Mathematics Education Library

More information about this series at http://www.springer.com/series/6276

John Monaghan • Luc Trouche
Jonathan M. Borwein

Tools and Mathematics

Instruments for learning

Springer

John Monaghan
Department of Mathematical Sciences
University of Agder
Kristiansand, Norway

Luc Trouche
Institut Français de l'Education
Ecole Normale Supérieure de Lyon
Lyon, France

Jonathan M. Borwein
Centre for Computer Assisted Research
Mathematics and its Applications
School of Mathematical
and Physical Sciences
University of Newcastle
Newcastle, Australia

ISSN 0924-4921 ISSN 2214-983X (electronic)
Mathematics Education Library
ISBN 978-3-319-02395-3 ISBN 978-3-319-02396-0 (eBook)
DOI 10.1007/978-3-319-02396-0

Library of Congress Control Number: 2015952621

Printed on acid-free paper

This Springer imprint is published by Springer Nature
The registered company is Springer International Publishing AG Switzerland

To Seymour Papert—for inspiration over the decades

Foreword

In our book *Windows on Mathematical Meanings: Learning Cultures and Computers*, Celia Hoyles and I wrote:

> The role of mathematics in underpinning social and economic life stretches back to the dawn of the industrial revolution and beyond. Every aspect of modern society is infused with the congealed mathematical labour of mathematicians, computer scientists, engineers and so on. Yet at the same time, this mathematics is increasingly invisible to those who merely share in, rather than construct, the artefacts of the culture ... It is mathematics which lies dormant inside the chips of vacuum cleaners, the warheads of missiles, and the graphical displays of news broadcasts. Even the simple exchange of goods and commodities, once relatively amenable to mathematisation, has been overwhelmed by the workings of global markets which are dominated by invisible mathematical forces that are increasingly out of control. (Noss & Hoyles, 1996, p. 253)

I am sure neither of the authors would claim any special prescience in predicting the state of affairs we find ourselves in some 20 years later. It would be banal, now, to force such an obvious point. In a world of CGI, mobile communications and automated drone strikes, the importance of mathematics in our lives has never been greater—or more greatly acknowledged. The world has truly changed, and with it, the role of mathematics has become a critical force.

In school, while the importance of mathematics is widely appreciated, it is often unclear what this implies or how this appreciation should manifest itself. Celia and I argued that the computer could have some special role in changing that. Centrally, our argument was to do with the possibility of employing computers to open up—to make visible—the knowledge that drives the world, not just how to improve the transmission of knowledge from teacher to student. Our friend and inspiration, Seymour Papert, at the 17th ICMI Study Conference on Digital Technologies and Mathematics Teaching and Learning: Rethinking the Terrain, encouraged mathematics educators present to acknowledge the dependence of mathematical expression on tools and the ways in which mathematics is shaped by their use. Papert challenged us to spend 'just 10 %' of our time focusing on the *what* rather on only the *how* of mathematics teaching and learning. A necessary—though far from sufficient—condition for change is for computers to be ubiquitous, to be freely available as a tool that is genuinely useful to accomplish worthwhile goals.

It is happening. Each of us—in the developed world at least—has a powerful computer in our pocket. Yesterday, I attended the launch of the BBC's *Micro Bit* computer, which—according to its designers—is a 'pocket-sized computer set to be given to about one million UK-based children in October'. The details don't matter: a free computer for every child is around the corner. And at the same time, the UK government has mandated the teaching of programming to all children—again, a massive opportunity for developing new ways to *use* mathematics to accomplish things that could not otherwise be done except by the very few.

So the publication of this book is timely in the extreme. Tools matter, and as the authors of the book make clear, tools—and especially computers—cannot sensibly be thought of as just passive artefacts that can be sprinkled on the mathematical terrain. The computer points to possibilities. It sets us thinking about ways of reasserting mathematics as a cultural practice, not merely a schoolish endeavour to achieve technological know-how for the few, but to educate in the broadest sense.

Of course the advent of free computers is just the start. Computers sealed in their boxes, programming courses that exclude the big ideas of mathematics and computer science, failure to adequately support teachers, curricula that just miss the point... I could make a very long list of possible negative outcomes. But I, like the authors of this book, am optimistic. We *can* seize the moment to change what we teach, to focus on Papert's 10 %. This book will help.

Richard Noss

Preface

Chapter 1 introduces the book, so we use this preface to comment on the genesis of this book and the development of ideas during its writing.

This book originates in discussions at conferences between John and Luc. We (meaning John and Luc at the moment) had similar backgrounds: a number of years in high school mathematics teaching and then university work in mathematics education. In both sites we'd done a lot of experimenting (both practical and research) using digital technology. We shared prejudices for the types of tools/systems we liked to use with our students: graph plotters, *Logo*, computer algebra systems and similar tools/systems which our students could use to express mathematical relationships rather than tools/systems that might go under the name of 'computer-aided instruction'. When we first met, in the later 1990s, we'd read papers of the other where we'd being trying to 'get a theoretical handle' on what went on in high school mathematics classrooms when teachers and students used these tools/systems. Our theoretical frameworks were not identical (John was influenced by activity theory and Luc by instrumentation theory), but there was plenty of room for productive dialogue; we were both also influenced by reading papers of and talking to Michèle Artigue and Jean-baptiste Lagrange who were, in turn, influenced by Yves Chevallard's anthropological theory of the didactic and Brousseau's theory of didactical situations.

In the early years of the new millennium, an interesting thing happened—we realised that much of our thought about digital technology and mathematics (education) was a special case of tool use and mathematics (education). This realisation probably came quite slowly, but once implanted the importance of tools (and later the distinction between artefacts, instruments and tools) in mathematical activity produced a way of viewing mathematics that was both simple and profound (at least for us): of course mathematical activity is not possible without tools; why are people writing about issues in learning and teaching mathematics and not mentioning the tools used in this learning and teaching? There were, of course, papers on tool use in learning and teaching mathematics, but few of these were research papers which paid attention to the *problématique* of tool use and were targeted at an international audience; and there was no book on tools and mathematics.

The first idea of this book was suggested by John to Luc and cemented over several days in charming Montpellier about 4 years ago. During our discussions in that visit, we sketched the structure of the book (quite similar to the final structure) and decided we wanted to work with a research mathematician. John and Luc had both read 'popular' papers by Jon and were intrigued by the 'experimental mathematics' he was putting forward. Working with a research mathematician would, we thought, temper naïve statements we might make about tools in the practice of professional mathematicians, and Jon clearly had a lot of interesting things to say about tools and mathematics. John approached Jon to write a chapter in the book, Jon agreed to be involved, and 'we' now becomes 'John, Luc and Jon'.

The first three chapters of this book to be completed were Chaps. 1–3 (more or less in the form they appear in the final version). This was to establish a base for the rest of the book. Jon's chapter on experimental mathematics and his own research was deemed especially important as something that could be referred to in the chapters to be written. The chapters which follow were written over several years with many drafts and many reviews and exchanges of ideas between the three of us. We three have quite a lot in common, but there are significant differences too. One set of exchanges concerned 'a common voice' in the chapters of this book, and it resulted in 'no, we should not attempt a common voice' in the chapters. Each 'solo' chapter thus represents the views of the author alone. This, we think, respects emerging theorising of scholarship on tools and mathematics.

Another set of exchanges concerned the 'encyclopaedic' scope of the book (which has actually been toned down a little since the proposal was sent to the publishers). We see the encyclopaedic nature of the book as both a strength and weakness. We have, however, tried to vary the 'scope' of that which we explore through the structure of the book. For example, with regard to the history of mathematics, Chap. 4 has a 'large scope', but Chap. 5 focuses down on a specific time and place in the history of mathematics. We acknowledge that we have tried to cover too much (though, as the Interlude between Parts C and D mentions, we have missed out some important issues). But this is the first book on the subject of tools and mathematics, and we are content to err on the side of 'surveying the scene' and leave it to others to fill in details we have missed.

We have written the book as a linear object to be read from cover to cover, but, of course, readers may not read it this way. We think that most chapters, however, work pretty well as 'stand-alone' chapters. Many chapters also have multiple references to other chapters.

There are people we'd like to thank for input on this book (reviewing chapters and technical help), and these include Naomi Borwein, Hazel Cathan, Nikolaos Fotou, Ghislaine Gueudet, Celia Hoyles, Stefan Lesnianski, Richard Noss, Christine Proust, Janine Rogalski, Kenneth Ruthven and Rudolf Straesser.

Kristiansand, Norway — John Monaghan
Lyon, France — Luc Trouche
Newcastle, Australia — Jonathan M. Borwein

Contents

About the Authors

John Monaghan is Professor of Mathematics Education at the University of Agder, Norway, and Emeritus Professor at the University of Leeds, UK.

Luc Trouche is Professor of Didactics of Mathematics at the Institut Français de l'Education, Ecole Normale Supérieure de Lyon, France.

Jonathan M. Borwein is Professor Laureate and Director at the Centre for Computer Assisted Research Mathematics and its Applications, School of Mathematical and Physical Sciences, University of Newcastle, Australia.

Part I
Setting the Scene

This book, on tools and mathematics with specific regard to mathematics education, is in four parts. Part I provides an introduction to the book and includes two chapters on tool use in mathematics and two chapters addressing tool use in the history of mathematics. Part II jumps forward in time, to developments in the second half of the twentieth century to the present, to consider 'modern' developments related to tools and mathematics and theoretical approaches concerned with understanding the learning mathematics (with tools). Part III considers four substantive issues related to tool use in mathematics education. Part IV looks to the future, a future we expect to be shaped, in part, by increased use of digital tools.

Part I has six chapters. Chapters 1 and 6, respectively, provide an introduction to the whole book and a review of issues arising in Chaps. 2–5. Chapters 2 and 3 focus on tool use in mathematics. Chapter 2 takes one task, bisecting an angle, and discusses mathematical and educational issues arising from doing this task with four different tools. Chapter 3 considers tool use in the life of a research mathematician with specific regard to the role of visual computing made possible by the affordances of modern computing environments. Chapters 4 and 5 focus on tool use in the history of mathematics. Chapter 4 presents an overview of sorts and looks at tool us in four periods of time (as well as tool use in prehistory). Chapter 5 looks at tool use at a particular moment in the development of mathematics and the learning of mathematics, 2000 BCE in Mesopotamia.

There is much more that can be said about tools and mathematics than is covered in the chapters in Part I, but we hope we have raised some crucial issue for the reader to reflect on the importance of, and the diversity of, tool use in mathematics.

Chapter 1
Introduction to the Book

John Monaghan and Luc Trouche

1.1 Introduction

This chapter sets the scene for the book. In Sect. 1.2 we state the purpose and scope of this book. In Sect. 1.3 we make two attempts at answering the question 'what is a tool?' In Sect. 1.4 we outline the structure of the book.

1.2 The Purpose and Scope of This Book

This book is an exploration of tools and mathematics and issues in mathematics education related to tool use. There is much that can be said of mathematics without explicit reference to tools and there is much that can be said of tools without focusing on mathematics. We will explore such things when a consideration of issues is relevant to *mathematics and tools*. Similarly, we will explore issues in education when a consideration of such issues is relevant to mathematics education and, in particular, tool use in mathematics education.

What is a tool? Answers to this question will emerge in the course of this book; at this point we simply say 'it is something you use or create to do something'. There are many related words (implement, instrument, utensil, artefact. . .) but these words often have different nuances for different people (we write more on this shortly). Tools have been and are omnipresent in our lives; go into your kitchen and look in wonder at the array of tools. Tool use is important in mathematics. In school mathematics certain tools are valued, for example compass, protractor and ruler, whilst other tools are controversial, for example the calculator. Chapter 3 shows how tools are important in the life of a research mathematician.

What is mathematics? That's a big question that we will skirt but, whatever it is, tool use seems important for mathematics. We have heard it said (by colleagues and by students) that 'mathematics is a tool-box'. What is usually meant when people

J. Monaghan et al., *Tools and Mathematics*, Mathematics Education Library,
DOI 10.1007/978-3-319-02396-0_1

say something like this is that doing mathematics involves selecting an appropriate algorithm to get an answer to a problem (or question or given task). This makes sense at an 'everyday level' and also characterises much of what is done in school mathematics: a student is set a problem, e.g. *solve* $x^2 - x - 2 = 0$, and selects an appropriate tool from the tool-box, e.g. 'factorise, $(x-2)(x+1)$, and find the zeros of the linear factors' or 'use the equation $x = \frac{1 \pm \sqrt{1^2 - 4\times 1\times(-2)}}{2\times 1}$'. But the view that mathematics is just a tool-box is a pretty impoverished view of mathematics. It does not account for the structure of mathematics or the purposes of mathematics (beyond solving an equation you have been told to solve) or rigour or how the different solution strategies (tools) are related or what the solver, in selecting the tool, focuses on in the question in order to select a tool. A view of mathematics that will emerge in this book is that many of the actions of doing mathematics involve selecting, using and creating tools but mathematics is more than just a tool-box.

This book will consider the past, the present, and the future. Mathematics has a long (and glorious) past and tool use is part of the history of mathematics. This book is not conceived as a history of mathematics and tools but it will explore selected periods of the history of mathematics with regard to tool use and the development of mathematics. The history of mathematics is also important to appreciate controversial issues in mathematics in the present day (early twenty-first century) such as the calculator debate. This book will explore 'issues in current mathematics education' where tool use is an important (and often controversial) factor; the clash between people's valuations of old and new tools is often one reason (amongst others) for controversies. This book will also tentatively consider the future of mathematics education and the role of new tools and new ways of using tools in this future. We say 'tentatively' because new tools often arise from serendipity and it is somewhat foolish to say that things will develop in this way. Consider the case of the hand-held digital calculator. Although the first such calculator was designed for doing arithmetic, this design was realised because of advances in electronics, which had nothing to do with educational interests, it was a case of 'we now have the technology to do this'.

A theme running through this book is the dialectical relationship between mathematics, tool use, networks of artefacts (objects) and social activity. Perhaps a tool can be used in isolation by a single person for his/her own benefit (e.g. spear fishing) but it is common for a set of artefacts/tools to be used in social activity. For example, the salesman who arrived at John's house to make an estimate for carpeting the rooms was involved in social intercourse, a sales transaction. He arrived with a set of artefacts, a tape measure, a set of charts on carpet widths and a cool electronic device that gave the width of the room at the touch of a button. In an educational setting consider a high school lesson on quadratic functions. The classroom will have a social structure which will conform, to some extent, with the school's social structure. The students will have a variety of artefacts on their desks (pencil, ruler, graph paper, calculator) which they may use, alone or together, at different times during the lesson. These networks of artefacts and the social structure interact—the teacher may direct the students to use their graph paper or the students even may have to ask if they can use their calculator. Even a single artefact may be viewed as a network of artefacts, e.g. a graphic calculator as an arithmetic calculator, a graph plotter and a machine with built-in statistical functions.

We have written this book because we are interested in tool use in mathematics and mathematics education and because we feel there is a gap in the literature. Having said this, we are aware of two twenty-first century books with the words 'mathematics' and 'tools' in the title: *Symbolising, modelling, and tool use in mathematics education* (Gravemeijer, Lehrer, van Oers, & Verschaffel, 2002); and *Tools of American mathematics teaching, 1800-2000* (Kidwell, Ackerberg-Hastings & Roberts, 2008). Gravemeijer et al. (2002) is a multi-authored publication resulting from a conference on symbolising and modelling in mathematics education. Tool use in mathematics education is mainly considered from the point of instructional design. Kidwell, Ackerberg-Hastings, and Roberts (2008) is an historical work—the first author is the curator of the mathematics collections at the National Museum of American History, Smithsonian Institution. The book traces tools used in American mathematics teaching under four headings: tools of presentation and general pedagogy; tools of calculation; tools of measurement and representation; and electronic technology and mathematical learning. Both of these books are important but we have set out to do something different: to examine the dialectic between tool use and doing mathematics; to explore the complexities of using tools in the learning and the teaching of mathematics; to consider philosophical positions regarding the aforementioned dialectic and complexities; to consider current issues in mathematics education with regard to tool use; and to speculate on future issues in mathematics education with regard to tool use.

The above we hope, sets out the overall scope of this book. Section 1.4 describes the parts and chapters of this book. We now turn to an initial attempt (attempts actually) at saying what a tool is.

1.3 What Is a Tool?

We gave a first, somewhat crude, definition above, 'it is something you use to do something'. We now refine our response to this question a little. We do this in two subsections, each allowing one author to state his position on tools. A word of explanation appears to be in order for presenting two views. We came together to write this book because we share an interest in tool use in mathematics. We have also, over the years, learnt through reading each other's work and in talking to each other about tool use in mathematics. But certainly learning through reading someone else's work and talking to them is only possible if the two (or more) discussants differ to some degree in their knowledge-base and/or their interpretation. We choose to celebrate these differences rather than hide them.

1.3.1 *John's Attempt to Address This Question*

Rather than give a direct definition of the form 'a tool is . . .', I choose to make four distinctions related to tools. I follow this with a comment on mathematical tools and on the academic language surrounding discourse on the use of tools in current mathematics education.

I first make a distinction between an artefact and a tool. An artefact is a material object, usually something that is made by humans for a specific purpose, e.g. a pencil. An artefact becomes a tool when it is used by an agent, usually a person, to do something. The compass becomes a tool when it is used to draw a circle (which is its intended purpose); the same artefact becomes a different tool when it is used to stab someone. This establishes, for me, an irreducible bond between agent, purpose and tool; it is not possible to comment on a tool, for example, 'is it a good tool?', without considering the user and the purpose. After being used as a tool (for whatever purpose), the compass returns to being an artefact. The materiality of an artefact is not just that open to touch. An algorithm, e.g. for adding two natural numbers, is an artefact and it is material in as much as it written and can be programmed into a computer.

Secondly, I distinguish between an artefact/tool and ways of using the artefact/tool. I illustrate this with an example, the algorithm referred to above. There are a number of ways of enacting the traditional algorithm. Two ways are shown below in the example 27 + 36.

$$\begin{array}{r} 27 \\ +_{1}\underline{36} \\ 63 \end{array}$$

20	7	
30	6	
50	13	63

All the readers can see in the print above is a summary of the physical actions, which will be executed in some temporal sequence. Behind these physical actions are intentions, understandings and routines with regard to ways of using the algorithm.

My third distinction is between 'mental representations'[1] of artefact/tool use and material actions in artefact/tool use, but this distinction comes with an interrelationship: to carry out material actions with an artefact/tool you need some form of mental representation, which may be quite crude, of how to act with the artefact/tool, but actions with the artefact/tool will provide feedback to the user which may change the mental representation.

---

[1] I use this term reluctantly and because I cannot think of a better one. I could have used the terms 'schemas', or 'scripts' or 'mental models' but these all come, to me, with more theoretical baggage that 'mental representations'. In using the latter term I simply wish to communicate that 'there must be something in the user's mind before s/he uses the artefact/tool' without getting into issues in the philosophy of mind.

My fourth distinction is between signs and tools. Signs, like tools, are artefacts but a sign signifies/points to something whereas a tool does something. Having said this, concatenations of signs can function as tools. On their own the artefacts '(','), '×', '+', 'x', 'y', 'a', 'b' and '2' are signs but the artefact composed of these signs, $(x+a) \times (x+b) = x^2 + (a+b) \times x + a \times b$, can function as a tool. In mathematics education concatenations of signs such as $y = mx + c$ are called 'representations' and $y = mx + c$ is referred to as an algebraic representation (which has related graphical, numeric, and natural language representations). Representations can function as tools (they can be used to do something, usually something mathematical) but they also have non-tool functions, e.g. to signify/point to a mathematical object.

Is there such a thing as a 'mathematical tool'? My answer is that there are simply artefacts which become tools in use, though there is a sense in which the compass is a mathematical tool when it is used to draw a circle but it is not a mathematical tool when it is used to stab someone. When artefacts are used for mathematical purposes they generally (artefacts such as pencils are exceptions) incorporate mathematical features, as we have seen in the examples above, e.g. a compass encapsulates the equidistant relationship between the centre of a circle and points on the circumference of that circle.

Prefixing the word 'tool' with an adjective, as in 'mathematical tool', is common in academic discourse. Examples include cultural tool, semiotic tool, cognitive tool, ICT tool, dragging tool. My comments in the paragraph above apply to these terms also. This subdivision of tools can be useful for communication, e.g. the term 'ICT tool' does locate a certain subset of tools. But in general I find such categorisations of tools unnecessary and, at times, ontologically confusing. A tool which is designed as a cognitive tool only lives up to its name if, in practice, it transforms the cognition of the user. A dragging tool is obviously an ICT tool and is also often a semiotic tool and a cognitive tool. And what of cultural tools? A culture could be defined as a co-ordinated set of artefacts (most of which function as tools), so is there any tool which is not a cultural tool?

1.3.2 Luc's Attempt to Address This Question

I would like to define a tool from one epistemological position, specified by four dualities.

Using the word 'artefact' constitutes, indeed, an essential epistemological stance that means: seeing the trace of humans on anything. Most of the objects, symbolic as well as materials we use (in mathematics or elsewhere) are saturated in history and culture, and, when using them, each human situates herself in a world of culture. We inherit from our predecessors human knowledge encapsulated in the objects they have created.

First duality, acting (performing a task, solving a problem) is both a process of *using* and *creating* artefacts. John has given several examples of using artefacts but creating artefacts, sometimes by serendipity, sometimes consciously, is also a necessity for action. A compass, or any given artefact, was once created and this historical act of creation reappears every day in mathematics classes, even if we are not aware of this.

Second duality, the relation between a user and an artefact consists in a double shaping process. In one sense, the artefact *shapes the way* the user is acting (see John examples in Chap. 2: the way we bisect an angle depends on the set of available artefacts); in the reverse sense, the user *shapes*, along the course of her/his situated action, *the artefact* that s/he appropriates. That is not obvious with very simple artefacts (as a book for bisecting an angle), but we will make it clear for more complex artefacts (for example, calculators) keeping the trace of their users and usages.

Third duality, the process of using an artefact is both a process of *producing* something and a process of *constructing* knowledge. The first process is generally visible (we do bisect—when we succeed—an angle with a compass, there is a tangible result), the second process is, at once, not (completely) visible, neither by the user nor by the observer (we do learn something, even if we do not succeed in performing the task, from the complex system artefact-task-social context of action). But creation goes on, for example the creation of an ad hoc measuring tool by students (see text above Fig. 2.6).

The last duality consists in an essential distinction between an *artefact* and an *instrument*. Like John I will situate an artefact as a starting point, something available for a user and an action oriented by a goal. When an artefact has been appropriated by a user, I will name *instrument* the mixed entity composed of the artefact and the associated knowledge (both the knowledge on the artefact, and the knowledge on the task constructed when using this artefact). For example, a compass, as an object I can see on a table, is an artefact. John's compass, incorporated in his activity over time, for performing various tasks, became an instrument.

These four dualities finally constitute four windows on a same world of artefacts seen as lived entities. And what about tools? To me, a tool is a thing somewhere on the way from artefact to instrument.

1.4 The Structure of the Book

This book is in four Parts, I to IV. Part I is called 'Setting the scene'. It includes this introductory chapter followed by a chapter in which we execute one task with four different tools; this allows us to raise a number of aspects about tool use in a concrete manner. These first two chapters have been deliberately written without references to the literature. We provide a large number of references in subsequent chapters but we thought it would be useful to force ourselves to develop our arguments *from first principles* (without reference to secondary sources) in these two opening chapters.

We now introduce the rationale for Chap. 3, an account of the life of a working research mathematician who makes considerable use of digital tools. This book was conceived by John Monaghan and Luc Trouche through face-to-face and distance communication over several years. Both John and Luc are mathematics educators who like a strong dose of mathematics in their mathematic education but they are not research mathematicians. We felt the input of a research mathematician into this book was important and Jon Borwein was an obvious person to approach; Jon is a prominent experimental mathematician with a strong interest in educational matters. Jon was initially recruited for a single chapter but his involvement ended up being influential in the overall shaping of this book. Jon labels himself 'Homo Habilis Mathematicus', with reference to the co-evolution of tools and our species (see Sect. 4.2). His chapter includes an exposition of what experimental mathematics is (and the essential role that tool use plays in determining the nature of experimental mathematics) and the role of tools in a part of his working life (visual theorems) circa 2014.

Chapters 4 and 5 consider the place of tools in the history of mathematics. The co-history of tools and mathematics is vast and merits a specialist encyclopaedia and these chapters do not attempt to be comprehensive. Chapter 4 introduces five aspects of tool use that we consider important. (1) Tool use and phylogenesis, the co-evolution of tools and our species. To set the record straight on 'man the tool user' we also consider tool use in the non-human animal world. (2) Tools of ancient Greek mathematics. The 'obvious' tools here, we think, are straight edges and compasses. We argue, however, following the work of a prominent historian of ancient Greek mathematics, that 'not so obvious' tools, the lettered diagram and the mathematical language, were a part of the development of mathematics in this culture. (3) An ancient Indian algorithm for computing square roots. This algorithm was selected because it represents a different ancient approach to that of ancient Greek deductions and has links with Chap. 3. (4) The use of the abacus for calculations. We chose this to illustrate the mutual support of hand, mind, and artefact in tool use. (5) A time and place (sixteenth century Europe) which witnessed a rapid development in tools for calculations; our intention in this section is to consider the statement 'tools for doing mathematics beget further tools for doing mathematics'. Chapter 4 is broad but Chap. 5 specialises in one time and place, 2000 BCE in Mesopotamian. This period and place illustrate the role of artefacts in both the development and in the teaching and learning of mathematics; it is proposed that the process of creating artefacts and the process of creating mathematics feed one another. The study of this period-place also suggests that new artefacts for doing mathematics co-exist with old artefacts for doing mathematics (something we may recognise in current mathematical practices). The creation and use of artefacts for doing mathematics in this period-place also reveals two aspects of masters of calculations, the teacher and the scholar, and these masters used co-ordinated systems of artefacts in both aspects of their working lives.

Part I ends with Chap. 6, a reflection/discussion of issues arising in Chaps. 2–5.

Part II jumps forward in time to developments in the second half of the twentieth century to the time of writing (2014). Chapter 7 opens this Part with a survey of intellectual and technological developments with regard to the development of digital artefacts and mathematics; scholarly work (e.g. theories of learning) with particular emphasis on work which considers tool use; and the ascent of mathematics education as an academic field of inquiry. Chapters 8–10 present 'case histories' of work, which impinges on an understanding of tool use in doing and learning mathematics, in the field of mathematics education in this period: constructionism; activity theory; and French didactics of mathematics. We were aware from an early stage in the planning of this book that it was unrealistic to attempt to cover every theoretical framework employed in mathematics education research and we chose the three areas that we thought would be most productive in terms of tool use. It only emerged in the course of writing that constructionism and activity theory were not as fruitful, with regard to understanding tool use in the doing and learning of mathematics, as we expected prior to writing—but this is interesting in itself (but you can judge this for yourself).

The origins of constructionism are a man, Seymour Papert, and a programming language, *Logo*. Papert's book *Mindstorms* is a visionary text on a Piagetian base. Papert's ideas were central in debates in the 1980s on the role of programming in mathematics education and 'which programming language is the best'. Fifteen years after *Mindstorms* its successor appeared, Noss & Hoyles' *Windows on mathematical meanings*. The Piagetian base had disappeared and was replaced by a hybrid social constructivist/sociocultural approach, but the vision of new forms of engaging in mathematics remained. Constructionism today continues as sociocultural approach with a vision of human–machine interaction and design for mathematical activity. As such it has the potential to make an interesting case history of tool use and mathematics.

Chapter 9 considers activity theoretic approaches. Western academic mathematics education came to embrace activity theoretic approaches quite late (significant work only from 1990) but then the seeming importance of such approaches really took off. Activity theory has many forms and the 'generations' and variant foci of these forms are our first consideration. This establishes a basis for a consideration of the differing foci in activity theoretic studies in mathematics education in the last 20 years.

Chapter 10 considers the development of French didactics of mathematics education. The aim is to consider this field with regard to tool use (which we think we do) but to do this the work on tool use needs to be set in a wider setting of mathematical, intellectual, and institutional developments in France. The chapter has four themes: the roots of French didactics of mathematics; two important theoretical frameworks (Brousseau's theory of didactical situations and Vergnaud's theory of conceptual fields); Chevallard's anthropological theory of didactics; and approaches dedicated to artefacts and resources in mathematics education.

Part II ends with Chap. 11, a reflection/discussion of issues arising in Chaps. 7–10.

Part III considers selected issues in tool use in mathematics education. There are many issues regarding tool use in mathematics education which could be considered. We have selected four that we consider to be particularly important: the curriculum (and assessment and policy); the calculator debate; mathematics in the real world; and the mathematics teacher and digital technology.

Chapter 12 considers curriculum and assessment policies with regard to the integration of digital technologies into the learning and teaching of mathematics. With regard to curricula it focuses on interrelationships between tools and mathematics curricula and argues that: school mathematics develops from 'really used' tools; the development of tools is related to the implemented as well as the intended curricula. Assessment is viewed as a 'problem area' and the final section of the chapter uses French policy on assessment as a case study to examine tensions that can arise.

Chapter 13 considers the longstanding debate on the place and value of the calculator in the learning, teaching and assessment of mathematics. The calculator has inspired one of (if not the most) controversial debates regarding tool use in mathematics education. After an introduction to the issues, actors and claims, this debate will be viewed by a framework informed by Wertsch's ten theses on mediational means. The chapter ends with a consideration of the future of this debate.

Chapter 14 focuses on mathematics in the real world and the problem of linking this mathematics with school mathematics. This leads us to address questions concerning the nature of mathematics. We consider the use of tools in leisure and in working practices. Tool use is omnipresent in out-of-school mathematics but school mathematics privileges specific tools. The chapter considers the problem of 'suspension of sense making' in school mathematics and opportunities for using real-life artefacts to link in-school mathematics to out-of-school mathematical activities. The increasing presence of digital technology in everyday life and work opens up new opportunities (and problems) for linking in-school to out-of-school mathematical activities.

Chapter 15 considers the teacher with regard to mathematical knowledge and the use of technology. The teacher, once jokingly referred to as something that could be replaced by teaching machines, is arguably more important in classrooms where digital technology is a central feature than those without. But mathematics teachers, en masse, are often reluctant to enact deep integration of digital technology in their classrooms—why is this? A consideration of this question will include a critical review of attempts to categorise forms of teacher knowledge and what teacher should do.

Between Parts III and IV we insert an *Interlude* in which we reflect on issues in mathematics education; the issues we have written chapters on and issues to which we have not devoted a chapter.

Part IV looks to the future, a future we expect to be shaped, in part, by increased use of digital artefacts. We explore three themes: tasks and tools; games; and connectivity. Chapter 17 considers tasks and tools. Task design in mathematics education is so very important but as a subject of study it is in its infancy. Using different tools for what appears to be the same task involves the person doing the task in different mathematical processes (as was seen in Chap. 2)—what are the implications of this for the design of tasks with digital tools and for the design of digit tools themselves?

Chapters 18 and 19 look to emerging forms of activity made available through the use of digital tools: connectivity and games. Chapter 18 considers the range of games; artefacts and tools in games and gameplay; and research on games and mathematics. The final section looks to future development. Chapter 19 considers the different meanings and the potential of 'connectivity': connecting students in their learning and connecting teachers in their professional development. It also discusses the concept itself with regard to the future of mathematics education. The book ends with an *Epilogue*, a reflection on matters for further thought and action.

References

Gravemeijer, K., Lehrer, R., van Oers, B., & Verschaffel, L. (2002). *Symbolizing, modelling and tool use in mathematics education*. Dordrecht: Kluwer.

Kidwell, P. A., Ackerberg-Hastings, A., & Roberts, D. L. (2008). *Tools of American mathematics teaching, 1800-2000*. Baltimore: The John Hopkins University Press.

Chapter 2
Doing Mathematics with Tools: One Task, Four Tools

John Monaghan

2.1 Introduction

In this second introductory chapter I illustrate a variety of mathematical and educational issues arising from doing a single task with different tools. I will bisect an angle using four tools, a straight edge and compass, a protractor, a dynamic geometry system and a book, and reflect on what was done. The last one, a book, sounds a bit strange but I hope you will soon appreciate it as a nice mathematical tool (in some respects). You may like to stop and do this task before you read on.

2.2 Bisecting an Angle with a Straight Edge and a Compass

A compass has mathematical beauty in as much as it encapsulates the equidistant relationship between the centre of a circle and points on the circumference of that circle. Mathematics is many things and one of these is *activity-with-relationships* and here is a tool that captures a simple, but fascinating and historically important, relationship between pairs of points in a classic geometric figure. Properties of a circle, however, are not essential in bisecting an angle with a compass (though the whole circle can certainly be used); the compass is used to provide two equidistant line segments from the vertex of the angle and another two (intersecting) equidistant line segments from the end points of the initial line segments. The straight edge also has a built-in mathematical feature, it is linear. Figure 2.1a shows a standard school construction with all arc lengths (generating line segments) of the same length. Figure 2.1b shows a slight variation, the radius of the second pair of arcs is longer than the radius of the first pair of arcs.

Over the years I have asked hundreds of pre-service mathematics teachers why this construction bisects the angle. Only a tiny handful have ever explained why. Do you know why? Our answer depends on seeing congruent triangles. These may be

J. Monaghan et al., *Tools and Mathematics*, Mathematics Education Library,
DOI 10.1007/978-3-319-02396-0_2

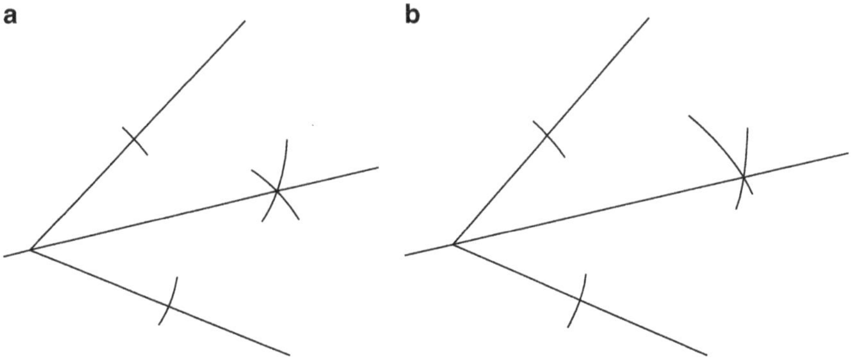

Fig. 2.1 (**a**) Straight edge and compass bisection of an angle where pairs of arcs have the same radii. (**b**) Straight edge and compass bisection of an angle where pairs of arcs have different radii

more obvious if we redraw Fig. 2.1b to emphasise different mathematical objects (and relationships), see Fig. 2.2. Line segments *AB* and *AC* are identical because they have been constructed via arcs which have the same radii. Similarly, line segments *BD* and *CD* are identical. *AD* is a common side. Thus triangles *ABD* and *ACD* are congruent (side, side, side). Thus $\angle BAD = \angle CAD$.

I now comment on aspects of mathematics, tools, actions and thought involved in this construction. Regarding tools: (1) There are two main physical tools used in this construction, a straight edge and a compass. I opened this section by praising the properties of the compass but the compass is of no use in this construction without the straight edge (I have underlined the word 'this' because the theorem of Mohr-Mascheroni proves that any construction made using a compass and straight edge can be constructed using a compass alone). This, to me, illustrates an important aspect of tool use (in mathematics or elsewhere), tools are rarely used in isolation, they are almost always used with other tools (though mathematicians, as in the case of the Mohr-Mascheroni theorem, value economy). Note that we also need a tool for making marks, typically a pencil. (2) Neither the compass nor the straight edge was designed to tackle the task of bisecting an angle, they just happen to be useful. Design, including the design of tools, is an important part of mathematics education but tools can be adapted for purposes other than their primary design purpose (to draw a straight line and to draw a circle in the case of the straight edge and the compass). (3) The user needs to know how to use the tools as physical actions must be enacted to perform the task. The straight edge and, particularly, the compass require quite advanced motor control to achieve a result approximating to the ideal of lines without width and equal line segments. (4) The user also needs to have an intention to use the straight edge and the compass for particularly ends. Related to this, the mind (whatever that is), something internal to the user, and the tools, things that are usually external to the user, need to be co-ordinated.

I now consider the educational aspects of mathematics in this task. (1) The task is 'isolated'. Why should anyone except a mathematician want to bisect an angle? Well, if that person is a student, then the task might be set as an exercise; this task

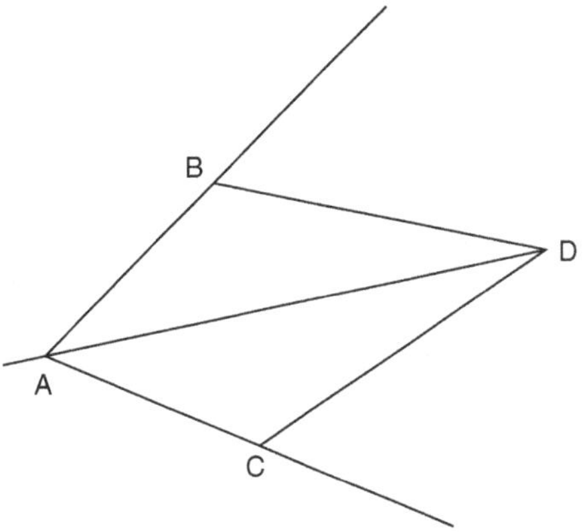

Fig. 2.2 Figure 2.1b redrawn to focus on congruent triangles

may be set for the student to 'learn some mathematics' or it might be set in order for the student to practice using a compass. Or the task might be a necessary step in solving a bigger task, such as drawing the inscribed circle of a triangle. Indeed, an interesting question is 'to what problem is a bisected angle a solution?' (2) As mentioned above, the reason why this construction bisects the angle may not be clear even to people preparing to teach mathematics. (3) Line segments *BD* and *CD* (Fig. 2.2) are not explicit in either Fig. 2.1; the mathematical relationships which the compass makes explicit and the mathematical relationships which may aid the proof are subtlety different.

Before proceeding with the next tool for doing this task I go on to consider a related task, trisecting an angle. In general, this task cannot be done with a straight edge and a compass. Some angles can be trisected using a straight edge and a compass; for example, an angle for which a straight edge and a compass construction is possible for one third of the angle (so 90° can be trisected using a straight edge and a compass because 30° can be constructed in this way but 60° cannot be trisected using a straight edge and a compass because 20° cannot be constructed in this way). I will not go into details, which require a little bit of university level algebra, but the interested reader can find explanations on another tool, the internet.

Some modern straight edges are clear plastic rulers with a line down the middle, a result of the production process. Using this straight edge (not the numbers, which themselves would allow an arbitrary angle to be trisected) we can trisect an arbitrary angle. I refer to Fig. 2.3a.

Place the straight edge with the central line over the top of one of the line segments forming the angle (position *A* in Fig. 2.3a) and draw the two parallel lines by tracing along the straight edges. Now place the straight edge in position *B*, with its top left corner meeting one side of the line segment forming the angle, the central axis running through the intersection point of the angle and the top right corner meeting the previously drawn line. Mark the three point *C*, *D* and *E* as shown.

I now consider points *C*, *D* and *E* together with the vertex of the angle, *O*, and the point *F* such than $\angle OFE = 90^{\text{o}}$ (see Fig. 2.3b).

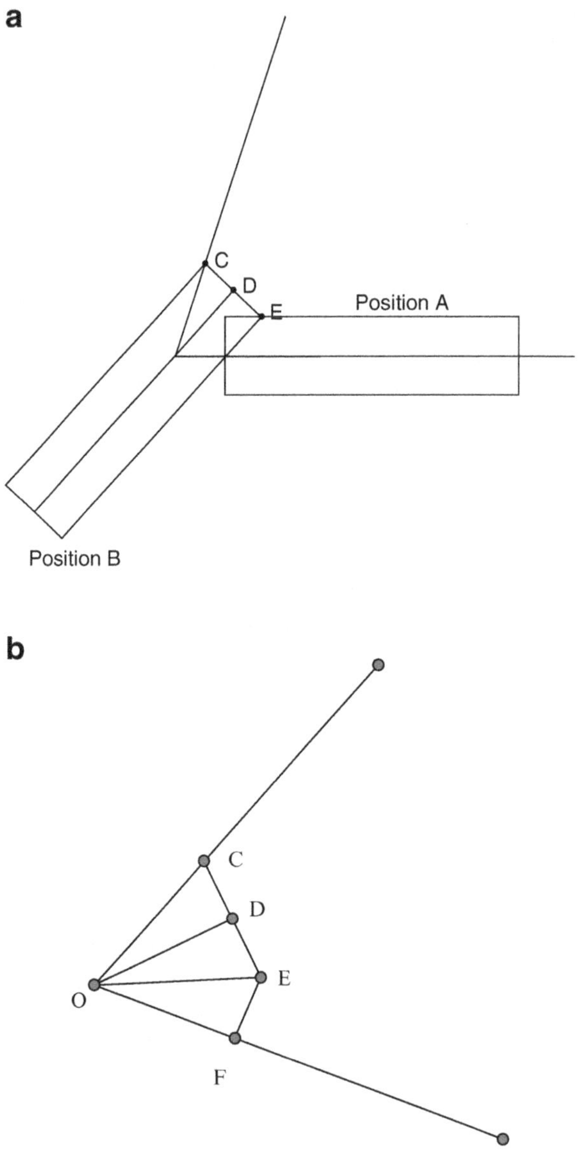

Fig. 2.3 (**a**) Two positions of the straight edge with a *line down* the *middle*. (**b**) The relationships between points marked on (**a**)

Since *CD*, *DE* and *EF* have been constructed using the same ruler, they are equal in length.

Since $\angle CDO$ and $\angle ODE$ are right angles, and segment *OD* is common, triangles *CDO* and *EDO* are congruent (side, angle, side).

Therefore $\angle COD = \angle EOD$.

Similarly, since $\angle ODE$ and $\angle OFE$ are right angles and segment *OE* is common, $DO = FO$ by Pythagoras' theorem.

Therefore triangles *ODE* and *OFE* are congruent (side, side, side).
As a result $\angle DOE = \angle FOE$ and $\angle COF$ has been trisected.

I went off on this little tangent on trisecting an angle to show how a little change in one tool, a visible central axis, changed the mathematics that was possible with the tool; note, however, that this feature, the central axis, is itself a mathematical feature, a line of symmetry. Other points of interest are: (1) the compass was not used in this trisection; (2) the change in the tool was the product of an accident, not of design (the production process caused a visible central axis).

2.3 Bisecting an Angle with a Protractor

The protractor is generally circular or semi-circular in shape but, unlike the compass, it does not rely on circle properties for its use (a circular shape is merely convenient). Figure 2.4 shows a standard way to do this with a semi-circular protractor. The centre of the protractor is placed at the vertex of the two lines (generally aligning a zero of the protractor with one of the lines) and reading off the degree where the other line crosses the protractor.

As before I comment on aspects of mathematics, tools, actions and thought involved in this construction. I first note that the protractor can be regarded a single tool or ‘two tools in one’, a straight edge and an angle measuring tool. It is also a tool designed for the purpose at hand (to measure angles). However, to complete the task, another tool is, arguably, needed: a tool to divide the numeric value of the measured angle by 2. As with the straight edge and compass the user needs to: perform physical actions to enact to the task (though these are arguably simpler than those needed for the straight edge and compass construction); know how to use the tool; have an intention to use it for the particular end; and internal thought and external tool need to be co-ordinated (I shall not keep repeating these last three points in this subsection).

With regard to the educational aspects of mathematics in this task I first note the incorporation of arithmetic into the solution method. Arithmetic was not a feature of the straight edge and compass solution. This is obvious but sometimes the obvious is worthy of comment. To labour this point, arithmetic is not needed to solve this geometric task (the straight edge and compass construction shows that it can be posed and solved entirely within geometry); arithmetic is brought into the solution because the tool, the protractor, essentially uses numbers. The use of arithmetic, and manually reading a scale, also brings a theoretical inaccuracy (an approximation) to the task. The straight edge and compass construction is also open to manual error (which may be, depending on the user, be greater than the error using a protractor) but the straight edge and compass construction is ‘ideal’ in a way that the protractor solution is not. This raises an aesthetic issue, this ‘construction’ is, to me, aesthetically inferior to the straight edge and compass construction (but this ‘inferior’ method does come with a greater range of

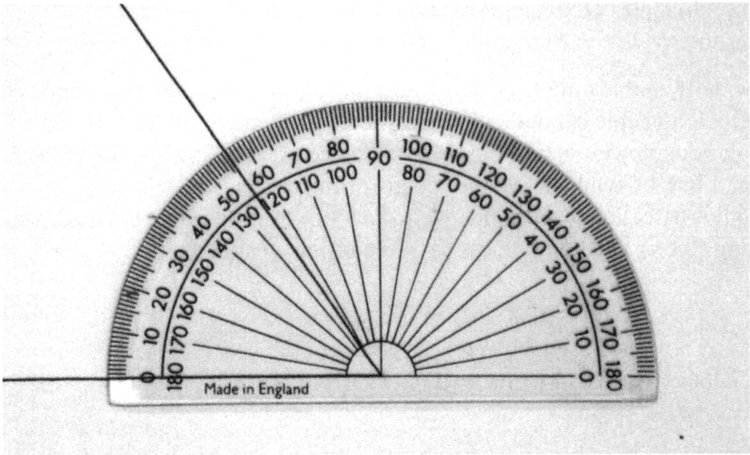

Fig. 2.4 Positioning a protractor on the vertex of two lines to measure the angle

application, with the protractor there is no problem in trisecting or n-secting an angle). The aesthetic dimension of the agent-tool dyad is, I hold, worthy of note.

The protractor solution does, however, 'beat' the straight edge and compass construction with regard to transparency of the solution: the angle is bisected because the degrees in the angle are halved. Is this a proof? Well, with a few surrounding statements, I believe it can be turned into a proof! Furthermore, unlike the straight edge and compass construction, seeing why the protractor solution works does not require 'seeing' what is implicit (such as line segments *BD* and *CD* in Fig. 2.2).

Is one tool 'better' than the other? I hedge my answer, yes and no: 'yes' if we appeal to a value judgement such as mathematical aesthetics or transparency with regard to reason; 'no' in as much as the answer depends on the valuation you adopt. Value judgements, closely linked to aesthetics, are something we cannot ignore when we consider tool use in mathematics. I now move on to a tool that invokes extreme valuations, the computer or, more precisely, dynamic geometry systems.

2.4 Bisecting an Angle with a Dynamic Geometry System

Dynamic geometry systems (DGS) perform digital manipulations of 'digital geometric objects' (points, lines, polygons, circles, etc.). There are many different versions, with some similarities and some individual features. With this caveat I present a method of bisecting an angle using *GeoGebra*, a popular freeware DGS at the time of writing. *GeoGebra* also has a graph plotter, a basic computer algebra system and a spreadsheet-like facility but I focus only on its DGS (which is remarkably similar to that of the DGS *Cabri*).

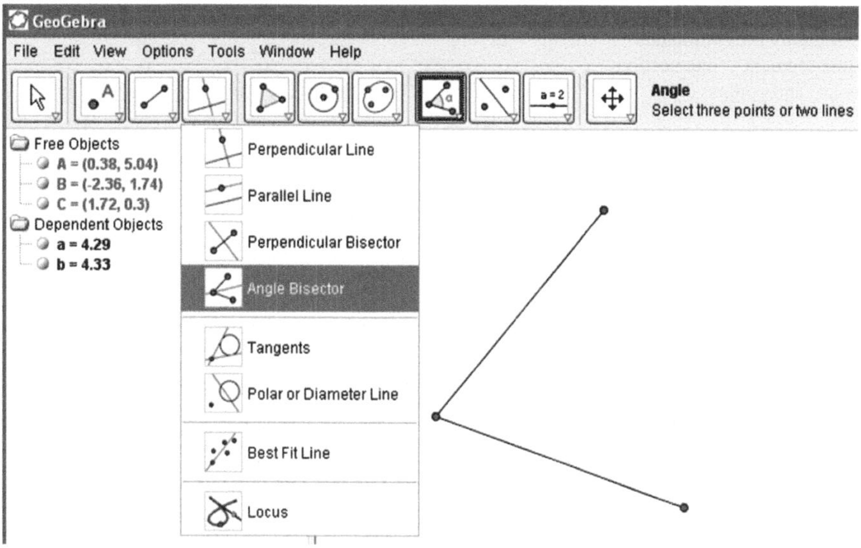

Fig. 2.5 A *GeoGebra* screenshot at the point at which the user has clicked on the drop-down menu including 'Angle Bisector'

I start my explanation at the point at which the lines meeting at a point and forming the angle to be bisected have been 'drawn[1]' and the user has clicked on the drop-down menu including 'Angle Bisector' (see Fig. 2.5). Once the user has clicked on 'Angle Bisector' two actions can be used to produce the angle bisector: clicking on the two line segments; clicking on the three points (in an order in which the vertex is the middle point selected). Using the 'Angle Bisector' command is the expected method in this DGS but other methods of producing a bisector are possible, including methods 'imitating' the straight edge and compass and the protractor methods; information and communications technology (ICT) often allows a number of solution methods.

Again, I comment on aspects of mathematics, tools, actions and thought involved in using this tool. The DGS could be considered as a set of tools, a toolbox, and I have selected one tool from this set. It is possible to co-ordinate the use of some DGSs with other digital tools but, in educational use, a DGS is generally used as a self contained system. Although 'geometry' is the G in DGS and the surface features are iconic, the system is digital, i.e. numeric. The method, the algorithm, the DGS uses to perform the construction, i.e. bisecting an angle, is not transparent to the user (this is sometimes referred to as a 'black box'); this lack of transparency is deeper than the lack of transparency of the straight edge and compass

[1] Perhaps we should replace the action word 'draw' with a computer equivalent, say, 'C-draw', as the physical action of drawing on pencil & paper and on most (at the time of writing) ICT environments, differ.

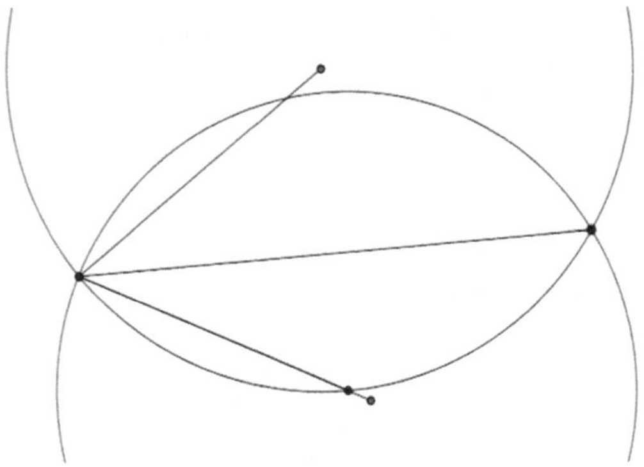

Fig. 2.6 Bisecting an angle using only the commands 'line segment', 'intersection point' and compass

construction of the angle bisector, one would have to find out about the algorithm behind the DGS construction and how it was implemented in the specific DGS.

Another transparency aspect of DGS (and many other mathematical software systems) is that the teacher can choose what functions are available for the student or not. For example, s/he could hide the angle-bisector command, and other related commands. In such a case it might be expected that teachers could force students to employ geometric constructions. For example, with only the commands 'line segment', 'intersection point' and compass, we can bisect an angle as shown in Fig. 2.6. But students can be very inventive and I have found students who, in such a situation, measure the given angle, then draw an approximate bisecting line, measure the angle made by that line and then move the line until the second angle measures half the first one.

It could be argued that the mathematical educational value of this ICT-construction as a stand-alone task is virtually non-existent; the user has merely learnt that a specific sequence of clicking actions results in the appearance of an angle bisector. But potential mathematical educational value exists through the action of 'dragging'. In the straight edge and compass construction the geometric Fig. 2.2 is static, it does not move relative to the paper on which it is constructed. The D in DGS is 'dynamic' and once a dependent DGS object is constructed (the angle bisector above is a dependent object, it depends on the line segments which depend on the points) it is dragged when independent objects are dragged. This feature has the potential to help users 'see' dependencies (which are mathematical relationships) in constructions through the recognition of mathematical invariances.

A potential mathematical educational value also exists when this task is a sub-task of a larger mathematical task such as constructing the inscribed circle to a triangle (see Fig. 2.7a). The speed and accuracy with which angle bisectors of the angles of a triangle can be constructed both arguably aid the user in not getting bogged down in a sub-task of a larger task. Dragging can (when the geometric object is constructed with appropriate mathematical relationships), again, help users recognise mathematical invariances (see Fig. 2.7b). A further potential mathematical educational value can also exist in constructing the dependencies. For example in the static straight edge and compass construction of the inscribed circle

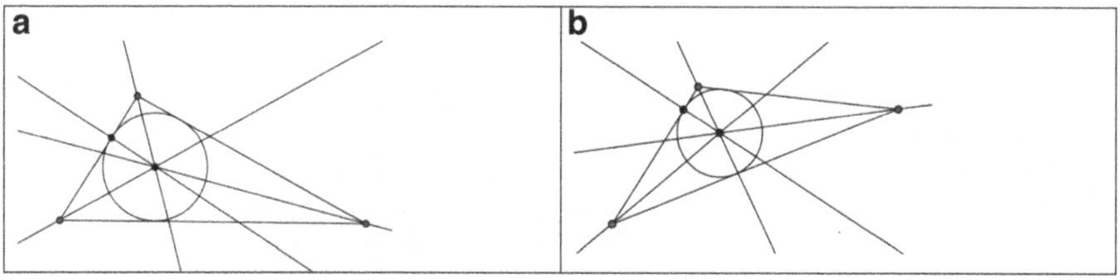

Fig. 2.7 (**a**) A DGS construction of the inscribed *circle* to a *triangle*. (**b**) The result of 'dragging' (**a**)

to a triangle, the in-centre is located and the inscribed circle can be drawn with the compass 'by eye', i.e. so that it looks like the inscribed circle. This method will not result in dependencies which will successfully drag in a DGS; a radial point must be constructed on a line segment of the triangle so that the line through this point and the in-centre is perpendicular to the line segment the point lies on, if the construction is to successfully drag (this point can be seen in Figs. 2.7a, b).

But these are potential mathematical educational values, they won't necessarily be realised in practice. One aspect of realising these values is in users linking actions with mathematical relationships.

2.5 Bisecting an Angle with a Book

I end with a description of a book as a mathematical tool to illustrate that an artefact can be used for a very different purpose to that which it was designed and, related to this, to illustrate that a tool to do mathematics does not need to be designed as a mathematical tool. I first note that a book could be used to imitate a compass: using one side of the book, use one end of this side as the point of the compass and the other end of this side, together with a pencil, as the pencil end of the compass. This is not how I will use a book in this example, I will use it as a kind of set square. You need a large sheet of paper. I will explain Figs. 2.8a–d.

Figure 2.8a simply shows a book placed near the angle to be bisected. Figure 2.8b shows the first construction line being drawn: the book is positioned so that one corner of the book is at the vertex of the angle to be bisected and one side of the book placed against one of the two rays describing the angle. This last procedure is repeated with the second ray in Fig. 2.8c. Figure 2.8d shows the two construction lines and the angle bisector (which joins the vertex to the point where the two construction lines meet).

I have tried this, and other 'constructions by the book', construction with groups of in-service teachers and most of them were able to explain why the construction worked. The reason for this transparency with regard to the user seeing why the construction works, I assume, is linked to the greater transparency of Fig. 2.2, compared to Fig. 2.1, as discussed above.

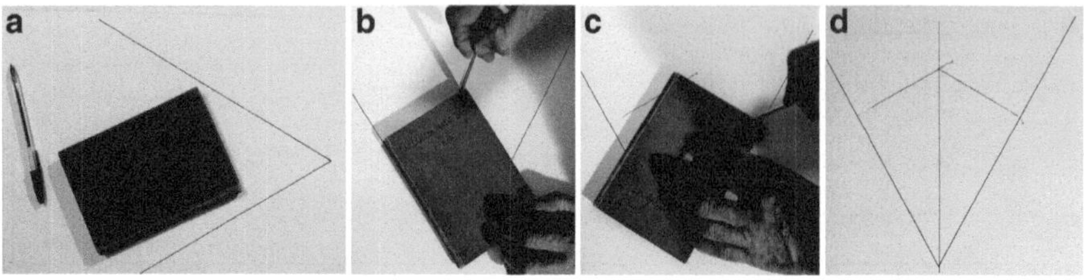

Fig. 2.8 Steps in bisecting an angle using a book

I am not suggesting wholesale advocacy of a new method of teaching geometrical constructions but I must confess to having a certain affection for this unusual mathematical tool. Constructions using a book can be quite interesting. For example, the 'A series' of books, i.e. A4, have an inbuilt mathematical property, the ratio of their sides is 2:√2. This property can be used to construct an octagon using an A series book but the construction of an octagon by any book is not possible.

The examples in this chapter were chosen to show different tools being used to perform a task, not to exhaust all possibilities. There are other means and tools available to bisect an angle. Indeed it is possible to construct the bisector without an obvious tool, by folding the sheet on which the angle is drawn. The tool in this case is the sheet of paper coupled with knowledge of an axis of symmetry.

Chapter 3
The Life of Modern Homo Habilis Mathematicus: Experimental Computation and Visual Theorems

Jonathan M. Borwein

3.1 Introduction

> The computer has in turn changed the very nature of mathematical experience, suggesting for the first time that mathematics, like physics, may yet become an empirical discipline, a place where things are discovered because they are seen.—David Berlinski[1]

In this chapter I want to talk, both generally and personally, about the use of tools in the practice of modern research mathematics. To focus my attention I have decided to discuss the way I and my research group members have used tools primarily computational (visual, numeric and symbolic) during the past 5 years. When the tools are relatively accessible I shall exhibit details; when they are less accessible I settle for illustrations and discussion of process.

Long before current graphic, visualization and geometric tools were available, John E. Littlewood, 1885–1977, wrote in his delightful Miscellany:

> A heavy warning used to be given [by lecturers] that pictures are not rigorous; this has never had its bluff called and has permanently frightened its victims into playing for safety. Some pictures, of course, are not rigorous, but I should say most are (and I use them whenever possible myself). (Littlewood, 1953, p. 53)

Over the past 5 years, the role of visual computing in my own research has expanded dramatically. In part this was made possible by the increasing speed and storage capabilities—and the growing ease of programming—of modern multi-core computing environments (Borwein, Skerritt, & Maitland, 2013). But, at least as much, it has been driven by my group's paying more active attention to the possibilities for graphing, animating or simulating most mathematical research activities.

[1] In "Ground Zero: A Review of The Pleasures of Counting, by T. W. Koerner," 1997.

J. Monaghan et al., *Tools and Mathematics*, Mathematics Education Library,
DOI 10.1007/978-3-319-02396-0_3

3.1.1 Who I Am and How I Got That Way

In my academic lifetime, tools went from graph paper, log tables, slide rules and slipsticks to today's profusion of digital computational devices. Along the way came the CURTA, HP programmable calculators, TI calculators, and other transitional devices not to mention my grandfather's business abacus. When a radically new tool has come along, it can be adapted very quickly as was the case with the use of log-tables in the early seventeenth century after Brigg's 1616 improvement of Napier's 1614 logarithms and the equally rapid abandonment of slide-rule in the 1970s after 350 years of ubiquity. I feel obliged to record that well into the 1980s business mathematics texts published compound interest tables with rates up to 5 % when mortgage rates were well over 20 %.

Let me next reprise material I wrote for a chapter for the 2015 collection *The Mind of a Mathematician* (Borwein, 2012).

I wish to aim my scattered reflections in generally the right direction: I am more interested in issues of creativity á la Hadamard (Borwein, Liljedahl, & Zhai, 2010) than in Russell and foundations, or Piaget and epistemology. . .and I should like a dash of "goodwill computing" thrown in. More seriously, I wish to muse about how we work, what keeps us going, how the mathematics profession has changed and how "plus ça change, la plus ça reste pareil",[2] and the like while juxtaposing how we perceive these matters and how we are perceived. Elsewhere, I have discussed at length my own views about the nature of mathematics from both an aesthetic and a philosophical perspective (see, e.g., Gold & Simons, 2008; Sinclair, Pimm, & Higginson, 2007).

I have described myself as 'a computer-assisted quasi-empiricist'. For present more psychological proposes I will quote approvingly from Brown (2009, p. 239):

> . . .Like Ol' Man River, mathematics just keeps rolling along and produces at an accelerating rate "*200,000 mathematical theorems of the traditional handcrafted variety . . .annually.*" Although sometimes proofs can be mistaken—sometimes spectacularly—and it is a matter of contention as to what exactly a "proof" is—there is absolutely no doubt that the bulk of this output is correct (though probably uninteresting) mathematics.—Richard C. Brown

I continued: Why do we produce so many unneeded results? In addition to the obvious pressure to publish and to have something to present at the next conference, I suspect Irving Biederman's observations below plays a significant role.

> "While you're trying to understand a difficult theorem, it's not fun," said Biederman, professor of neuroscience in the USC College of Letters, Arts and Sciences. . . ."But once you get it, you just feel fabulous." . . .The brain's craving for a fix motivates humans to maximize the rate at which they absorb knowledge, he said. . . ."I think we're exquisitely tuned to this as if we're junkies, second by second."—Irving Biederman[3]

---

[2] For an excellent account of the triumphs and vicissitudes of Oxford mathematics over eight centuries, see Fauvel, Flood, and Wilson (1999). The description of Haley's ease in acquiring equipment (telescopes) and how he dealt with inadequate money for personnel is by itself worth the price of the book.

[3] Discussing his article in the *American Scientist* at www.physorg.com/news70030587.html.

Mathematical tools are successful especially when they provide that rapid 'fix' of positive reinforcement. This is why I switched from competitive chess to competitive bridge. Being beaten was less painful and quicker, reward was more immediate.[4]

In Borwein (2012) I again continued: Take away all success or any positive reinforcement and most mathematicians will happily replace research by adminstration, more and (hopefully better) teaching, or perhaps just a favourite hobby. But given just a little stroking by colleagues or referees and the occasional opiate jolt, and the river rolls on. For a fascinating essay on the modern university in 1990 I recommend Giametti (1990).

The pressure to publish is unlikely to abate and qualitative measurements of performance[5] are for the most part fairer than leaving everything to the whim of one's Head of Department. Thirty-five years ago my career review consisted of a two-line mimeo "your salary for next year will be ..." with the relevant number written in by hand.

At the same time, it is a great shame that mathematicians have a hard time finding funds to go to conferences just to listen and interact. Csikszentmihalyi (1997) writes:

> [C]reativity results from the interaction of a system composed of three elements: a culture that contains symbolic rules, a person who brings novelty into the symbolic domain, and a field of experts who recognize and validate the innovation. All three are necessary for a creative idea, product, or discovery to take place.—Mihalyy Csikszentmihalyi

We have not paid enough attention to what creativity is and how it is nurtured. Conferences need audiences and researchers need feedback other than the mandatory "nice talk" at the end of a special session. We have all heard distinguished colleagues mutter a stream of criticism during a plenary lecture only to proffer "I really enjoyed that" as they pass the lecturer on the way out. A communal view of creativity requires more of the audience.

And the computer as provider of tools can often provide a more sympathetic and caring, even better educated, audience.

3.1.2 What Follows

We first discuss briefly in Sect. 3.2 what is meant by a *visual theorem*. In Sect. 3.3 we talk about *experimental computation* and then turn to *digital assistance*. In a key Sect. 3.4 we examine a substantial variety of accessible examples of these three concepts. In Sect. 3.5 we discuss simulation as a tool for pure mathematics.

In the final three sections, we turn to three more sophisticated sets of case studies. They can none-the-less be followed without worrying about any of the more complicated formulae. First in Sect. 3.6 comes *dynamic geometry* (iterative

[4] I played twice against Cambridge on losing Oxford bridge teams.

[5] For an incisive analysis of citation metrics in mathematics, I thoroughly recommend the IMU report and responses at: http://openaccess.eprints.org/index.php?/archives/417-Citation-Statistics-International-Mathematical-Union-Report.html.

reflection methods Aragon & Borwein, 2013) and *matrix completion problems*[6] (applied to *protein conformation* Aragon, Borwein, & Tam, 2014) (see Case Studies I). In Sect. 3.7 for the second set of Case Studies, we then turn to numerical analysis (see Case Studies II). I end in Sect. 3.8 with description of recent work from my group in *probability* (behaviour of short random walks Borwein & Straub, 2013; Borwein, Straub, Wan, & Zudilin, 2012) and *transcendental number theory* (normality of real numbers Aragon, Bailey, Borwein, & Borwein, 2013).

3.1.3 Some Early Conclusions

I have found it is often useful to make some conclusions early. So here we go.

1. Mathematics can be done *experimentally* (Bailey & Borwein, 2011a) (it is fun) using computer algebra, numerical computation and graphics: *SNaG* computations. Tables and pictures are experimental data but you cannot stop thinking.
2. Making mistakes is fine as long as you learn from them, and keep your eyes open (conquering fear).
3. You cannot use what you do not know and what you know you can usually use. Indeed, you do not need to know much before you start research in a new area (as we shall see).
4. Tools can help democratize appreciation of and ability in mathematics.

3.2 Visual Theorems and Experimental Mathematics

In a 2012 study *On Proof and Proving* (ICMI, 2012), the International Council on Mathematical Instruction wrote:

> The latest developments in computer and video technology have provided a multiplicity of computational and symbolic tools that have rejuvenated mathematics and mathematics education. Two important examples of this revitalization are experimental mathematics and visual theorems.

3.2.1 Visual Theorems

By a *visual theorem*[7] I mean a picture or animation which gives one confidence that a desired result is true; in Giaquinto's sense that it represents "coming to believe it in an independent, reliable, and rational way" (either as discovery or validation) as

---

[6] See http://www.carma.newcastle.edu.au/jon/Completion.pdf and http://www.carma.newcastle.edu.au/jon/dr-fields11.pptx.

[7] See http://vis.carma.newcastle.edu.au/.

described in Bailey and Borwein (2011b). While we have famous pictorial examples purporting to show things like all triangles are equilateral, there are equally many or more bogus symbolic proofs that '$1+1=1$'. In all cases 'caveat emptor'.

Modern technology properly mastered allows for a much richer set of tools for discovery, validation, and even rigorous proof than our precursors could have ever imagined would come to pass—and it is early days. That said just as books on ordinary differential equations have been replaced by books on *dynamical systems*, the word *visual* now pops up frequently in book titles. Unless ideas about visualization are integrated into the text this is just marketing.

3.2.2 On Picture-Writing

The *ordinary generating function* associated with a sequence $a_0, a_1, \ldots, a_n, \ldots$ is the formal series[8]

$$A(x) := \sum_{k=0}^{\infty} a_k x^k \tag{3.1}$$

while the *exponential generating function* is

$$A(x) := \sum_{k=0}^{\infty} a_k \frac{x^k}{k!}. \tag{3.2}$$

Both forms of generating function are ideally suited to computer-assisted discovery.

George Pólya, in an engaging eponymous *American Mathematical Monthly* article, provides three compelling examples of converting pictorial representations of problems into generating function solutions (Pólya, 1956):

1. *In how many ways can you make change for a dollar?*

   This leads to the (US currency) *generating function*

   $$\sum_{k=1}^{\infty} P_k x^k = \frac{1}{(1-x^1)(1-x^5)(1-x^{10})(1-x^{25})(1-x^{50})},$$

   which one can easily expand using a *Mathematica* command,

```
Series[1/((1-x)*(1-x^5)*(1-x^10)*(1-x^25)*(1-x^50)), {x, 0, 100}]
```

   to obtain $P_{100} = 292$ (242 for Canadian currency, which lacks a 50 cent piece). Pólya's illustration is shown in Fig. 3.1.

[8] In computational cases we often use only the initial segment of the series and so we do not care whether it converges or not.

We look at a related generating function for counting additive partitions in Example 3.4.6.

2. *Dissect a polygon with n sides into $n - 2$ triangles by $n - 3$ diagonals and compute D_n, the number of different dissections of this kind.*

   This is illustrated in Fig. 3.2 and leads to the fact that the generating function for $D_3 = 1, D_4 = 2, D_5 = 5, D_6 = 14, D_7 = 42, \ldots$

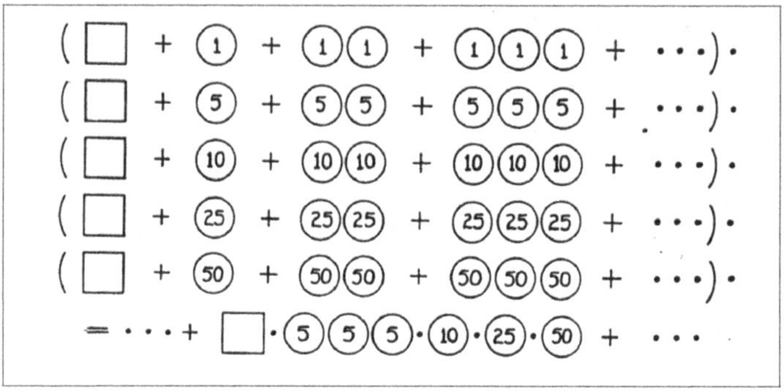

Fig. 3.1 Pólya's illustration of the change solution (courtesy Mathematical Association of America)

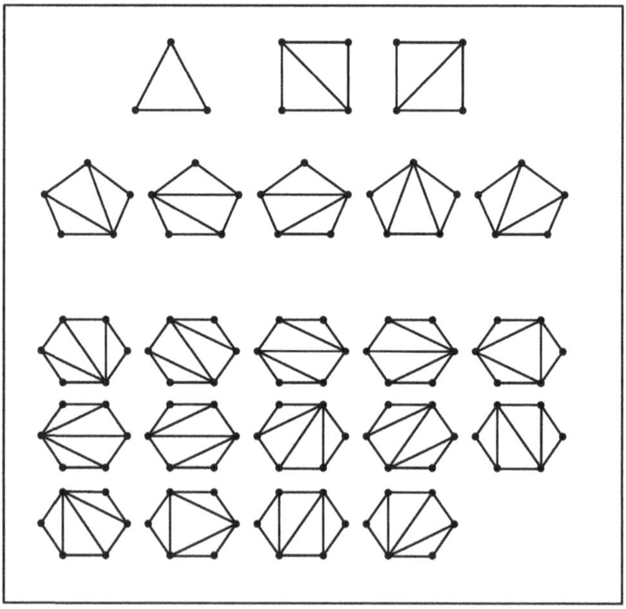

Fig. 3.2 The first few sets of dissections

$$D(x) = \sum_{k=1}^{\infty} D_k x^k$$

satisfies

$$D(x) = x\,[1 + D(x)]^2,$$

whose solution is therefore

$$D(x) = \frac{1 - 2x - \sqrt{1 - 4x}}{x}.$$

The *Mathematica* command

```
Series[((1 - 2 x) - Sqrt[1 - 4 x])/x, {x, 0, 10}]
```

returns

$$2x + 4x^2 + 10x^3 + 28x^4 + 84x^5 + 264x^6 + 858x^7 \\ + 2860x^8 + 9724x^9 + 33592x^{10} + O(x^{11}).$$

with list of coefficients

$$\{0, 2, 4, 10, 28, 84, 264, 858, 2860, 9724, 33592\}$$

and D_{n+2} turns out to be the nth Catalan number $\binom{2n}{n}/(n+1)$. This can be discovered using Sloane's wonderful *Online Encyclopedia of Integer Sequences* as illustrated in Fig. 3.3. Note that we only used the first six non-zero terms and had four left to 'confirm' our experiment.

3. *Compute T_n, the number of different (rooted) trees with n knots.*[9]

   This is a significantly harder problem so we say less:

   The ordinary generating function of the T_n becomes a remarkable result due to Cayley, namely

[9] Roots are now more commonly called vertices or nodes. For rooted labeled trees (and hence labeled trees): http://www.math.ucla.edu/~pak/hidden/papers/Moon-counting_labelled_trees.pdf is Moon's monograph with a nice discussion of the history of Cayley's formula, including the fact that Cayley himself acknowledged that Borchardt had proved it earlier, and that it appeared without proof in a work of Sylvester. A more modern (if not necessarily more relevant or accurate) reference is http://en.wikipedia.org/wiki/Cayley.

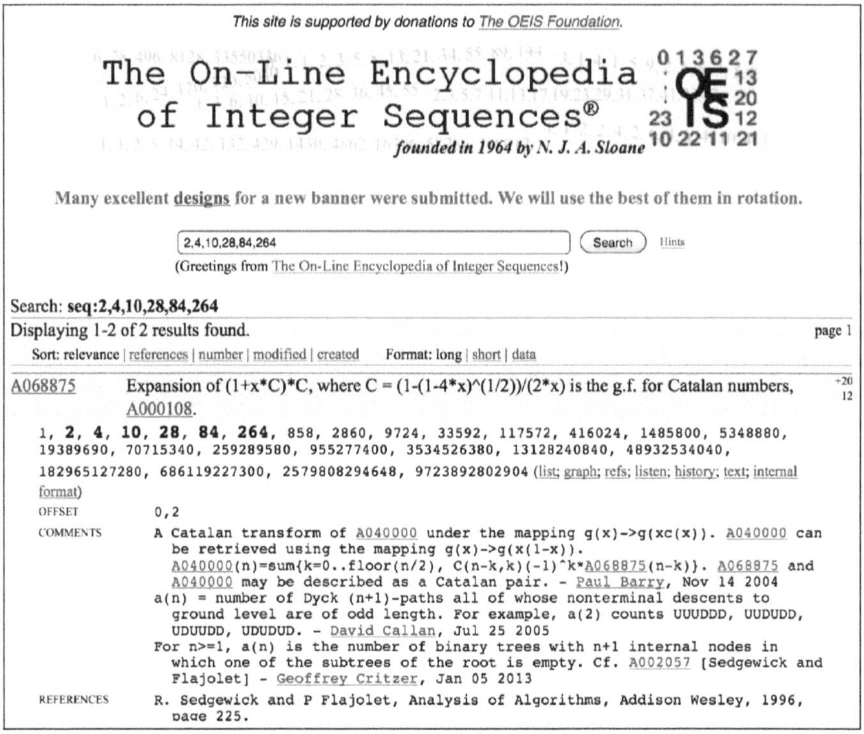
This site is supported by donations to The OEIS Foundation.

The On-Line Encyclopedia of Integer Sequences®

founded in 1964 by N. J. A. Sloane

Many excellent designs for a new banner were submitted. We will use the best of them in rotation.

2,4,10,28,84,264 Search Hints

(Greetings from The On-Line Encyclopedia of Integer Sequences!)

Search: **seq:2,4,10,28,84,264**

Displaying 1-2 of 2 results found. page 1

Sort: relevance | references | number | modified | created Format: long | short | data

A068875 Expansion of (1+x*C)*C, where C = (1-(1-4*x)^(1/2))/(2*x) is the g.f. for Catalan numbers, A000108. +20 12

1, **2, 4, 10, 28, 84, 264**, 858, 2860, 9724, 33592, 117572, 416024, 1485800, 5348880, 19389690, 70715340, 259289580, 955277400, 3534526380, 13128240840, 48932534040, 182965127280, 686119227300, 2579808294648, 9723892802904 (list; graph; refs; listen; history; text; internal format)

OFFSET 0,2

COMMENTS A Catalan transform of A040000 under the mapping g(x)->g(xc(x)). A040000 can be retrieved using the mapping g(x)->g(x(1-x)). A040000(n)=sum{k=0..floor(n/2), C(n-k,k)(-1)^k*A068875(n-k)}. A068875 and A040000 may be described as a Catalan pair. - Paul Barry, Nov 14 2004

a(n) = number of Dyck (n+1)-paths all of whose nonterminal descents to ground level are of odd length. For example, a(2) counts UUUDDD, UUDUDD, UDUUDD, UDUDUD. - David Callan, Jul 25 2005

For n>=1, a(n) is the number of binary trees with n+1 internal nodes in which one of the subtrees of the root is empty. Cf. A002057 [Sedgewick and Flajolet] - Geoffrey Critzer, Jan 05 2013

REFERENCES R. Sedgewick and P Flajolet, Analysis of Algorithms, Addison Wesley, 1996, page 225.

Fig. 3.3 Using https://oeis.org/ to identify the Catalan numbers

$$T(x) = \sum_{k=1}^{\infty} T_k x^k = x \prod_{k=1}^{\infty} (1 - x^k)^{-T_k}, \tag{3.3}$$

where remarkably the product and the sum share their coefficients. This can be used to produce a recursion for T_n in terms of $T_1, T_2, \ldots, T_{n-1}$, which starts: $T_1 = 1, T_2 = 1, T_3 = 2, T_4 = 4, T_5 = 9, T_6 = 20, \ldots$.

In each case, Pólya's main message is that one can usefully draw pictures of the component elements—(a) in pennies, nickels dimes and quarters (plus loonies in Canada and half dollars in the USA), (b) in triangles and (c) in the simplest trees (with the fewest knots).

That said, I often find it easier to draw pictures from generating functions rather than go in the other direction. In any event, Pólya's views on heuristic reasoning and his books on problem solving (Pólya, 1981, 1990) remain as engaging, if idiosyncratic, today as when first published.[10]

[10] I mention also Klein and Grothendieck's *desin d'enfant* see www.ams.org/notices/200307/what-is.pdf.

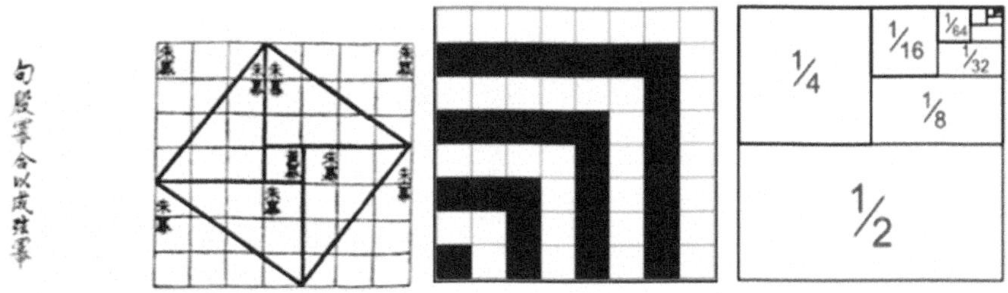

Fig. 3.4 Three classical proofs without words

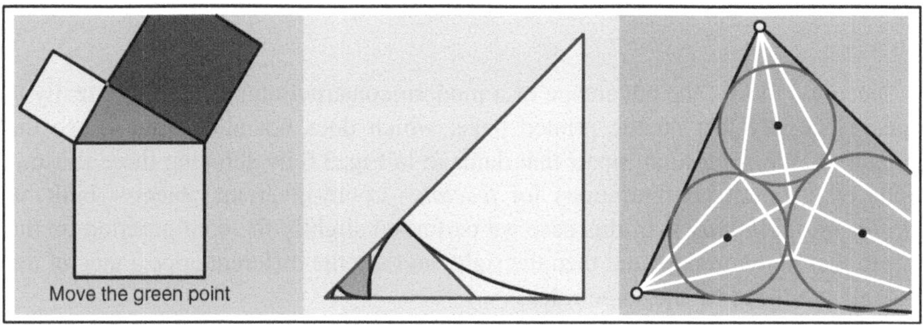

Fig. 3.5 Three modern proofs without words

3.2.2.1 Proofs Without Words

In Fig. 3.4 we reproduce three classic *proofs without words*—though most such proofs benefit from a few words of commentary. In Fig. 3.5 we display three modern (dynamic geometry) proofs without words from http://cinderella.de/files/HTMLDemos/Main.html.[11]

Figure 3.4 shows from left to right the following three results:

1. Pythagoras theorem;
2. $1 + 3 + 5 + (2n - 1) = n^2$;
3. $1/2 + 1/4 + 1/8 + \cdots = 1$.

The Pythagorean proof is from the *Zhou Bi Suan Jing* which dates from the Zhou Dynasty (1046 BCE–256 BCE), and is one of the oldest recorded.

Figure 3.5 shows from left to right the following three results:

1. Pythagoras theorem;
2. $\sqrt{2}$ is irrational as suggested by Tom Apostol[12];
3. How to inscribe three tangent circles in a triangle.

[11] See also http://www.usamts.org/About/U_Gallery.php.

[12] Assume the large triangle is the smallest 45° right-angled triangle with integer sides. The complement of the brown kite is a smaller such triangle.

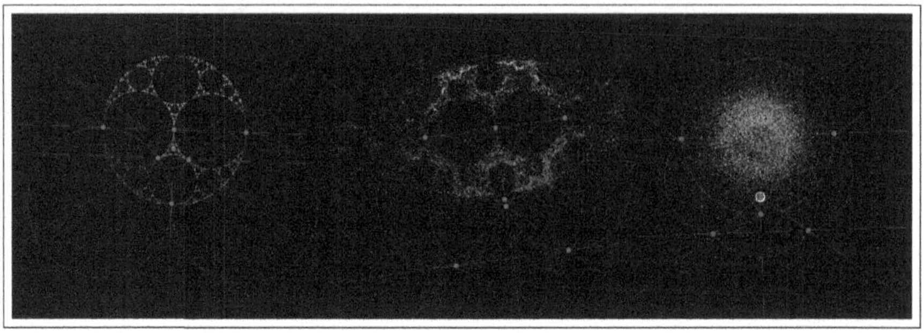

Fig. 3.6 Three fractals generated by different Apollonian configurations

Needless to say, the advantage of a modern construction—when there really is one—is largely lost on the printed page, which does not allow one to see the dynamics. We somewhat repair that damage in Fig. 3.6 by showing three illustrations of different configurations for *fractals*—zoom-invariant objects—built on *circles of Apollonius*.[13] In this case we perturbed slightly the configuration on the left to that in the middle and then the right, and see the different appearance of the fractals produced by the same rules.

3.3 Experimental Mathematics

The same ICMI study (2012), quoting (Borwein & Devlin, 2008, p. 1), says enough about the meaning of *experimental mathematics* for our current purposes:

> Experimental mathematics is the use of a computer to run computations—sometimes no more than trial-and-error tests—to look for patterns, to identify particular numbers and sequences, to gather evidence in support of specific mathematical assertions that may themselves arise by computational means, including search.
>
> Like contemporary chemists—and before them the alchemists of old—who mix various substances together in a crucible and heat them to a high temperature to see what happens, today's experimental mathematicians put a hopefully potent mix of numbers, formulas, and algorithms into a computer in the hope that something of interest emerges.

3.3.1 Experimental Mathodology

I originally mistyped 'mathodology' intending 'methodology', but I liked the mistake and have kept it. We started (Borwein & Devlin, 2008) with Justice Potter Stewart's famous 1964 comment on pornography: "I know it when I see it."

[13] See http://en.wikipedia.org/wiki/Circles_of_Apollonius.

A bit less informally, by *experimental mathematics* I intend, as discussed in Borwein and Bailey (2008) and elsewhere:

1. Gaining insight and *intuition*;
   - We illustrate this repeatedly below by drawing many simple functions. Almost always, as in Example 3.4.3, we see things in a picture that were not clear in our mind's eye.
   - Sometimes, as in Example 3.4.2, a new pattern jumps out that we were not originally intent on studying. By contrast, in Example 3.4.9 we show how the computer can tell you things, such as that a number is algebraic, that you can then verify but probably would never find.
2. *Discovering* new relationships;
   - Computers generate patterns we might well not see by hand. See Examples 3.4.2, 3.4.5, 3.4.6 and 3.4.10.
3. *Visualizing* math principles;
   - Computers allow one to switch representations easily. This can be like drawing a curtain open as in Example 3.3.4 or Example 3.4.6.
4. *Testing* and especially *falsifying* conjectures;
   - See Example 3.4.1 where we conclude one equality is invalid and are led to a proof of why another similar looking one holds.
   - Examples 3.4.16 and 3.4.17 underscore that seemingly compelling patterns can fail to be hold. Learning how to trust one's judgement is a subtle context-dependent matter.
5. *Exploring* a possible result to see if it *merits* formal proof;
   - In a traditional Lemma–Theorem–Corollary version of deductive mathematics, one has to prove every step of a chain of arguments to get to the end. Suppose there are six steps in a complicated result, and the third is a boring but hard equation, whose only value is that it leads to step six. Then it is appropriate to *challenge* step six a lot, before worrying about proving step three.
6. *Suggesting* approaches for formal proof;
   - For me this connotes computer-assisted or computer-directed proof and is quite far from *Formal Proof* as was the topic of a special issue of the *Notices of the AMS* in December 2008.
   - See Examples 3.4.1 and 3.4.11 which look at how our tools change both induction and integration.
7. *Computing* replacing lengthy hand derivations;
   - Example 3.3.1 discusses this for matters like taking roots, or factoring large numbers.

- It also the case that many computations that used to be too lengthy to perform by hand are no longer so. For instance, the *Maple* fragment

$$\mathrm{add}(\mathrm{ithprime}(k), k = 1..100000);$$

returned the sum of the first 10^5 primes, 62660698721, in 0.171 s. Adding a million took much longer though! My preference on tests, rather than banning calculators or computers, is to adapt the questions to make them computationally aware.[14]

8. *Confirming* analytically derived results.

- I illustrate this in Example 3.4.12 by confirming some exact results knowing only their general structure.

All of these uses play a central role in my daily research life. We will see all of these eight notions illustrated in the explicit examples of Sect. 3.3.2 and of Sect. 3.4.

3.3.2 When Science Becomes Technology

What tools we choose to use—and when—is a subtle and changeable issue.

Example 3.3.1 (When Science Becomes Technology). We 'unpack' methods when we want to understand them or are learning them for the first time. Once we or our students have mastered a new tool we 'repack' it. For instance,

$$2^{2^7} + 1 = 340282366920938463463374607431768211457$$

which factors as

$$(59649589127497217)(5704689200685129054721).$$

If we are teaching or taking a course in factorization methods, we may well want to know 'how' this was done. In most contexts, we are happy to treat the computer as a reliable tool and to take the answer without further introspection.

In like fashion,

$$t := 1.259921049894873164767210607 28\ldots = \sqrt[3]{2}$$

will be computed by most packages to the displayed precision. We can confirm this since

[14] Though how to stop things like a student scanning a question and then going to the toilet to consult Wolfram Alpha is a never-ending issue.

$$t^3 = 2.0000000000000000000000000001\ldots$$

If we wish to understand what the computer has done—probably by *Newton's method*, we must go further, but if we only wish to use the answer that is irrelevant. The first is science or research, the second is technology. ◇

The William Lowell Putnam competition taken each year by the strongest North American undergraduate students has conventionally had one easy question (out of 12) based on the current year.

Example 3.3.2 (A 1998 Putnam Examination Problem). The problem was
Let N be the positive integer with 1998 decimal digits, all of them 1; that is, $N = 1111\ldots 11.$ Find the thousandth digit after the decimal point of $\sqrt{N}$.

This can be done by brute force

```
> evalf[10](sqrt(add(10^k,k=0..1997))/10^1000;
```

which is not what the posers had in mind. ◇

Example 3.3.3 (A 1995 Putnam Examination Problem). The problem requests
Evaluate:

$$\sqrt[8]{2207 - \cfrac{1}{2207 - \cfrac{1}{2207 - \ldots}}} \tag{3.4}$$

Express your answer in the form $(a + b\sqrt{c})/d$, where a, b, c, d are integers.

Proof. If we call the *repeated radical* above α, the request is to solve for

$$\alpha^8 = 2207 - \cfrac{1}{2207 - \cfrac{1}{2207 - 1/\alpha^8}},$$

and a `solve` request to a CAS will return

$$\left(\frac{2207}{2} + \frac{987\sqrt{5}}{2}\right)^{1/8} = \frac{3 + \sqrt{5}}{2}. \tag{3.5}$$

We may determine the last reduction in many ways (1) via *identify*, (2) using the *inverse symbolic calculator (ISC)*, (3) using a *resolvent* computation to find the quadratic polynomial satisfied by α as given by Eq. (3.5), or (4) by repeatedly computing the square root. Indeed `identify` will return the answer directly from (3.4) which already agrees with the limit to 20 places. □

```
> with(numtheory): cfrac(Pi, 6)=cfrac(Pi, 6, 'quotients');
```

$$3+\cfrac{1}{7+\cfrac{1}{15+\cfrac{1}{1+\cfrac{1}{292+\cfrac{1}{1+\cfrac{1}{1+\ldots}}}}}} = [3, 7, 15, 1, 292, 1, 1, \ldots]$$

Fig. 3.7 The simple continued fraction for π (L) in compact form (R)

With access to computation the problem becomes too straight-forward. ◇

I next recall some continued fraction notation. Figure 3.7 shows the two most common ways of writing a *simple or regular continued fraction*—in this case for π. For any $\alpha > 0$, this represents the process of going from

$$\alpha \to \alpha' := \frac{1}{\alpha - \lfloor \alpha \rfloor}$$

and repeating the process, while recording the integer part $\lfloor \alpha \rfloor$ each time. This is usually painful to do by hand but is easy for our computer.

Example 3.3.4 (Changing Representations). Suppose I wish to examine the numbers

$$\alpha := 0.697774657964007982006790592552$$

and

$$\beta := 0.920016900019100086226595799993.$$

As floating point numbers there is nothing to distinguish them; but the *Maple* instruction `convert(alpha,confrac);` returns the *simple continued fraction* for α in compact form (Borwein and Bailey, 2008)

$$\alpha = [0, 1, 2, 3, 4, 5, 6, 7, 8, 9, 10, 11, 12, 13, 14, 15, 16, \ldots]$$

while `convert(beta,confrac);` returns

$$\beta = [0, 1, 11, 1, 1, 94, 6, 2, 9, 2, 1, 5, 1, 6, 7, 3, 4, 24, 1, 8, 1, 2, 1, 2, 1, \ldots].$$

So, in this new representation, the numbers no longer look similar. As described in Borwein and Bailey (2008), Borwein and Devlin (2008), and Bailey and Borwein (2011a), continued fractions with terms in arithmetic progression are well studied, and so there are several routes now open to discovering that $\alpha = I_1(2)/I_0(2)$ where for $\nu = 0, 1, 2, \ldots$

Fig. 3.8 The golden mean $a + b : a = a : b$

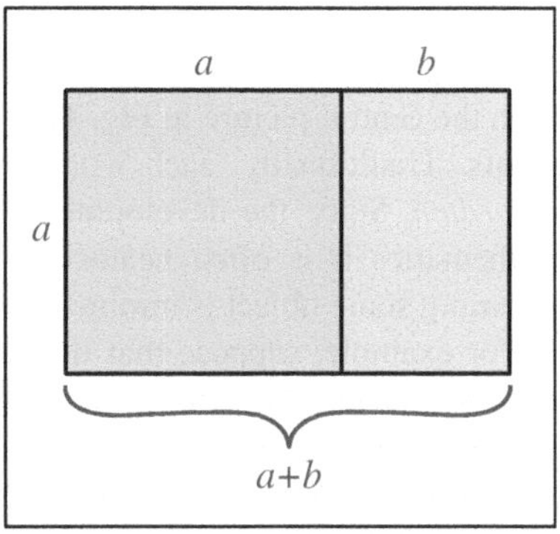

$$I_\nu(2z) = z^\nu \sum_{k=0}^{\infty} \frac{(z^2)^k}{k!(\nu + k)!}.$$

For instance, on May 23, 2014, entering `"continued fraction" "arithmetic progression"` into *Google* returned 23,700 results of which the first http://mathworld.wolfram.com/ContinuedFractionConstant.html gives the reader all needed information, as will the use of the *ISC*. My purpose here was only to show the potential power of changing a representation. For example, the continued fraction of the irrational *golden mean* $\frac{\sqrt{5}+1}{2} = 1.6180339887499\ldots$ is $[1, 1, 1, \ldots]$. Figure 3.8 illustrates the golden mean, and also provides a proof without words that it is irrational as we discuss further in the next section.

It is a result of Lagrange that an irrational number is a quadratic if and only if it has a non-terminating but eventually repeating simple continued fraction. So quadratics are to continued fractions what rationals are to decimal arithmetic. This is part of their power. ♢

As the following serious quotation makes clear, when a topic is science and when it is technology is both time and place dependent.

> A wealthy (15th Century) German merchant, seeking to provide his son with a good business education, consulted a learned man as to which European institution offered the best training."If you only want him to be able to cope with addition and subtraction," the expert replied, "then any French or German university will do. But if you are intent on your son going on to multiplication and division – assuming that he has sufficient gifts – then you will have to send him to Italy."[15]

[15] Quoted from p. 577 of George Ifrah, "The Universal History of Numbers: From Prehistory to the Invention of the Computer", trans. from French, John Wiley, 2000. This was also quoted a century ago by Tobias Dantzig.

3.3.2.1 Minimal Configurations

Both the central picture in Fig. 3.5 and the picture in Fig. 3.8 illustrate irrationality proofs. Traditionally, each would have been viewed as showing a *reductio ad absurdum*. Since the development of modern set theory and of modern discrete mathematics it is often neater to view them as deriving a contradiction from assuming some object is minimal.

For example, suppose that the $(a + b) \times a$ rectangle was the smallest integer rectangle representing the golden mean in Fig. 3.8, then the $a \times b$ rectangle cannot exist. Because of the geometric simplicity of this argument, it is thought that this may be the first number the Pythagoreans realized was irrational. Figure 3.5, by contrast, illustrates a reductio. If we continue, we will eventually get to an impossibly small triangle with integer sides. A clean picture for minimality is shown in Fig. 3.9.

Example 3.3.5 (Sylvester's Theorem, Bailey & Borwein, 2011b). The theorem conjectured by Sylvester in the late nineteenth century establishes that *given a finite set of non-colinear points in the plane there is at least one 'proper' line through exactly two points.* The first proof 40 years later was very complicated. Figure 3.10 shows a now-canonical minimality proof.

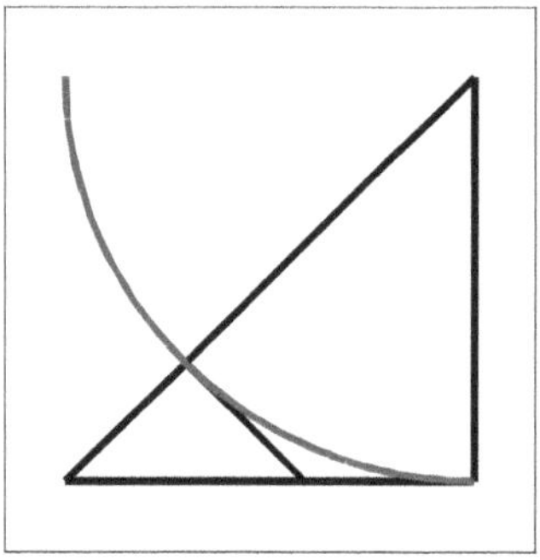

Fig. 3.9 A minimal configuration for irrationality of $\sqrt{2}$

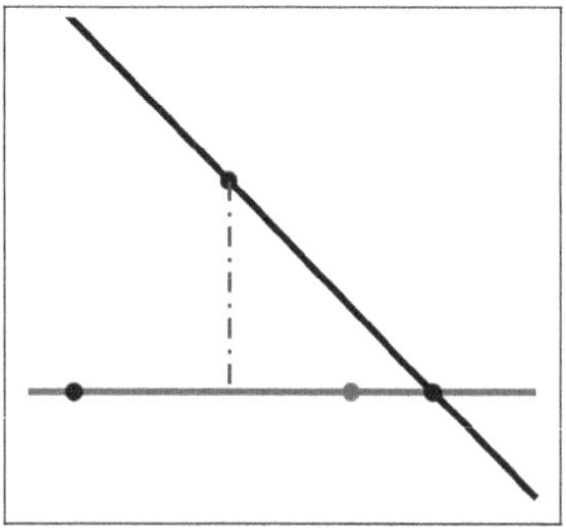

Fig. 3.10 A minimal configuration for Sylvester's theorem

The objects used in this picture are pairs (L,p) where L is a line through at least two points of the set and p is the closest point in the set but not on the line. We consider the $(\overline{L},\overline{p})$ with $\overline{p}$ closest to $\overline{L}$. We assert that $\overline{L}$ (the red horizontal line) has only two points of the set on it. If not two points lie on one side of the projection of $\overline{p}$ on $\overline{L}$. And now the black line L_0 through $\overline{p}$ and the farther point on L, and p_0 the red point nearer to the projection constructs a configuration (L_0,p_0) violating the minimality of $(\overline{L},\overline{p})$.

Subtle, ingenious and impossible to grasp without a picture! Here paper and coloured pencil are a fine tool. ◇

3.3.3 *Mathematical Discovery (or Invention)*

Giaquinto's attractive encapsulation: "In short, *discovering* a truth is coming to believe it in an independent, reliable, and rational way" Giaquinto (2007, p. 50) has the satisfactory consequence that a student can discover results whether known to the teacher or not. Nor is it necessary to demand that each dissertation be original (only that the results should be independently discovered).

Despite the conventional identification of mathematics with deductive reasoning, Kurt Gödel (1906–1978) in his 1951 Gibbs Lecture said: "If mathematics describes an objective world just like physics, there is no reason why inductive methods should not be applied in mathematics just the same as in physics". He held this view until the end of his life despite—or perhaps because of—the epochal deductive achievement of his incompleteness results.

Also, one discovers that many great mathematicians from Archimedes and Galileo—who apparently said "All truths are easy to understand once they are discovered; the point is to discover them."—to Gauss, Poincaré, and Carleson have emphasized how much it helps to "know" the answer. Two millennia ago Archimedes wrote to Eratosthenes[16] "For it is easier to supply the proof when we have previously acquired, by the method, some knowledge of the questions than it is to find it without any previous knowledge". Think of the *Method* as an ur-precursor to today's interactive geometry software—with the caveat that, for example, *Cinderella* actually does provide certificates for much Euclidean geometry.

As 2006 Abel Prize winner Lennart Carleson describes in his 1966 ICM speech on his positive resolution of Luzin's 1913 conjecture (about the pointwise convergence of Fourier series for square-summable functions) after many years of seeking a counterexample he decided none could exist. The importance of this confidence is expressed as follows:

> The most important aspect in solving a mathematical problem is the conviction of what is the true result. Then it took 2 or 3 years using the techniques that had been developed during the past 20 years or so.

[16] Introduction to his long-lost and recently re-constituted *Method of Mechanical Theorems*.

3.3.4 Digital Assistance

By *digital assistance* I mean use of *artefacts* as:

1. *Modern Mathematical Computer Packages*—symbolic, numeric, geometric or graphical. Symbolic packages include the commercial computer algebra packages *Maple* and *Mathematica*, and the open source SAGE. Primarily numeric packages start with the proprietary MATLAB and public counterparts *Octave* and *NumPy*, or the statistical package (*R*). The dynamic geometry offerings include *Cinderella, Geometer's Sketchpad, Cabri* and the freeware *GeoGebra.*
2. *Specialized Packages* or *General Purpose Languages* such as Fortran, C++, Python, CPLEX, PARI, SnapPea and MAGMA.
3. *Web Applications* such as: Sloane's Encyclopedia of Integer Sequences, the ISC,[17] Fractal Explorer, Jeff Weeks' Topological Games, or Euclid in Java.[18]
4. *Web Databases* including Google, MathSciNet, ArXiv, GitHub, Wikipedia, MathWorld, MacTutor, Amazon, Wolfram Alpha, the DLMF (Olver, Lozier, Boisvert, & Clark, 2012) (all formulas of which are accessible in MathML, as bitmaps, and in TE X) and many more that are not always so viewed.

All entail *data-mining* in various forms. Franklin (2005) argues Steinle's "exploratory experimentation" facilitated by "widening technology", as in pharmacology, astrophysics, medicine and biotechnology, is leading to a reassessment of what legitimates experiment; in that a "local model" is not now prerequisite. Sørenson (2010) cogently makes the case that *experimental mathematics*—as 'defined' above—is following similar tracks.

> These aspects of exploratory experimentation and wide instrumentation originate from the philosophy of (natural) science and have not been much developed in the context of experimental mathematics. However, I claim that e.g., the importance of wide instrumentation for an exploratory approach to experiments that includes concept formation also pertain to mathematics.

In consequence, boundaries between mathematics and the natural sciences and between inductive and deductive reasoning are blurred and getting more so. (See also Avigad, 2008.) I leave unanswered the philosophically vexing if mathematically minor question as to whether genuine *mathematical experiments* (as discussed in Borwein & Bailey, 2008) exist even if one embraces a fully idealist notion of mathematical existence. They sure feel like they do.

---

[17] Most of the functionality of the ISC is built into the "identify" function *Maple* starting with version 9.5. For example, identify(4.45033263602792) returns $\sqrt{3} + e$. As always, the experienced will extract more than the novice.

[18] A cross-section of such resources is available through www.carma.newcastle.edu.au/jon/portal.html and www.experimentalmath.info.

3.3.5 The Twentieth Century's Top Ten Algorithms

The modern computer itself, being a digital repurposable tool, is quite different from most of its analogue precursors. They could only do one or two things. The digital computer, of course, greatly stimulated both the appreciation of and need for algorithms and for algorithmic analysis.[19] These are what allows the repurposing. This makes it reasonable to view substantial mathematical algorithms as tools in their own right.

At the beginning of this century, Sullivan and Dongarra could write "Great algorithms are the poetry of computation", when they compiled a list of the ten algorithms having "the greatest influence on the development and practice of science and engineering in the twentieth century".[20] Chronologically ordered, they are:

- #1. 1946: **The Metropolis Algorithm for Monte Carlo**. Through the use of random processes, this algorithm offers an efficient way to stumble toward answers to problems that are too complicated to solve exactly.
- #2. 1947: **Simplex Method for Linear Programming**. An elegant solution to a common problem in planning and decision-making.
- #3. 1950: **Krylov Subspace Iteration Method**. A technique for rapidly solving the linear equations that abound in scientific computation.
- #4. 1951: **The Decompositional Approach to Matrix Computations**. A suite of techniques for numerical linear algebra.
- #5. 1957: **The Fortran Optimizing Compiler**. Turns high-level code into efficient computer-readable code.
- #6. 1959: **QR Algorithm for Computing Eigenvalues**. Another crucial matrix operation made swift and practical.
- #7. 1962: **Quicksort Algorithms for Sorting**. For the efficient handling of large databases.
- #8. 1965: **Fast Fourier Transform (FFT)**. Perhaps the most ubiquitous algorithm in use today, it breaks down waveforms (like sound) into periodic components.
- #9. 1977: **Integer Relation Detection**. A fast method for spotting simple equations satisfied by collections of seemingly unrelated numbers.
- #10. 1987: **Fast Multipole Method**. A breakthrough in dealing with the complexity of n-body calculations, applied in problems ranging from celestial mechanics to protein folding.

[19] The discussion in Guin, Ruthven, and Trouche (2005, Chap. 3) regarding the computer science issues arising when using *Maple* bears rereading a decade later.

[20] From "Random Samples", *Science* page 799, February 4, 2000. The full article appeared in the January/February 2000 issue of *Computing in Science & Engineering*.

I observe that eight of these ten winners appeared in the first two decades of serious computing, and that Newton's method was apparently ruled ineligible for consideration.[21] Most of the ten are multiply embedded in every major mathematical computing package. The last one is the only one that occurs infrequently in my own work.

Just as layers of software, hardware and middleware have stabilized, so have their roles in scientific and especially mathematical computing. When I first taught the simplex method more than 30 years ago, the texts concentrated on 'Y2K'-like tricks for limiting storage demands.[22] Now serious users and researchers will often happily run large-scale problems in MATLAB and other broad spectrum packages, or rely on *CPLEX* or, say, *NAG* library routines embedded in *Maple*.

While such out-sourcing or commoditization of scientific computation and numerical analysis is not without its drawbacks, I think the analogy with automobile driving in 1905 and 2005 is apt. We are now in possession of mature—not to be confused with 'error-free'—technologies. We can be *fairly* comfortable that *Mathematica* is sensibly handling round-off or cancellation error, using reasonable termination criteria and the like. Below the hood, *Maple* is optimizing polynomial computations using tools like Horner's rule, running multiple algorithms when there is no clear best choice, and switching to reduced complexity (*Karatsuba* or *FFT*-based) multiplication when accuracy so demands. Though, it would be nice if all vendors allowed as much peering under the bonnet as *Maple* does.

3.3.6 Secure Knowledge Without Proof

Given real floating point numbers

$$\beta, \alpha_1, \alpha_2, \ldots, \alpha_n,$$

Helaman Ferguson's *integer relation method*—see #9 of Sect. 3.3.5 above—called unhelpfully *PSLQ*, finds a nontrivial linear relation of the form

$$a_0\beta + a_1\alpha_1 + a_2\alpha_2 + \cdots + a_n\alpha_n = 0, \tag{3.6}$$

where a_i are integers—if one exists and provides an *exclusion bound* otherwise. This method is very robust. Given adequate precision of computation (Borwein and Bailey, 2008) it very rarely returns spurious relations.

[21] It would be interesting to construct a list of the ten most influential earlier algorithms.

[22] 'Y2K' was geek-speak for *the Year 2000* when there was concern that a trick used to save a storage bit decades earlier was going to crash all computers. It turned out to be much less serious, but who knew?

If $a_0 \neq 0$, then (3.6) assures β is in the rational vector space generated by

$$\{\alpha_1, \alpha_2, , , \ldots, , \alpha_n\}.$$

Moreover, as a most useful special case, if $\beta := 1, \alpha_i := \alpha^i$, then α is *algebraic* of *degree n* (see Example 3.4.9).

Quite impressively here is an unproven 2010 integer relation discovery by Cullen:

$$\frac{2^{11}}{\pi^4} \stackrel{?}{=} \sum_{n=0}^{\infty} \frac{(\frac{1}{4})_n (\frac{1}{2})_n^7 (\frac{3}{4})_n}{(1)_n^9} (21 + 466n + 4340n^2 + 20632n^3 + 43680n^4) \left(\frac{1}{2}\right)^{12n}. \quad (3.7)$$

We have no idea why it is true but you can check it to almost any precision you wish. In Example 3.4.10 we shall explore such discoveries.

3.3.7 Is 'Free' Software Better?

I conclude this section by commenting on open-source versus commercial software. While free is very nice, there is no assurance that most open source projects such as *GeoGebra* (based on *Cabri* and now very popular in schools as replacement for *Sketchpad*) will be preserved when the founders and typically small core group of developers lose interest or worse. This is still an issue with large-scale commercial products but a much smaller one.

I personally prefer *Maple* to *Mathematica* as most of the source code is accessible, while *Mathematica* is entirely sealed. This is more of an issue for researchers than for educators or less intense users. Similarly, *Cinderella* is very robust, unlike *GeoGebra*, and mathematically sophisticated—using Riemann surfaces to ensure that complicated constructions do not crash. That said, it is the product of two talented and committed mathematicians but only two, and it is only slightly commercial. In general, software vendors and open source producers do not provide the teacher support that has been built into the textbook market.

3.4 A Dozen or So Accessible Examples

Modern graphics tools change traditional approaches to many problems. We used to teach calculus techniques to allow graphing of even reasonably simple functions. Now one should graph to be guided in doing calculus.

Example 3.4.1 (Graphing to Do Calculus). Consider a request in a calculus text to compare the function given by $f(y) := y^2 \log y$ (red) to each of the functions given by

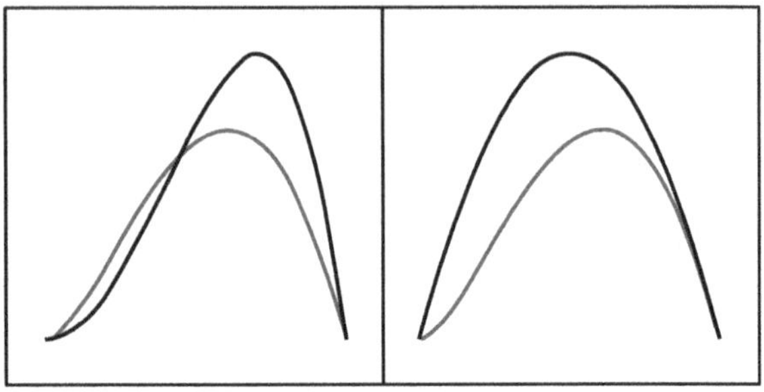

Fig. 3.11 The functions f and h (L) and f and g (R)

$g(y) := y - y^2$ and $h(y) := y^2 - y^4$ for $0 \leqslant y \leqslant 1$; and to prove any inequality that holds on the whole unit interval.

The graphs of f, g are shown in the left of the picture in Fig. 3.11, and the graphs of f, h to the right. In any plotting tool we immediately see that f and g cross but that $h \geqslant f$ appears to hold on $[0, 1]$. Only in a neighbourhood of 1 is there any possible doubt. Zooming in—as is possible in most graphing tools—or re-plotting on a smaller interval around 1 will persuade you that $f(y) > h(y)$ for $0 < y < 1$. This is equivalent to $k(x) := \log(x) - 1 + 1/x > 0$ and so that is what you try to prove. Now it is immediate that $k'(x) < 0$ on the interval and so k strictly decreases to $k(1) = 0$ and we are done. □

Likewise, *computer algebra systems* (CAS) now make it possible to find patterns which we prove *ex post facto by* induction. Before CAS many of these inductive statements might have been inaccessible.

Example 3.4.2 (Induction and Computer Algebra). We all know how to show

$$\sum_{k=1}^{N} k = \frac{n(n+1)}{2}$$

with or without induction. But what about

$$\sum_{k=1}^{N} k^5 = ?$$

Consider the following three lines of *Maple* code.

```
> S:=(n,N)->sum(k^n,k=1..N):
> S5:=unapply(factor(simplify(S(5,N))),N);
> simplify(S5(N)-S5(N-1));
```

The first line defines the sum $\sum_{k=1}^{N} k^n$. The second finds this sum for $n = 5$ and makes it into a function of N. We obtain:

$$\sum_{k=1}^{N} k^5 = \frac{1}{12} N^2 \left(2 N^2 + 2 N - 1\right)(N + 1)^2. \tag{3.8}$$

The third line proves this by induction—on checking that $S5(1) = 1$. The proof can of course be done by hand. Jakob Bernoulli (1655–1705) invented his *Bernoulli numbers*[23] and associated polynomials in part to evaluate such sums. Indeed, using the same code with $N = 10$ we arrive at a proof that

$$\sum_{k=1}^{N} k^{10} = \frac{N(2N+1)(N+1)\left(N^2+N-1\right)\left(3N^6+9N^5+2N^4-11N^3+3N^2+10N-5\right)}{66}$$

and so that

$$\sum_{k=1}^{100} k^{10} = 959924\underline{142434241}924250,$$

and

$$\sum_{k=1}^{1000} k^{10} = 914099242\underline{4242434242}41924242500.$$

This later computation by Bernoulli is accounted as the first case of real computational number theory. Likewise

$$\sum_{k=1}^{10000} k^{10} = 9095909924242424\underline{1424242434242424241}924242425000.$$

We finish with interior palindromes in each of the three sums centered at the '3' and leave its explanation and other apparent patterns to the reader. Of course, unlike Bernoulli, we could simply have added the three sums without finding the closed form but then we would know much less. □

Large matrices often have structure that pictures will reveal but which numeric data may obscure.

Example 3.4.3 (Visualizing Matrices). The picture in Fig. 3.12 shows a 25 × 25 *Hilbert* matrix on the left and on the right a matrix required to have 50 % sparsity and non-zero entries random in [0, 1].

The 4 × 4 Hilbert matrix in *Maple* is generated by `with(LinearAlgebra); HilbertMatrix(4);` which code produces

[23] If you are unfamiliar with them, just ask *Maple*, *Mathematica* or *Wikipedia*.

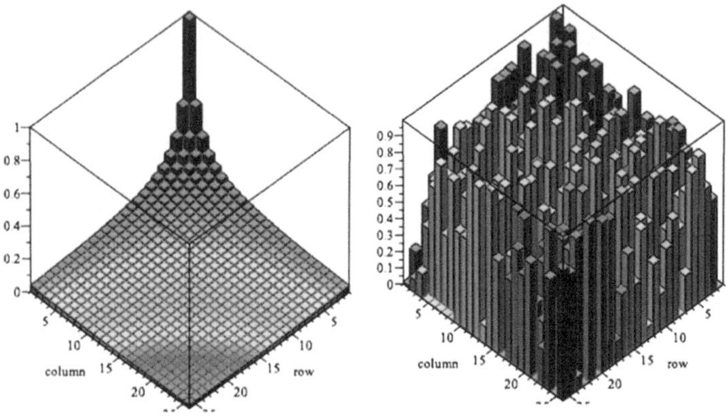

Fig. 3.12 The Hilbert matrix (L) and a sparse random matrix (R)

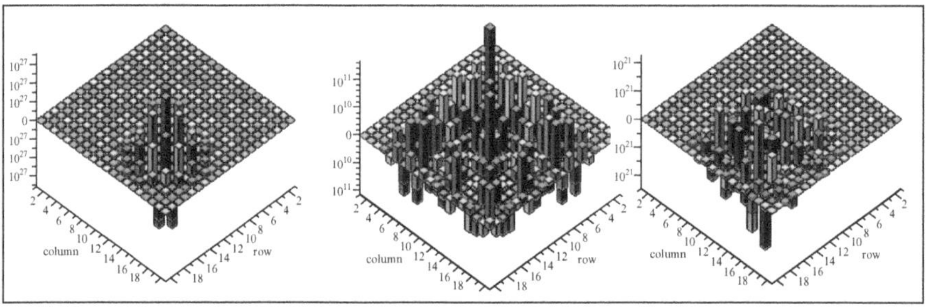

Fig. 3.13 Inverse 20 × 20 Hilbert matrix (L) and 2 numerical inverses (R)

$$\begin{bmatrix} 1 & 1/2 & 1/3 & 1/4 \\ 1/2 & 1/3 & 1/4 & 1/5 \\ 1/3 & 1/4 & 1/5 & 1/6 \\ 1/4 & 1/5 & 1/6 & 1/7 \end{bmatrix}$$

from which the general definition should be apparent. Hilbert matrices are notoriously unstable numerically. The picture on the left of Fig. 3.13 shows the inverse of the 20 × 20 Hilbert matrix when computed *symbolically and so exactly*. The picture in the middle shows the enormous *numerical errors* introduces if one uses 10 digit precision, and the right shows that even if one uses 20 digits, the errors are less frequent but even larger.

Representative *Maple* code for drawing the symbolic inverse is:

```
> with(plots):
> matrixplot(MatrixInverse(HilbertMatrix(20)),
  heights = histogram, axes = frame, gap = .2500000000,
color = proc (x, y) options operator, arrow; sin(y*x) end proc);}
```

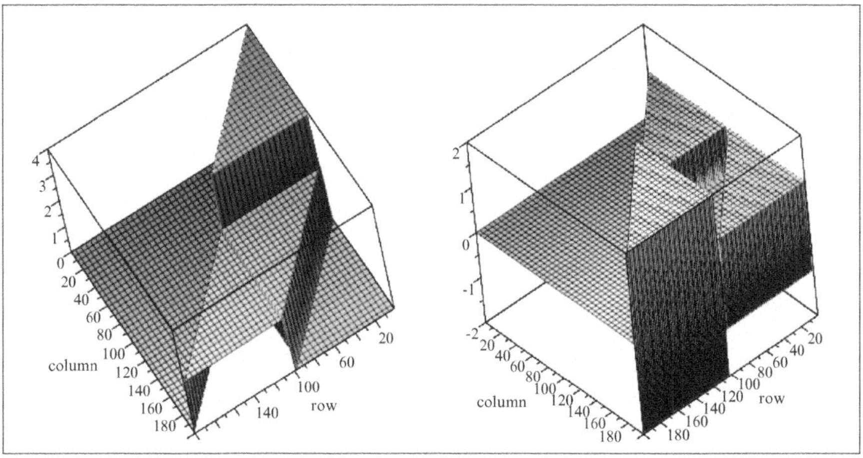

Fig. 3.14 The matrix $Q(100)$ (L) continuing the pattern in (3.9) and its inverse (R)

It is very good fun to play with pictures of very large matrices constructed to have complicated block structure. Consider the sequence of $2n \times 2n$ matrices $Q(n)$, with entries only $0, 1, 2, 4$ which start

$$\begin{bmatrix} 2 & 1 \\ 0 & 1 \end{bmatrix}, \quad \begin{bmatrix} 2 & 2 & 1 & 0 \\ 0 & 2 & 2 & 1 \\ 0 & 0 & 1 & 1 \\ 0 & 0 & 0 & 1 \end{bmatrix}, \quad \begin{bmatrix} 2 & 2 & 2 & 1 & 0 & 0 \\ 0 & 2 & 4 & 2 & 1 & 0 \\ 0 & 0 & 2 & 2 & 2 & 1 \\ 0 & 0 & 0 & 1 & 1 & 1 \\ 0 & 0 & 0 & 0 & 1 & 2 \\ 0 & 0 & 0 & 0 & 0 & 1 \end{bmatrix} \tag{3.9}$$

We cannot possibly present $Q(100)$ as a symbolic or numerical matrix but Fig. 3.14 visually shows everything both about the matrix and its inverse. □

Let us continue with a different exploration of matrices.

Example 3.4.4 (Abstract Becomes Concrete). Define, for $n > 1$ the $n \times n$ matrices $A(n), B(n), C(n), M(n)$ by

$$A_{kj} = (-1)^{k+1} \binom{2n-j}{2n-k}, \quad B_{kj} = (-1)^{k+1} \binom{2n-j}{k-1},$$

$$C_{kj} = (-1)^{k+1} \binom{j-1}{k-1}$$

(for $k, j = 1, \ldots, n$) and set $M := A + B - C$. For instance,

$$M(7) = \begin{bmatrix} 1 & 0 & 0 & 0 & 0 & 0 & 0 \\ 26 & -12 & 9 & -7 & 5 & -3 & 1 \\ 156 & -78 & 55 & -42 & 30 & -18 & 6 \\ 572 & -286 & 176 & -120 & 80 & -46 & 15 \\ 1430 & -715 & 385 & -220 & 126 & -65 & 20 \\ 2574 & -1287 & 627 & -297 & 135 & -56 & 15 \\ 3432 & -1716 & 792 & -330 & 120 & -36 & 7 \end{bmatrix}.$$

In my research (Borwein, Bailey, & Girgensohn, 2005, §3.3), I needed to show $M(n)$ was invertible. After staring at numerical examples without much profit, I decided to ask *Maple* for the *minimal polynomial* of $M(10)$ using

```
> MP:=LinearAlgebra[MinimalPolynomial]: MP(evalm(M(10)),t);
```

and was surprised to get $t^2 + t - 2$. (One way to write B in *Maple* is

```
> B:=n->matrix(n,n,(i,j)->(-1)^(j+1)*binomial(2*n-j,i-1));
```

and there are many other formats.) I got the same answer for $M(30)$ and so I *knew* $M(n)^2 + M(n) = 2I$ for all $n > 1$ or equivalently that

$$M^{-1} = \frac{M + I}{2}.$$

But why? I decided to explore A, B, C with the same tool and discovered that A and C satisfied $t^2 = 1$ and B satisfied $t^3 = 1$. This led me to realize that A, B, C generated the symmetric group on three elements and so to a computer discovered proof that M was as claimed.

As an illustration of the robustness of such discoveries, if we change the $i = 1$, $j = 10$ entry in $M(10)$ to $\varepsilon \neq 0$ from 0, we find the minimal polynomial is far from as simple: $t^4 + 2\,t^3 - 3\,t^2 - 4\,t + 4 - (252\,t^2 + 252\,t - 504)\varepsilon$, which also shows the discontinuity at $\varepsilon = 0$. Similarly, for the 5×5 Hilbert matrix we get

$$-\frac{1}{266716800000} + \frac{61501\,t}{53343360000} - \frac{852401\,t^2}{222264000} + \frac{735781\,t^3}{2116800} - \frac{563\,t^4}{315} + t^5 = 0.$$

The constant term is of course giving minus the determinant. When I was a student characteristic and minimal polynomials seemed to be entirely abstract and matrix decompositions were in their infancy. Now they are technology. ◇

Example 3.4.5 (Hardy's Taxi-Cab). Hardy when visiting Ramanujan in hospital in 1917 remarked that his taxi's number, 1729, was very dull. Ramanujan famously replied that it was very interesting being the smallest number expressible as a sum of two cubes in two distinct ways (not counting sign or order):

$$1729 = 12^3 + 1^3 = 10^3 + 9^3.$$

Let us ask "what is the second such number? As in Sect. 3.2.2, we can look at a generating function—this time for cubes. The coefficients of $C^2(q)$ will be 0 when n is not the sum of two cubes, 1 when $n = 2m^3$, 2 when $n = m^3 + k^3$ for $k \neq m$. If there are two distinct representations, the coefficient will be 4. The *Maple* fragment

```
> C:=convert((add(q^(n^3),n=1..20)^2),polynom):C-(C mod 4):
```

outputs $4\,q^{4104} + 4\,q^{1729}$ which both proves Ramanujan's assertion and finds that the second example is $4104 = 15^3 + 9^3 = 2^3 + 16^3$. If we change 20–25 in our code, we uncover the third such number. Alternatively entering just the first two into the *OIES* produces sequence A001235 consisting of the 'taxi-cab numbers': 1729, 4104, 13832, 20683, 32832, 39312, 40033, 46683, 64232, 65728, 110656, 110808, ... ◇

Example 3.4.6 (Euler's Pentagonal Number Theorem). The number of *additive partitions* of n, $p(n)$, is generated by

$$P(q) = 1 + \sum_{n \geqslant 1} p(n) q^n = \prod_{n \geqslant 1} (1 - q^n)^{-1}. \tag{3.10}$$

Thus $p(5) = 7$ since

$$\begin{aligned} 5 &= 4+1 = 3+2 = 3+1+1 = 2+2+1 \\ &= 2+1+1+1 = 1+1+1+1+1, \end{aligned}$$

as we ignore "0" and permutations. Additive partitions are mathematically less tractable than multiplicative ones as there is no analogue of unique prime factorization nor the corresponding structure.

Partitions provide a wonderful example of why Keith Devlin calls mathematics "the science of patterns". They do sometimes enter the school curriculum through the back-door in the guise of *Cuisenaire rods* (or réglets), as illustrated by a staircase in Fig. 3.15.

Fig. 3.15 A 10×10 Cuisenaire staircase

Formula (3.10) is easily seen by expanding $(1-q^n)^{-1}$ and comparing coefficients. A modern computational temperament leads to:

> *Question:* How hard is $p(n)$ to compute—in 1900 (for MacMahon the "father of combinatorial analysis") or in 2015 (for *Maple* or *Mathematica*)?

Answer: The famous computation by Percy MacMahon of $p(200) = 3972999029388$ at the beginning of the twentieth century, done symbolically and entirely naively from (3.10) in *Maple* on a laptop, took 20 min in 1991 but only 0.17 s in 2010, while the many times more demanding computation

$$p(2000) = 4720819175619413888601432406799959512200344166$$

took just 2 min in 2009 and 40.7 s in 2014.[24] Moreover, in December 2008, the late Richard Crandall was able to calculate $p(10^9)$ in 3 s on his laptop, using the Hardy-Ramanujan-Rademacher 'finite' series for $p(n)$ along with FFT methods. Using these techniques, Crandall was also able to calculate the probable primes $p(1000046356)$ and $p(1000007396)$, each of which has roughly 35,000 decimal digits.[25]

Such results make one wonder when easy access to computation discourages innovation: *Would Hardy and Ramanujan have still discovered their marvelous formula for p(n) if they had powerful computers at hand?* The *Maple* code

```
N:=500; coeff(series(1/product(1-q^n,n=1..N+1),q,N+1),q,N);
```

Twenty-five years ago computing $P(q)$ in *Maple* was very slow, while taking the series for the reciprocal of the series for

$$Q(q) = \prod_{n\geqslant 1}(1-q^n)$$

was quite manageable!

Why? Clearly the series for Q must have special properties. Indeed it is *lacunary*:

$$\begin{aligned} Q(q) = 1 - q - q^2 + q^5 + q^7 - q^{12} - q^{15} + q^{22} + q^{26} - q^{35} - q^{40} + q^{51} \\ + q^{57} - q^{70} - q^{77} + q^{92} + O(q^{100}). \end{aligned} \tag{3.11}$$

[24] The difficulty of comparing timings and the growing inability to look under the hood (bonnet) in computer packages, either by design or through user ignorance, means all such comparisons should be taken with a grain of salt.

[25] See http://fredrikj.net/blog/2014/03/new-partition-function-record/ for a lovely description of the computation of $p(10^{20})$, which has over 11 billion digits and required knowing π to similar accuracy.

This lacunarity is now recognized automatically by *Maple*, so the platform works much better, but we are much less likely to discover Euler's gem:

$$\prod_{n=1}^{\infty}(1-q^n) = \sum_{n=-\infty}^{\infty}(-1)^n q^{n(3n+1)/2}.$$

If we do not immediately recognize the *pentagonal numbers*, $(3(n+1)n/2)$, then Sloane's online *Encyclopedia of Integer Sequences*[26] again comes to the rescue with abundant references to boot.

This sort of mathematical computation is still in its reasonably early days but the impact is palpable. □

Example 3.4.7 (Ramanujan's Partition Congruences). Ramanujan had access to the first 200 values of $p(n)$ thanks to MacMahon's lengthy work which the following *Maple* snippet reconstructs near instantly:

```
> N:=200:L200:=
sort([coeffs(convert(series(1/product(1-q^n,n=1..N+1),q,N
+1),polynom))]);
```

The list starts $1, 1, 2, 3, \mathbf{5}, \mathbf{7}, \mathbf{11}, 15, 22, 30, 42, 56, 77, 101, 135, 176, 231, 297, 385, 490, 627\ldots$ with $p(0) := 1$, and Ramanujan noted various modular patterns. Namely $p(5n+4)$ is divisible by 5, and $p(7n+5)$ is divisible by 7. This is hard to see from a list but a little software can help. The snippet below reshapes the beginning of a list of $n \times m$ or more entries into an $n \times m$ matrix:

```
> reshape:=proc (L, n, m) local k;
    linalg[matrix](n, m, [seq(L[k], k = 1 .. m*n)])
  end proc
```

For instance, `> reshape(L200 mod 5, 8,20)` produces the first 160 entries of the list with 20 columns in each of 8 rows as

$$\begin{bmatrix} 1 & 1 & 2 & 3 & 0 & 2 & 1 & 0 & 2 & 0 & 2 & 1 & 2 & 1 & 0 & 1 & 1 & 2 & 0 & 0 \\ 2 & 2 & 2 & 0 & 0 & 3 & 1 & 0 & 3 & 0 & 4 & 2 & 4 & 3 & 0 & 3 & 2 & 2 & 0 & 0 \\ 3 & 3 & 4 & 1 & 0 & 4 & 3 & 4 & 3 & 0 & 1 & 3 & 4 & 1 & 0 & 1 & 3 & 4 & 0 & 0 \\ 2 & 0 & 1 & 4 & 0 & 3 & 0 & 4 & 0 & 0 & 3 & 0 & 3 & 4 & 0 & 4 & 1 & 3 & 4 & 0 \\ 1 & 2 & 0 & 4 & 0 & 2 & 2 & 3 & 4 & 0 & 3 & 4 & 2 & 2 & 0 & 4 & 4 & 0 & 1 & 0 \\ 2 & 1 & 4 & 0 & 0 & 4 & 1 & 4 & 4 & 0 & 1 & 3 & 1 & 3 & 0 & 1 & 3 & 1 & 3 & 0 \\ 0 & 1 & 2 & 1 & 0 & 2 & 2 & 0 & 0 & 0 & 0 & 4 & 4 & 2 & 0 & 1 & 0 & 1 & 1 & 0 \\ 0 & 3 & 0 & 2 & 0 & 4 & 4 & 3 & 2 & 0 & 3 & 2 & 1 & 4 & 0 & 2 & 4 & 4 & 2 & 0 \end{bmatrix}.$$

We now see only zeroes in the columns congruent to 4 modulo 5 and discover the first congruence $5|p(5n+4)$. Similarly, `> reshape(L200 mod 7, 8,21)` reveals

[26] A fine model for twenty-first century databases, it is available at https://oeis.org/.

$$\begin{bmatrix}
1 & 1 & 2 & 3 & 5 & 0 & 4 & 1 & 1 & 2 & 0 & 0 & 0 & 3 & 2 & 1 & 0 & 3 & 0 & 0 & 4 \\
1 & 1 & 2 & 0 & 5 & 0 & 0 & 1 & 1 & 4 & 3 & 5 & 0 & 4 & 1 & 1 & 0 & 3 & 0 & 0 & 0 \\
2 & 2 & 2 & 3 & 5 & 0 & 0 & 2 & 1 & 4 & 0 & 0 & 0 & 0 & 3 & 2 & 2 & 3 & 5 & 0 & 4 \\
2 & 2 & 2 & 3 & 5 & 0 & 4 & 3 & 2 & 4 & 6 & 5 & 0 & 0 & 2 & 2 & 4 & 3 & 5 & 0 & 0 \\
3 & 3 & 6 & 6 & 3 & 0 & 1 & 3 & 3 & 4 & 3 & 5 & 0 & 0 & 4 & 3 & 4 & 6 & 5 & 0 & 1 \\
5 & 3 & 6 & 6 & 3 & 0 & 0 & 5 & 4 & 6 & 6 & 3 & 0 & 4 & 5 & 4 & 6 & 6 & 5 & 0 & 0 \\
6 & 5 & 1 & 2 & 3 & 0 & 1 & 5 & 5 & 1 & 6 & 3 & 0 & 4 & 0 & 5 & 1 & 5 & 1 & 0 & 1 \\
0 & 6 & 3 & 2 & 1 & 0 & 4 & 1 & 6 & 3 & 5 & 1 & 0 & 1 & 1 & 6 & 3 & 2 & 1 & 0 & 1
\end{bmatrix}$$

and 'discovers' the second congruence $5|p(7n + 5)$ for all $n \geqslant 0$. The third congruence $6|p(11n + 6)$ for all $n \geqslant 0$ can be discovered by appropriate reshaping—and if wished confirmed by taking more terms. These partition congruences are discussed and the first two proved in Borwein and Borwein (1987, §3.5). ◇

Maple has since version 9.5 had a function called 'identify'. It takes many tools such as PSLQ (Sect. 3.3.6) and attempts to predict an answer for a floating point number. A related *ISC* is on-line at http://isc.carma.newcastle.edu.au/. This lets you enter a real number or a *Maple* expression and ask the computer "What is that number?"

Another excellent example of how packages are changing mathematics is the *Lambert W function* (Borwein et al., 2005), whose remarkable properties and development are very nicely described in a fairly recent article by Hayes (2005), *Why W?* Informally, $W(x)$ solves

$$x = W(x)e^{W(x)}.$$

As a real function, its domain is $(-1/e, \infty)$. We draw W and the quite similar log function on the left of Fig. 3.16. Its use can be traced back to Lambert (1728–1727), and W as a notation was used by Pólya and Szegö in 1925. However, this very useful non-elementary function only came into general currency after it was named

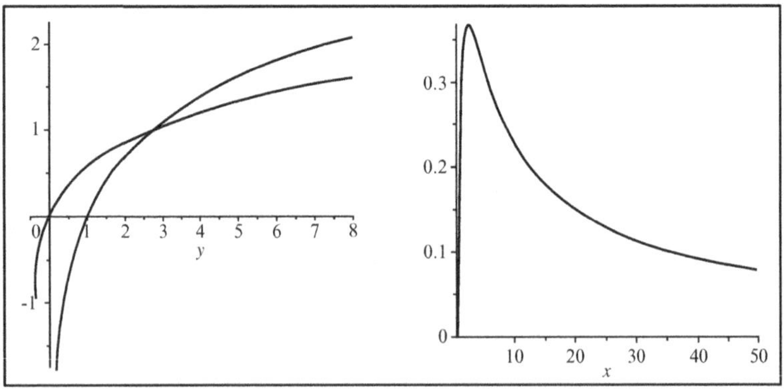

Fig. 3.16 (L) W and log (R) $(\log x)/x$

and then implemented in both *Maple* and *Mathematica*. It is hard to use or popularize an un-named function. Now most CAS know the expansion

$$W(z) = \sum_{k=1}^{\infty} \frac{(-k)^{k-1}}{k!} z^k$$

with radius of convergence 1/e.

Example 3.4.8 (Solving Equations with W). We first look at $x^y = y^x$.

(a) Let us fix $x > 0$ and try to solve

$$x^y = y^x \qquad \text{for } y > 0. \tag{3.12}$$

Of course, we seek a non-trivial solution with $x \neq y$ such as $x = 2, y = 4$. The *Maple* `solve` command returns

$$y(x) = \left(\frac{-x}{\log x}\right) W\left(\frac{-\log x}{x}\right). \tag{3.13}$$

This may confuse initially more than help. If we take logarithms in (3.12) and rearrange, we are trying to solve

$$\frac{\log y}{y} = z := \frac{\log x}{x} \tag{3.14}$$

for $y > 0$.

The right of Fig. 3.16 shows that the function $(\log x)/x$ is positive only on $1 < x < \infty$ and then and only then has two solutions—except for $x = e$ where the maximum of $1/e$ occurs—and now in *Maple* `solve(log(x)/x=z,x)` returns $-W(-\log z)/z$. This solution is shown on the left in Fig. 3.17 where z implicitly must satisfy $0 < z < 1/\mathrm{e}$. This yields (3.13), shown in the center of

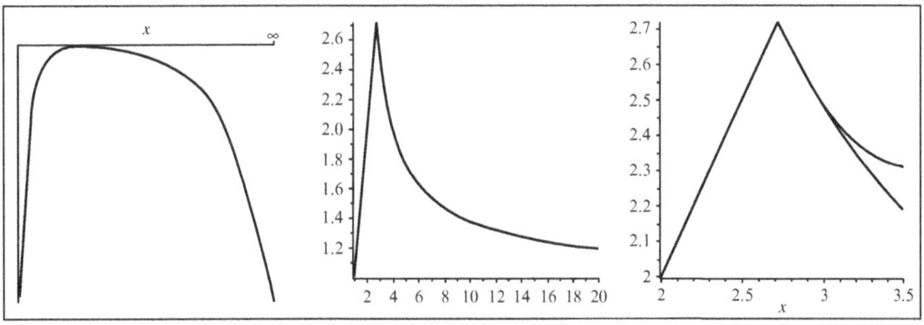

Fig. 3.17 (L) Solution to (3.13). (M) Solution to (3.12). (R) Quadratic Taylor approximation around e

Fig. 3.17, where we know now that $x > 1$ is requisite. For instance, $y_3 := -3W(-(\log 3)/3)/\log 3 = 2.478052685\ldots \neq 3$ solves $3^{y_3} = (y_3)^3$.

Now, we may not know W but our computer certainly now does. For instance, `identify(0.56714329040978)` returns $W(1)$ and the Taylor series for $y(x)$ around e starts

$$y(x) = \mathrm{e} - \mathrm{sign}\,(x - \mathrm{e})(x - \mathrm{e}) + \frac{5\,\mathrm{e}^{-1}(\mathrm{sign}(x - \mathrm{e}) + 1)}{6}(x - \mathrm{e})^2 + O\left((x - \mathrm{e})^3\right) \tag{3.15}$$

as shown on the right of Fig. 3.17. For $x > \mathrm{e}$, $y(x) = 1/(3\mathrm{e})(11\,\mathrm{e}^2 - 13\,x\mathrm{e} + 5\,x^2)$ while for $x < \mathrm{e}$ we get the trivial solution x.

(b) A parametric form of the solution is $x = r^{1/(r-1)}, y = rx = r^{r/(r-1)}$, for $r > 1$. Equivalently with $r = 1 + 1/s$, where $s > 0$, we have $x = (1 + 1/s)^s$, $y = (1 + 1/s)^{s+1}$. This is shown in Fig. 3.18.

(c) *Repeated exponentiation.* How many distinct meanings may be assigned to *product towers* for the n-fold exponentiation

$$x{*}^{x}_{n} = x^{x^{x^{\cdot^{\cdot^{\cdot^{x}}}}}}\ ?$$

Recursions like $x_1 = t > 0$ and $x_n = t^{x_{n-1}}$ for $n > 0$ have been subjected to considerable scrutiny. We can check that the solution to $t^x = x$ is $t \mapsto -W(-\log t)/\log t$ which solution exists for $t \in [e^{-e}, e^{1/e}]$. ◇

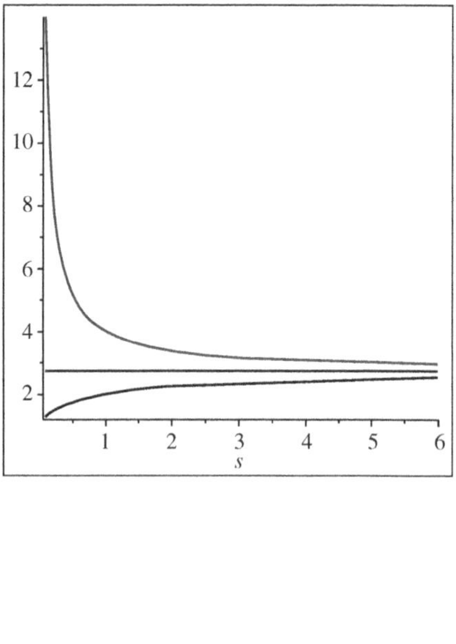

Fig. 3.18 Parametric solutions of (3.12) separated by $y = e$

Example 3.4.9 (Finding Algebraic Numbers). Both *Maple* and *Mathematica* have algorithms that can predict if a number is algebraic and even find its minimal polynomial. We described this a little further in Sect. 3.3.6. For instance, using `identify`—with no tuning of parameters—as with

```
> Digits:=20:a:=evalf(7^(1/2)+3^(1/2));identify(a);
```

returns

$$\sqrt{7}+\sqrt{3}$$

but

```
> Digits:=20:a:=evalf(2^(1/3)+3^(1/2));identify(a);
```

returns

$$2.9919718574637504583$$

meaning *Maple* could not identify the surd from 20 digits. The ISC at http://isc.carma.newcastle.edu.au/advancedCalc runs tuned algorithms and will identify the constant as shown in Fig. 3.19.

However:

```
> Digits:=30:a:=evalf(2^{1/3}+3^{1/2});identify(a);
```

returns

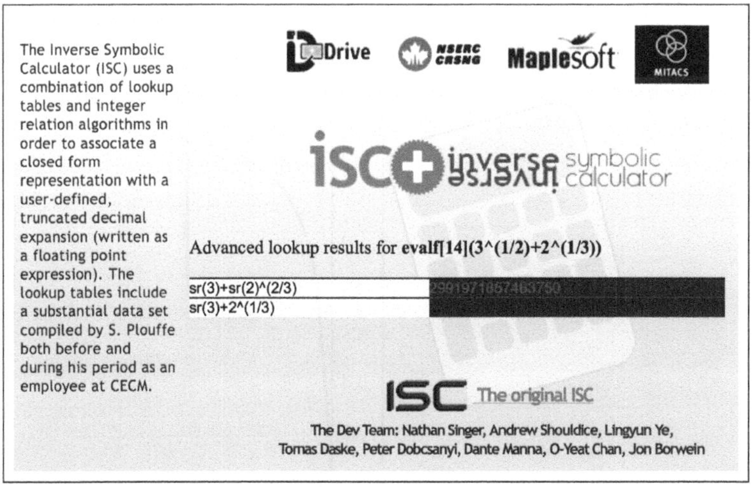

Fig. 3.19 Identifying $2^{1/3}+3^{1/2}$

$$Root\ Of\left(\_Z^6 - 9\_Z^4 - 4\_Z^3 + 27\_Z^2 - 36\_Z - 23, index = 1\right)$$

which allows us to recover $2^{1/3} + 3^{1/2}$. For example, we can factor in Q($\sqrt{3}$) using the command:

```
> factor(_Z^6-9*_Z^4-4*_Z^3+27*_Z^2-36*_Z-23,sqrt(3));
```

which yields

$$-\left(3\ \sqrt{3}\_Z^2 + \_Z^3 + 3\ \sqrt{3} + 9\_Z - 2\right)\left(3\ \sqrt{3}\_Z^2 - \_Z^3 + 3\ \sqrt{3} - 9\_Z + 2\right).$$

The quadratic formula now applies to determine that the only real roots of the sextic polynomial are $2^{1/3} \pm \sqrt{3}$. □

We can do more exciting things of this kind.

Example 3.4.10 (What is that Number?). Let us illustrate it for the integral

$$\mathcal{R}(a) := \mathcal{R}(a,a) = 2\int_0^1 \frac{t^{1/a}}{1+t^2}\,\mathrm{d}t \tag{3.16}$$

whose origins are described in Sect. 3.5. We plot $R(a)$ in Fig. 3.20. (We used the hypergeometric form given below. *Maple* will find this form if you input (3.16).) Note that the graph is consistent with the fact that $\mathcal{R}$ increases to the blue asymptote $\mathcal{R}(\infty) = \frac{\pi}{2}$. You may be able to evaluate some other values by hand.

Most CAS will answer that the values of $\mathcal{R}(1/m)$, for $1 \leqslant m \leqslant 6$, are

$$\log 2, \quad 2 - \frac{\pi}{2}, \quad 1 - \log 2, \quad -\frac{4}{3} + \frac{\pi}{2}, \quad -\frac{1}{2} + \log 2, \quad \frac{26}{15} - \frac{\pi}{2}.$$

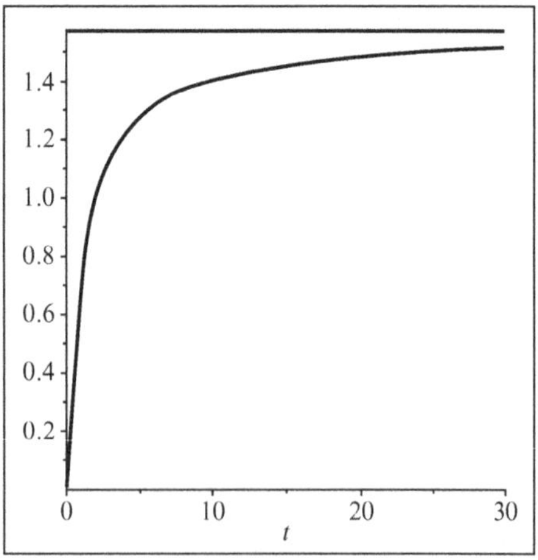

Fig. 3.20 The function $\mathcal{R}(a)$ for $0 < a < 30$

We then check that $\mathcal{R}(1/7) = \frac{5}{6} - \log 2$ and $\mathcal{R}(1/8) = -\frac{152}{105} + \frac{\pi}{2}$. From this it should be plausible that

$$(-1)^n \mathcal{R}\left(\frac{1}{2n+1}\right) = \log 2 + a_n$$
$$(-1)^n \mathcal{R}\left(\frac{1}{2n}\right) = \frac{\pi}{2} + b_n$$

for rational numbers a_n, b_n. As experimental confirmation of this conjecture we can check that

$$\mathcal{R}\left(\frac{1}{21}\right) = -\frac{1627}{2520} + \log 2, \quad \mathcal{R}\left(\frac{1}{20}\right) = -\frac{22128676}{14549535} + \frac{\pi}{2}.$$

If we ask the computer for $\mathcal{R}(2)$, we get a complicated (ostensibly complex) expression that simplifies to $\frac{\sqrt{2}}{4}\left(2\,\pi + \log\left(17 - 12\,\sqrt{2}\right)\right)$. If we try PSLQ for $a := 2, 2/3, 2/5, \ldots$ we discover that each such sum evaluates in terms of three *basis* vectors:

$$1, \sqrt{2}\pi \quad \text{and} \quad \sqrt{2}\log(1+\sqrt{2}).$$

For instance,

$$\mathcal{R}\left(\frac{11}{2}\right) = \frac{164}{45} - \frac{\pi}{\sqrt{2}} - \sqrt{2}\,\log\left(1+\sqrt{2}\right).$$

If, however, we leave out the constant term '1', we find

$$\mathcal{R}(2) = \frac{\pi}{\sqrt{2}} - \sqrt{2}\,\log\left(1+\sqrt{2}\right)$$

but have no such luck as we need that pesky constant term.

Actually, Borwein, Crandall, and Fee (2004) give a closed form for every instance of $\mathcal{R}\left(\frac{p}{q}\right)$ with p, q positive integers. □

We turn to another example where the CAS provides a proof that we would not have been likely to arrive at without the current tools.

Example 3.4.11 (π Is Not 22/7). Even *Maple* or *Mathematica* 'knows' this since

$$0 < \int_0^1 \frac{(1-x)^4 x^4}{1+x^2}\,\mathrm{dx} = \frac{22}{7} - \pi, \tag{3.17}$$

though it would be prudent to ask 'why' it can perform the integral and 'whether' to trust it?

Fig. 3.21 The integrand in (3.17)

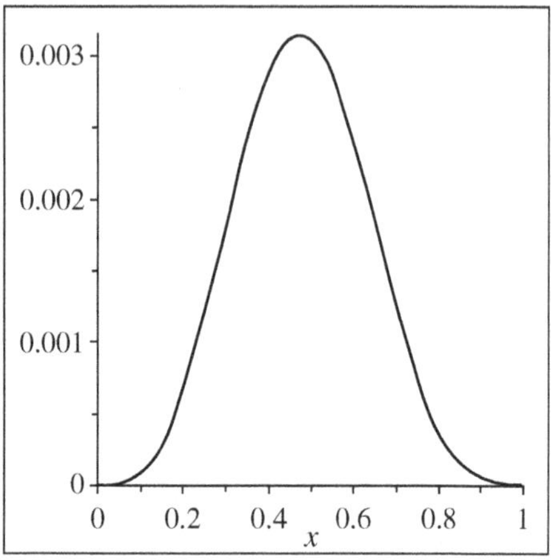

1. Assuming we trust our software, the integrand is strictly positive on $(0, 1)$, see Fig. 3.21, and so the answer in (3.17) is an *area* which is necessarily strictly positive, despite millennia of claims that π is 22/7.
2. Quite accidentally, 22/7 is one of the early continued fraction approximation to π—and is why it is a pretty reasonable approximation to π. These commence:

$$3, \frac{22}{7}, \frac{333}{106}, \frac{355}{113}, \dots$$

   but no one has found a way to replicate (3.17) for these other fractions. Some coincidences are just that—happenstance. Similarly, there is no good reason why $e^{\pi} - \pi = \underline{19.99909997}918947576\ldots$, but it is most impressive on a low precision calculator.
3. We turn to *proving* π is not $\frac{22}{7}$ with computational help. In this case, taking the *indefinite* integral provides immediate reassurance. We obtain

$$\int_0^t \frac{x^4(1-x)^4}{1+x^2}\,\mathrm{dx} = \frac{1}{7}\,t^7 - \frac{2}{3}\,t^6 + t^5 - \frac{4}{3}\,t^3 + 4\,t - 4\,\arctan(t)$$

   as differentiation and simplification—by hand or by computer—easily confirms. Now the *Fundamental theorem of calculus* **proves** (3.17). A traditional proof would probably have developed the partial fraction expansion and thence performed the integral.
4. One can take this idea a bit further. Note that

$$\int_0^1 x^4(1-x)^4 \mathrm{d}x = \frac{1}{630}. \tag{3.18}$$

Hence

$$\frac{1}{2}\int_0^1 x^4(1-x)^4 \mathrm{d}x < \int_0^1 \frac{(1-x)^4 x^4}{1+x^2}\,\mathrm{d}x < \int_0^1 x^4(1-x)^4 \mathrm{d}x.$$

5. Combine this with (3.17) and (3.18) to derive: $223/71 < 22/7 - 1/630 < \pi < 22/7 - 1/1260 < 22/7$ and so re-obtain Archimedes' famous

$$3\frac{10}{71} < \pi < 3\frac{10}{70}. \tag{3.19}$$

Note that by not cancelling the zeros on the right it is much easier to see that $1/7 > 10/71$. All rules must be broken occasionally.

Even without using (3.19), a glance at Fig. 3.21 shows how small the error is. Indeed the maximum occurs at 1/2 with maximum value a tiny 1/320.

Never Trust References In 1971 Dalziel published this development in *Eureka*, a Cambridge student magazine of the period. Integral (3.17) was earlier on the 1968 *William Lowell Putnam* examination, an early 1960s Sydney honours exam, and traces back to a 1944 paper by the self-same Dalzeil who opted not to reference it 27 years later.[27] The message here is that what might appear to be a primary source may well not be, and even the author may not necessarily tell you the whole truth. □

The take away from Examples 3.4.2 and 3.4.11 is that whenever a CAS can do a definite sum, product or integral, it is well worth seeing if it can perform the corresponding indefinite one.

I have built a little function 'pslq' in Maple that when input data for PSLQ predicts an answer to the precision requested but checks it to ten digits more (or some other precision). This makes the code a real experimental tool as it predicts and confirms. One of my favourite uses of it is to quickly check answers for a lecture in cases where I know the general form of an answer but cannot remember all the details.

Example 3.4.12 (Preparing for Class). In all serious computations of π from 1700 (by John Machin) until 1980 some version of a *Machin formula* (Borwein and Bailey, 2008) was used. This is a formula which writes

$$\begin{aligned}\arctan(1) = a_1 \cdot \arctan\left(\frac{1}{p_1}\right) + a_2 \cdot \arctan\left(\frac{1}{p_2}\right)\\ + \cdots + a_n \cdot \arctan\left(\frac{1}{p_n}\right)\end{aligned} \tag{3.20}$$

for rational numbers $a_1, a_2, \ldots, a_n$ and integers $p_1, p_2, \ldots, p_n > 1$. When combined with the Taylor series for arctan, namely

[27] I am certainly guilt of some such sins herein.

$$\arctan(x) = \sum_{n=0}^{\infty} \frac{(-1)^n}{2n+1} x^{2n+1}.$$

This series when combined with (3.20) allows one to compute $\pi = 4\arctan(1)$ efficiently, especially if the values of p_n are not too small.

For instance, Machin found

$$\pi = 16 \arctan\left(\frac{1}{5}\right) - 4 \arctan\left(\frac{1}{239}\right) \tag{3.21}$$

while Euler discovered

$$\arctan(1) = \arctan\left(\frac{1}{2}\right) + \arctan\left(\frac{1}{5}\right) + \arctan\left(\frac{1}{8}\right). \tag{3.22}$$

The code in Fig. 3.22 used 20 digits to confirm (3.22) to 30 digits. The input is a *Maple* or numeric real, followed by a list of basis elements, and the third variable is the precision to use. The code in Fig. 3.23 used 20 digits to likewise confirm (3.21) to 30 digits.

The code in Fig. 3.24 used 20 digits to find another relation and confirm it to 30 digits. This is what happens if you mistype 1/3 for 1/5.

```
> pslq(arctan(1),[arctan(1/2),arctan(1/5), arctan(1/8)],20);
```

[1, 1, 1, 1], "Error is", 0., "checking to", 30, *places*

$$\frac{1}{4}\pi = \arctan\left(\frac{1}{2}\right) + \arctan\left(\frac{1}{5}\right) + \arctan\left(\frac{1}{8}\right)$$

Fig. 3.22 Finding equation (3.22)

```
> pslq(Pi,[arctan(1/5), arctan(1/239)],20);
```

[1, 16, -4], "Error is", $2.8\ 10^{-30}$, "checking to", 30, *places*

$$\pi = 16 \arctan\left(\frac{1}{5}\right) - 4 \arctan\left(\frac{1}{239}\right)$$

Fig. 3.23 Finding equation (3.21)

```
> pslq(arctan(1),[arctan(1/2),arctan(1/3), arctan(1/8)],20);
```

[1, 1, 1, 0], "Error is", $-1.\ 10^{-30}$, "checking to", 30, *places*

$$\frac{1}{4}\pi = \arctan\left(\frac{1}{2}\right) + \arctan\left(\frac{1}{3}\right)$$

Fig. 3.24 Finding an unexpected equation

```
> pslq(arctan(1),[arctan(1/2),arctan(1/5), arctan(1/9)],20);
```

$$[42613, 72375, 22013, -40066], \text{"Error is"}, 2.31604649037\ 10^{-15}, \text{"checking to"}, 30, \mathit{places}$$

$$\frac{1}{4}\pi = \frac{72375}{42613}\arctan\left(\frac{1}{2}\right) + \frac{22013}{42613}\arctan\left(\frac{1}{5}\right) - \frac{40066}{42613}\arctan\left(\frac{1}{9}\right)$$

Fig. 3.25 When no relation exists

If, however, as in Fig. 3.25, you use 1/9 instead of 1/8 you get a 'mess'.

This shows that when no relation exists the code will often find a very good approximation but will use very large rationals in the process. So it diagnoses failure both because it uses very large coefficients and because it is not true to the requested 30 places. □

We next find the limit of an interesting *mean iteration*—an idea we take up again in Example 3.8.1. Recall that a *mean* M is any function of positive numbers a and b which always satisfies $\min\{a,b\} \leqslant M(a,b) \leqslant \max\{a,b\}$. A mean is *strict* if $M(a,b) = a$ or $M(a,b) = b$ implies $a = b$ as is true for the *arithmetic* mean $A(a,b) := \frac{a+b}{2}$, the *geometric* mean $G(a,b) := \sqrt{ab}$ or the *harmonic* mean $H(a,b) := \frac{2ab}{a+b}$. Every mean clearly is *diagonal* meaning that $M(a,a) = a$.

Theorem 3.4.13 (Invariance Principle, Borwein & Borwein, 1987). *Suppose M,N are means and at least one is strict. The mean iteration given by* $a_{n+1} = M(a_n, b_n)$ *and* $b_{n+1} = N(a_n, b_n)$ *is such that the limit* $L(a,b) = \lim_n a_n = \lim_n b_n$ *exists and is necessarily a mean. Moreover, it is the unique continuous and diagonal mapping satisfying for all n:*

$$L(a_n, b_n) = L(a_{n+1}, b_{n+1}). \tag{3.23}$$

Proof. We sketch the proof (details may again be found in Borwein & Borwein, 1987, Chap. 8). One first checks that the limit, being a pointwise limit of means is itself a mean and so is continuous on the diagonal. The principle follows since, L being diagonal satisfies

$$\lim_n a_n = L(\lim_n a_n, \lim_n b_n) = L(a_{n+1}, b_{n+1}) = L(a_n, b_n) = \cdots = L(a,b),$$

as asserted. □

A simple but satisfying application of Theorem 3.4.13 is to show that with $a_0 := a > 0, b_0 := b > 0$, the mean iteration

$$a_{n+1} = \frac{a_n + b_n}{2}, \quad b_{n+1} = \frac{2a_n b_n}{a_n + b_n}$$

converges quadratically to $\sqrt{ab}$.

Fig. 3.26 The function $x \mapsto L(x)\log x$ on $[1/3, 3]$

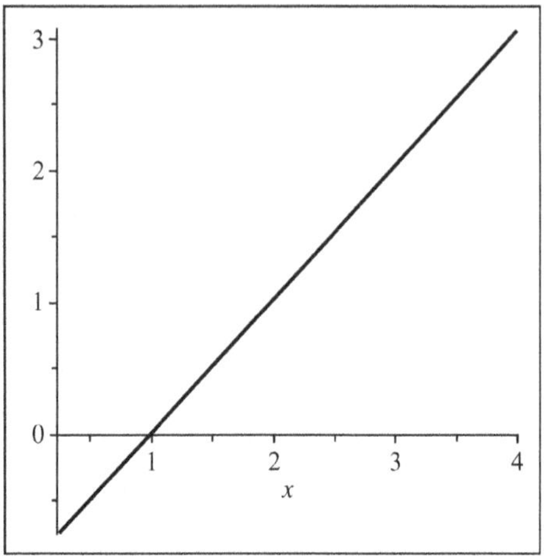

Example 3.4.14 (Finding a Limit). Consider the iteration that takes positive numbers $a_0 := a, b_0 := b$ and repeatedly computes the mixed arithmetic-geometric means:

$$a_{n+1} = \frac{a_n + \sqrt{a_n b_n}}{2}, \quad b_{n+1} = \frac{b_n + \sqrt{a_n b_n}}{2}. \tag{3.24}$$

In this case convergence is immediate since $|a_{n+1} - b_{n+1}| = |a_n - b_n|/2$. The following *Maple* function will compute the Nth step of the iteration to the precision of the environment.

```
L:=proc(a0,b0,n) local a,b,c,k;a:=evalf(a0);b:=evalf(b0); for
k to n do
c:=evalf((a+sqrt(a*b))/2);b:=evalf((b+sqrt(a*b))/2);a:=c;
od;a;end;
```

If we set the precision at 14 digits and try `identify(L(2,1,50));` we get $1/\log 2$ and `identify(L(3,1,50));` gives $2/\log 3$. After checking that $x = 4$ gives $3/\log 4$ and 5 behaves similarly, it seems worthwhile considering $\log(x)L(x, 1)$. We only want a few digits so we plot $F(x) := L(x, 1, 5)\log x$ on $[1/3, 3]$. The result in Fig. 3.26 is a straight line and strongly supports the *conjecture* that $L(x, 1) = (x - 1)/\log x$ in which case $L(a, b) = b(a/b - 1)/\log(a/b) = (a - b)/(\log a - \log b)$. Even dull plots can be interesting. ◊

We are ready to prove our conjecture.

Example 3.4.15 (Carlson's Logarithmic Mean). Consider the iterations with $a_0 := a > 0, b_0 := b > a$ and

$$a_{n+1} = \frac{a_n + \sqrt{a_n b_n}}{2}, \qquad b_{n+1} = \frac{b_n + \sqrt{a_n b_n}}{2},$$

for $n \geqslant 0$. If asked for the limit, you might make little progress. But suppose we have just completed Example 3.4.14. Then we can see that answer is the *logarithmic mean*

$$\mathcal{L}(a,b) := \frac{a-b}{\log a - \log b},$$

for $a \neq b$ and a (the limit as $a \to b$) when $a = b > 0$. We check that

$$\mathcal{L}(a_{n+1}, b_{n+1}) = \frac{a_n - b_n}{2\,\log \frac{a_n + \sqrt{b_n a_n}}{b_n + \sqrt{b_n a_n}}} = \mathcal{L}(a_n, b_n),$$

since $2\log \frac{\sqrt{a_n}}{\sqrt{b_n}} = \log \frac{a_n}{b_n}$. The invariance principle of Theorem 3.4.13 then confirms that $\mathcal{L}(a,b)$ is the limit. In particular, for $a > 1$,

$$\mathcal{L}\left(\frac{a}{a-1}, \frac{1}{a-1}\right) = \frac{1}{\log a},$$

which quite neatly computes the logarithm (slowly) using only arithmetic operations and square roots. ◇

And finally, we look at two examples that emphasize that no initial pattern is a proof. They involve the highly oscillatory sinc function

$$\mathrm{sinc}(x) := \frac{\sin x}{x}$$

which is plotted in Fig. 3.27.

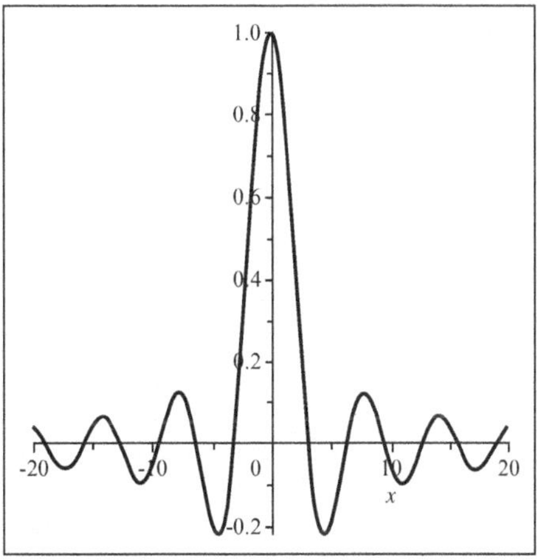

Fig. 3.27 The sinc function

Example 3.4.16 (Inductive Reasoning Has Its Limits). Consider

$$J_n := \int_{-\infty}^{\infty} \operatorname{sinc}(x) \cdot \operatorname{sinc}\left(\frac{x}{3}\right) \cdots \operatorname{sinc}\left(\frac{x}{2n+1}\right) \, dx.$$

Then—as *Maple* and *Mathematica* are able to confirm—we have the following evaluations:

$$\begin{aligned}
J_0 &= \int_{-\infty}^{\infty} \operatorname{sinc}(x) \, dx = \pi, \\
J_1 &= \int_{-\infty}^{\infty} \operatorname{sinc}(x) \cdot \operatorname{sinc}\left(\frac{x}{3}\right) dx = \pi, \\
&\vdots \\
J_6 &= \int_{-\infty}^{\infty} \operatorname{sinc}(x) \cdot \operatorname{sinc}\left(\frac{x}{3}\right) \cdots \operatorname{sinc}\left(\frac{x}{13}\right) dx = \pi.
\end{aligned}$$

As explained in detail in Borwein et al. (2005, Chap. 2), the seemingly obvious pattern is then confounded by

$$\begin{aligned}
J_7 &= \int_{\infty}^{-\infty} \operatorname{sinc}(x) \cdot \operatorname{sinc}\left(\frac{x}{3}\right) \cdots \operatorname{sinc}\left(\frac{x}{15}\right) dx \\
&= \frac{467807924713440738696537864469}{467807924720320453655260875000} \pi < \pi,
\end{aligned}$$

where the fraction is approximately $0.99999999998529\ldots$ which, depending on the precision of calculation used, numerically might not even be distinguished from 1.

These integrals now called the *Borwein integrals* have gathered a life of their own as illustrated in Fig. 3.28 and http://oeis.org/A068214/internal. □

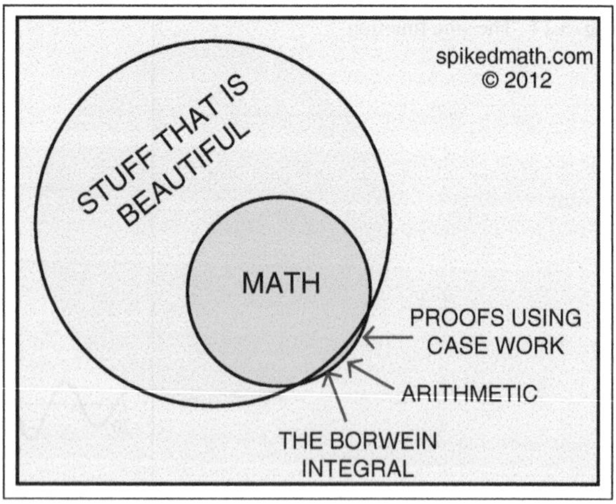

Fig. 3.28 What is beauty?

In case this caution against inductively jumping to conclusions was not convincing, consider the next example.

Example 3.4.17 (Inductive Reasoning Really Has Its Limits). The following "student's dream" identity of a sum equalling an integral again engages the sinc function:

$$\sum_{n=-\infty}^{\infty} \operatorname{sinc}(n) \operatorname{sinc}(n/3) \operatorname{sinc}(n/5)\cdots\operatorname{sinc}(n/23) \operatorname{sinc}(n/29)$$
$$= \int_{-\infty}^{\infty} \operatorname{sinc}(x) \operatorname{sinc}(x/3) \operatorname{sinc}(x/5)\cdots\operatorname{sinc}(x/23) \operatorname{sinc}(x/29)\,\mathrm{d}x, \tag{3.25}$$

where the denominators range over the odd primes up to 29, was first discovered empirically.

Provably, the following is true: The analogous "sum equals integral" identity remains valid for '29' replaced by any one of more-than-the first 10176 primes but stops holding after some larger prime, and *thereafter* the "sum less the integral" is positive but *much less than one part in a googolplex.* An even stronger estimate is possible assuming the *generalized Riemann hypothesis* (GRH) (Baillie, Borwein, & Borwein, 2008). What does it mean for two formulas to differ by a quantity that can never be measured in our assumed-to-be finite universe? □

3.5 Simulation in Pure Mathematics

Pure mathematicians have not frequently thought of simulation as a relevant tool though it has a long lineage. An early and dubious example of simulation of π is called *Buffon's needle.* The Comte de Buffon (1700–1778) was an early vegetarian and his claimed result is much too good—it gets too accurate a result for the sample size (Fig. 3.29).

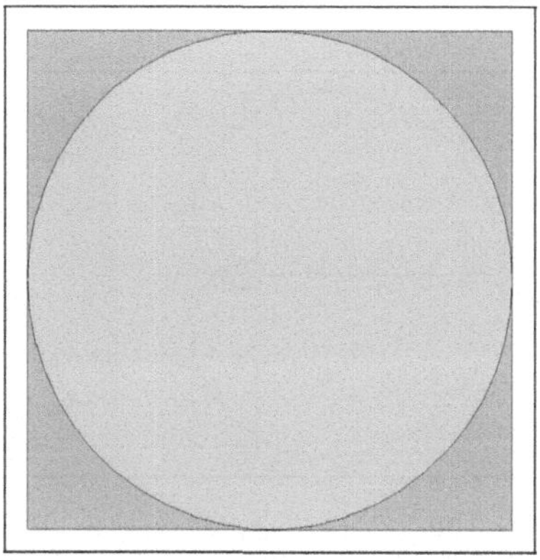

Fig. 3.29 Simulating π

3.5.1 *Monte Carlo Simulation of π*

Stanislaw Ulam (1900–1984) can be considered the inventor of modern *Monte Carlo* sampling methods—named for the casino parlours of that city. See also the first of our top ten algorithms in Sect. 3.3.5. Such simulations were crucial during the Manhattan project when early computers were inadequate to perform the needed computations, even though they intrinsically are not efficient. One expects to need n^2 measurements to get an accuracy of $O(1/n)$. An easy illustration is simulation of π.

Example 3.5.1 (Why a Serial God Should Not Play Dice). Consider inscribing a circle in a square pan of side one, and sprinkling a fine particle (e.g., salt or grain) and counting the proportion of particles that fall in the circle. It should approximate $\pi/4$ as that is the area of the circle.

If one can pour all the particles at once and uniformly over the square, this is a fast and parallel method of estimating π. But if one has to do this in serial it is painfully slow. One can do this at the computer by selecting pairs of pseudo-random numbers in the square $0 < x, y < 1$ and counting how often $x^2 + y^2 \leqslant 1$. Four times that proportion should converge to π. ◇

Despite the slowness of the method, in the early days of personal computers I implemented this on each new desktop or laptop. It was a terrible way to compute π but a great way to test the random number generator. I would start the program and after a few thousand trials would have roughly 3.14. When I looked again the next morning I might have converged to 3.57... or some such because the built-in random number generator was far from random.

3.5.2 *Finding a Region of Convergence*

The cardioid at the left of Fig. 3.30 was produced by a scatter plot while trying to determine for which complex numbers $z = b/a$ an improper continued fraction due to Ramanujan, $\mathcal{R}(a, b)$, converged. It is given for complex numbers a and b by

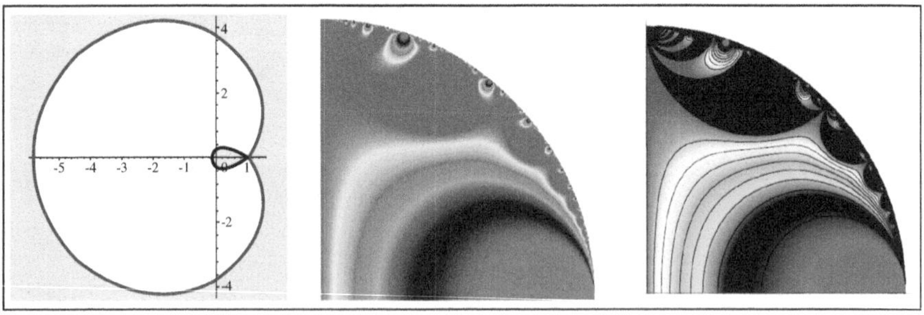

Fig. 3.30 (L) cardioid discovered by simulation. (M) and (R) a fractal hidden in $\mathcal{R}$

$$\mathcal{R}(a,b) = \cfrac{a}{1+\cfrac{b^2}{1+\cfrac{4a^2}{1+\cfrac{9b^2}{1+\ddots}}}}, \tag{3.26}$$

see Borwein et al. (2005, Ex. 53, p. 69).

As often I first tried to compute $R(1,1)$ and had little luck[28]—it transpires that for $a = b \in \mathbb{R}$ convergence is $O(1/n)$ but is geometric for $a \neq b$. So what looks like the simplest case analytically is the hardest computationally. We did eventually determine from highly sophisticated intermediate steps that:

Theorem 3.5.2 (Four Formulae for $\mathcal{R}(a,a)$). *For any $a > 0$*

$$\begin{aligned}\mathcal{R}(a,a) &= 2a\sum_{k=1}^{\infty}\frac{(-1)^{k+1}}{1+(2k-1)a}\\ &= \frac{1}{2}\left(\psi\left(\frac{3}{4}+\frac{1}{4a}\right)-\psi\left(\frac{1}{4}+\frac{1}{4a}\right)\right)\\ &= \frac{2a}{1+a}\,{}_2F_1\left(\left.\begin{matrix}\frac{1}{2a}+\frac{1}{2},1\\ \frac{1}{2a}+\frac{3}{2}\end{matrix}\right| -1\right)\\ &= 2\int_0^1 \frac{t^{1/a}}{1+t^2}\,\mathrm{d}t.\end{aligned}$$

Here ${}_2F_1$ is the hypergeometric function defined in (3.35). If you do not know the ψ or Ψ ('psi') function, you can easily look it up once you can say 'psi'. Notice that

$$\mathcal{R}(a,a) = 2\int_0^1 \frac{t^{1/a}}{1+t^2}\,\mathrm{d}t$$

now allows us to evaluate $R(1,1) = \log 2$ as discussed in Example 3.4.10.

The development of this theory exploited modular and theta functions. We used the square counting *theta functions* $\theta_3(q) := \sum_{n=-\infty}^{\infty} q^{n^2}$ and $\theta_4(q) := \theta_3(-q)$. The pictures on the right of Fig. 3.30 shows the level sets of the modulus of the ratio $\theta_4(q)/\theta_3(q)$ for $q := re^{i\theta}$ in the first quadrant; black regions have modulus exceeding one. From this simple recipe comes beautiful fractal complexity.

After making no progress analytically, Crandall and I decided to take a somewhat arbitrary criterion for convergence and colour yellow the points for which the fraction seemed to converge. Treating the iteration implicit in (3.26) as a black box, we sampled one million starting points and reasoned that a few thousand

[28] I could see that $\mathcal{R}(1,1) = 0.693\ldots$ as is suggestive of $\log 2 = 0.6931471806\ldots$.

mis-categorizations would not damage the experiment. Figure 3.30 is so precise that we could identify the cardioid. It is the points where

$$\sqrt{|ab|} \leqslant \frac{|a+b|}{2}$$

and since for positive a, b the fraction satisfies

$$\mathcal{R}\left(\frac{a+b}{2}, \sqrt{ab}\right) = \frac{\mathcal{R}(a,b)+\mathcal{R}(b,a)}{2}$$

this gave us enormous impetus to continue our eventually successful hunt for a rigorous proof (Borwein & Crandall, 2004; Borwein, Borwein, Crandall, & Mayer, 2007).

Example 3.5.3 (Digital Assistance, arctan(1) and a Black-Box). Consider for integer $n > 0$ the sum

$$\sigma_n := \sum_{k=0}^{n-1} \frac{n}{n^2+k^2}.$$

The definition of the Riemann sum means that

$$\begin{aligned} \lim_{n\to\infty} \sigma_n &= \lim_{n\to\infty} \sum_{k=0}^{n-1} \frac{1}{1+(k/n)^2}\,\frac{1}{n} \\ &= \int_0^1 \frac{1}{1+t^2}\,\mathrm{d}t = \arctan(1). \end{aligned} \tag{3.27}$$

Even without being able to do this *Maple* will quickly tell you that

$$\sigma_{10^{14}} = 0.78539816339746\ldots$$

Now if you ask for 100 billion terms of most slowly convergent series, a computer will take a long time. So this is only possible because *Maple* knows

$$\sigma_N = -\frac{i}{2}\Psi(N-iN) + \frac{i}{2}\Psi(N+iN) + \frac{i}{2}\Psi(-iN) - \frac{i}{2}\Psi(iN)$$

using the imaginary i, and it has a fast algorithm for our new friend the psi function. Now `identify`(0.78539816339746) yields $\frac{\pi}{4}$.

We can also note that

$$\tau_n := \sum_{k=1}^{n} \frac{n}{n^2+k^2}$$

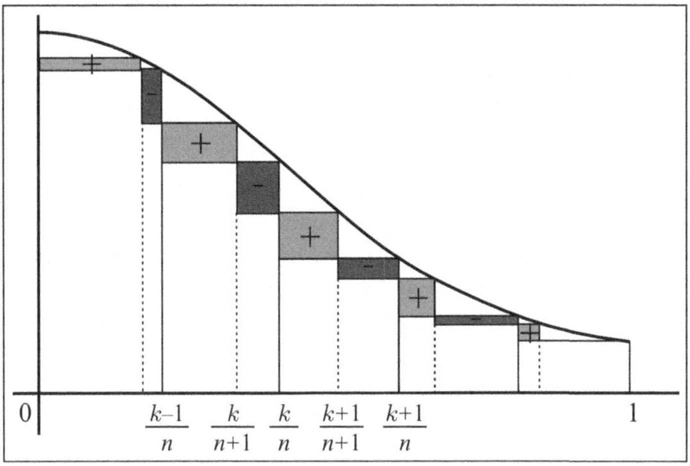

Fig. 3.31 Difference in the lower Riemann sums for $1/(1+x^2)$

is another Riemann sum. Indeed, $\sigma_n - \tau_n = \frac{1}{2n} > 0$. Moreover, experimentally it *seems* that $\tau)$ increases and σ_n decreases to $\pi/4$.

If we enter "monotonicity of Riemann sums" into Google, one of the first entries is http://elib.mi.sanu.ac.rs/files/journals/tm/29/tm1523.pdf which is a 2012 article (Szilárd, 2012) that purports to show the monotonicity of the two sums. The paper goes on to prove that if f: [0, 1] $\to R$ is continuous, concave and decreasing then $\tau_n := \frac{1}{n}\sum_{k=1}^{n} f\left(\frac{k}{n}\right)$ increases and $\sigma_n := \frac{1}{n}\sum_{k=0}^{n-1} f\left(\frac{k}{n}\right)$ decreases to $\int_0^1 f(x)\,\mathrm{d}x$. Moreover, if f is convex and decreasing, then instead σ_n increases and τ_n decreases.

All proofs are based on looking at the rectangles which comprise the difference between τ_{n+1} and τ_n as in Fig. 3.31 (or the corresponding sums for σ_n). This is

$$\sum_{k=1}^{n}\left\{\frac{(n+1-k)}{n+1}f\left(\frac{k}{n+1}\right)+\frac{k}{n+1}f\left(\frac{k+1}{n+1}\right)-f\left(\frac{k}{n}\right)\right\}. \tag{3.28}$$

In the easiest case, each bracketed term

$$\delta_n(k) := \frac{(n+1-k)}{n+1}f\left(\frac{k}{n+1}\right)+\frac{k}{n+1}f\left(\frac{k+1}{n+1}\right)-f\left(\frac{k}{n}\right)$$

has the same sign for all n and $1 \leqslant k \leqslant n$ as happens for concave or convex and decreasing (for increasing consider $-f$).

But in Szilárd (2012) the author mistakenly asserts this for $1/(1+x^2)$ which has an inflection point at $1/\sqrt{3}$. It appears, on checking in a CAS, that $\delta_n(k)+\delta_n(n-k) \geqslant 0$ which will repair the hole in the proof. Indeed, this suggests we consider $g(x) :=$

$\frac{f(x)+f(1-x)}{2}$ which for $f(x) := 1/(1+x^2)$ is concave on $[0, 1]$ and has the same value for (3.28). The details of a correct result based on *symmetric Riemann sums* are to be found in Borwein, Borwein, and Sims (2015). What a fine example of digital assistance in action! ◇

I conclude this section by saying that most of my more sophisticated research computing is an admixture of tools like the ones above—used appropriately and in context. In the remainder of this chapter we look at mathematics originating in my recent research. Details are given in the references but a reader who knows some secondary school algebra, geometry and calculus should be able to follow the broad brushes of what follows. We now turn to three sets of more sophisticated case studies. Remember in each case the pictures are central.

3.6 Case Studies I: Dynamic Geometry

Dynamic or interactive geometry packages take points and lines as primitive objects—usually in two dimensions—and add various conic sections and the like. Once positioned the entire construction is moveable. Thence, the qualitative 'generic' properties of a configuration often become clear very quickly. In Cinderella one can work in various geometries: Euclidean, hyperbolic spherical and more. One can also export a construction as a Java html object useable in a web page. For example, http://www.carma.newcastle.edu.au/jon/lm.html will illustrate much of the next section's discussion and many additional features.

3.6.1 *Case Study Ia: Iterative Reflections*

Let $S \subset R^m$. The (nearest point or metric) *projection* onto S is the (set-valued) mapping, $P_S x := \operatorname{argmin}_{s \in S} \parallel s - x \parallel$. The *reflection* with respect to S is then the (set-valued) mapping, $R_S := 2P_S - I$. The projections and reflection are illustrated in Fig. 3.32 for a convex set (where they are unique) and a non-convex set where they need not be.

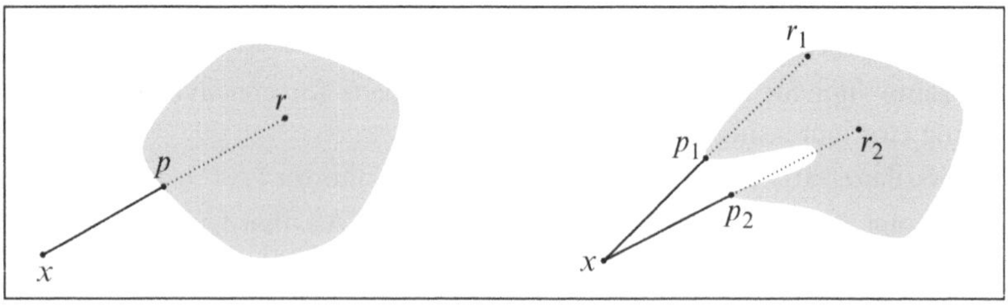

Fig. 3.32 Projections and reflections for a convex set (L) and for a non-convex set (R)

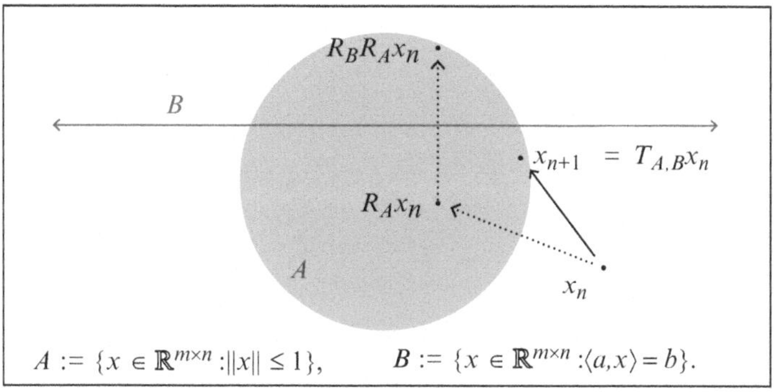

Fig. 3.33 One step of the Douglas–Rachford method

Iterative projection methods have a long and successful history going back to von Neumann, Wiener and many others. The basic model (Aragon and Borwein, 2013; Aragon et al., 2014) finds a point in $A \cap B$ assuming information about the projections on A and B individually is accessible. Precisely we repeatedly compute

$$x_{n+1} := S_{A,B}x_n \quad \text{where} \quad S_{A,B} := P_B P_A.$$

The corresponding reflection methods are more recent and often appear more potent.

Theorem 3.6.1 (Douglas–Rachford (1956–1979)). *Suppose $A, B \subset R^m$ are closed and convex. For any $x_0 \in R^{\mathrm{m}}$ define*

$$x_{n+1} := T_{A,B}x_n \quad \text{where} \quad T_{A,B} := \frac{I + R_B R_A}{2}.$$

If $A \cap B \neq \emptyset$, then $x_{\mathrm{n}} \to x$ such that $P_A x \in A \cap B$. Else if $A \cap B = \emptyset$, then $\| x_n \| \to \infty$.

In Fig. 3.33 we illustrate one step of 'reflect-reflect-average' as Douglas–Rachford's method is also called below.[29]

The method also can be applied to a good model for *phase reconstruction*, namely for B affine and A a boundary 'sphere'. In this case we have some few local convergence results and even fewer global convergence results; but much positive empirical evidence—both numeric and geometric—using tools such as *Cinderella*, *Maple* and SAGE.

Is Fig. 3.34 showing a "generic visual theorem" establishing global convergence off the (provably chaotic) y-axis? Note the *error*—scattered red points—from using 'only' 14 digit computation.

[29] See also http://www.carma.newcastle.edu.au/jon/reflection.html and http://carma.newcastle.edu.au/jon/expansion.html.

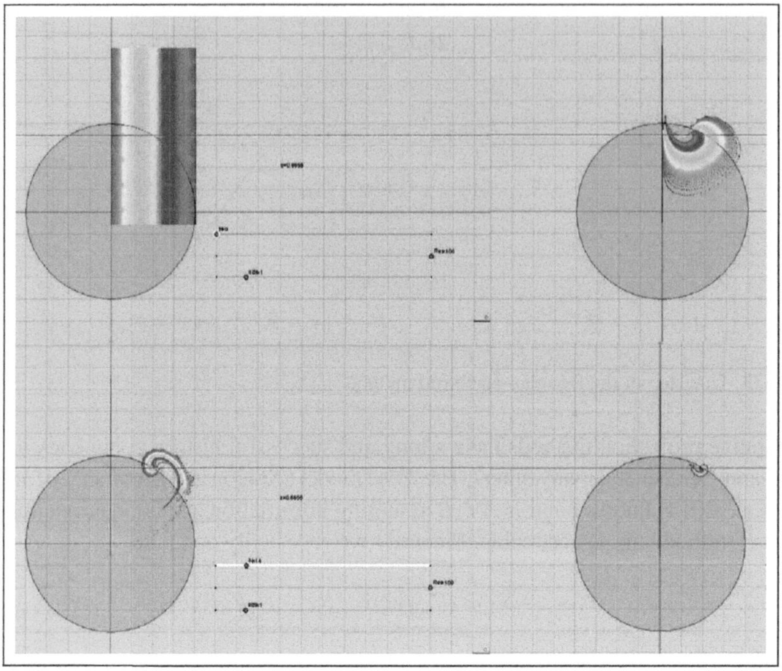

Fig. 3.34 Trajectories of a *Cinderella applet* showing 20,000 starting points coloured by distance from y-axis after $0, 7, 14, 21$ steps

Figure 3.35 illustrates that what we can *prove* (L) is frequently less than what we can *see* (R). There is nothing new here. The French academy stopped looking at attempts to solve the three classical ruler-and-compass *construction* problems of antiquity—trisection of an angle, doubling the cube, and squaring the circle—centuries before they were proven impossible during the nineteenth century.[30]

It is quite striking that an algorithm based on three simple operations of high-school geometry can so effective solve complicated real-world problems.

3.6.2 Case Study Ib: Protein Conformation

We need three concepts. First, a *matrix completion problem* starts with a fixed class of matrices $\mathcal{A}$ (say doubly stochastic, symmetric or positive semidefinite) and seeks a matrix $A \in \mathcal{A}$ consistent with knowledge of some *prescribed* subset of its entries. Of course this is not always possible.

[30] Indeed, changing the tools slightly makes all three constructions possible.

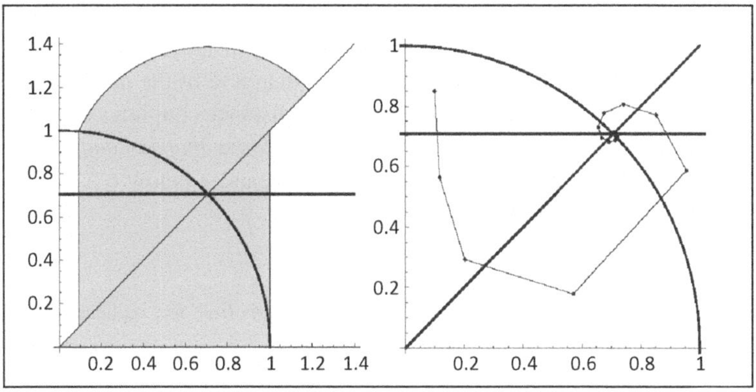

Fig. 3.35 Proven region of convergence in *grey*

Second, a *distance matrix*, with respect to *metric* d on a set X, is a symmetric square $n \times n$ matrix (a_{ij}) with real entries $a_{ij} := d^2(p_i, p_j)$ for points $p_1, p_2, \ldots, p_N \in X$. It is *Euclidean* if $X = \mathbb{R}^N$ and $d(x,y) = \| x - y \|$ is the metric induced by the Euclidean norm (Gower, 1985).[31] Note that $a_{ii} = 0$ for any distance matrix.

$$A := \begin{bmatrix} 0 & ?? & \frac{4}{9} & \frac{3}{4} \\ ?? & 0 & \frac{10}{9} & \frac{3}{4} \\ ?? & \frac{10}{9} & 0 & \frac{19}{36} \\ ?? & ?? & \frac{19}{36} & 0 \end{bmatrix} \qquad B := \begin{bmatrix} 0 & 2 & 4/9 & 3/4 \\ 2 & 0 & \frac{10}{9} & 3/4 \\ 4/9 & \frac{10}{9} & 0 & \frac{19}{36} \\ 3/4 & 3/4 & \frac{19}{36} & 0 \end{bmatrix}. \tag{3.29}$$

Expression (3.29) shows a partial Euclidean matrix A (left) and a completion B (right) based on the four points given as columns

$$p_1 := \begin{bmatrix} 1 \\ 0 \\ 1 \end{bmatrix}, \quad p_2 := \begin{bmatrix} 0 \\ 1 \\ 1 \end{bmatrix}, \quad p_3 := \begin{bmatrix} 1/3 \\ 0 \\ 1 \end{bmatrix}, \quad p_4 := \begin{bmatrix} 1/2 \\ 1/2 \\ 1/2 \end{bmatrix}.$$

Finally, *proteins* are large biomolecules comprising multiple amino acid chains. For instance, RuBisCO (responsible for photosynthesis) has 550 amino acids (making it smallish). Proteins participate in virtually every cellular process and their structure predicts how functions are performed. NMR spectroscopy (the

[31] This paper uses a different normalization: $a_{ij} = -d^2(p_i, p_j)/2$.

Nuclear Overhauser effect, a coupling which occurs through space, rather than chemical bonds) can determine a subset of interatomic distances without damage to the sample (under 6 Å typically constituting less than 8 % of the distances).

Reconstructing a protein given only these short distance couplings can profitably be viewed as a non-convex *low-rank Euclidean distance matrix completion* problem with points in $\mathbb{R}^3$. We use only interatomic distances below 6 Å and use our reflection method to predict the other distances.

We illustrate with a numerical table.

Six proteins from a protein database: average (maximum) errors from five replications

Protein	# Atoms	Rel. error (dB)	RMSE	Max error
1PTQ	40	−83.6 (−83.7)	0.0200 (0.0219)	0.0802 (0.0923)
1HOE	581	−72.7 (−69.3)	0.191 (0.257)	2.88 (5.49)
1LFB	641	−47.6 (−45.3)	3.24 (3.53)	21.7 (24.0)
1PHT	988	−60.5 (−58.1)	1.03 (1.18)	12.7 (13.8)
1POA	1067	−49.3 (−48.1)	34.1 (34.3)	81.9 (87.6)
1AX8	1074	−46.7 (−43.5)	9.69 (10.36)	58.6 (62.6)

Here

$$\text{Rel.error}(dB) := 10\log_{10}\left(\frac{\| P_{C_2}P_{C_1}X_N - P_{C_1}X_N\|^2}{\| P_{C_1}X_N\|^2}\right),$$

$$\text{RMSE} := \sqrt{\frac{\sum_{i=1}^{m} \| \widehat{p}_i - p_i^{true} \|_2^2}{\text{of atoms}}}, \quad \text{Max} := \max_{1\leq i\leq m} \| \widehat{p}_i - p_i^{true} \|_2.$$

The points $\widehat{p}_1, \widehat{p}_2, \ldots, \widehat{p}_n$ denote the best fitting of $p_1, p_2, \ldots, p_n$ when rotation, translation and reflection are allowed.

The numeric estimates above do not well-segregate good and poor reconstructions, as we discover by asking what the reconstructions *look* like. Two instances are shown in Figs. 3.36 and 3.37.

The picture of 'failure' suggests many strategies for greater success, and the method can be accelerated by lots of standard techniques now that we know it is promising.[32] The consequent set of more honed and successful results is described in Borwein and Tam (2012).

Moreover, there are many projection methods, so it is fair to ask *why use Douglas–Rachford*? The sets of images below in Figs. 3.38 and 3.39 show the striking difference in the methods of averaged alternating reflections and that of alternating projections. Yet the method of alternating projections works very well

[32] Video of the first 3000 steps of the 1PTQ reconstruction is at http://carma.newcastle.edu.au/DRmethods/1PTQ.html.

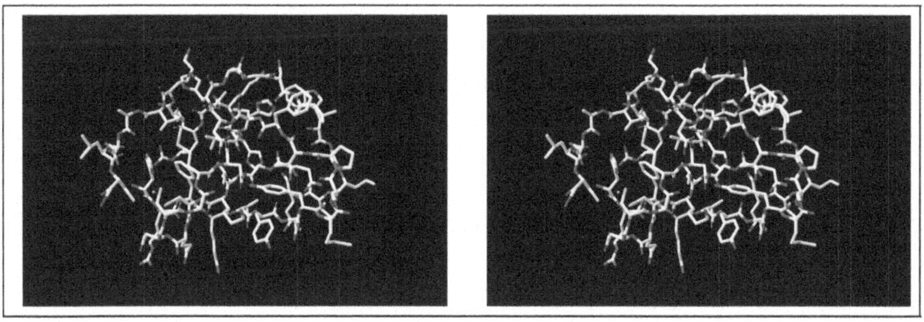

Fig. 3.36 1PTQ (actual) and 5000 DR-steps. Error of -83.6 dB (perfect)

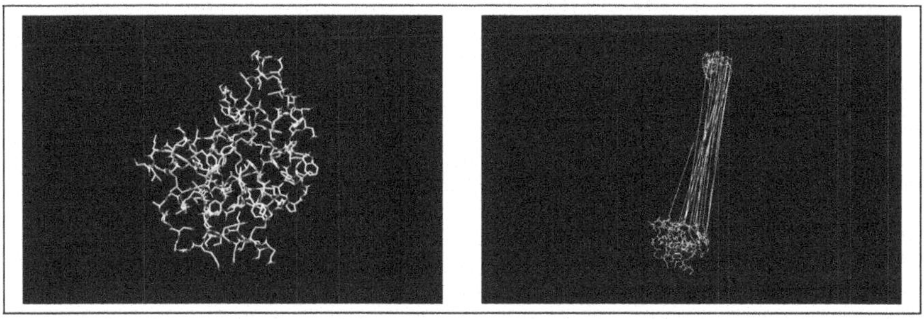

Fig. 3.37 1POA (actual) and 5000 DR-steps. Error of -49.3 dB (mainly good!)

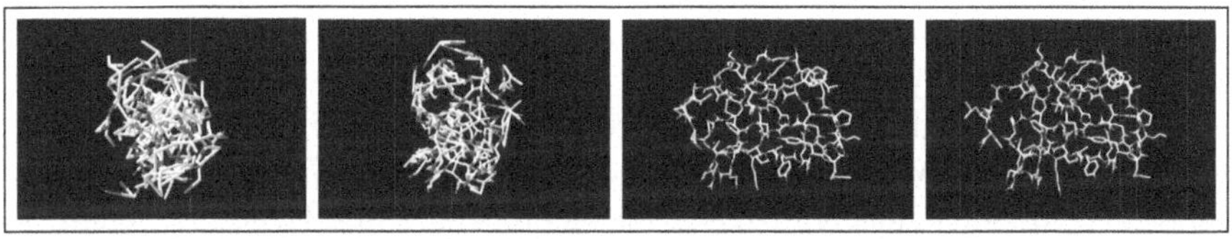

500 steps, -25 dB. 1000 steps, -30 dB. 2000 steps, -51 dB. 5000 steps, -84 dB.

Fig. 3.38 Douglas–Rachford projection method reconstruction

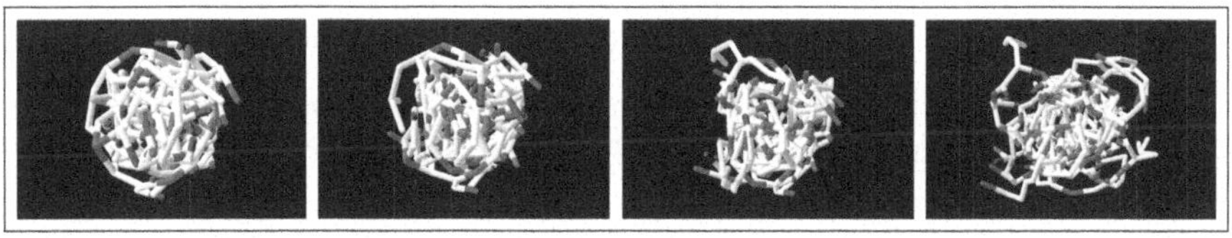

500 steps, -22 dB. 1000 steps, -24 dB. 2000 steps, -25 dB. 5000 steps, -28 dB.

Fig. 3.39 Alternating projection method reconstruction

for optical abberation correction (originally on the Hubble telescope and now on amateur telescopes attached to laptops). And we still struggle to understand why and when these methods work or fail on different non-convex problems.

3.7 Case Studies II: Numerical Analysis

Famously, in 1962 Richard Hamming wrote in *Numerical Methods for Scientists and Engineers*:

> The purpose of computing is insight, not numbers.

This is even more true 50 years on. We turn to three examples of problems arising in numerical analysis.

3.7.1 *Case Study IIa: Trefethen's 100 Digit Challenge*

In the January 2002 issue of *SIAM News*, Nick Trefethen presented ten diverse problems used in teaching *modern* graduate numerical analysis students at Oxford University, the answer to each being a certain real number. Readers were challenged to compute ten digits of each answer, with a $100 prize to the best entrant. Trefethen wrote, "If anyone gets 50 digits in total, I will be impressed." To his surprise, a total of 94 teams, representing 25 different nations, submitted results. Twenty received a full 100 points (10 correct digits for each problem). Bailey, Fee and I quit contentedly at 85 digits!

The problems and solutions are dissected most entertainingly in Bornemann, Laurie, Wagon, and Waldvogel (2004) and are online at http://mathworld.wolfram.com/Hundred-DollarHundred-DigitChallengeProblems.html. Quite full details on the contest and the now substantial related literature are beautifully recorded on Bornemann's website http://www-m3.ma.tum.de/m3old/bornemann/challengebook/. We shall examine the two final problems.

> **Problem #9**. The integral $I(a) = \int_0^2 [2 + \sin(10\alpha)]x^\alpha \sin\left(\frac{\alpha}{2-x}\right) \mathrm{d}x$ depends on the parameter α. What is the value $\alpha \in [0, 5]$ at which $I(\alpha)$ achieves its maximum?

The function $I(\alpha)$ is expressible in terms of a *Meijer- G function*. See Fig. 3.40. This is a special function, invented in 1936, with a solid history that we use below. While knowledge of this function was not common among contestants, *Mathematica* and *Maple* both will figure this out; help files or a web search then quickly informs the scientist. This is another measure of the changing environment. It is usually a good idea—and not at all immoral—to *data-mine*, and find out what your favourite one of the 3Ms knows about your current object of interest. For example, Fig. 3.41 shows the beginning of *Maple*'s help file.

Fig. 3.40 $I(\alpha)$?

$$I(\alpha) = 4\sqrt{\pi}\,\Gamma(\alpha)\, G_{2,4}^{3,0}\left(\frac{\alpha^2}{16}\,\middle|\, \begin{matrix} \frac{\alpha+2}{2}, \frac{\alpha+3}{2} \\ \frac{1}{2}, \frac{1}{2}, 1, 0 \end{matrix}\right)[\sin(10\,\alpha) + 2].$$

MeijerG - Meijer G function

Calling Sequence

```
MeijerG([as, bs], [cs, ds], z)
```

Parameters

Description

- The Meijer G function is defined by the inverse Laplace transform

$$\text{MeijerG}([as, bs], [cs, ds], z) = \frac{1}{2\,\pi\,\mathrm{I}} \oint_L \frac{\Gamma(1 - as + y)\,\Gamma(cs - y)}{\Gamma(bs - y)\,\Gamma(1 - ds + y)}\, z^y \mathrm{d}y$$

where

$$as = [a1, ..., am], \Gamma(1 - as + y) = \Gamma(1 - a1 + y) \ldots \Gamma(1 - am + y)$$

$$bs = [b1, ..., bn], \Gamma(bs - y) = \Gamma(b1 - y) \ldots \Gamma(bn - y)$$

$$cs = [c1, ..., cp], \Gamma(cs - y) = \Gamma(c1 - y) \ldots \Gamma(cp - y)$$

$$ds = [d1, ..., dq], \Gamma(1 - ds + y) = \Gamma(1 - d1 + y) \ldots \Gamma(1 - dq + y)$$

and L is one of three types of integration paths $L_{\gamma + \infty \mathrm{I}}$, L_{∞}, and $L_{-\infty}$.

Contour L_{∞} starts at $\infty + \mathrm{I}\,\phi 1$ and finishes at $\infty + \mathrm{I}\,\phi 2 (\phi 1 < \phi 2)$.

Contour $L_{-\infty}$ starts at $- \infty + \mathrm{I}\,\phi 1$ and finishes at $- \infty + \mathrm{I}\,\phi 2 (\phi 1 < \phi 2)$.

Contour $L_{\gamma + \infty \mathrm{I}}$ starts at $\gamma - \infty$ and finishes at $\gamma + \infty \mathrm{I}$.

All the paths L_{∞}, $L_{-\infty}$, and $L_{\gamma + \infty \mathrm{I}}$ put all $cj + k$ poles on the right and all other poles of the integrand (which must be of the form $aj - 1 + k$) on the left.

- The classical notation used to represent the MeijerG function relates to the notation used in Maple by

$$G_{pq}^{mn}\left(z \,\middle|\, \begin{matrix} a_1, \ldots, a_n, a_{n+1}, \ldots, a_p \\ b_1, \ldots, b_m, b_{m+1}, \ldots, b_q \end{matrix}\right) = \text{MeijerG}([[a_1, \ldots, a_n], [a_{n+1}, \ldots, a_p]], [[b_1, \ldots, b_m], [b_{m+1}, \ldots, b_q]], z)$$

Note: See Prudnikov, Brychkov, and Marichev.

Fig. 3.41 *Maple* help file for Meijer-G

This is a function that only a software package could love, let alone define correctly, and it hums beneath the surface of a great many special function computations. Another excellent example of how packages are changing mathematics is the *Lambert W function*, already met in Example 3.4.8.

Problem #10. A particle at the center of a 10×1 rectangle undergoes Brownian motion (i.e., 2-D random walk with infinitesimal step lengths) till it hits the boundary. What is the probability that it hits at one of the ends rather than at one of the sides?

Bornemann starts his remarkable solution by exploring *Monte-Carlo methods*, which are shown to be impracticable. A tour through many areas of pure and applied mathematics produces huge surprises. Using *separation of variables* on a related PDE on a general $2a \times 2b$ rectangle, we learn that

$$p(a,b) = \frac{4}{\pi} \sum_{n=0}^{\infty} \frac{(-1)^n}{2n+1} \operatorname{sech}\left(\frac{\pi(2n+1)}{2} \frac{a}{b}\right). \qquad (3.30)$$

Equation (3.30) is very efficient computationally since sech decays exponentially. Thence, using only the first *three* terms we obtain

$$p(10,1) = 0.00000038375879792512261034071331862048391007930055940724\ldots$$

to fifty flamboyant places. Thus, (3.30) is also a great example of learning to read formula. It may look foreboding but it is not and one can quickly apprehend its power.

Equation (3.30) and other delights ultimately lead to *elliptic integrals* and *modular functions* and results in a *proof* that the answer is $p = \frac{2}{\pi} \arcsin(k_{100})$ where

$$k_{100} := \left(\left(3 - 2\sqrt{2}\right)\left(2 + \sqrt{5}\right)\left(-3 + \sqrt{10}\right)\left(-\sqrt{2} + \sqrt[4]{5}\right)^2\right)^2,$$

is an example of a so-called *singular value* which were much beloved by Ramanujan.

In general for an $a \times b$ rectangle $p(a,b) = \frac{2}{\pi} \arcsin\left(k_{(a/b)^2}\right)$.

No one (except perhaps harmonic analysts) anticipated a closed form—let alone one like this. This analysis can be extended to some other shapes, and the computation has been performed by Nathan Clisby for self-avoiding walks.

3.7.2 *Case Study IIb: Algorithms for Polylogarithms*

The classical *polylogarithm* of order s is defined by

$$\begin{aligned}\mathrm{Li}_s(z) &= z + \frac{z^2}{2^s} + \frac{z^3}{3^s} + \cdots + \\ &= \sum_{k=1}^{\infty} \frac{z^k}{k^s}.\end{aligned} \tag{3.31}$$

In particular $\mathrm{Li}_1(x) = -\log(1-x)$ and $\mathrm{Li}_s(1) = \zeta(s) = 1 + 1/2^s + 1/3^s + \cdots$ is the famous *Riemann zeta* function. While (3.31) is only guaranteed for $|z| < 1$ the functions Li_s and $\zeta(s)$ may be continued analytically with many wonderful properties (Olver et al., 2012; Borwein et al., 2005).

For small z and most s it is easy to sum $\mathrm{Li}_s(z)$ to high precision—as required in our experimental mathematical-physics studies—from (3.31) but as we approach the radius of convergence of 1 this becomes impracticable. Remarkably we have the following result which is best as the modulus increases.

Theorem 3.7.1 (Polylogarithms). *For $s = n$ a positive integer,*

$$\mathrm{Li}_n(z) = \sum_{m=0}^{\infty}{}' \zeta(n-m)\frac{\log^m z}{m!} + \frac{\log^{n-1} z}{(n-1)!}\Big(H_{n-1} - \log(-\log z)\Big). \tag{3.32}$$

For any complex order s *not a positive integer,*

$$\mathrm{Li}_s(z) = \sum_{m\geq 0} \zeta(s-m)\frac{\log^m z}{m!} + \Gamma(1-s)(-\log z)^{s-1}. \tag{3.33}$$

Here $H_n := 1 + \frac{1}{2} + \frac{1}{3} + \cdots + \frac{1}{n}$ are the *harmonic numbers* and, and $\sum'$ avoids the singularity at $\zeta(1)$. In (3.32), $|\log z| < 2\pi$ precludes use when $|z| < e^{-2\pi} \approx 0.0018674$. For small $|z|$, however, it suffices to use the definition.

We found that (3.31) was faster than (3.32) whenever $|z| < 1/4$, for precisions from 100 to 4000 digits. We illustrate this for Li_2 in Fig. 3.42. Timings show microseconds required for 1000 digit accuracy as the modulus goes from 0 to 1 with blue showing superior performance of (3.32). The regions record trials of random z, such that $-0.6 < \mathrm{Re}\,(z) < 0.4$, $-0.5 < \mathrm{Im}\,(z) < 0.5$. We do not yet have an explanation for the wonderful regularity of the 'eggs' (drawn in *Mathematica*) of Fig. 3.42 but it seems a general phenomenon for all orders s and variable precisions. We may never be able to prove this but we can use it in our algorithm design.

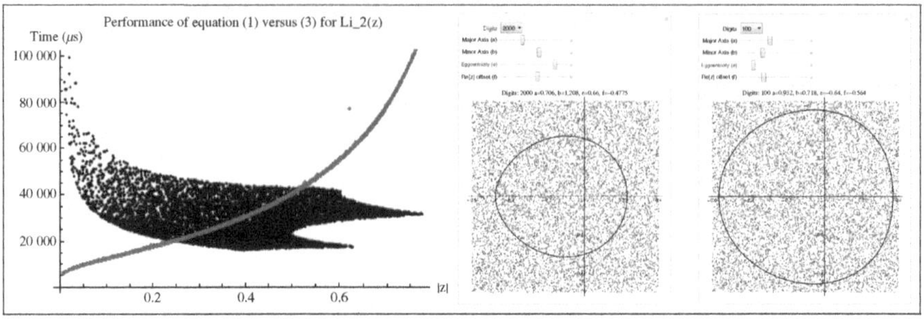

Fig. 3.42 L: timing (3.32) (*blue*) and (3.31) (*red*) for Li_2. M: *blue* region where (3.32) is faster for 100 digits. R: region for 2000 digits

3.8 Case Studies III: Randomish Walks

I have no satisfaction in formulas unless I feel their arithmetical magnitude. Baron William Thomson Kelvin.[33]

The first 'random walk' appears to have been drawn on the base-ten digits of π by John Venn (1834–1923) in 1870. He ignored the digits '8' and '9' and assigned the digits 0 through 7 to the vertices of a regular octagon.[34] The modern study started with questions by Pearson in (1905).

3.8.1 Case Study IIIa: Short Walks

The final set of studies expressedly involve random walks. Our group, motivated initially by multi-dimensional quadrature techniques for higher precision than Monte Carlo can provide, looked at the moments and densities of n-step walks of unit size with uniform random angles (Borwein & Straub, 2013; Borwein et al., 2012). Intensive numeric-symbolic and graphic computing led to some striking new results for a century old problem. Here we mention only two. Let p_n be the radial density of the n-step walk ($p_n(x) \sim \frac{2x}{n} e^{-x^2/n}$) so that

$$W_n(s) := \int_0^n t^s p_n(t)\, \mathrm{d}t$$

is the *moment* function and $W_n := W_n(1)$ is the expected distance travelled in n-steps. The direct definition of $W_n(s)$, for $\mathrm{Re}\, s > -1$, is given by

33 In Lecture 7 (7 Oct 1884), of his Baltimore Lectures on *Molecular Dynamics and the Wave Theory of Light*.

34 See www.theguardian.com/science/alexs-adventures-in-numberland/gallery/2014/mar/14/pi-day-pi-transformed-into-incredible-art-in-pictures/.

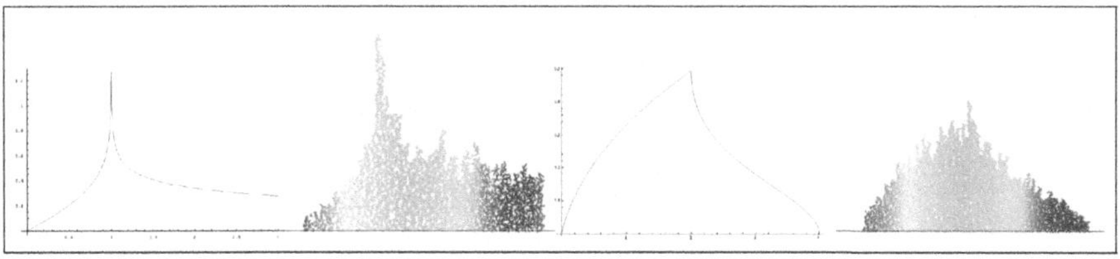

Fig. 3.43 The densities and simulations for p_3 (L) and for p_4 (R)

$$W_n(s) = \int_{[0,1]^n} \left| \sum_{k=1}^{n} e^{2\pi x_k i} \right|^s \mathrm{d}(x_1, \ldots, x_{n-1}, x_n)$$
$$= \int_{[0,1]^{n-1}} \left| 1 + \sum_{k=1}^{n-1} e^{2\pi x_k i} \right|^s \mathrm{d}(x_1, \ldots, x_{n-1}).$$

In particular $W_1 = 1, W_2 = 4/\pi$.

We show the radial densities for three and four step walks in Fig. 3.43 and draw W_4 in the complex plane in Fig. 3.45. These are hard to draw before good analytic expressions such as (3.36) and (3.40).

3.8.1.1 The Three-Step Walk

After learning a good way to compute p_3 numerically (using Bessel functions), we soon discovered, from symbolic and numeric computation, that

$$\sigma(x) := \frac{3-x}{1+x}$$

is an *involution* on $[0,3]$ since

$$\sigma(\sigma(x)) = \frac{\left(3 - \frac{3-x}{x+1}\right)}{\left(1 + \frac{3-x}{x+1}\right)} = x,$$

and σ exchanges $[0,1]$ with $[1,3]$ and leaves 1 fixed. Moreover,

$$p_3(x) = \frac{4x}{(3-x)(x+1)}\, p_3(\sigma(x)). \tag{3.34}$$

Equation (3.34) implies that

$$\frac{3}{4} p_3'(0) = p_3(3) = \frac{\sqrt{3}}{2\pi}, \quad p(1) = \infty,$$

as we see in the picture.

We then found and proved that in terms of the Gauss *hypergeometric* function (Olver et al., 2012, 15.2.1) which is the analytic continuation of the sum (convergent for $|z|<1$)

$$ {}_2F_1\left(\begin{matrix}a,b\\c\end{matrix}\middle|z\right) = 1+\frac{ab}{c}z+\frac{a(a+1)b(b+1)}{c(c+1)2!}z^2+\cdots \tag{3.35}$$

the density of a three-step walk is

$$p_3(\alpha) = \frac{2\sqrt{3}\alpha}{\pi\,(3+\alpha^2)}\,{}_2F_1\left(\begin{matrix}\frac{1}{3},\frac{2}{3}\\1\end{matrix}\middle|\frac{\alpha^2(9-\alpha^2)^2}{(3+\alpha^2)^3}\right). \tag{3.36}$$

Example 3.8.1 (A Cubically Convergent Mean Iteration, Borwein & Bailey, 2008). I had earlier in 1991 used the same hypergeometric form appearing in (3.36)

$$ {}_2F_1\left(\begin{matrix}\frac{1}{3},\frac{2}{3}\\1\end{matrix}\middle|s^3\right) = \sum_{n\geq 0}\frac{(3n)!}{(n!)^3}\left(\frac{s}{3}\right)^{3n}$$

and knew it had remarkable properties. These translated into the result that, for $0<\alpha<3$, we can compute

$$p_3(\alpha) = \frac{2\sqrt{3}}{\pi}\,\frac{\alpha}{\mathrm{AG}_3\left(3+\alpha^2, 3\sqrt[3]{(1-\alpha^2)^2}\right)},$$

where AG_3 is the *cubically convergent* mean iteration given by

$$\mathrm{AG}_3(a,b) := \lim_n a_n = \lim_n b_n$$

with

$$a_{n+1} = \frac{a_n+2b_n}{3}, \qquad b_{n+1} = \sqrt[3]{b_n\cdot\frac{a_n^2+a_nb_n+b_n^2}{3}},$$

starting with $a_0=a, b_0=b$.

For instance, computing $\mathrm{AG}_3(2, 1)$ to 80 places we see that the number of correct digits triples with each step.

$a_1 = \underline{1.3}33333333333333333333333333333333333333333333333333333333333$
33333333333333333333

$$
\begin{aligned}
a_2 &= \underline{1.3286793}7953253152436989510838861457299650489371918427831842415082768902340248 26\\
a_3 &= \underline{1.32867937794645800860304513}102055519425852580042240482420772826100481718474853 45\\
a_4 &= \underline{1.3286793779464580086030451309577743646584603595070873284535}45700752292898721468 8\\
a_5 &= 1.32867937794645800860304513095777436465846035950708732845354570075229289872146 89
\end{aligned}
$$

◇

From (3.38) below, we eventually proved the stunning closed form:

$$W_3 = \frac{16\sqrt[3]{4}\,\pi^2}{\Gamma(\frac{1}{3})^6} + \frac{3\Gamma(\frac{1}{3})^6}{8\sqrt[3]{4}\,\pi^4},$$

in terms of $\pi, 4^{1/3}$ and $\Gamma(1/3)$ where the *Gamma function* is defined by

$$\Gamma(x) := \int_0^\infty t^{x-1} e^{-t}\, \mathrm{d}t$$

for $x > 0$.

Example 3.8.2 (The Bohr–Mollerup Theorem). More usefully, by the *Bohr–Mollerup theorem*, Γ is the unique function, G, mapping positive numbers to positive numbers that satisfies (1) $G(1) = 1$, (2) $G(x+1) = xG(x)$ and (3) is *logarithmically convex*: $\log G$ is convex on $(0, \infty)$. In particular, Γ agrees with the factorial at integers—in that $\Gamma(n+1) = n!$. This result allows one to automate proofs of many interesting facts such as the fact that the *beta function* is a ratio of Gamma functions:

$$B(a,b) := \int_0^1 t^{a-1}(1-t)^{b-1}\, \mathrm{d}t = \frac{\Gamma(a)\Gamma(b)}{\Gamma(a+b)}. \tag{3.37}$$

This is usually proved by change of variable in a two-dimensional integral. Instead, we define $G(a) := \Gamma(a+b)B(a,b)/\Gamma(b)$ and check that G satisfies the three conditions of the Bohr–Mollerup theorem, see Borwein and Bailey (2008, §5.4). ◇

3.8.1.2 The Four-Step Walk

Crucially, for Re $s > -2$ and s not an odd integer the corresponding *moment functions* (Borwein & Straub, 2013), W_3, W_4 have Meijer-G representations

$$W_3(s) = \frac{\Gamma(1+\frac{s}{2})}{\sqrt{\pi}\ \Gamma(-\frac{s}{2})}\, G^{21}_{33}\left(\begin{matrix} 1,1,1 \\ \frac{1}{2}, -\frac{s}{2}, -\frac{s}{2} \end{matrix} \middle| \frac{1}{4}\right), \tag{3.38}$$

$$W_4(s) = \frac{2^s}{\pi} \frac{\Gamma(1+\frac{s}{2})}{\Gamma(-\frac{s}{2})}\, G^{22}_{44}\left(\begin{matrix} 1, \frac{1-s}{2}, 1, 1 \\ \frac{1}{2}, -\frac{s}{2}, -\frac{s}{2}, -\frac{s}{2} \end{matrix} \middle| 1\right). \tag{3.39}$$

Surprisingly, from (3.39) we ultimately got a *modular closed form*:

$$p_4(\alpha) = \frac{2}{\pi^2} \frac{\sqrt{16-\alpha^2}}{\alpha}\, \mathrm{Re}\,{}_3F_2\left(\begin{matrix} \frac{1}{2}, \frac{1}{2}, \frac{1}{2} \\ \frac{5}{6}, \frac{7}{6} \end{matrix} \middle| \frac{(16-\alpha^2)^3}{108\,\alpha^4}\right). \tag{3.40}$$

Let me emphasize that we do not need to know about the Meijer-G function to exploit (3.38) and (3.39). We need only read the help file we sampled in Fig. 3.41. We illustrate this in Figs. 3.44 and 3.45. We leave it to the reader to consider which representation carries more information.

As an illustration of the growing complexity of short walks we mention the question of which paths will return to the starting point in exactly n steps. For $n=2$ or 3 this is easy. For two steps one must retrace the first step and for 3 steps the path must be an equilateral triangle. What about 4 and 5 steps?

Fig. 3.44 W_3 plotted by *Mathematica* from the Meijer-G representation (3.38). Each point is coloured by argument. *Black* is a zero and *white* is a pole (infinity). These can only occur where all four quadrants meet

Fig. 3.45 W_4 plotted by *Mathematica* from the Meijer-G representation (3.39). Each quadrant is coloured differently. *Black* is a zero and *white* is a pole (infinity). These can only occur where all four quadrants meet

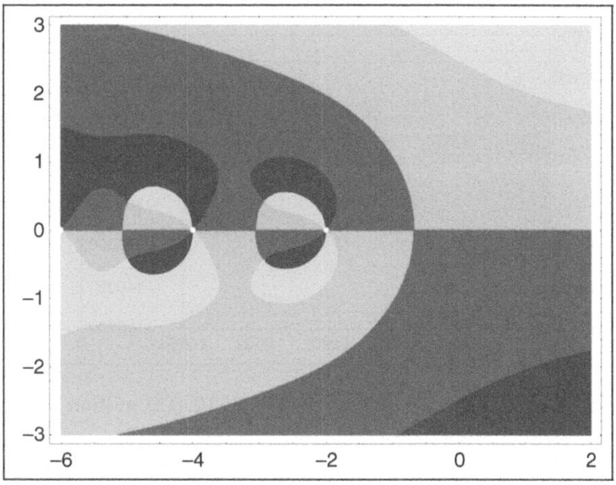

Fig. 3.46 A 108 Gigabit walk on Pi

3.8.2 Case Study IIIb: Number Walks

Our final studies concern representing base-b representations of real numbers as planar walks. For simplicity we consider only binary or hex numbers and use two bits for each direction: $0 = \text{right}$, $1 = \text{up}$, $2 = \text{left}$, and $3 = \text{down}$ (Aragon et al., 2013). From this idea we eventually produced the 100-billion-step walk on the hexadecimal-digits of π shown in Fig. 3.46. The colours move through the spectrum (ROYGBIV and back to red.) We believe this to be the largest mathematical illustration ever made. The picture in Fig. 3.46 can be explored on line at http://gigapan.org/gigapans/106803.

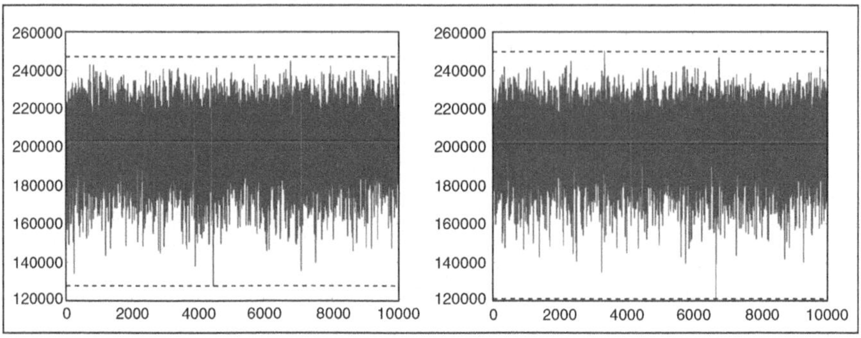

Fig. 3.47 Number of points visited by 10,000 million-step base-4 random walks (L) and by equally many walks on π (R)

This idea also allows us to compare the statistics of walks on any real number to those for pseudo-random walks[35] of the same length. For now in Fig. 3.47 we illustrate only the similarity between the number of points visited by 10,000 million-step pseudo-random walks and for 20 billion bits of π chopped up into 10,000 million-step walks.

All the statistics we have considered in Aragon et al. (2013) and elsewhere have π behaving very randomly even though it is not, and even though it is not yet proven normal in any base.

3.8.3 Case Study IIIc: Normality of Stoneham Numbers

A real constant α is *b-normal* if, given $b \geq 2$, every m-long string of digits appears in the base-b expansion of α with precisely the expected limiting frequency $1/b^m$. Borel showed that almost all irrational real numbers are b-normal in any base but no really explicit numbers (e.g., $e, \pi, \sqrt{2}, \zeta(3)$) have been proven normal. The first number proven 10-normal was the *Champernowne number*

$$C_{10} := 0.12345678910111121314\ldots$$

which comes from concatenating the natural numbers. This number is clearly far from random but as noted it is normal. What do the pictures tell us?

To complete our final study we shall detail the visual discovery of the next theorem. It concerns the *Stoneham numbers*, first studied by Stoneham 40 years ago, which are defined by $\alpha_{b,c} := \sum_{n=1}^{\infty} \frac{1}{c^n b^{c^n}}$ (Fig. 3.48).

[35] Python uses the *Mersenne Twister* as generator with a period of $2^{19937} - 1 \approx 10^{6002}$.

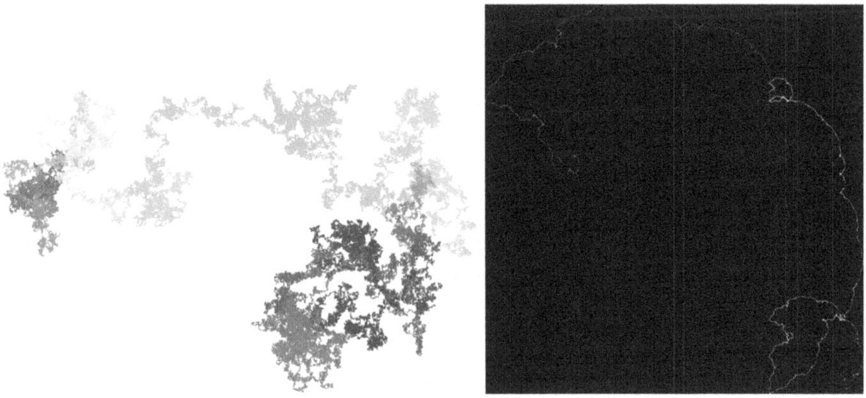

Fig. 3.48 A pseudo-random walk (L) and a walk on C_{10} (R)

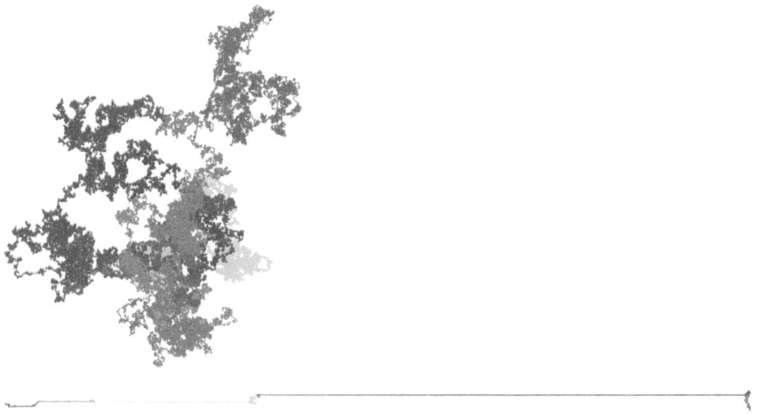

Fig. 3.49 $\alpha_{2,3}$ is 2-normal (*top*) but 6-nonnormal (*bottom*). Is seeing believing?

Theorem 3.8.3 (Normality of Stoneham Constants). *For coprime pairs b,c $\geq$ 2, the constant $\alpha_{b,c}$ is b-normal, while if $c < b^{c-1}$, $\alpha_{b,c}$ is bc-nonnormal.*

Since $3 < 2^{3-1} = 4, \alpha_{2,3}$ is 2-normal but 6-nonnormal. This yields the first concrete transcendental to be shown normal in one base yet abnormal in another. Our final Fig. 3.49 illustrates this result.

There are clearly too many '0's base six (equivalently, too many steps to the right). This is what we ultimately proved.

What is less obvious is that while the shape base-two of $\alpha_{2,3}$ looks like that of a random number, some colours are missing. Indeed, as was discovered from animations that can be viewed at http://walks.carma.newcastle.edu.au/, the walk repeats itself and overwrites large portions!

3.9 Conclusion

In Pólya (1981) George Pólya wrote much that rings very true in the presence of our current tools:

- This "quasi-experimental" approach to proof can help to de-emphasis a focus on rigor and formality for its own sake, and to instead support the view expressed by Hadamard when he stated "The object of mathematical rigor is to sanction and legitimize the conquests of intuition, and there was never any other object for it."[36]
- Intuition comes to us much earlier and with much less outside influence than formal arguments which we cannot really understand unless we have reached a relatively high level of logical experience and sophistication. Therefore, I think that in teaching high school age youngsters we should emphasize intuitive insight more than, and long before, deductive reasoning (Pólya, 1981, p. 2–128).
- In the first place, the beginner[37] must be convinced that proofs deserve to be studied, that they have a purpose, that they are interesting (Pólya, 1981, p. 2–128).
- The purpose of a legal proof is to remove a doubt, but this is also the most obvious and natural purpose of a mathematical proof. We are in doubt about a clearly stated mathematical assertion, we do not know whether it is true or false. Then we have a problem: to remove the doubt, we should either prove that assertion or disprove it (Pólya, 1981, p. 2–120).

We will do well to heed these observations and to think about the many ways our experimental mathodology meshes with them.

Acknowledgements Thanks are due to many of my collaborators and students for comments and input into this work. I especially mention Matt Tam, Nathalie Sinclair, Neil Calkin and Ben Brawn for careful reading of versions of this manuscript.

References

Aragon, F., Bailey, D. H., Borwein, J. M., & Borwein, P. B. (2013). Walking on real numbers. *Mathematical Intelligencer, 35*(1), 42–60.

Aragon, F., & Borwein, J. M. (2013). Global convergence of a non-convex Douglas-Rachford iteration. *Journal of Global Optimization, 57*(3), 753–769.

Aragon, F., Borwein, J. M., & Tam, M. (2014). Douglas-Rachford feasibility methods for matrix completion problems. *ANZIAM Journal*. Online, 28 pages. Available at http://arxiv.org/abs/1308.4243.

Avigad, J. (2008). Computers in mathematical inquiry. In P. Mancuso (Ed.), *The philosophy of mathematical practice* (pp. 302–316). Oxford: Oxford University Press.

Bailey, D. H., & Borwein, J. M. (2011a). Exploratory experimentation in mathematics. *Notices of the AMS*. See www.ams.org/notices/201110/rtx111001410p.pdf.

Bailey, D. H., & Borwein, J. M. (2011b). Exploratory experimentation and computation. *Notices of the AMS, 58*(10), 1410–1419.

[36] J. Hadamard, in E. Borel, *Leçons sur la theorie des fonctions*, 3rd ed. 1928, quoted in Pólya (1981, p. 2–127).

[37] We are all beginners most of the time.

Baillie, R., Borwein, D., & Borwein, J. M. (2008). Some sinc sums and integrals. *American Mathematical Monthly, 115*(10), 888–901.

Bornemann, F., Laurie, D., Wagon, S., & Waldvogel, J. (2004). *The SIAM 100-digit challenge: A study in high-accuracy numerical computing*. Philadelphia: SIAM.

Borwein, J. M. (2012). The Future of Mathematics: 1965 to 2065. Pp 313–330 in *Celebrating a Century of Advancing Mathematics* (MAA Centenary Volume), Stephen Kennedy et al editors, 2015. Print ISBN: 978-0-88385-588-1 and Electronic ISBN: 978-1-61444-522-7.

Borwein, J. M., & Bailey, D. H. (2008). *Mathematics by experiment: Plausible reasoning in the 21st century* (extended 2nd edn.). Wellesley: AK Peters Inc.

Borwein, J. M., Bailey, D. H., Girgensohn, R. (2005). *Experimentation in mathematics: Computational paths to discovery*. Wellesley: AK Peters.

Borwein, J. M., & Borwein, P. B. (1987). *Pi and the AGM: A study in analytic number theory and computational complexity*. New York: Wiley.

Borwein, D., Borwein, J., Crandall, R., & Mayer, R. (2007). On the dynamics of certain recurrence relations. *Ramanujan Journal (Special issue for Richard Askey's 70th birthday), 13*(1–3), 63–101.

Borwein, D., Borwein, J. M., & Sims, B. (2015). Monotonicity of symmetric Riemann sums. Preprint. See https://www.carma.newcastle.edu.au/jon/riemann.pdf.

Borwein, J. M., & Crandall, R. E. (2004). On the Ramanujan AGM fraction. Part II: The complex-parameter case. *Experimental Mathematics, 13*, 287–296.

Borwein, J. M., Crandall, R. E., & Fee, G. (2004). On the Ramanujan AGM fraction. Part I: The real-parameter case. *Experimental Mathematics, 13*, 275–286.

Borwein, J., & Devlin, K. (2008). *The computer as crucible: An introduction to experimental mathematics*. Wellesley: AK Peters.

Borwein, P. B., Liljedahl, P., & Zhai, H. (2010). *Creativity and mathematics*. MAA Spectrum Books. Providence: American Mathematical Society.

Borwein, J. M., Skerritt, M., & Maitland, C. (2013). Computation of a lower bound to Giuga's primality conjecture. *Integers, 13*. Online September 2013 at #A67, http://www.westga.edu/~integers/cgi-bin/get.cgi.

Borwein, J. M., & Straub, A. (2013). Mahler measures, short walks and logsine integrals. *Theoretical Computer Science, 479*(1), 4–21.

Borwein, J. M., Straub, A., Wan, J., & Zudilin, W., (with an Appendix by Don Zagier). (2012). Densities of short uniform random walks. *The Canadian Journal of Mathematics, 64*(5), 961–990.

Borwein, J. M., & Tam, M. K. (2012). Reflection methods for inverse problems with applications to protein conformation determination. August 2015. For Springer volume on CIMPA school *Generalized Nash Equilibrium Problems, Bilevel programming and MPEC*, New Delhi.

Brown, R. C. (2009). *Are Science and Mathematics Socially Constructed?* Singapore: World Scientific.

Csikszentmihalyi, M. (1997). *Creativity: Flow and the psychology of discovery and invention*. New York: HarperCollins.

Fauvel, J., Flood, R., & Wilson, R. (Eds.) (1999). *Oxford figures. 800 years of the mathematical sciences*. Oxford: Oxford University Press.

Franklin, L. R. (2005). Exploratory experiments. *Philosophy of Science, 72*, 888–899.

Giametti, A. B. (1990). *A free and ordered space: The real world of the University*. New York: Norton.

Giaquinto, M. (2007). *Visual thinking in mathematics. An epistemological study*. Oxford: Oxford University Press.

Gold, B., & Simons, R. (2008). *Proof and other dilemmas*. MAA Spectrum Books. Providence: American Mathematical Society.

Gower, J. C. (1985). Properties of Euclidean and non-Euclidean distance matrices. *Linear Algebra and Its Applications, 67*, 81–97.

Guin, D., Ruthven, K., & Trouche, L. (Eds.). (2005). *The didactical challenge of symbolic calculators: Turning a computational device into a mathematical instrument*. New York: Springer. http://www.springerlink.com/content/978-0-387-23158-7/#section=512733&page=1.

Hayes, B. (2005). Why W? *American Scientist, 93*, 1004–1008.

ICMI. (2012). In G. Hanna, M. de Villiers (Eds.), *Proof and proving in mathematics education*. The 19th ICMI Study. New ICMI Study Series (Vol. 15). New York: Springer.

Littlewood, J. E. (1953). *A mathematician's miscellany*. London: Methuen; Littlewood, J. E. (1986). In J. E. Littlewood & B. Bollobás (Eds.), *Littlewood's miscellany*. Cambridge: Cambridge University Press.

Olver, F. W. J., Lozier, D. W., Boisvert, R. F., & Clark, C. W. (2012). *NIST digital handbook of mathematical function*. Cambridge: Cambridge University Press. Online at http://dlmf.nist.gov.

Pólya, G. (1956). On picture-writing. *American Mathematical Monthly, 63*, 689–697.

Pólya, G. (1981). *Mathematical discovery: On understanding, learning and teaching problem solving* (Combined Edition). Hoboken, NJ: Wiley.

Pólya, G. (1990). *Mathematics and plausible reasoning* (Vols. I, II). Princeton: Princeton University Press (orig. published 1968).

Sinclair, N., Pimm, D., & Higginson, W. (Eds.). (2007). *Mathematics and the aesthetic: New approaches to an ancient affinity*. CMS Books in Mathematics. New York: Springer.

Sørenson, H. K. (2010). Exploratory experimentation in experimental mathematics: A glimpse at the PSLQ algorithm. In B. Löwe & T. Müller (Eds.), *Philosophy of mathematics: Sociological aspects and mathematical practice (PhiMSAMP)* (pp. 341–360). London: College Publications.

Szilárd, A. (2012). Monotonicity of certain Riemann type sums. *Teaching of Mathematics, 15*(2), 113–120.

Chapter 4
Tools, Human Development and Mathematics

John Monaghan

4.1 Introduction

As stated in Chap. 1, the chapters in Part I of this book serve as introductions to different aspects of tool use in mathematics. Chapters 2 and 3 are set in the present and concern, respectively, tool use in mathematics education and tool use in research mathematics. This chapter and the next consider the past. The place of tools in the history of mathematics is much too large a topic to cover in two chapters, so we are selective. A principal rationale for the next chapter is to illustrate that learning mathematics in ancient Mesopotamia was learning to use a set of material and symbolic artefacts and that this learning had internal (mental) and embodied aspects. The purpose of this chapter is to raise a number of issues that I consider important with regard to tool use and mathematics: to locate tool use in the development of the human species (phylogenesis, Sect. 4.2); to show that tool use in a mathematical culture, ancient Greek mathematics, may go beyond the obvious tools (Sect. 4.3); to examine an algorithm from ancient Indian mathematics that bears some resemblances to Jon's experimental mathematics described in Chap. 3 (Sect. 4.4); to illustrate the mutual support of hand, mind and artefact in expert use of an abacus (Sect. 4.5); and to examine a period (sixteenth-century Europe) where there was a rapid advance in the development of mathematical tools (Sect. 4.6). This chapter closes with a discussion (Sect. 4.7).

This chapter is a reflection on tools in pre-history and history by a mathematics educator; I am not an anthropologist or an historian. My knowledge in the field I write upon is limited and a further limitation is my Western background. But I believe that my specialist academic niche provides a background for an interpretation of tool use in the past and I strongly believe that we must say something of the past in this book before we consider tool use in mathematics in more recent times. My preparation for writing this chapter includes a lifetime of 'dabbling' in the history of mathematics, numerous articles I have read in the course of writing this chapter and four books: Fauvel and Gray (1987)—a selection of readings

J. Monaghan et al., *Tools and Mathematics*, Mathematics Education Library,
DOI 10.1007/978-3-319-02396-0_4

(translated into English) from the history of mathematics with accompanying commentary from the editors; Gibson and Ingold (1993)—an edited collection of articles by anthropologists, archaeologists, linguists and psychologists on tools, language and cognition in human evolution; Netz (1999)—an historical consideration of the 'shaping of deduction' in ancient Greek mathematics. The fourth book is the instruction manual that came with my Chinese abacus (*Suan-pan*); I cannot reference because it does not name the publisher or year of publication.

4.2 Tool Use and Phylogenesis

Before considering the joint development of tools and the human species I note that humans are not the only species to use tools. As this section concerns the fields of animal behaviour and then anthropology, I use definitions from these fields of animal behaviour: tool use is 'the external deployment of an unattached environmental object to alter more efficiently the form, position, or condition of another object' (Beck, 1980, p. 10); tool-making is 'any modification of an object by the user or conspecific so that the object serves more efficiently as a tool' (Beck, 1980, p. 11). This definition is different to the ones offered in Chap. 1 but it seems sensible to work with definitions in these fields in this section.

Apes, especially chimpanzees, have been the focus of many studies of animal tool use/making. Pruetz and Bertolani (2007) documents chimpanzees using sticks to hunt bush babies. Boesch (1993) documents how chimpanzees collect nuts and carry them to the root of a tree (which serves as an anvil) where they crack the nuts using a wooden club or stone. This is premeditated (thoughtful) behaviour as both nuts and clubs are carried to the anvil. It is also a traditional/cultural behaviour as this practice is regional (not species-wide) and chimpanzees in regions practising such nut cracking actively teach their young how to do it. Recognition of the fact that humans are not the only species to use and make tools has been slow. In 1778 Benjamin Franklin coined the term *man the tool-maker* and it was really only Jane Goodall's 1963 photographs of chimpanzees making tools that dispelled the myth that humans are unique in their tool-making (see Gibson, 1993a). But then our recognition of similarities between animal and human behaviours is a mere 150 years old (Darwin's *The Descent of Man* was published in 1871) and this recognition has been gradual; for instance Vygotsky, writing in 1930, on tool use said 'It is a means by which *human* external activity is aimed at mastering and triumphing over nature' (Vygotsky, 1978, p. 55, my italics). The twentieth century witnessed the development of the study of animal behaviour and tool use and tool-making in many species has been documented.

Although apes have been a major focus of attention, tool use/making is not the preserve of primates. Weir, Chappell, and Kacelnik (2002) report research that 'raise the possibility that these birds may rival nonhuman primates in tool-related cognitive possibilities' (Weir et al., 2002, p. 981). The researchers conducted trials where a New Caledonian crow was presented with food in a bucket in a transparent tube beyond the reach of her beak and a straight thin wire. In nine out of ten valid trials she bent the wire (using sticky tape available or by holding one end with her feet) to make a hook to raise the bucket and successfully get the food. In the wild, New Caledonian Crows 'are renowned for their complex tool-oriented behaviour, which involves both tool use and manufacture' (Kenward, Rutz, Weir, & Kacelnik, 2006, p. 1329) but the crow in the Weir et al. (2002) experiment was captive and had 'little exposure to and no prior training with pliant material' (Weir et al., 2002, p. 981). So much for the derogatory term 'bird brained'! I now move on to tool use and human development, phylogenesis.

The Descent of Man (Darwin, 1879) suggests that modern humans, *Homo sapiens*,[1] are descended from apelike beings but this descent was partially ordered, not linear. Current knowledge on the family tree of our possible ancestors (see Tattersall & Schwartz, 2000) over the last 5 million years shows branches that die out and possible branches where current knowledge does not have definitive evidence for evolutionary links between species. In the lower half of this 5 million year time line is the genus *Australopithecus* and in the upper half is the genus *Homo*. Both genuses have species which have died out and there are species for which the genus is disputed. For example, there is debate (see Miller, 2000) whether the species *Homo habilis*, celebrated by Jon in his title for Chap. 3 of this book, is from the genus *Australopithecus* or the genus *Homo*.

Australopithecines had slightly larger cranial capacities than modern chimpanzees, walked upright, had complex social structures but there is no evidence that they used tools; this does not mean that there was no tool use as fossil evidence from wooden tools would be hard to come by. The earliest known tools, dated 2.6 million years ago, are pebble tools (also known as Oldowan tools), a class of tools which were stones shaped by other stones for the purpose of cutting or pounding. These were made and used by early *Homo* species, who had larger cranial capacities than Australopithecines.

But it should not be assumed that tool use/making resulted from the larger brains of early hominids. Experts in the field have, for many years, argued for multiple and complementary factors in hominid development. The anthropologist Sherwood Washburn was an early advocate for the importance of tools in human development. Washburn (1960) argues for the interrelations of tool use/making, brain size and the development of the hand in hominid evolution. Washburn (1959) is careful to use the word 'speculations' in his consideration due to the incomplete nature of the fossil record. Washburn believes that 'the form of the human hand *is the result* of the new selective pressures which came in with the use of tools' (Washburn, 1959, p. 24) and that 'the tripling in the size of the brain came after man was a tool user' (Washburn, 1959, p. 25); evidence for the latter point being that 'When the brain increased in size, the area for the hand increased vastly more than that for the foot' (Washburn, 1959, p. 27). Washburn (1959) concludes that 'it was bipedalism which started man on his separate evolutionary career. But tool use was nearly as

[1] *Homo sapiens* is a specie in the genus *Homo* in the family *Hominidae*.

early. . . . Tools changed the whole pattern of life bringing in hunting, cooperation, and the necessity for communication and language' (Washburn, 1959, p. 31).

But Washburn may be overstating the case for tools and Dunbar (1993) argues that 'Primates are, above all, social animals' (Dunbar, 1993, p. 661), that evolution in large groups depends on developing time-efficient methods of social bonding and that language alone allows for this. Thus, the development of language and social organisation was an important catalyst for the increase in brain size in human evolution. Gibson's (1993b) argument subsumes both those of Washburn and Dunbar and argues for the interdependence of tool use, language, social structure and information processing in human evolution. Her argument is worth an expanded summary but we must remember that all of these arguments, though evidence-based, are speculations.

Gibson (1993b) seeks to explain why 'the linguistic and technological achievements of apes fall far short of human achievements' (Gibson, 1993b, p. 251). Her overall argument is in three parts:

1. *Humans possess a greater information processing capacity than apes and apply this to tool use, language and social behaviour*. Apes are capable of symbolic gestures but 'the content of ape communications is "information sparse"' (Gibson, 1993b, p. 253) and 'apes rarely exhibit advanced planning of a series of tool-using schemes' (Gibson, 1993b, p. 255).
2. *Tool use, language and social behaviour are mutually interdependent in humans but not in apes*. In humans: technology is inextricably linked with social structures; tool use is cooperative and social; language and technology are interdependent (paraphrase of pp. 256–257). Apes use tools and gestures and have a social structure but these do not exhibit interdependence. For instance, a social group of apes may take to the trees when a predator appears and throw branches at the predator but they do not co-ordinate their branch throwing activity.
3. *Tool use and language are genetically canalised to appear early in the early years of a human child but not in apes*. Gibson compares young children's and chimpanzees' tool-using and symbolic capacities and posits a canalisation in humans leading to logicomathematical understanding and classification skills. The evidence-base for this claim appears weaker than for the other claims.

As a mathematician interested in tool use/making I am drawn to Gibson's account but anthropology is a field with many interpretations of the place of tools, language and social behaviour in human evolution. One thing is certain, however, that early *Homo sapiens* did not spontaneously start using/making tools; tool use is a part of human evolution. I now leave species other than *Homo sapiens* and look at their pre-history.

The development of our species 100–150,000 years ago is what anthropologists call a *speciation event*. The dominant hypothesis concerning *Homo sapiens' speciation event* is that it took place in central Africa and spread around the world over 100,000 years. In the decades following *The Descent of Man* (Darwin, 1879) it was assumed (by those who gave credence to Darwin's work) that something akin to a linear descent occurred (that one species replaced another) but in reality different hominid species co-existed. Whilst wishing to avoid teleological arguments that our species is the *end of the line* of descent, we are the only remaining species of the *Homo* genus and we are the ones writing and reading this book. How *Homo sapiens* came about is another area of debate; Darwin was a gradualist who emphasised continuities in evolution but some, e.g. Tattersall (2002), believe the evidence points to saltational change (i.e. sudden large-scale mutation).

Early *Homo sapiens* made and used pebble tools and stone tool use dominated most of our 150,000 years on earth. There is a popular division of the periods of *Homo sapiens* by the material used for their tools: the stone age, the bronze age and the iron age. There are a number of problems with this division: modern humans still use stone tools; the global diffusion of metallurgy was not uniform over time; it is a rather Eurocentric division (e.g. Japan experienced the bronze age and the iron age simultaneously); and it implies that people who remained in the stone age long after others had gone beyond it were primitive (in the pejorative sense of the word). Further to this, each regional 'age' had stages. Clarke (1969) proposes five *modes* of stone technologies progressing from pebble tools to ground stone tools such as axes, used for clearing forests for agriculture. Wood and bone tools were used with as well as stone tools.

The origin of tool use for protomathematical purposes is shrouded in mystery. Fauvel and Gray (1987) cite an African (modern Zaire) bone artefact dated 9000–6500 BC as 'among the earliest evidence for protomathematical activity' (Fauvel & Gray, 1987, p. 5). The bone has series of notches in columns, for example: column 1 has 11, 13, 17 and 19 (prime numbers); column 3 has 11, 21, 19, 9 ($10 \pm 1, 20 \pm 1$). Its purpose is not known but it is suggested that it could have been used as an early lunar phase counting tool. Similarly, the rocks at Stonehenge, built between 3000 and 2000 BC in modern England, could have been used to predict celestial events (eclipses, solstices and equinoxes). There is a theory (see Fauvel & Gray, 1987, pp. 8–13) that around this time and place there was a standard unit of length, the *megalithic yard*, but this is another area of controversy and one where archaeologists appear sceptical of historians of mathematics who appear to look for evidence to support their theories.

I have effectively finished the section on tool use and phylogenesis, for the humans who constructed the bone artefact mentioned above and Stonehenge were, physically, virtually the same as humans today. The area I have briefly surveyed is full of controversy due to the patchwork nature of archaeological finds and the difficulty in interpreting what has been found. But controversy and difficulties in interpreting 'evidence' aside, work by thousands of archaeologists, anthropologists (biological and social), primatologists and animal behaviour; researchers over the last 150 years clearly shows us that tool use is something that is not just human and was an important aspect of the development of our species. I now move on from pre-history to history.

4.3 Ancient Greece

To mathematicians, the mathematics (geometry) developed by the ancient Greeks is one of the wonders of our history. But, as Fauvel and Gray (1987, p. 46) note,

> ... virtually nothing survives which was physically written by the mathematicians of ancient Greece ... The most substantial source describing the development of geometry ... is a passage from Proclus (fifth century AD). This is believed to have been largely based on the lost *History of Geometry* by Eudemus (late fourth century BC) ...

Greek geometry developed over a relatively short period, the first significant figure was Thales (624–548 BCE[2]) and the most significant later figure was Archimedes (287–212). Euclid is a significant name but he (or she[3]?) appears to simply have methodologically compiled results (his/her *Elements*[4])—this is not to belittle this achievement. Very little is known about Euclid but the *Elements* were written about 300 BCE and were, thereafter, a reference point for mathematicians (a point I shall use in the discussion below). I shall focus on Euclid's *Elements* in this section as it serves my purpose to present tool use in ancient Greek mathematics in a reasonably short number of pages.

The most noted artefacts associated with ancient Greek geometry are the straight edge and the compass[5] and the use of these artefacts as tools is implicit in the opening proposition in the *Elements*: *On a given finite straight line to construct an equilateral triangle* (Heath, 1926, p. 241) Fig. 4.1 shows a lettered diagram which accompanies the construction and proof of this proposition. *AB* is the given straight line and circles of radius *AB* are drawn with centres at the points *A* and *B*. The circles meet at *C* and *ABC* forms an equilateral triangle. It is clear that a straight edge and a compass are required for this construction.

The straight edge and the compass are a powerful pair of mathematical tools but I will not consider them further here because: this pair of tools was considered in Chap. 1; whilst the construction is very nice, the wonder of Greek mathematics is more than a set of nice constructions, it includes a corpus of theorems which were proved[6] and I explore tool use in the proof of theorems.

To enable an exploration of tool use in the proof of theorems in a limited space, I look at one of Euclid's proposition in depth. A theorem that suits my purpose is

[2] Before common era, a term preferred by scholars to BC (but virtually identical in terms of dates).

[3] Almost all the mathematicians of ancient Greece were upper class males.

[4] Thirteen strictly sequenced books of definitions, postulates, common notions and propositions.

[5] Evidence suggests that ancient Greeks also 'used pebbles for calculations on abaci ... but in a marginal role ... never at the centre of mathematical activity' (Netz, 1999, pp. 63–64).

[6] Without detracting from the wonder of Greek mathematics, there are mathematical problems with its definitions and proofs. We do not consider these here. The interested reader may consult Netz (1999).

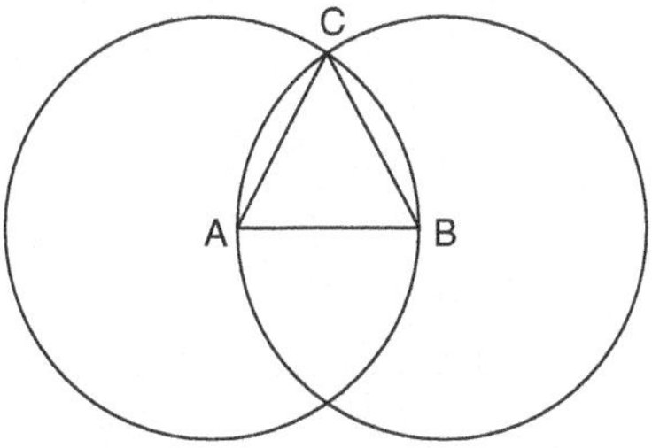

Fig. 4.1 Lettered diagram accompanying Proposition 1, Book 1 (Heath, 1926, p. 46)

Proposition 5 from Book II.[7] The following is Euclid's statement, construction (with diagram) and proof of Proposition 5 from Book II. Much of this is taken from Netz (1999, pp. 9–11) which differs from established English language expositions, for example Heath (1926), by placing text intended to aid reading, but which was not in the original, in <pointy brackets>. I follow Netz in this manner to draw attention to 'formulae' which I discuss in the commentary following Euclid's work.

<**Enunciation**>
If a straight line is cut into equal and unequal <segments>, the rectangle contained by the unequal segment of the whole, with the square on the <line> between the cuts, is equal to the square on the half.

<**Setting Out**>
For let some line, the <line> *AB*, be cut into equal <segments> at the <point> *C*, and into unequal <segments> at the <point> *D*.

<**Goal**>
I say that the rectangle contained by the <lines> *AD*, *DB* together with the square on the <line> *CD*, is equal to the square on the <line> *CB*.

<**Construction**>
For on the <line> *CB*, let a square be set up[8], the <square> *CEFB*
and let the <line> *BE* be joined,
and, through the <point> *D*, let the <line> *DG* be drawn parallel to either of the <lines> *CE*, *BF*,
and, through the <point> *H*, again let the <line> *KM* be drawn parallel to either of the <lines> *AB*, *EF*,
and again, through the <point> *A*, the <line> *AK* be drawn parallel to either of the <lines>[9] *CL*, *BM*.

<**Diagram**>
See Fig. 4.2.

---

[7] My reference here could be (Heath, 1926, p. 382) but I shall use Netz (1999) to describe the proof of this proposition.

[8] This makes implicit reference to Proposition I.46. We return to implicit references and expected knowledge in our discussion of the 'the tool box' after the proof.

[9] This makes implicit reference to Proposition I.31.

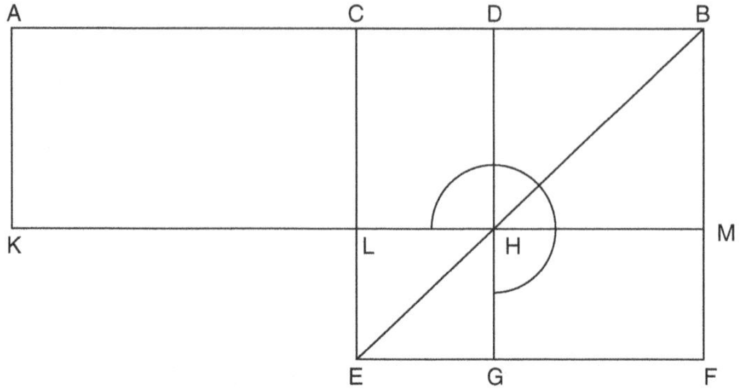

Fig 4.2 Diagram central to Proposition 5 from Book II

<Proof>

<1> And since the complement[10] of *CH* is equal to the complement of *HF*

<2> let the <square> DM be added <as> common

<3> therefore the whole *CM* equals the whole *DF*

<4> But the <area> *CM* is equal to the <area> *AL*[11]

<5> since the <line> *AC*, too, is equal to the <line> *CB*

<6> therefore the <area> AL, too, is equal to the <area> *DF*.

<7> Let the <area> *CH* be added <as> common

<8> therefore the whole *AH* is equal to the gnomon[12] *NOP*

<9> But the <area> *AH* is the <rectangle contained> by the <lines> *AD*, *DB*

<10> for the <line> *DH* is equal to the <line> *DB*

<11> therefore the gnomon *NOP*, too, is equal to the <rectangle contained> by the <lines> *AD*, *DB*

<12> Let the <area> *LG* be added <as> common

<13> which is equal to the <square> on the <line> *CD*

<14> therefore <the sum of> the gnomon *NOP* and the <area> *LG* are equal to <the sum of> the rectangle contained by the <lines> *AD*, *DB* and the square on the <line> *CD*

<15> but the gnomon *NOP* and the area *LG*, <as a> whole, is the square *CEFB*

<16> which is <the square> on the <line> *CB*

<17> therefore the rectangle contained by the <lines> *AD*, *DB*, with the square on the <line> *CD*, is equal to the square on the <line> *CB*.

[10] The *complement* of a parallelogram would be expected to be known to readers. Line <1> makes implicit reference to Proposition I.43.

[11] Implicit reference to Proposition I.36.

[12] The gnomon is defined in Definition II.2.

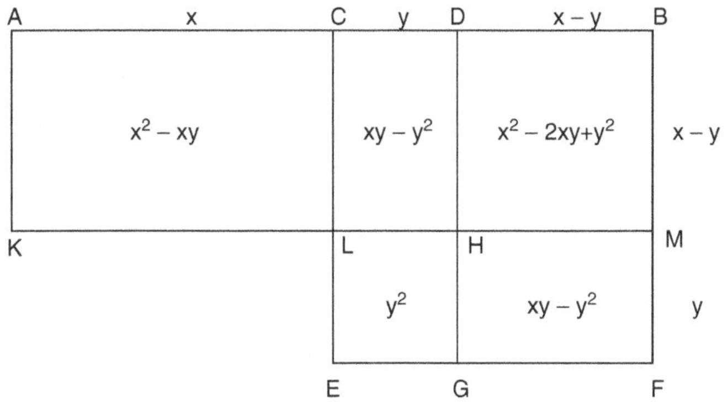

Fig. 4.3 Figure 4.2 with algebraic notation

<Conclusion>
Therefore if a straight line is cut into equal and unequal <segments>, the rectangle contained by the unequal segment of the whole, with the square on the <line> between the cuts, is equal to the square on the half; which it was required to prove.

4.3.1 *Discussion of This Proof*

As a twenty-first century mathematics educator I note that the argument is geometrical and the proof is much easier (to me) using algebra; writing length *AC* as x and length *CD* as y gives (Fig. 4.3):

The Proposition then states that $(x^2 - xy) + (xy - y^2) + y^2$ is equal to $(xy - y^2) + (x^2 - 2xy + y^2) + y^2 + (xy - y^2)$; which is straightforward to show by collecting like terms.

But this ancient Greek proof was written 2400 years ago and the Greek mathematicians did not have algebra. Although Euclid's Elements is not quite the paragon of rigour it is sometimes put up to be, the logic (the necessity of the conclusion) of Proposition II.5 is, mathematically, quite beautiful because nothing is superfluous. We, as mathematicians, might say that this is to be expected because it is mathematics but the Greek mathematicians did not have the history and culture that we do, they were pioneers, they were establishing cultural expectations for us.

4.3.2 *Tools*

I now use Proposition II.5 to discuss themes related to tool use raised in Netz[13] (1999, p. 89), 'Greek mathematical deduction was shaped by two tools: the lettered

[13] 'Netz' in the following pages refers to 'Netz (1999)'.

diagram and the mathematical language'. I first outline what Netz means by this and then consider his use of the term 'tools'.

Between the death of Thales and the birth of Plato the Greeks developed an alphabet, lettered scripts, media approximating to pen and paper and *the epistle* as a means of textual distance communication. The number of mathematicians in ancient Greece was not large and they were geographically dispersed and epistles were used by mathematicians to communicate their results to fellow mathematicians. When an audience was at hand, the media for communication probably varied greatly (sand, wax tablets, wooden tablets, wooden tablets painted white). 'Aristotle used the lettered diagram in his lectures' (Netz, p. 15) but the medium used is not known. Unlike modern whiteboards, it would not have been easy to erase text, so lettered diagrams would have been prepared in advance for oral communication of results. It is reasonable to believe that rulers (straight edges) and compasses were used in the preparation of these lettered diagrams.

Euclid entered a world where lettered diagrams accompanied expositions of mathematical results and in Proposition II.5 (and all Euclid's works) we can see the interdependence of lettered diagram and text. In the seven-part structure of the proof of Proposition II.5 above (enunciation, setting out, goal, construction, diagram, proof, conclusion) this interdependence can be seen in the central five parts. Consider, for example, the setting out: *For let some line, AB, be cut into equal <segments> at C and into unequal <segments> at D*. The diagram includes this line and the four points *A*, *B*, *C* and *D*. The specific line in the diagram is a general line (*let some line*) in the text. This generality is important for Euclid is presenting a deduction of a general theorem. The diagram clearly does not make sense without the text but the text does also not make sense without the diagram. For example, in line 8 of the proof, *therefore the whole AH is equal to the gnomon NOP*, one needs the diagram to locate the gnomon *NOP*.

To the modern mathematician, the insertion of letters into the diagram may appear natural but this was over 2000 years ago and was original (though we do not know the origin). Lettered diagrams would have been crucial for distant written communication; in an oral communication of the setting out one could point and say 'this point'. The letters are references (signifiers) and they refer to (signify) geometrical objects. In terms of the definition of a tool in Sect. 1.3.1, the lettered diagram in Proposition II.5 is a semiotic tool for the communication of a mathematical result to a reader not physically present; it is a material artefact that is used for a specific purpose. This semiotic tool is not just a *teaching aid* for, as mentioned above, it is impossible to follow the proof without recourse to the lettered diagram and, further to this, the logic of the proof depends on the lettered diagram, for example '<3> therefore the whole *CM* equals the whole *DF*'.

The second of Netz' tools is the mathematical language. Netz breaks this down into the mathematical lexicon and the use of 'formulae'. The following discussion of this compresses 81 pages of detailed argument and evidence and, of course, does not do justice to Netz' scholarship (but it serves my purpose here).

With regard to the lexicon, Netz argues that Greek mathematics is 'tiny, strongly skewed towards particular objects ... and is invariant within works and between authors' and it has 'very few synonyms, and even fewer homonyms' (p. 108), i.e. it 'operates on the principle of one-concept-one-word' (p. 113). The mathematical language uses repeated phrases, employing this minimalist lexicon, to highlight relationships between concrete objects. Repetition can be seen in II.5: in the *construction* the repeated use of 'let the ... be drawn'; and, in the *proof*, 'let the ... be added'. We can also see minimalist descriptions of relationships between concrete objects: in the *construction*, 'through the D, let the DG be drawn parallel to either of the CE, BF'; and, in the *proof*, 'therefore the gnomon *NOP* and the *LG* are equal to the rectangle contained by the *AD*, *DB* and the square on the *CD*'.

This minimalist lexicon is clearly a material cultural artefact. Netz argues that this minimalist lexicon facilitates 'self-regulating conventionality' (p. 113). Mathematics educators may view this in terms of *communities of practice* for it is, I feel, easy to imagine a novice of the time, who wants to become a mathematician, appropriating and reproducing this minimalist lexicon.

Regarding 'formulae' Netz (p. 132) writes, 'I count a group of words as formulaic if it is semantically marked OR it is *very* markedly repeated'. In II.5, 'let the ... be drawn' is marked by repetition and 'the *DG*' is semantically marked. Netz list five types of semantic marking in Greek mathematics: object formulae (such as 'the *DG*'); construction formulae (such as 'let the ...'); second-order formulae (such as 'I say that ...'); argumentation formulae (such as 'therefore ...'); and predicate formulae (which includes relations, such as 'the *CM* is equal to the *AL*').

Netz goes on to argue that there are formulae within formulae. This can be seen in II.5, 'therefore the gnomon *NOP*, too, is equal to the <rectangle contained> by the <lines> *AD*, *DB*'. Further to this, there is structure in these nested formulae. For example, using a nested functional representation, we could represent the line of the *proof* cited in the last sentence as 'argumentation formula(predicate formula (object formula, object formula))'. In terms of tool_definition_1 in Chapter, formulae are material artefacts that are used for a specific purpose, they are semiotic tools for communicating the structure of mathematical relationships.

The text of II.5 can thus be viewed as a highly structured artefact where the structure reflects the logical properties of the objects. This, together with the minimalist lexicon assists an appreciation of the global logic of the proof. For example, referring to the numbered statements in the *proof*, the argument can represent by:

<1> & <2> ∴ <3>[14]
<5> ∴ <4>
<3> & <4> ∴ <6>
<6> & <7> ∴ <8>
<10> ∴ <9>

[14] I have represented this using a symbol for 'therefore' instead of representing this as '<1> & <2> → <3>' as I am far from certain that implication in terms of mathematical logic (suggested by the '→' sign) is how the Greeks understood the relationship between '<1>, <2> and <3>'.

<8> & <9> ∴ <11>
<11> & <12> & <13>∴ <14>
<14> & <15> & <16>∴ <17>

4.3.3 Oral and Written Mathematics; Communities of Practice

I end this section on Greek mathematics with a consideration of modern issues in mathematics education, oral and written mathematics and communities of practice.

Putting mathematics aside for the moment, in reading about Greek 'high culture', I am struck by the importance of oral argument. For instance: the politics of the Greek city states were conducted in public arenas where *oration* was the means of argument; the dialogues of Plato are written (though somewhat contrived) forms of spoken discourse. But where, if at all, is the oral in Greek mathematics? I consider formulae. As Netz (see p. 128) notes, there are parallels between what he calls formulae in Greek mathematics and formulae in Homeric plays, where illiterate singers coped 'with the necessity of singing long stretches of metrical text without a script' by developing a tool, formulae, 'short phrases of given metrical shapes'. Moreover, Netz hypothesises that:

> propositions originated in many ways, but the most common was to draw a diagram, to letter it, accompanied by an oral dress rehearsal—an internal monologue perhaps—corresponding to the main argument; and then proceed to write down the proposition (p. 86)

But perhaps positing an oral/written duality is an 'artefact' of a twenty-first century mind. As Netz (p. 163) notes, if the Greeks wrote '$A+B=C+D$' in English, then it would be 'THEAANDTHEBTAKENTOGETHERAREEQUAL TOTHECANDTHED'

Netz rejects the duality and writes:

> Greek mathematical formulae are post-oral, but pre-written. They no longer rely on the aural; they do not yet rely on the layout. They are neutral: rather than oral or written, they are simply artefacts of language. (pp. 163–164)

I commented above on *communities of practice* and how a novice mathematician might appropriate and reproduce the minimalist lexicon; this is true of formulae too. Returning to Chap. 1 and the *irreducible bond* between agent, tool and purpose I note that a tool, be it a compass or a formulae in Greek times, is useless to do, or to advance, mathematics without someone using the tool for a purpose. But to understand Proposition II.5, one needs more than tools. One also needs to know the implicit references alluded to in footnotes <14> to <18>. Netz, following the

Japanese scholar Saito, regards such implicit references as part of 'the tool-box[15]' of Greek mathematics. Netz (p. 216) provides an eloquent definition of *the tool-box*:

> Every starting-point or argument whose truth is not obvious from the diagram or from some other intuitive basis must reflect a more specialised knowledge. When there is no explicit reference, this more specialised knowledge is assumed to be known to the audience. Sometimes this is easily secured: the relevant piece of knowledge is known to the audience because it was recently proved, in the same treatise . . . In other cases, the result invoked is not proved in the same treatise. Such results are the tool box.

Netz appears to be describing a part of the craft knowledge of a 'community of practice' in the sense that Lave and Wenger (1991) use this term, with various layers of participation (and knowledge), from novice to master. One might quibble about terms here for Netz writes 'it is clear that different persons must have internalised it [the tool box] to varying degrees' (p. 217) whilst Lave and Wenger write 'membership in communities of practice, like participation, can be neither fully internalised nor fully externalised' (p. 54) but the parallels are striking. Kanes and Lerman (2008, p. 320) state that 'The nature and role of artefacts and tools is hazy' in Lave and Wenger's exposition of communities of practice but it is clear that artefacts and tools were an essential part of the ancient Greek community of mathematical practice.

4.4 Ancient Indian Square Roots

In this section I consider an ancient Indian algorithm for computing square roots.[16] This section is based on Bailey and Borwein (2012). I first remind the reader that an algorithm is, by the definitions of artefact and tool given in Sect. 1.3.1, an artefact which, when it is used to do something, becomes a tool. I present the algorithm and its modern day version before discussing its origin and significance.

The text of the algorithm (in an English translation and with words not in the original added in square brackets) is:

[1] In the case of a non-square [number], subtract the nearest square number; divide the remainder by twice [the root of that number].
[2] Half the square of that [that is, the fraction just obtained] is divided by the sum of the root and the fraction and subtract [from the sum].
[3] [The non-square number is] less [than the square of the approximation] by the square [of the last term]. (Bailey & Borwein, 2012, pp. 649–650)

---

[15] I use the term Saito coined and Netz followed but do not scrutinise the term with regard to Chap. 1 definitions of tools. It can be assumed that the terms 'tool box' in this chapter and 'tool-box' in Chap. 1 refer to different things.

[16] The mathematical community considered also constructed algorithm for computing cube roots but I restrict my focus to square roots in this section.

The first two sentences present the algorithm; the third sentence provides a means to check the answer. The first two sentences in modern notation, with q as the number of which we are to find the square root and x_0 as the initial approximation, are:

$$a_n = \frac{q - x_n^2}{2x_n}$$
$$x_{n+1} = x_n + a_n - \frac{a_n^2}{2(x_n + a_n)}$$

Using this algorithm to find $\sqrt{123}$ with an initial approximation of 11, we calculate:

$$a_0 = \frac{123 - 11^2}{22} = \frac{1}{11}, \quad x_0 + a_0 = 11 + \frac{1}{11}$$
$$\frac{a_0^2}{2(x_0 + a_0)} = \frac{1/121}{2\left(11 + \frac{1}{11}\right)} = \frac{1}{2684}$$
$$x_1 = x_0 + a_0 - \frac{a_0^2}{2(x_0 + a_0)} = 11 + \frac{1}{11} - \frac{1}{2684} = \frac{29,767}{2684}$$

Approximating $\frac{29,767}{2684}$ and $\sqrt{123}$ on my TI-92 calculator gives the same answer to 10 significant figures, 11.09053651—the algorithm is pretty accurate.

The algorithm was found in an ancient mathematical text known as the Bakhshali manuscript. It was found in 1881 in the village of Bakhshali in the present day Pakistan.

> Among the topics covered in this document, at least in the fragments that have been recovered, are solutions of systems of linear equations, indeterminate (Diophantine) equations of the second degree, arithmetic progressions of various types, and rational approximations of square roots. (Bailey & Borwein, 2012, p. 648)

Indian/Hindu mathematics is one of the milestones in the history of mathematics though it is ignored by Fauvel and Gray (1987). In the early centuries of the common era (CE), Indian mathematicians introduced place value notation with a symbol for zero but the early traces from physical artefacts are sketchy (see Section 2 of Bailey and Borwein (2012) for details). The date of the Bakhshali manuscript is debated (with claims between the third and twelfth century CE, the seventh century being likely—see Section 4, Bailey and Borwein (2012) for details).

The Bakhshali manuscript was much more adventurous than my example of calculating $\sqrt{123}$ suggests. Bailey and Borwein (2012) present an example, arising from an analysis of additive series:

Find an accurate rational approximation to the solution of $\frac{3x^2}{4} + \frac{3x}{4} = 7000$.

The answer is $\frac{\sqrt{336009}-3}{6}$. To compute $\sqrt{336009}$ the manuscript author takes an initial estimate, x_0, as 579 and obtains (as x_1) $\frac{507533837627467432719 36}{7250483394675000000}$, which agrees with $\sqrt{336009}$ to 12 significant digits.

The Bakhshali manuscript did not, of course, use Arabic symbols and the modern square root sign but 'The digits are written left-to-right, and fractions are written as one integer directly over another (although there is no division bar). Zeroes are denoted by large dots' (Bailey & Borwein, 2012, p. 648). Bailey and Borwein (2012) goes on to prove that the Bakhshali square root algorithm is quartically convergent.

The reason I include a section on this ancient Indian tool in this chapter is to contrast ancient Greek tools (considered in the previous section) which lead to deduction being a pillar of Western mathematics, with tools for experimental mathematics that Jon laid out in Chap. 3. Bailey and Borwein (2012, pp. 655–656) state this succinctly:

> The Greek heritage that underlies much of Western mathematics, as valuable as it is, may have unduly predisposed many of us against experimental approaches that are now facilitated by the availability of powerful computer technology. In addition, more and more documents are now accessible for careful study—from Chinese, Babylonian, Mayan, and other sources as well. Thus a renewed exposure to non-Western traditions may lead to new insights and results, and may clarify the age-old issue of the relationship between mathematics as a language of science and technology, and mathematics as a supreme human intellectual discipline.

4.5 Abaci

I have wondered whether a very early mathematical act was a shepherd placing one stone in a pile for each sheep entering a field and, later, taking one stone away for each sheep leaving—one-to-one correspondence in order to ensure that all sheep are accounted for. This is sheer speculation but there is evidence that stones/peebles were used in early counting/arithmetic as we have seen in the section on Greek mathematics. Indeed, the word 'calculate' is derived from the Latin word 'calx', a stone and refers to the practice of using small stones as counters. Given the cognitive demands on memory and the scarcity of writing materials it is hardly surprising that stones (or notches or fingers or ...) were used to aid counting and arithmetic. If counting is sufficient to satisfy the demands of an activity, then the medium is not particularly important as long as it is manageable. But if the demands of the activity involve arithmetic, then a more advanced semiotic system is needed.

De Solla Price (1984) presents a survey of calculating machines with particular emphasis on their role in the development of astronomy and notes different developmental paths for commercial arithmetic and scientific calculation:

> Astronomy needed ingenious mathematical constructions and mechanical devices, but the keeping of accounts demanded but little elaboration of the primitive method of laying out pebbles and shells. (De Solla Price, 1984, p. 34)

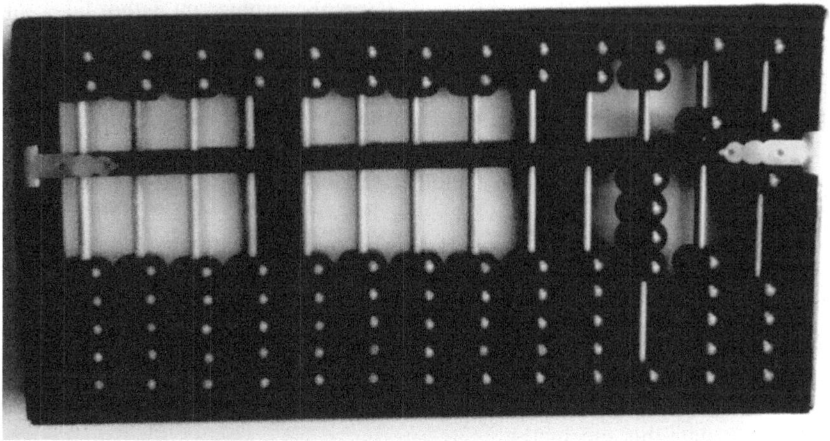

Fig. 4.4 A 13 column *Suan-pan*

Further to this the Greeks and Romans used the Babylonian sexagesimal system (including their multiplication tables) for scientific calculations, not pebble mathematics.

The word 'abacus' comes from the Greek αβαξ (a'bax) which is probably derived from the Semitic word for dust (see Smith, 1958, p. 156), referring to a table covered with dust or sand where marks could be drawn and erased. The dust abacus gave way to tables with ruled lines on which counters were placed (Smith, 1958, p. 157). The origin of the counter abacus is obscure and Smith (1958, p. 159) states that there is 'some reason for believing that this form of the abacus originated in India, Mesopotamia, or Egypt'.

Two abaci that remain in use today are the Chinese *Suan-pan* (dating from the late twelfth century) and the Japanese *soroban* (which appeared after the *Suan-pan*). I shall consider the *Suan-pan*. Figure 4.4 shows a 13 column *Suan-pan*. Note the 2 by 13 bead top array and the 5 by 13 bead lower array. The *soroban* is similar to the *Suan-pan* but the top array has only one bead on each column and the bottom array has four beads on each column. Prior to a *Suan-pan* calculation, the top (respectively lower) beads are placed against the top (respectively bottom) of the frame. Beads are brought into operation by moving them towards the middle bar. The *Suan-pan* represents denary place value in the standard written manner, left-to-right, with columns representing powers of 10. Any column can represent units but, as is common, I will take the unit column to be the rightmost column. In each column the top (respectively bottom) beads represent 5 (respectively 1) of the column power of ten; Fig. 4.4 thus represents the number 456.

The *Suan-pan* can be used for addition, subtraction, multiplication and division (and, with some ingenuity, squares, cubes, square roots and cube roots) but, for reasons of space, I will only deal with addition. To perform $3+4$ on the *Suan-pan*, first move 3 lower beads in the unit column to the middle. We have 4 to add on to this but only 2 more lower beads. Move these 2 to the middle. We now have 5 in the middle, replace these by 1 top bead by moving the 5 lower beads to the bottom of the frame and moving one top bead to the middle. We can now move two lower beads towards the middle to complete the addition. To perform 6 + 7 we move beads in a similar manner but in this calculation (which I will not describe in detail) we need to go to a new (left) column (when a column has 2 top beads to the middle, replace them by 1 bottom bead to the middle in the adjacent left column). The performance of more complex calculations (even 24 + 36) is going to get quite tricky and this is where the Chinese *secrets* (hints or action guides) are taught/learnt to make things manageable. I present the addition *secrets* in shorthand where: *lower 5* means 'move 1 top bead down to the middle'; *cancel* means 'replace beads that have already been moved to the middle to the neutral position'; *raise* means 'move a lower and/or top beads towards the middle'; and *forward 10* means 'move one lower bead in the next column towards the middle'. The 17 *secrets* for addition (in three sets) are shown in Table 4.1 below. I explain the first row (the explanation for the other rows are similar).

1	lower 5, cancel 4	(1 = 5 − 4)

The number on the left is the result of a hand movement. Then the shorthand 'lower 5, cancel 4' is listed. The addition sum for the first *secret* of each set is displayed in brackets.

Once the novice abacus user is familiar with these *secrets*, the addition sums above become:

3 + 4	raise 3; lower 5, cancel 1
6 + 7	raise 6; raise 2, cancel 5, forward 10
24 + 36	raise 24; second column—lower 5, cancel 2; first column—cancel 4, forward 10

Table 4.1 The 17 *secrets* for addition

1	Lower 5, cancel 4	(1 = 5 − 4)
2	Lower 5, cancel 3	
3	Lower 5, cancel 2	
4	Lower 5, cancel 1	
1	Cancel 9, forward 10	(1 = 10 − 9)
2	Cancel 8, forward 10	
3	Cancel 7, forward 10	
4	Cancel 6, forward 10	
5	Cancel 5, forward 10	
6	Cancel 4, forward 10	
7	Cancel 3, forward 10	
8	Cancel 2, forward 10	
9	Cancel 1, forward 10	
6	Raise 1, cancel 5, forward 10	(6 = 10 − 5 + 1)
7	Raise 2, cancel 5, forward 10	
8	Raise 3, cancel 5, forward 10	
9	Raise 4, cancel 5, forward 10	

From a mathematics education perspective there are a number of interesting features of such use:

- The physical artefact is extremely complicated to use without the *secrets*, which are specialised modes of action with the physical artefact. One could say the abacus as a tool is a combination of the physical artefact and the *secrets* (semiotic artefact).
- Related to the above point, Master abaci calculators go beyond memorising these *secrets*, they are automated bodily (hand) operations. Master abaci calculators can do addition sums such as 83,492 + 239,497 + 23,098 very quickly but they do not invoke 'number bonds', instead each digit becomes a specific hand movement on a specific column.
- The secrets are, traditionally, what the Master teaches the Novice.
- Unlike the standard written algorithm for addition, addition on a *Suan-pan* begins with the most significant digit.

4.6 Tools for Calculation in Europe Circa 1600

The need to perform accurate multi-digit arithmetic calculations was a reality in sixteenth century Europe, not least in astronomy (as the De Solla Price (1984) citation in the last section evidences). In this section I look at three calculation tools that developed in fairly quick succession: prosthaphaeresis, logarithms and slide rules. The need to calculate accurately in astronomy was twofold: European ships were exploring the world and ships' navigators charted courses using trigonometry and the positions of specific stars and planets; astronomical theory was developing, e.g. Tycho Brahe (1546–1601) was refining the Copernican theory of planetary motion which involved detailed empirical observations and calculations of the position of specific stars and planets.

Trigonometric functions and tables have a history that predates this period and is not Europe-centred but 'tables of the decimal sine, cosine, tangent and cosecant functions were introduced in Western mathematics about 1450' (Rosińska, 1987, p. 419). Trigonometric tables are clearly a useful tool for, and were widely used by, sixteenth century astronomers.[17] But in the sixteenth century it was discovered that they could be used to speed up arithmetic calculations by a method called prosthaphaeresis (from Greek words for addition and subtraction). I shall explain, via an example (see Table 4.2), the principles of this method before discussing it further. I shall use the identity

$$\cos a \cos b \equiv \frac{1}{2}(\cos(a+b) + \cos(a-b)) \tag{4.1}$$

[17] I drop the prefix 'European' for the remainder of this section.

Table 4.2 Calculating 123 × 456 using $\cos a \cos b \equiv \frac{1}{2}(\cos(a+b) + \cos(a-b))$

Example	To calculate 123 × 345 using (4.1)	
Step 1	Rewrite the digits of the numbers to be multiplied as numbers between −1 and 1 (and keep a note of powers of 10 lost in this process)	0.123 × 0.345 Note that $10^3 \times 10^3$ has been lost
Step 2	Find the inverse cosines of the numbers to be multiplied	$\cos^{-1}(.123) \approx 82.935^\circ$ $\cos^{-1}(.345) \approx 69.818^\circ$
Step 3	Substitute the angles obtained in step 2 into (4.1), use trigonometric tables to find the cosines and calculate (using only addition and division by 2)	½ (cos(152.753) + cos(13.117)) ≈0.084867 ÷ 2 = 0.042434
Step 4	Put back the powers of 10 lost in calculation	$0.042434 \times 10^3 \times 10^3 = 42{,}434$

in my example but prosthaphaeresis also uses other trigonometric identities. I shall work to 5 significant figures in my example but the accuracy of this method depends on the accuracy of the trigonometric tables and 7 figure trigonometric tables were available in the sixteenth century. Note that sixteenth century astronomers would have worked in degrees.

As a mathematician interested in tool use I think this is pretty cool. The steps form an algorithm (a tool) which uses another tool (trigonometric tables) to simplify a calculation. My example used three digit numbers but the power of this algorithm can be appreciated if we imagine numbers with more than three digits. With repeated use the algorithm can be executed quickly. The identities for the products of sines, and of cosines, were published in 1588 but Thoren (1988) states that the identity for the product of sines was discovered/invented in 1510. It is clear that Tycho Brahe used prosthaphaeresis before 1588.

The next step in the 'tools beget tools' claim I am putting forward is the introduction of logarithms, which is largely due to one person, John Napier, a Scottish landlord and an amateur mathematician, with the publication in 1614 of *Mirifici Logarithmorum Canonis Descriptio*. It appears (see Pierce, 1977) that Napier learnt of prosthaphaeresis and began playing with the correspondence between arithmetic and geometric progressions (APs and GPs). This can generate logarithmic functions; I shall call one, made up for introductory purposes here, *lg*, in the example below.

If we put the AP 2, 4, 6, . . . in term-wise correspondence with the GP 3, 9, 27, . . . we can define: $2 = lg(3)$, $4 = lg(9)$, $6 = lg(27)$, . . . *lg* obeys the laws of logarithms. For example, $lg(3 \times 9) = lg(3) + lg(9) = 2 + 4 = 6$. Similar logarithmic function can be set up for other APs and GPs. Like prosthaphaeresis it allows the calculation of products by way of addition. Note that this origin of logarithms is not connected with the inverses of exponential operations as it is in the modern definition of a logarithm.[18]

[18] $a = \log_b c \Leftrightarrow c = b^a, b > 0$ and $b \neq 1$.

The example I introduced is not useful for calculations because of the large 'gaps' between the terms of the GP. Napier got over this problem by using a GP with the first term 10^7 and a common ration of 0.9999999 (these numbers were chosen as he was inspired by studying prosthaphaeresis and the best trigonometrical tables he had were given to 7 decimal places). Napier used this logarithmic function (which does not really have a base and, for which, the logarithm of 1 is not 0) to construct a *table of radicals*; he made up the term 'logarithm' from Greek, logos (ratio) and arithmos (number).

Napier's publication came to the notice of the mathematician Henry Briggs and the two met, in 1615. Napier and Briggs agreed to modify Napier's logarithms so that $\log(1) = 0$ and $\log(10) = 1$, i.e. to what we now know as base 10 logarithms. The results were published in 1617 and the use of base 10 logarithmic tables quickly caught on; within 10 years publications were appearing in other European countries. Indeed, Pierce (1977, p. 26) writes, 'it has been postulated that logarithms literally lengthened the life span of astronomers, who had been sorely bent and broken early by the masses of calculations their art required' and mathematical folklore has it that Gauss memorised his table of logarithms.

Slide rules quickly followed in the wake of Napier and Brigg's logarithms as the mathematical principles are the same for both. E Gunter designed a 'logarithmic line of numbers' in 1620 and W Oughtred designed the first proper slide rule in 1622 (see Smith, 1958 for details). The twentieth century slide rule that the reader may be familiar with has many special built-in functions (a bit like a modern scientific calculator) but if we restrict ourselves to multiplication, the principle of operation is the same as Oughtred's slide rule. I illustrate this in Fig. 4.5 which shows how a slide rule is used to multiply 12×23.

The two logarithmic scales are identical. The one of the lower scale is aligned with 21 on the upper scale. One then reads along to 23 on the lower scale and reads the number vertically above it on the upper scale (which is just over 280).

Note: the leftmost number of a logarithmic line is 1, not 0, because $10^0 = 1$; a slide rule works on the digits, i.e. it does not differentiate between 12, 1.2, 0.12, etc.—the user must calculate the order of magnitude of the answer; the answer is rarely accurate to more than three significant figures (the calculation above is only accurate to two significant figures).

Like tables of logarithms, slide rules became popular. My chain of tools for calculation in Europe circa 1600 is now complete: trigonometric tables begat prosthaphaeresis which inspired Napier's logarithms and tables of radicals which begat base 10 tables of logarithms which begat slide rules. I end this section with educational comments.

It was not long before the use of tables of logarithms moved from scientific communities to elementary arithmetic, evidenced by a reference to Brigg's table of logarithms in a 1646 edition of the early mathematics educator Robert Recorde's *Ground of Artes* (see Smith, 1958, p. 518). By the time that the electronic calculators appeared, adolescent schoolchildren throughout the world took their log

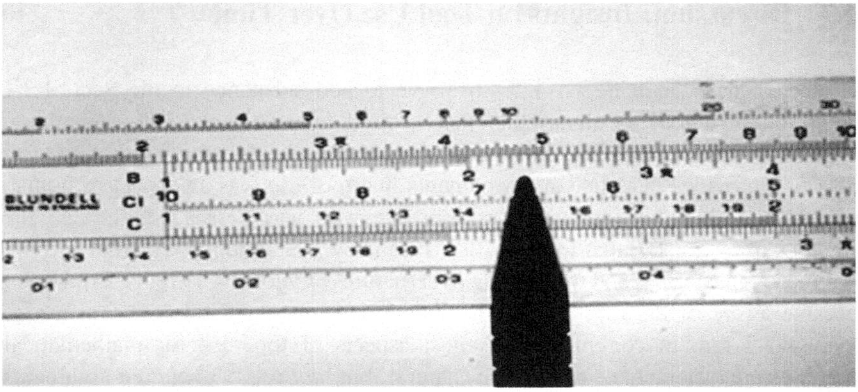

Fig. 4.5 Using a slide rule to calculate (estimate) 21×23

tables to mathematics lessons for calculations other than practice on standard written algorithms. It is instructive to reflect on the techniques that these almost modern schoolchildren were taught. I present an example, $120 \div 0.45$, using the once common (to schoolchildren) four-figure tables of logarithms.

Children were taught to find the characteristic and mantissa (terms that Briggs introduced in 1624). The mantissas of 120 and 0.45 are 0792 and 6532 and the characteristics are 2 and -1 (representing 10^2 and 10^{-1}). This makes the logarithms of the two respective numbers 2.0792 and -0.3468 but -0.3468 is not suited for the use of logarithmic table, so it is conceived as $-1+0.6532$, written as $\bar{1}.6532$ and read as 'bar 1 point 6532'. Whilst this is not difficult for a mathematician to understand, it is difficult for a 11-year-old child to appreciate that it is useful to consider 0.45 as $10^{-1+0.6532}$, and it can be presumed that the method was often taught without a mathematical justification. To perform $120 \div 0.45$ one calculates $2.0792 - \bar{1}.6532 = 2.426$, then looks up 426 in the table of antilogarithms and obtain the digits 2667 and then use the characteristic, 2, to get the answer, correct to 4 significant figures, 266.7.

As with the *Suan-pan*, once the techniques of using logarithmic tables are mastered, they are quickly, and usually accurately, performed. Using logarithmic tables is a written method of calculation in that the logarithms and antilogarithms of numbers are written down. Further to this, hand calculations are also employed (such as subtracting logarithms in performing a division).

Prior to the introduction of the electronic calculator, secondary students were also taught how to use a slide rule and they were popular with scientists and engineers. Indeed, I worked as a bridge designer in the 1970s and I used to carry a slide rule with me 'on site' in case of non-trivial calculations/estimations were needed. Some professions such as aviation science had specialist slide rules designed for their specific needs. The slide rule does not require written methods, like the *Suan-pan* it only requires hand–eye co-ordination (and a knowledge of the artefact and appropriate techniques).

4.7 Discussion: Insights on Tool Use Over Time

I selected the foci in Sects. 4.2–4.6 above to add substance to the Sect. 1.3.1 definition of a tool as an artefact that is used by an agent to do something. Section 4.2 shows that the agent does not need to be human. Further to this, Sect. 4.2 provides evidence and arguments that tool use was an aspect of human evolution. I find this interesting as someone who was persuaded, through the writings of Vygotsky, that tool use 'is a means by which human external activity is aimed at mastering, and triumphing over, nature' (Vygotsky, 1978, p. 55). Until I became aware of the role of tools in human development I rejected tool use as 'natural'. I remain convinced that most aspects of tool use for mathematical purposes are 'artefactual' rather than 'natural' but *tool use in evolution* arguments have tempered my views, there does appear to be something natural (hard-wired into human biology) is using (at least some) tools. Further to this, Gibson's (1993b) argument regarding the interdependence of tool use, language, social structure and information processing in human evolution appears to have something important to say about protomathematics and tool use: that humans have a proclivity for tool use, language, social structure and information processing (which includes logicomathematical understanding) but none of these four attributes on their own (or in pairs or in threes) is sufficient to explain our problem-solving abilities.

Section 4.3 introduces tools that may not strike one initially as tools. Before I read Netz (1999) I did not consider the tools of ancient Greek mathematics to be anything other than the compass and the straight edge. Netz makes a strong case for the lettered diagram and the mathematical language being tools of ancient Greek mathematics. Netz's claim about language here is quite specific, it concerns the mathematical lexicon and the use of what he calls 'formulae', and should not be confused with 'poetic claims' such as 'language is the tool of thought'. He also makes a case that deduction (perhaps the most important legacy of ancient Greek mathematics) was shaped by these two tools. But these tools are not sufficient on their own and Netz brings Saito's tool-box into his argument, the craft knowledge of a community of mathematicians through which the tool use makes sense. These themes (tools, language and community) have commonalities with Gibson's (1993b) four components. Both Gibson and Netz are trying to understand tool use which developed over significant periods of time (rather than a moment in time, which is often the time frame of mathematics education researchers). Gibson behoves us to pay due regard to language, social structure and information processing in consideration of tool use. The tools Netz considers are intertwined with a specialist language and the 'shaping of deduction' he outlines depends on a community which can make sense of specific tools through their joint practice which includes knowledge of the *tool box*.

Section 4.4 allows us to view mathematical activity (and the place of tools in mathematical activity) under different lens. The 'shaping of deduction' that Netz documents also shaped a view of mathematics as a deductive science. Bailey and Borwein (2012, p. 655) argue that this heritage, 'as valuable as it is, may have unduly predisposed many of us against experimental approaches'. Joseph (2010, p. xiii) goes further and states:

> A Euro centric approach to the history of mathematics is intimately connected with the dominant view of mathematics ... as a deductive system .. [s]ome of the most impressive work in Indian and Chinese mathematics ... involve computations and visual demonstrations that were not formulated with reference to any formal deductive system.

Similar (but not identical) views are expressed by some mathematics educators in their consideration of the place of tools in mathematical activity:

> the premise that Western culture establishes, in the range of human practices, a structural opposition between activities considered to be 'manual' and activities considered to be 'intellectual'. This opposition is not neutral. Western cultural axiology prioritises activities of 'the spirit' ... over the work 'of the hand' Bosch and Chevallard (1999, p. 89 translation by M Bosch & J Monaghan)

The consideration of abaci in Sect. 4.5 provides complementary insights into tool use. Although an abacus can be used, by a novice, as an external tool to perform simple calculations, its use in the hands of a master for complex calculations depends on the physical artefact being used in concert with the *secrets*. There are, effectively, two distinct tools here corresponding with the novice (without *secrets*) and the master (with *secrets*) 'ways of using the artefact' (cf. Sect. 1.3.1). The *secrets*, moreover, are directly linked to body movements and the tool use by a master irreducibly combines manipulating the artefact, dexterity and information processing. This, I posit, is true of physical tools in general.

I bring this chapter to a close with a consideration of timescales and a comparison of Sects. 4.5 and 4.6. Although abaci, in their various forms, developed over millennia, the *Suan-pan* (and the *soroban*) reached their current state many centuries ago. This stands in stark contrast to the rapid rise and fall of tools for calculations *circa* 1600 outlined in Sect. 4.5. The history of tool use (in general and in mathematics in particular) appears to be that old tools are replaced by new tools; tools such as *Suan-pan* and the *soroban* are exceptions. I think De Solla Price (1984, p. 34; cited above) hints at a reason for this general rule and the exceptions:

> Astronomy needed ingenious mathematical constructions and mechanical devices, but the keeping of accounts demanded but little elaboration of the primitive method of laying out pebbles and shells. (De Solla Price, 1984, p. 34)

Scientific development, past and present, requires new tools and also often provides the means to develop these new tools. The digital age we are now in has parallels to the sixteenth century with new tools, digital and algorithmic rising and being replaced. One of the reasons for writing this book was to try to better understand mathematics and mathematics education in this digital age and an awareness of the history of tools is useful in this search for an understanding.

References

Bailey, D. H., & Borwein, J. M. (2012). Ancient Indian square roots: An exercise in forensic paleo-mathematics. *The American Mathematical Monthly, 119*(8), 646–657.

Beck, B. B. (1980). *Animal tool behavior: The use and manufacture of tools by animals*. New York: Garland STPM Press.

Boesch, C. (1993). Aspect of transmission of tool use in wild chimpanzees. In K. R. Gibson & T. Ingold (Eds.), *Tools, language and cognition in human evolution* (pp. 171–183). Cambridge: Cambridge University Press.

Bosch, M., & Chevallard, Y. (1999). La sensibilité de l'activité mathématique aux ostensifs. Objet d'étude et problématique. *Recherches en Didactique des Mathématiques, 19*(1), 79–124.

Clarke, G. (1969). *World history: A new outline* (2nd ed.). Cambridge: Cambridge University Press.

Darwin, C. (1879). *The descent of man* (2nd ed.). St Ives: Penguin Classics (2004).

Dunbar, R. I. M. (1993). Coevolution of neocortical size, group size and language in humans. *Behavioral and Brain Sciences, 16*, 661–736.

Fauvel, J., & Gray, J. (1987). *The history of mathematics: A reader*. Milton Keynes: The Open University.

Gibson, K. R. (1993a). General introduction: Animal minds, human minds. In K. R. Gibson & T. Ingold (Eds.), *Tools, language and cognition in human evolution* (pp. 3–19). Cambridge: Cambridge University Press.

Gibson, K. R. (1993b). Tool use, language and social behavior in relationship to information processing capacities. In K. R. Gibson & T. Ingold (Eds.), *Tools, language and cognition in human evolution* (pp. 251–269). Cambridge: Cambridge University Press.

Gibson, K. R., & Ingold, T. (Eds.). (1993). *Tools, language and cognition in human evolution*. Cambridge: Cambridge University Press.

Heath, T. L. (1926). *The thirteen books of Euclid's Element* (Vol. 1). London: Cambridge University Press.

Joseph, G. G. (2010). *The crest of the peacock: Non-European roots of mathematics*. Princeton: Princeton University Press.

Kanes, C., & Lerman, S. (2008). Analysing concepts of communities of practice. In A. Watson & P. Winbourne (Eds.), *New directions for situated cognition in mathematics education* (pp. 30–328). New York: Springer.

Kenward, B., Rutz, C., Weir, A. A. S., & Kacelnik, A. (2006). Development of tool use in New Caledonian crows: Inherited action patterns and social influences. *Animal Behaviour, 72*, 1329–1343.

Lave, J., & Wenger, E. (1991). *Situated learning: Learning as situated peripheral participation*. Cambridge: Cambridge University Press.

Miller, J. M. A. (2000). Craniofacial variation in *Homo habilis*: An analysis of the evidence for multiple species. *American Journal of Physical Anthropology, 112*(1), 103–128.

Netz, R. (1999). *The shaping of deduction in Greek mathematics*. Cambridge: Cambridge University Press.

Pierce, R. C. (1977). A brief history of logarithms. *Two-Year College Mathematics Journal*, 22–26.

Pruetz, J. D., & Bertolani, P. (2007). Savanna chimpanzees, *Pan troglodytes verus*, hunt with tools. *Current Biology, 17*, 412–417.

Rosińska, G. (1987). Tables of decimal trigonometric functions from ca. 1450 to ca. 1550. *Annals of the New York Academy of Sciences, 500*(1), 419–426.

Smith, D. E. (1958). *History of Mathematics, Volume II*. New York: Dover Publications.

Solla Price, D. (1984). A history of calculating machines. *Micro, IEEE, 4*(1), 22–52.

Tattersall, I. (2002). The case for saltational events in human evolution. In T. J. Crow (Ed.), *The speciation of modern Homo sapiens: Proceedings of the British Academy* (Vol. 106, pp. 49–59). Oxford: Oxford University Press.

Tattersall, I., & Schwartz, J. H. (2000). *Extinct humans*. New York: Perseus books.
Thoren, V. E. (1988). Prosthaphaeresis revisited. *Historia Mathematica, 15*, 32–39.
Vygotsky, L. S. (1978). *Mind in Society*. Cambridge, MA: Harvard University Press.
Washburn, S. L. (1959). Speculations on the interrelations of the history of tools and biological evolution. *Human Biology, 31*(1), 21–31.
Washburn, S. L. (1960). Tools and human evolution. *Scientific American, 203*, 63–75.
Weir, A. S., Chappell, J., & Kacelnik, A. (2002). Shaping of hooks in New Caledonian crows. *Science, 297*, 981.

Chapter 5
The Development of Mathematics Practices in the Mesopotamian Scribal Schools

Tablets and tokens, lists and tables, wedges and digits, a complex system of artefacts for doing and learning mathematics, 2000 years BCE

Luc Trouche

5.1 Introduction

This chapter proposes a view on a particular moment in the learning of mathematics, 2000 BCE in Mesopotamia: a moment particular regarding the medium, with the development of writing and of systems of signs; particular regarding the development of mathematics, with the development of a sexagesimal positional numerical system, and of associated algorithms; particular regarding the places dedicated to learning, with the development of scribal schools; and, last but not least, particular regarding the supports, with the use of clay tablets 'still alive' today.

I will look at this particular moment through the eyes of a contemporary researcher on mathematics education, aware of the difficulty of looking at the past through the eyes of the present, and of the interest of enriching the present didactical questions by an historical lighting.

5.2 A Critical Moment

The period of Mesopotamian mathematics is certainly a critical one: 'The development of scribal schools in the late third millennium and the early second millennium in Mesopotamia corresponds to a switch in the medium used for the accumulation and transmission of knowledge, from memorisation, the medium became essentially written during this period' (Proust, 2012a, p. 161). This switch could be compared to another major one that of the translation from paper to digital era (see Chaps. 2, 11, 13 and 17). This critical period is also a privileged one: 'Concerning Mesopotamian scribal schools, the situation is exceptionally favourable, due to the huge quantity of school tablets handed down to us. No other educational system of the distant past is as well documented as that of Mesopotamia' (Proust, 2012a, p. 162). This situation is due to the material used

J. Monaghan et al., *Tools and Mathematics*, Mathematics Education Library,
DOI 10.1007/978-3-319-02396-0_5

for building the tablets: 'The conservation of the unskilled writings of students is partially accidental. It is due primarily to the nature of the writing medium, the clay, a nearly indestructible material. It also ensues from the reuse of dry and waste tablets as construction material. Trapped in walls, floors or foundations of houses, tablets produced by students and subsequently discarded have escaped other forms of destruction' (Proust, 2012a, p. 163): 4000 years after, clay tablets are still alive, speaking to whom is able to understand them...

I will evoke[1] here four aspects of this rich mathematics teaching context: the computation practices and their support; the set of artefacts necessary for doing computations; the persistence of old artefacts (from the pre-writing period) in the new context of scribal schools; the algorithms for calculating the reciprocal[2] of a regular number, evidencing, in this context, the mastering of a complex and efficient system of artefacts.

5.3 The Computation Practices and Their Support in Scribal Schools

In this section, I will situate the importance of scribal schools as an essential structure for learning/teaching writing,[3] the importance of writing as an essential means for communicating and thinking, and I evidence the importance of artefacts used for writing and computing. These three elements are interrelated: the scribes were the persons mastering the art of writing, essential for writing and reading administrative texts, or for calculating area and taxes; the Sumerian name for 'tablet' is DUB, for 'scribe' is DUB.SAR, meaning 'the one who writes on tablets'; for 'scribal school' is É.DUB.BA, meaning 'the house of the tablets'. The schools are well described by Veldhuis (1997) in his study of Elementary education at Nippur (one of the main cities in this area for this period). From a number of literary texts, scribal schools appear as an institution supported by aristocracy, focusing on

---

[1] We would like here to greatly thank Christine Proust, historian of mathematics specialist of this period, for her precious advices, particularly about relevant references, and her careful re-reading of this chapter; Ghislaine Gueudet, for her re-reading on an advanced version of the chapter.

[2] Reciprocal of x stands here for $1/x$.

[3] The question of 'who was allowed to attend a scribal school?' is essential to evaluate the scope of writing in society. Veldhuis (1997, p. 27) gives some information about it: 'Admittance was not restricted to members of clerical families. This is shown [...] by two kinds of evidence. First, the teacher was not paid by state or temple, but by the parents of the pupil. Payment by the parents is attested in the literary text called Schooldays. Payment by state or temple [...] would have left traces in official documents, which is not the case. Second, a few girls attended school. Both points are [...] indications of a certain freedom of choice, and a non-mechanistic procedure for admission. One must admit, however, that this freedom of choice must have been restricted to the happy few'. This suggests that the transition from memorisation to writing concerns a quite restricted sphere of the Mesopotamian society, that is not the case for translation era from paper to digital support.

the art of writing, and where the learning of computation is essential, see for example the following text where the king Šulgi describes his childhood:

> When I was young I learned at school the scribal art on the tablets of Sumer and Akkad.
> Among the highborn no one could write like me.
> Where people go for instruction in the scribal art there I mastered completely subtraction, addition, calculating, and accounting.
> The fair Nanibgal Nisaba[4] provided me lavishly with knowledge and understanding.
> I am a meticulous scribe who does not miss a thing! (Veldhuis, 1997, p. 24)

Scribal schools appeared with the development of writing as an essential support for communicating. We know from Goody (1977) the importance of writing for cognitive and intellectual development. Speech has no spatial aspect, but writing has. The writing conditions knowledge into formats in one dimension (list) or two dimensions (tables), leading to what Goody names a 'graphic reason'.[5]

The spatial aspect in this period took the form of *clay tablets* (see an example Fig. 5.1), containing texts, lists and tables. Veldhuis (1997, p. 28) distinguished, for the tablets coming from Nippur and concerning elementary learning, four types: Type I tablets are large tablets containing a long text, continuously and densely inscribed on the obverse and on the reverse[6]; Type II tablets contain different texts on the obverse and on the reverse. On the obverse, a model was noted in an archaic style by a master,[7] and copied once or twice by a student; the copies were sometimes traced and erased repeatedly.[8] On the reverse, a dense text was written by heart by a student; Type III tablets are small rectangular tablets containing a short extract, often a multiplication table; Type IV tablets are small square or round tablets, containing a short exercise.

The set of lists and tables to be learnt constitutes the basis of the Mesopotamian curriculum, as it has been reconstructed by the historians (Table 5.1).

Students began by learning metrological lists[9] and finished by learning the table of roots. The analysis of the structure of clay tablets (see Fig. 5.1) evidences a part

[4] Nisaba is the patroness of the scribal schools and the goddess of writing and mathematics.

[5] Bachimont (2010) oppose this 'graphic reason' (linked to the writing era) to the 'digital reason' of the digital era. The digital reason allows the gathering in the same space of heterogeneous contents, and a multidimensional writing and reading (thanks to hyperlinks). Bachimont underlines the essential function of the supports of knowledge: they are not only the *consequence*, but also the *cause* of knowledge.

[6] Obverse and reverse stand, for the Assyriologists, for front and back of the tablet.

[7] We use the term of master following Proust's choice: 'Since we ignore the exact nature of the scholars' charge, I prefer to refer to them as 'masters' rather than as 'teachers', a term which could implicitly suggest that teaching at the elementary level was their unique activity' (Proust 2012a, 2012b, p. 163). The persons learning in scribal schools are called in this paper 'students', for reasons of facility, instead of apprentice scribes.

[8] In order to erase signs impressed in wet clay, scribes simply rub them lightly with their finger. Tablets bear often fingerprints and erased signs covered by others.

[9] The metrological lists are enumerations of measures of weight, area or length. The metrological tables consist of the same items in the same order, but each measurement is associated with a number written in sexagesimal place value notation: they constitute tables of conversion between quantities and 'pure' numbers.

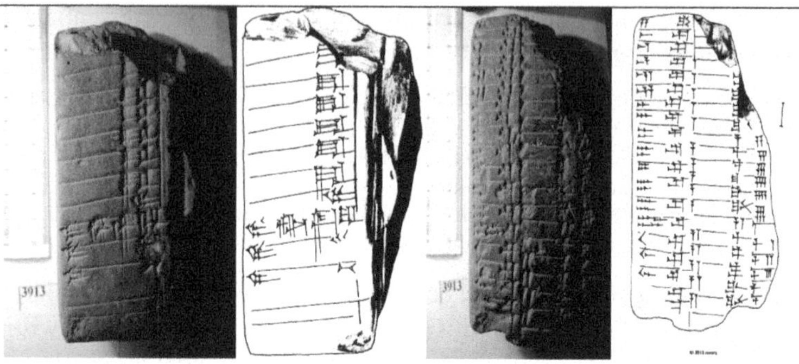

The obverse contains a Sumerian lexical list, including mathematical terms regarding volume calculations. The reverse contains a list of measures of capacity. The right side of the tablet, which contained student copies, is lost. Note the characteristic appearance of the fracture, which results from the fact that the right columns have been written and erased several times, becoming thinner and forming a ledge.

Fig. 5.1 School tablet (Type II) from Nippur, courtesy Istanbul Archaeological Museum (Proust, 2012a, p. 168)

Table 5.1 Mathematical curriculum in Nippur (Proust, 2012a, p. 170)

Metrological lists	Capacity list
	Weight list
	Surface list
	Length list
Metrological tables	Capacity table
	Weight table
	Surface table
	Length table
	Height table
Division/multiplication tables	Reciprocal table
	Multiplication tables
	Square table
Tables of roots	Square root table
	Cubic root table

allocated to the master (resp. to the student) and gives access to the mode of learning lists and tables: 'In a first step, the students learnt to write short excerpts, reproducing a model on the obverse of tablets, then they memorised the pronunciation, they recited the excerpt, and, in the last step, they reproduced by heart a large part of the list by writing it on the reverse of a tablet. Learning therefore inextricably combined writing and memorisation' (Proust, 2012a, p. 171).

Let us analyse this crucial importance of the clay tablets (in addition of the conservation for historians that I have mentioned in the introduction).

It appears clearly that the nature of this writing support conditions the student's work: the still fresh clay allows the student to write and erase what s/he wants to change (see Fig. 5.1). The dimensions of the tablets of Type IV (from 6 to 8 cm^2), dedicated to the work at home are called 'im-šu' (meaning tablets for hand) allow the student to bring them at home.[10] Kramer gave access to a text evidencing the importance of the tablet for student's work:

> Schoolboy, where did you go from earliest days
> I went to school.
> What did you do in school?
> I read my tablet, ate my lunch,
> prepared my tablet, wrote it, finished it; then
> my prepared lines were prepared for me
> (and in) the afternoon, my hand copies were prepared for me.
> Upon the school's dismissal, I went home,
> entered the house, (there) was my father sitting.
> I spoke to my father of my hand copies, then
> read the tablet to him, (and) my father was pleased;
> truly I found favour with my father. (Kramer, 1949, p. 205)

This text shows that one of the first things that a student had to do at school was the preparation of the tablet.[11] This tablet was also an essential support for the interaction between master and student, and between the student and his/her father.

For deepening this analysis, we have to consider, instead of *one* artefact, a *duo of artefacts*: a clay tablet, support of the writing and a *calame*[12] (in sumerian GI.DUB. BA, in akkadian *qan tuppi*(*m*), meaning 'reed of/for tablet'). Unlike tablets, no calame has been found till now. The existence of this artefact is attested by literary texts:

> You who speak as sweet as honey, whose name suits the mouth, longed-for husband of Inana, to whom Enki gave broad wisdom as a gift! Nisaba, the woman radiant with joy, the true woman, the scribe, the lady who knows everything, guides your fingers on the clay: she makes them put beautiful wedges on the tablets and adorns them with a golden stylus. Nisaba generously bestowed upon you the measuring rod, the surveyor's gleaming line, the yardstick, and the tablets which confer wisdom. (ETCSL, 2-5-5-2)

The existence of calames and their properties are also attested by the shape of their traces on the tablets themselves. It was probably a piece of reed (Proust, 2007, p. 81), sometimes of bone or ivory, of wood or of metal, especially pointed or rounded at first, then with a flat triangular form, or beveled thereafter. The incision of this artefact in fresh clay makes it difficult to draw lines and curves and encourages the user to draw short segments. This gave the Mesopotamian cuneiform writings a distinctive appearance (see Fig. 5.1). One must first plant a tip,

---

[10] The importance of such handheld device for appropriation by students is certainly crucial, as evidenced, in a recent period, for the purpose of mathematics teaching, by the use of handheld calculators (Trouche & Drijvers, 2010).

[11] The making of clay tablets, particularly those used in schools, was an important aspect of the technology of writing in this period and this geographical era. Bread of fine clay for tablets have been found, stored in jars (Suse), or cavities (palace of Mari). Clay is an abundant material in the Mesopotamian alluvial plain. But the clay used for writing had to be very pure. They had to be degreased and refined to prevent them from cracking as they dry (Charpin, 2002, p. 408).

[12] *Calame* (*pen* in Arabic) has been chosen by some historians to translate the Akkadian name. Other translation used: *stylus*.

giving the shape of a wedge, and then draw a line generally following (certain signs being simple wedges). The incision of signs on a malleable media finally gives not a flat writing like that obtained with ink and paper, but an embossed writing, and signs should be read with lighting that allows the reader to identify all incisions in order to avoid misinterpretation.[13]

The most used Mesopotamian numeration system, following the system used, in this region, before the writing era, was sexagesimal. In mathematical texts, the numbers are made of sequences of digits following a positional principle in base 60: each sign noted in a given place represents 60 times the same sign noted in the previous place (on its right).[14] Using this duo of artefacts for writing numbers, easily and without ambiguity, leads to the introduction of a minimum number of well-contrasted signs, actually two signs were enough: ones (vertical wedges 𒁹) and tens (oblique wedges 𒌋),[15] concatenated to represent the 59 digits used in the sexagesimal system (as 0 did not exist in this period). Proust (2007) presents the usual layout of these numbers, aggregated by a maximum three figures, to allow for rapid reading (see Table 5.2).

Several wedges are thus combined for writing numbers, with precise rules:

- If vertical wedges are written at the right of oblique wedges, they are at the same position; for example 𒌋𒈫 stands for 12 in our numeration system.[16]
- If vertical wedges are written at the left of oblique wedges, they are at an upper position, for example 𒈫𒌋 stands for 130 (2 × 60 + 10). It is transcribed by the historians as 2.10.
- The concatenation has to be considered carefully: 𒈫 stands for 2, and 𒁹 𒁹 stands for 1.1 (i.e. 61 in our decimal positional system).

We are now able to analyse an exercise written on a tablet (Fig. 5.2).

---

[13] Lavoie (1994) analyses also the importance of the artefact for writing in another context: the passage of the quill of goose to the quill of iron in the primary schools, at the beginning of the twentieth century, in Québec.

[14] Among the Mesopotamian versions of sexagesimal numeration systems, there is only one which is positional, and this is this one which has been developed/used in the scribal schools.

[15] One can hypothesis that these two figures are the written transpositions of token used for computing before the writing era, see Fig. 5.5.

[16] In the Old Babylonian period, the cuneiform writing did not allow to distinguish 12 and 10.2. This ambiguity of the notation created errors, and was corrected in later period by the use of a new sign, 𒑱 to denote the absence of digit. In this improved system, 𒌋 𒈫 stands for 12, and 𒌋 𒑱 𒈫 stands for 10.2.)

Table 5.2 The ergonomic display of the numbers (Proust, 2007, p. 74)

Units	1	2	3	4	5	6	7	8	9
Tens	10	20	30	40	50				

Tablet UM 29-15-192 (Neugebauer & Sachs 1984)

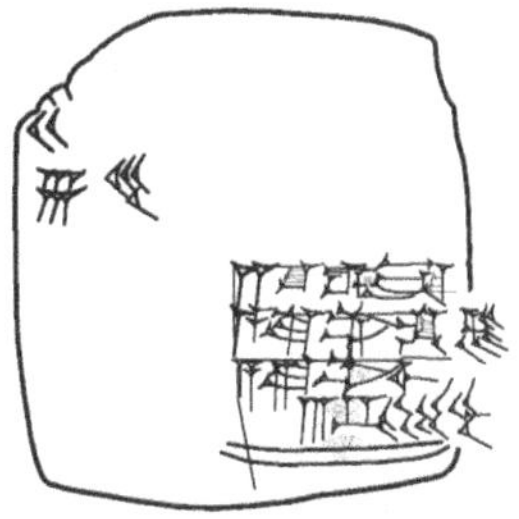

Hand copy made by Proust, personal communication

UM 29-15-192 -Transcription	Translation	Interpretation
[2]0 20 6.40	20 x 20 = 6.40	
2 šu-si ib$_2$-si$_8$ -------------------- a-ša$_3$-bi en-nam -------------------- a-ša$_3$-bi igi- 3-gal$_2$ še-kam ==============	2 šu-si the side of the square What is its area? Its surface is 1/3 še [a *šu-si* (= a finger) is a length measuring unit a *še* (= a grain) is an area measuring unit]	2 šu-si → 20 20 x 20 = 6.40 6.40 → 1/3 še

Fig. 5.2 A tablet (type IV), its picture, hand-made copy, translation and interpretation (Proust, 2007, p. 193)

The layout of the tablet (Fig. 5.2) shows two distinct places: a place for *computation* (in the upper left area), following the positional system, and a place for *quantification* (in the lower right area), giving the text of the problem in terms of unit of measure and area. Then the student's work can be reconstituted:

– In the lower right place, *using* the metrological tables of lengths for converting length measurements in numbers.

- In the upper left place, *using* the multiplication tables for making the multiplication.[17]
- Back to the lower right place, *using* the metrological tables of area for converting the number in area measurement.

When I say '*using* metrological, or multiplication tables', it has to be understood in a large way: I have explained above the importance of reading and memorisation of such tables, fundamental elements of scribal school practices. For performing such computations (Fig. 5.2), students had certainly to mobilise memorised results from their learning of tables.

I have described, in this section, a set of artefacts used for learning mathematics: symbolic artefacts (as the sexagesimal positional numeration system), written artefacts (as the wedges), material artefacts, some of them have been preserved for us (as the clay tablets of different types), but evidence for some of them are suggested by their traces (calames). Are we sure that this enumeration is exhaustive? We will see in the following section that the answer is probably no.

5.4 Evidencing Computing Artefacts Complementing the Usage of Tablets and Memory

For Proust (2012a, p. 173), 'the resources of the masters [. . .] might have included a complex system of written texts, memorised texts, calculation devices and various communicational processes, but only the written artefacts reached us. We have then to reconstruct a rich environment from truncated evidence'. This reconstruction can rely on three arguments: the necessity of artefacts outside of the tablets for doing intermediate computations, the interpretation of frequent similar errors in the tablets, and the persistence of artefacts coming from the pre-writing era.

Firstly, the necessity of artefacts dedicated to these intermediate computations, for too big multiplications. Some tablets (see Fig. 5.3) show indeed important multiplications without any intermediate results.

We could imagine that such intermediate computation, supported by a given algorithm, could have been made on a 'draft tablet', but such tablet had never been found. We could also imagine that this draft could have been made on the tablet bearing the problem itself, then erased: a careful analysis of the tablet suggests that this was not the case here. One possibility is the presence of an artefact dedicated to such computation, that is, not a clay tablet, but made of a material, which vanished over time due to its nature (lexical evidence suggests that this device was made of wood, cf. Lieberman, 1980).

[17] An application 'mesocalc' has been developed par Mélès for doing such computations, which is useful for a better understanding of this system and a reading of the tablets: http://baptiste.meles.free.fr/site/mesocalc.html#multiplication.

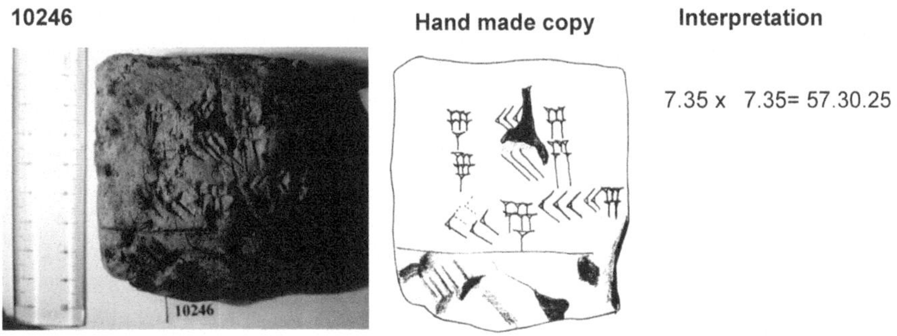

Fig. 5.3 An example of computation difficult to do mentally (Proust, 2007, p. 168)—picture: courtesy Archaeological Museums of Istanbul

The second argument for the use of such a disappeared artefact comes from the careful analysis of the display of the computation on the tablets. Proust (2000) presents a table of successive doublings of the initial term 2.5 (see Table 5.3).

Something strange appears line 21, i.e. as soon as the writing of the number exceeds 5 positions: this writing is split into two parts, separated by a sign (a vertical wedge and an oblique one), and these two parts are separately doubled. This writing of the big numbers in two parts needs afterwards a *reattachment*, taking into account the relative sexagesimal positions of the digits, and this reattachment could explain a number of errors found through tablets during the whole cuneiform history. For example, the following error has been discovered in a tablet (300 BCE), about the computation of the reciprocal of 1.16.53.12.11.15 (Proust, 2000, p. 4). The result displayed is 46.49.19.**54.58**.53.20, instead of 46.49.19.**40.14.48**.53.20 (the curious reader could use the application Mesocalc, see Footnote 13, to check it. . .). The error could derived from a wrong reattachment of separate reciprocal computation of two parts of the number, leading to 46.49.19.**40 and 14.48**.53.20: instead of concatenate these two numbers, the scribe had added the two proximate digits, 40 and 14, giving 54 (otherwise an error of copy could explain the writing of 58 instead of 48). The repetition of such error could be explained by two computations using an artefact other than a tablet (and therefore without writing), and then performing a mental operation of reattachment of the two numbers leading to the written result (which is sometimes an error).

The third argument suggesting the existence of a non written artefact is the persistence of ancient artefacts, *the tokens*, during the period of the cuneiform writing. This persistence can be supported by the large presence of these artefacts in the Ancient Near East just before—and during—the use of writing on clay tablets. Tokens, that were small objects (Fig. 5.4), made of clay, modelled into many shapes such as cones, spheres, cylinders, disks and tetrahedrons are used for counting. Studying them brings us to a period 5000 years before the present day:

Table 5.3 The successive doubling of 2.5: 2.5 times 2 makes 4.10, etc. (Proust, 2000, p. 300)

Line	Obverse of the tablet	Line	Reverse of the tablet
1	2.5	21	10 + 6.48.53.20
2	4.10	22	12 + 13.37.46.40
3	8.20	23	40 + 27.15.33.20
4	16.40	24	1.20 + 54.31.6.40
5	33.20	25	2.**40** + **1**.49.2.13.20
6	1.6.40	26	5.**20** + **3**.38.4.26.40
7	2.13.20	27	10.**40** + **7**.16.8.53.20
8	4.26.40	28	21.**20** + **14**.32.17.46.40
9	8.53.20	29	42.**40** + **29**.4.35.33.20
10	17.46.40	30	1.25.**20** + **58**.9.11.6.40
11	35.33.20	31	2.50.**40**(+)1.**56**.18.22.13.20
12	1.11.6.40	32	5.41.**20**(+)3.**52**.36.44.26.40
13	2.22.13.20	33	11.22.**40**(+)7.**45**.13.28.53.20
14	4.44.26.40	34	22.45.**20**(+)15.**30**.26.57.46.40
15	9.28.53.20	35	45.30.**40**(+)31.**0**.53.55.33.20
16	18.57.46.40	36	1.13.1.**20**(+)1.2.**1**.47.51.6.40
17	37.55.33.20	37	
18	1.15.51.6.40	38	
19	2.31.42.13.20	39	
20	5.3.24.26.40	40	

Fig. 5.4 Complex tokens representing (*above*, from *right* to *left*) one sheep, one jar of oil, one ingot of metal, one garment, (*Below*, from *right* to *left*) one garment, one honeycomb, from Susa, Iran, ca. 3300 BC Courtesy Musée du Louvre, Département des Antiquités Orientales, Paris (Schmandt-Besserat, 2009, p. 148)

> Tokens started to appear in the Fertile Crescent of the Near East, from Syria to Iran, around 7500 BC. This means that counting coincided with farming, and in particular, the redistribution economy that derived from agriculture. Tokens were probably used to pool together community surpluses for the preparation of the religious festivals that constituted the lynchpin of the redistribution economy. The tokens helped leaders to keep track of the goods in kind collected and their redistribution as offerings to the gods and the various community needs. (Schmandt-Besserat, 2009, p. 146)

One could distinguish two major trends in the evolution of tokens:

- A first period of diversification, the tokens having to represent, on a symbolic and imaginative way, the variety of 'things' to be counted (Fig. 5.4):

> the number of token shapes, which was limited to about 12 around 7500 BC, increased to some 350 around 3500 BC, when urban workshops started contributing to the redistribution economy. Some of the new tokens stood for raw materials such as wool and metal while others represented finished products, among them textiles, garments, jewelry, bread, beer and honey. (Schmandt-Besserat, 2009, p. 148)

- A second period of abstraction, around 3000 BCE, linked to the emerging of writing (Fig. 5.5):

> plurality was no longer indicated by one-to-one correspondence. The number of jars of oil was not shown by repeating the sign for "jar of oil" as many times as the number of units to record. The sign for "jar of oil" was preceded by numerals—signs indicating numbers. Surprisingly, no new signs were created to symbolize the numerals but rather the impressed signs for grain took on a numerical value. The wedge that formerly represented a small measure of grain came to mean "1" and the circular sign, formerly representing a large measure of grain meant "10". (Schmandt-Besserat, 2009, p. 148)

The shape of the signs, sketched with a pointed calame, is obviously very close to the shape of the vertical and oblique wedge characteristics of the cuneiform writing, the vertical wedges standing for one, and the oblique wedges standing for ten. Nevertheless, it should be a mistake to imagine that the token had progressively vanished for leaving room to writing on clay tablets. Till now the researchers hypothesise that various forms of cohabitation had existed between token and clay tablets. Some traces of this cohabitation had been evidenced: material cohabitation as for these kinds of spherical envelop (Fig. 5.6) containing inside circular and wedge tokens, and keeping their traces on its surface[18]; symbolic cohabitation

[18] The interpretation of this envelop cannot be done out of its cultural environment. A modern eye could interpret this sphere full of token as a typical artefact for learning numbers in schools. At the opposite, these clay purses have been interpreted by historians as accounting artefacts: 'By 3300 BC, tokens were still the only accounting device to manage the redistribution economy that was now administered at the temple by priestly rulers. The communal offerings in kind for the preparation of festivals continued, but the types of goods, their amounts, and the frequency of delivery to the temple became regulated, and non-compliance was penalized. The response to the new challenge was the invention of envelopes where tokens representing a delinquent account could be kept safely until the debt was paid. The tokens standing for the amounts due were placed in hollow clay balls and, in order to show the content of the envelopes, the accountants created

Fig. 5.5 Pictographic tablet featuring an account of 33 measures of oil, (circular = 10, wedges = 1) from Godin Tepe, Iran, ca. 3100 BC Courtesy Dr. T. Cuyler Young, Royal Ontario Museum, Toronto, Canada (Schmandt-Besserat, 2009, p. 150)

as evidenced by a double system of computation on clay tablets (see Fig. 5.2, abstract numbers vs. quantities).

This hypothesis has been recently validated by a very important discovery. Excavations in South eastern Turkey have uncovered a corpus of tokens dating to the first millennium BCE:

> These tokens are found in association with a range of other artefacts of administrative culture—tablets, dockets, sealings and weights—in a manner which indicates that they had cognitive value concurrent with the cuneiform writing system and suggests that tokens were an important tool in Neo-Assyrian imperial administration. (MacGinnis, Willis Monroe, Wicke, & Matney, 2014, p. 289)

MacGinnis et al. (2014) show how these tokens, under different forms (Fig. 5.7) could intervene in working with tablets, for administrative purposes, in a complementary way.

> They represent a system of accounting that worked in conjunction with tablets to allow for a more flexible type of record keeping that could be achieved by the use of tablets alone. Specifically they provided a system of movable numbers that allowed for stock to be moved and accounts to be modified and updated without committing anything to writing. At the same time, because these tokens exist alongside a contemporary cache of administrative documents, they illustrate the concurrent use of clay tokens and tablets. (MacGinnis et al., 2014, p. 303)

---

markings by impressing the tokens on the wet clay surface before enclosing them' (Schmandt-Besserat, 2009, p. 149).

Fig. 5.6 Envelope, tokens and corresponding markings, from Susa, Iran (Courtesy Musée du Louvre, Département des Antiquités Orientales)

Fig. 5.7 A diversity of tokens intervening in computation (MacGinnis et al., 2014, p. 294)

It appears here two types of *conjunction* between tablets and token use: firstly the tokens provide a system of *movable numbers* allowing *updating without writing*. Secondly they could intervene for supporting, articulated with the use of tablets, a given computation. This articulation could be mastered by a single agent, or via the collaboration of different agents. MacGinnis et al. (p. 302) 'assume that under the trained scribes who wrote the cuneiform tablets were assistants helping to load and unload the grain and counting out transaction', using tokens. Finally, the computation results of a *flexible* combination of artefacts and agents.[19]

This discovery constitutes clearly a proof of the co-existence of written (clay tablets) and not written (token) artefacts for working with numbers during this period. The hypothesis of a device combining wood and clay token is still to be proved, beyond the different clues we have evoked in this section, but the reality of the system of artefacts, mainly tablets and tokens, supporting the practices and the learning of computation in scribal schools seems to be established. A last evidence comes from the analysis of a crucial algorithm, this of reciprocal computation, what we will examine in the next section.

5.5 Analysing the Algorithm for Calculating a Reciprocal, a Way for Entering the Spirit of Mesopotomian Computation

Calculating the reciprocal of A is essential for performing the division B/A as the multiplication $B \times A^{-1}$. Analysing the algorithm supporting this calculation opens an enlightening window on the Mesopotamian mathematics practices and knowledge. The following example is extracted from the tablet CBS 1215 (Fig. 5.9), which could come, according to an estimation of Proust (2012b), from the scribal schools of southern Mesopotamia, during the Old Babylonian period (beginning of the second millennium BCE). This is a multi-column tablet containing advanced mathematics (which is therefore out of the classification in four types, see Sect. 5.3). The existence of such tablets, in scribal schools, evidences the fact that the 'masters' (see Footnote 5) were not only teaching elementary mathematics, but worked also as scholars, for developing mathematics, exchanging texts between masters across the different schools.

I reconstitute below, in modern terms, the computation displayed on this tablet, analysed by Proust (2012b).[20]

The computation of a reciprocal only concerns regular numbers, i.e. in the sexagesimal numeration, numbers that are products of powers of 2, 3, and 5: only such numbers are present in the tablets displaying such a computation. The goal of the algorithm is to decompose the regular number at stake as the

(continued)

[19] We have to keep in mind that the context described by MacGinnis et al. is an administrative one. The computation, in such a context, can be based on highly specialised tasks assigned to different agents, as the learning is not an objective of this activity.

[20] For following the development of the computation, the reader could again use the application Mesocalc (see Footnote 14).

Table 5.4 Table of reciprocal of usual regular numbers, underlining the couples used in the computation of the reciprocal of 25.18.45

n	inv(n)
2	30
3	20
4	15
5	12
6	10
8	7.30
9	6.40
10	6
12	5
15	4
16	3.45
18	3.20
20	3
24	2.30
25	2.24
27	2.13.20
30	2
32	1.52.30
36	1.40
40	1.30
45	1.20
48	1.15
50	1.12
54	1.6.40
1.4	56.15
1.21	44.26.40

product (non unique) of regular numbers whose reciprocal is well known (this algorithm lies therefore on the property: 'the reciprocal of a product of numbers is the product of the reciprocals of these numbers').

The 'well known reciprocals' come from a table (Table 5.3) part of the curriculum (see Table 5.1, Sect. 5.2). Note that the digit 0 is not used in this sexagesimal numeration, therefore $2 \times 30 = 60$, i.e. 1.0, is noted as 1. The reciprocal of 2 is therefore 30.

Let us calculate, following the tablet (Fig. 5.9), the reciprocal of $A = 25.18.45$.

The second property supporting the algorithm is: 'if a regular number terminates the writing of A, then it is a regular factor in one decomposition of A'.

(continued)

Knowing that, we can now begin the computation (For following the development of the computation, the reader could again use the application Mesocalc (see Footnote 14)):

- First step, we isolate, in the final digits of A (thinking A as 25.15 + 3.45), a number present in the table (3.45), which reciprocal is 16 (see Table 5.4).
- Second step, we try to write A as a product of n and 3.45; the number n is therefore equal to $A \times 16$ (which is the reciprocal of 3.35), i.e. $n = 6.45\ldots$

And we apply again the same technic for 6.45.

- First step, we isolate, in the final digits of this number, a number present in the table: 45, which reciprocal is 1.20 (see Table 5.4).
- Second step, we try to write 6.45 as a product of m and 45; the number m is therefore equal to 6.45×1.20 (which is the reciprocal of 45), i.e. $m = 9$.

The number 9 is present in the table of reciprocals, here is therefore the end of the algorithm.

Finally, the number A has been written as a product of three numbers belonging to the table: $A = 3.45 \times 45 \times 9$, and the reciprocal of A is the product of the reciprocal of these three numbers:

$1/A = 16 \times 1.20 \times 6.40 = 2.22.13.20$.

Analysing the way of applying this algorithm (personal communication of C. Proust), we could again question the presence of hidden artefacts, mobilising tokens. Understanding the cutting of the number 25.18.45 is indeed easier if we have in mind[21] a 'token-based representation' instead of a written representation (Fig. 5.8), i.e. if we consider a number, not as a succession of digits (here three positions of wedges) but as a grouping of tokens.

This hypothesis of a hidden artefact is supported also by the examination of the corresponding tablet displaying the computation (Fig. 5.9) in a very ergonomic layout. This extract details the calculation of the reciprocal of 5.55.57.25.18.45 (the curious reader could try to apply the algorithm from this number). The computation I have presented above is just a part of this calculation (see inside the highlighted rectangular). Following our observations in the previous section, we can imagine that this kind of sophisticated computation is supported par 'an artefact outside the tablet'.

Finally, for performing efficient computations of this kind, the scribal school masters and advanced students had to combine a set of material and symbolic

[21] *Having in mind* could mean 'using tokens, eventually integrated in a wooden device, to assist the computation', as shown in Fig. 5.8; or 'keeping the memory of old practices of computation based on token'. Remember, in another context, the Bachelard's sentence, concerning the man of the twentieth century: 'Même chez l'homme nouveau, il reste des vestiges du vieil homme. En nous, le XVIIIe siècle continue sa vie sourde...' (Bachelard, 1934).

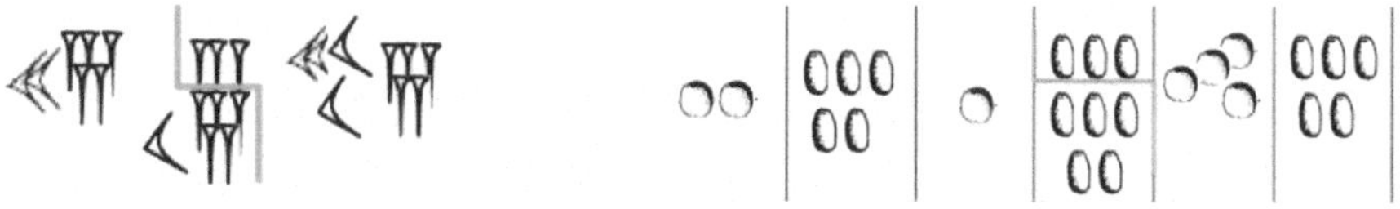

Fig. 5.8 Cutting of 25.18.45 in 25.15 + 3.35, in the sexagesimal numeration (*left*) and in a token representation (*right*)

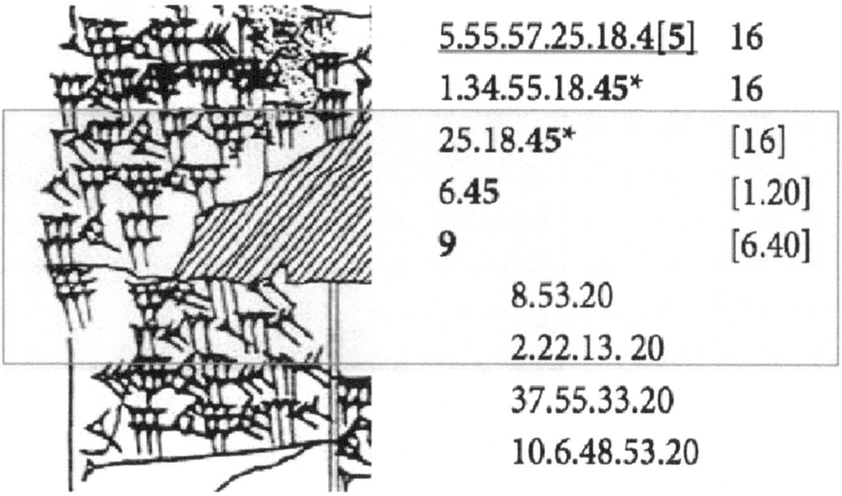

Fig. 5.9 Extract of tablet CBS 1215 (Sachs, 1947, copy Robson, 2000, p. 23), *left*, and its transcription, *right*, by Proust (2012b)

artefacts in their minds or/and in their hands. Learning to use these artefacts was, for them, a part of learning mathematics, the two modes of learning supporting one another (see the discussion of techniques and schemes in Chap. 10): conceptualisation and instrumentation are completely nested (Trouche, 2000).

5.6 Conclusion and Discussion

We have proposed in this chapter to have a look on a very rich period for the development of: mathematics; for learning mathematics; and for the learning on how mathematics was learnt and taught. We can draw, from this examination, supported by the historic research literature on this period, several observations.

Firstly, what appears clearly is *the importance of artefacts for supporting mathematics practices and learning*. We could say that the process of creating artefacts and the process of creating mathematics feed one another. The close analysis, from their traces on clay tablets, of the mathematics practices leads to

Fig. 5.10 The cohabitation between computation with Indian digits, and computation with abacus, during several centuries in France (Hébert, 2004)

the conjecture of the existence of disappeared wooden artefacts allowing to directly manipulate numbers through tokens. The combination of these artefacts allowed for the development of efficient methods of computation.

The second observation is that *new artefacts do not necessarily make old ones redundant*. We know from history that *phases of transition* between an old and a new artefact are *phases of cohabitation*, see the transition between abacus and Indian digits in France (Fig. 5.10), or the shorter transition between slide rule and calculator (Fig. 5.11). Once said that, it remains an important issue: is the use of old artefacts a brake, delaying the integration of new artefacts (i.e. does the death of the former condition the integration of the new artefact)? Or does the integration of a new artefact lead to the establishment of a new equilibrium in the conduct of computations? What we learn from the material of scribal schools, more than 1000 years after the invention of writing, is that writing did not replace, in schools, memorisation and that tablets did not replace tokens: on the contrary the combination of different means supporting calculations seems to have led to a constitution of a few articulated levels of mathematics practices: manipulating numbers through tokens, memorising tables and intermediate results, developing and using highly structured algorithms dedicated to specific mathematical tasks, expressed in a very few lines for saving place on clay tablets. In a very different context, that of the modern period of integration of powerful calculators in mathematics teaching, we

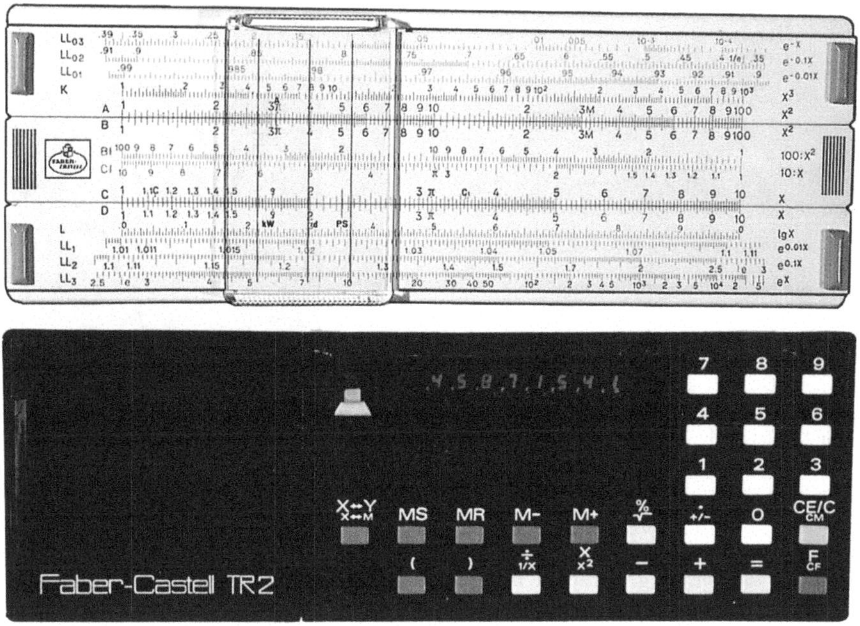

Fig. 5.11 The cohabitation between the slide rule and the calculators, two faces of a same artefact, during several years in France (Trouche, 2005)

find again what Artigue (2005) named 'the intelligence of computing', connected, for her, to three structuring abilities:

- The *relevant use of given repertoires*, in the case of scribal schools, the tables.
- The *flexibility of computing*, that is the ability of switching between several *frames*, *semiotic registers* (Mariotti and Maracci, 2012) or *points of view* (see Chap. 8), in the case of scribal schools the switching between the writing on clay tablets and the manipulation of tokens via wooden device; or the switching between computations on numbers, and computations on quantities; or the switching between a wedges-based representation and a token-based representation.
- The *ability to combine genericity and specificity*, that is the ability, for each kind of computation, to use both global properties of the computation, and specific properties linked to the domain, what we have observed in the case computing a reciprocal (see also, for the modes of reasoning beyond a given computation, Høyrup, 2002).

This intelligence of computation may also be developed by the combination of artefacts artificially reconstructed for pedagogical purpose: Maschietto and Soury-Lavergne (2013) evidenced the interest of introducing, for studying the *decimal* positional numeration system in primary schools, both a physical artefact (a reconstruction of the first calculator designed by Pascal in 1652) and its digital

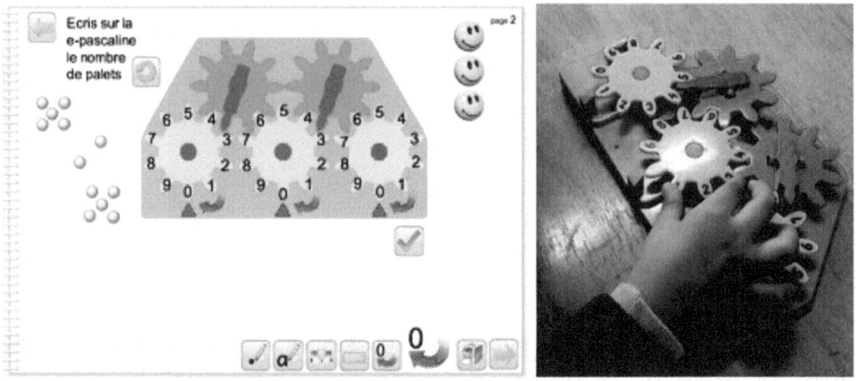

Fig. 5.12 A combination of two twin artefacts for learning mathematics, manipulation through a digital tablet on the one side, direct manipulation on the other side (Maschietto & Soury-Lavergne, 2013)

counterpart (see Fig. 5.12). There is on one side a tablet—here a digital one—allowing to turn the gear representing the units by clicking on an arrow, and on the other side a tangible device allowing to directly manipulate the gear. The digital tablet allows to combine two semiotic registers: digits (driven by the gears), and tokens (see Fig. 5.12, 12 tokens displayed on the screen). The tablet and the tangible device, even if they come from the same gear principle, lead to different gestures, and then different representations of the process of constructing numbers. These combinations of registers and gestures aim to support an essentially difficult transition in the learning of the decimal positional numeration (Bednarz & Janvier, 1984), the transition between the conception of 'a number as a sequence of digits read from left to right', and a conception of 'a number as a sequence of digits giving from right to left, the units, tens, hundreds, thousands, constituting the given number'. When history and didactics meet…

The third observation concerns the analysis of the masters 'resource system' (see Chap. 13), i.e. the set of 'things' which support their work in scribal schools. We have some information on this system, considering two faces of the masters' work: the face 'the master as a teacher', via the students' resources, a great variety of tables enlightening the curriculum and the way of learning basic computation; the face 'the master as a scholar', via tablets as CBS1215 (see Fig. 5.9), enlightening the type of sophisticated mathematical work the master could perform and share. There are a relatively few such tablets that had been found: that is easily understandable, as there are essentially tablets of unskilled students that had been trapped in wall, and that had been then preserved till now (see Sect. 5.2). The tablets integrating a rich mathematical content that had probably travelled in the whole region, from school to school, what could explain the very standardised character of the scribal school curriculum and teaching material. As for the missing artefacts that the historical research supports the existence, we could pledge the reality of missing 'lived resources' (Gueudet, Pepin, & Trouche, 2012) for scribal school masters, including tablets aiming to generate tablets of exercises, tablets describing the mode of combining artefacts, tablets describing the art of teaching, tablets with masters epistolary correspondence...

There are probably, on this subject, fruitful possible interactions between historians and researchers in mathematics education, the study of the masters resources in scribal schools enlightening the study of the master's resources in today schools, and vice versa, even if the contexts deeply differ.

References

Artigue, M. (2005). L'intelligence du calcul. *Actes de l'Université d'été "Le calcul sous toutes ses formes". Ministère de l'éducation nationale*, France (Internet link to be added).

Bachelard, G. (1934). *La formation de l'esprit scientifique. Contribution à une psychanalyse de la connaissance objective* (Vol. 5). Paris: Librairie philosophique J. Vrin, 1967.

Bachimont, B. (2010). Le numérique comme support de la connaissance: entre matérialisation et interprétation. In G. Gueudet & L. Trouche (Eds.), *Ressources vives. Le travail documentaire des professeurs en mathématiques* (pp. 75–90). Rennes: PUR.

Bednarz, N., & Janvier, B. (1984). La numération (première partie). Les difficultés suscitées par son apprentissage. *Grand N, 33*, 5–31. Retrieved from http://www-irem.ujf-grenoble.fr/revues/revue_n/fic/33/33n1.pdf.

Charpin, D. (2002). Esquisse d'une diplomatique des documents mésopotamiens *Bibliothèque de l'École des chartes, 160*, 487–511. Retrieved from http://www.persee.fr/web/revues/home/prescript/article/bec_0373-6237_2002_num_160_2_451101#.

Goody, J. (1977). *The domestication of the savage mind.* Cambridge: Cambridge University Press.

ETCSL. *The electronic text corpus of sumerian literature.* Oxford University. Retrieved August 2014, from http://etcsl.orinst.ox.ac.uk/cgi-bin/etcsl.cgi?text=t.2.5.5.2&display=Crit&charenc=gcirc&lineid=t2552.p2#t2552.p2.

Gueudet, G., Pepin, B., & Trouche, L. (Eds.). (2012). *From text to 'lived' resources: Mathematics curriculum materials and teacher development.* New York: Springer.

Hébert, E. (Dir.). (2004). *Instruments scientifiques à travers l'histoire. Actes du colloque Les instruments scientifiques dans le patrimoine: quelles mathématiques?* Rouen les 6, 7 et 8 avril 2001. Paris: Ellipses

Kramer, S. N. (1949). Schooldays: A sumerian composition relating to the education of a scribe. *Journal of the American Oriental Society, 69*(4), 199–215.

Høyrup, J. (2002). *Lengths, widths, surfaces. A portrait of old Babylonian algebra and its kin* (Studies and sources in the history of mathematics and physical sciences). Berlin: Springer.

Lavoie, P. (1994). *Contribution à une histoire des mathématiques scolaires au Québec: l'arithmé tique dans les écoles primaires (1800-1920).* thèse de doctorat, Faculté des sciences de l'éducation, Université de Laval, Québec.

Lieberman, S. J. (1980). On clay pebbles, hollow clays balls, and writing: A Sumerian view. *American Journal of Archaeology, 84*, 339–58.

MacGinnis, J., Willis Monroe, M., Wicke, D., & Matney, T. (2014). Artefacts of cognition: The use of clay tokens in a neo-Assyrian Provincial Administration. *Cambridge Archaeological Journal, 24*, 289–306. doi:10.1017/S0959774314000432.

Mariotti, M. A., & Maracci, M. (2012). Resources for the teacher from a semiotic mediation perspective. In G. Gueudet, B. Pepin, & L. Trouche (Eds.), *From text to 'lived' resources: Mathematics curriculum materials and teacher development* (pp. 59–75). Dordrecht: Springer.

Maschietto, M., & Soury-Lavergne, S. (2013). Designing a duo of material and digital artifacts: The pascaline and Cabri Elem e-books in primary school mathematics. *ZDM, The International Journal on Mathematics Education, 45*(7), 959–971.

Neugebauer, O., & Sachs, A. J. (1984). Mathematical and metrological texts. *Journal of Cuneiform Studies, 36*, 243–51.

Proust, C. (2000). Multiplication babylonienne: la part non écrite du calcul. *Revue D'histoire des Mathématiques, 6*, 293–303.

Proust, C. (2007). *Tablettes mathématiques de Nippur* (Varia Anatolica Vol. XVIII). Istanbul: IFEA, De Boccard.

Proust, C. (2012a). Master' writings and student's writings: School material in mesopotamia. In G. Gueudet, B. Pepin, & L. Trouche (Eds.), *From text to 'lived' resources: Mathematics curriculum materials and teacher development* (pp. 161–179). Dordrecht: Springer.

Proust, C. (2012b). Interpretation of reverse algorithms in several mesopotamian texts. In K. Chemla (Ed.), *The history of mathematical proof in ancient tradition* (pp. 384–422). Cambridge: Cambridge University Press.

Robson, E. (2000). Mathematical cuneiform tablets in Philadelphia. Part 1: Problems and calculations. *SCIAMVS, 1*, 11–48.

Sachs, A. J. (1947). Babylonian mathematical texts I. *Journal of Cuneiform Studies, 1*, 219–40.

Schmandt-Besserat, D. (2009). Tokens and writing: The cognitive development. *SCRIPTA, 1*, 145–154. The Hunmin Jeongeum Society.

Trouche, L. (2000). La parabole du gaucher et de la casserole à bec verseur, éléments de méthode pour une étude des processus d'apprentissage dans un environnement de calculatrices complexes. *Educational Studies in Mathematics, 41*(3), 239–264.

Trouche, L. (2005). *Des artefacts aux instruments: Une approche pour guider et intégrer les usages des outils de calcul dans l'enseignement des mathématiques.* Actes de l'Université d'été "Le calcul sous toutes ses formes". Ministère de l'éducation nationale, France. Retrieved from https://www.academia.edu/2744338/Trouche_L._2005_Des_artefacts_aux_instruments_une_approche_pour_guider_et_integrer_les_usages_des_outils_de_calcul_dans_lenseignement_des_mathematiques

Trouche, L., & Drijvers, P. (2010). Handheld technology for mathematics education: Flashback into the future. *ZDM–The International Journal on Mathematics Education, 42*(7), 667–681.

Veldhuis, N. C. (1997). *Elementary education at Nippur. The lists of trees and wooden objects.* PhD Rijksuniversiteit Groningen. Retrieved from http://dissertations.ub.rug.nl/faculties/arts/1997/n.c.veldhuis/.

Chapter 6
Discussions of Part I Chapters

Luc Trouche, Jonathan M. Borwein, and John Monaghan

6.1 Introduction

This chapter could be taken as a breath in this book, giving a space for a reader's reflective point of view on the previous chapters. It is one of the moments where the three authors have decided to question each other, from their own perspective (mathematical, cultural, historical, epistemological...). In this chapter, I have the initiative of this questioning. Here is the way I have proceeded: in the following sections, I ask my questions either to John or Jon, and each of them has a special place for developing his point of view, feeling free also to answer to a question that is not addressed to him! After receiving their answers, I do not propose to 'answer the answers', or provide a conclusion but provide some reflections arising from my colleagues' answers, wishing to empower the reader to enter the debate.

This chapter is organised in three sections. In the first one, I continue previous interactions I had with Jon and John during the process of writing the chapters. In the second one, I ask a set of new questions arising from re-reading the chapters once when the book was in a state of near completion. Then comes my conclusion.

6.2 Interactions with John and Jon Follow-Up

During the design of the book, two discussions were particularly interesting. The first one with John, about history, the second one with Jon, about proving 'graphical and numerical evidence'.

J. Monaghan et al., *Tools and Mathematics*, Mathematics Education Library,
DOI 10.1007/978-3-319-02396-0_6

6.2.1 Writing and Mathematics, a Dual Invention?

I had a discussion with John about the historical chapter: was it desirable to embrace the whole history of tools in a single chapter, or was it better to focus on specific periods? Finally, we decided to choose... both, with Chap. 4 dedicated to *Tool, human development and mathematics*, and Chap. 5 dedicated to the *Mesopotamian scribal schools*. Reading now these two chapters, I wonder if, with this organisation, we did not miss something. Actually, Chap. 4 skips from 'Tools use and phylogenesis' (Sect. 4.2) to 'Ancient Greece' (Sect. 4.3), i.e. from Prehistory to History. The essential element, distinguishing these two periods, is the invention of writing (that could not be reduced to some inscriptions on a bone). To what extent this invention is linked to mathematics? It is certainly a complex question:

- On the one hand, it seems to be clear that writing and information processing... including mathematics, came together, as stated by Nissen, Damerow, and Englund (1993): 'This innovation [the writing] was quite certainly more than a simple change in the means of storing information, or in the representation of language. Observing that at the end of the third millennium BC, during the so-called Ur III period, the human labour force was subjected to complete administrative control made possible through the developed techniques of writing (see Chap. 11), we must realise that this level of centralisation would have been impossible without the methods of information processing developed more than 1000 years earlier'.
- On the other hand, the most ancient texts that are known (called 'proto-cuneiforms') are written in an unknown language. The only part of them (constituting actually the main part of the corpus) that has been understood is composed of computations and have been deciphered by historian of mathematics: 'It may surprise some that the most important recent advances in the decipherment of the proto-cuneiform documents have been made by and in collaboration with mathematicians with no formal training in Assyriology, J. Friberg and P. Damerow. But remembering that the great majority of archaic texts are administrative records of the collection and distribution of grain, inventories of dairy fats stored in jars of specific sizes, and so on, that is, documents above all made to record in time quantifiable objects, it is reasonable to expect that such documents would contain, no less than the accounts of current institutions, evidence of mathematical procedures used in the archaic period and that they would thus contain the seeds of the mathematical thinking which developed during the third millennium'. (Englund, 1998, p. 111)[1]

So, the most ancient translated written texts seem to be mathematical texts because there only understandable part was about mathematics. Anyway, there seems to be a very strong relationship between the emergence of writing and the emergence of mathematics. Could you develop on that, John?

[1] Thanks to Christine Proust, historian of mathematics, who gave me the two references evoked in this section.

Emergence of Writing and Mathematics

John: As a mathematics educator my expertise is in the didactics of mathematics. Because mathematics comes with a long tradition (culture) and doing mathematics is a social and a cognitive activity my research is informed by work in the history of mathematics, philosophy, psychology and sociology but I am not an expert in these areas. I thus shy away from any pretence to have a definitive answer to the question on the co-emergence of writing and mathematics. But my reading in the history of mathematics provides support for this co-emergence.

In Sect. 4.3.2, informed by Netz (1999), I note that the development of ancient Greek mathematics was co-temporal with the development of an alphabet, lettered scripts and media approximating to pen and paper. And Netz (1999, p. 86) hypothesises that a common origin to mathematical propositions was 'to draw a diagram, to letter it, accompanied by an oral dress rehearsal—an internal monologue perhaps—corresponding to the main argument; and then proceed to write down the proposition'. Further to this, and with reference to your statement, Luc, that 'most ancient translated written texts seem to be mathematical texts because there only understandable part was about mathematics', Singh (2000), adds support to this statement. In a discussion about a twentieth century attempt to decipher an ancient Cretian text from 1450 to 1375 BCE, he writes, 'Many of the tablets seemed to contain inventories. With so many columns of numerical characters it was relatively easy to work out the counting systems, but the phonetic characters were far more puzzling'. (Singh 2000, p. 220)

6.2.2 *Proving (What Appears As) Numerical and Graphical Evidences*

In 2014, I had a very challenging discussion with Jon about a mysterious (to me) property. 20 years ago, I was studying with my students a Riemann sum (see T_n right side). This sum converges towards I, and the proof is quite easy to establish. Seeing the successive values of the sum (0.500000, 0.650000, 0.697436, 0.720294, 0.733732…), and supported by a strong visual support (the more numerous are the rectangle under the curve, the better they fit this curve), it seems quite obvious that this sum is increasing. Quite confident with that, I ask a student to prove it… and it appeared quickly that it was not so obvious! I try to prove it myself… and did not succeed… I ask a colleague, who also could not prove it. Finally, after some new attempts, I renounced, but I kept always in mind this open ended question… until I read the first version of Jon Chapter (Example 5.3), where this example was presented, from the point of view of the convergence, without questioning the monotonic point of view. I questioned Jon at once: what about the increasing nature of this sequence? There ensued an email interaction, see below.	$I = \int_0^1 \frac{1}{1+x^2} = \frac{\pi}{4}$ $S_n = \sum_{i=0}^{n-1} \frac{n}{i^2+n^2}$ $T_n = \sum_{i=1}^{n} \frac{n}{i^2+n^2}$

Luc to Jon: June 14th, 4:21

[...] to illustrate how a « simple » mathematical question could resist to a « naive » thinking. I have « in me » the following question, since I was a teacher in an university (see the formulas below). It is « visually obvious » that $T_n < I < S_n$, and that both S_n and T_n converge towards I. And the students can prove this. If we consider the table of values of the two sequences, it is again quite obvious that S_n is decreasing, and that T_n is increasing. *I* asked once a student: can you prove it? He was quite confident (at the beginning), me too... But he cannot do that... neither me. What should be the answer of an experimental mathematician facing this question?

Jon to Luc: June 14th, 5:08

I do not need to since I know it is a Riemann sum, but I could mention it.

Luc to Jon: June 14th, 6:30

Yes we do know that this sequence is a convergent one, and we do not need to know any thing regarding its variation but... My question is a kind of aesthetic one, the sequence appears to be decreasing, do I have some means to prove it or to explore further this 'potential' property? I have just read your Example 5.3, but you do not evoke the decreasing aspect of the sequence?

Jon to Luc: June 14th, 5:45 a.m

I will think on this....

Luc to Jon: Jun 14th, 8:24 p.m

Yes, it seemed to me very easy to prove that the Riemann sum was decreasing, as the function $1/(1+x^2)$ is decreasing... Of course no direct relationship between the behaviour of the function on [0, 1], and the behaviour of the corresponding Riemann sum. Actually it is easy to prove that the sequence $S(2^n)$ is decreasing, not at all the same easiness for $S(n)$...

Jon to Luc: June 14th, 8:36

Yes the powers of two are easy, but I think the monotonicity of the whole sequence is subtle...

Jon to Luc: June 19th, 13:37

Dear Luc, did this paper trigger your example? Szilard Andras (2012). Monotonicity of Certain Riemann-Type Sums. The teaching of mathematics, 15(2), 113–120.

Sadly, the argument at the top of page 16 has an error and the proof does not work, although it does work for the general results on convex and concave functions. But $1/(1+x^2)$ is neither and the sufficient condition fails despite the claimed proof to the contrary.

Interesting. I checked in Maple and got a different condition to check.... I found the article typing 'monotonicity of Riemann sums' into Google. Digital assistance in action!

Luc to Jon: Jun 19th, 15:31
Thank you Jon (and digital assistance!). Interesting for convex and concave function, but, if I have well understood, the proof doesn't work (top of the page 116) and it remains to be done? Will try to find extra digital assistance!

Jon to Luc: June 19th, 20:05
Yes it is unproven. Did you know this paper?

Luc to Jon: Jun 20th, 1:49
No, I didn't know. The paper is interesting, and the way of access too! Thank you.

As the reader could check, the current—and last—version of Jon's Chapter contains a complete proof of this property.

Then, my questions to Jon: could you tell the 'end of the story'? To what extent looking for proving a property is interesting for you, as the main result (the sequence at stake is convergent) is already achieved? Could you develop on your way of reflecting to a new problem?

What Is to Be Proven, and Why?

Jon: We eventually found a clean proof of much but not all that we wanted. In particular (https://www.carma.newcastle.edu.au/jon/riemann.pdf) we proved that[2]: if the function f is decreasing on the interval [0, 1] and its symmetrisation $f(x)+f(1-x)$ is concave, then T_n increases with n.

How much I care about a proof—once I know the result is true—is context dependent; if it is just a step on a route, then I have little interest unless the destination is interesting. This issue bedevils teaching proofs in classes as often the thing the student is asked to generalise and prove is intrinsically uninteresting. Why bother?

When I am given a new problem what do I do?

If it is a conjecture, I usually try to falsify it numerically, graphically or symbolically. Even if the question is not directly computational, I look for a consequence or a similar question which is. I find that the act of quantifying the problem sufficiently to play with it in Maple or Mathematica is enormously helpful. It forces a deeper understanding of the question, of unintended ambiguity and much else.

(continued)

[2] Interestingly we cannot prove the decreasing behaviour by our methods for the increasing case.

If the computations lead to a counter-example, then great. If not, and they add reassurance that the conjecture is probably true, I then let things slosh around in my head for a few days. I hunt for similar things I have seen or topics it reminds me of. I probably also then ask one of my network of collaborators if they know more.

If after that no progress comes, I try to judge if more effort is needed. After over 40 years, I trust my gut feelings. If my mind wants to keep worrying about the problem, I let it. And so it goes. This can be a long process. One of my post doctoral students and I are currently making sense and making an article out of a set of overheads from a talk I gave in 1983 but never turned into a paper.

6.3 And Some Fundamental Questions

Re-reading the chapters in this part of the book, I realised how complex were the questions arising from the consideration of tools, mathematics and learning. My reflections developed in two directions: What is mathematics? In the digital area, what links can be established between mathematics and computer science?

6.3.1 Mathematics, What Else?

Chapter one opens with these two essential questions addressed by John and me: what is a tool? What is mathematics? We can read in the first page of this chapter '[...] the view that mathematics is just a tool-box is a pretty impoverished view of mathematics'. Re-reading that, I realised that this sentence (written by John and me) witnessed perhaps an impoverished view on tools, and on tool-boxes, and perhaps a view that is not in line with the purpose of the book, this purpose giving tools a great potential for doing, evolving, thinking... What should be your comment, John, on that?

Mathematics, As a Tool-Box

John: The statement 'mathematics is just a tool-box' appears to ignore essential dimensions of what mathematics is, for example semiotic and social dimension, and what is involved in doing mathematics. Doing mathematics is usually directed at an outcome (an answer, a proof, a construction, etc.) in which tool use is essential but there is 'doing mathematics' prior to this outcome. For example, in extending and validating the pattern $1+2=3$, 4

(continued)

$+5+6=7+8, 9+10+11+12=13+14+15$ there is 'recognition': the numbers are consecutive; the terms on either side of the equal sign have patterns (the number of terms on the left hand side is 1 more than the right hand side, the number of numbers increases by 1 each time); the leftmost number of the nth equation is n^2. At this point we can bring in a tool-box of sorts and formulate a conjecture in algebraic terms.

Further to this, 'mathematics is a tool-box' is a metaphor. Metaphors are important. They can be things of beauty in the literary arts and it is very difficult to communicate (especially in instruction) without using metaphors but 'scientific inquiry' should, I hold, try to eliminate metaphors whenever possible. The term 'tool' is often used metaphorically. Vygotsky (1978, p. 53) wrote of this, 'Expressions such as "the tongue is the tool of thought" . . . are usually bereft of any definite content and hardly mean more than what they really are: simple metaphors . . .' I've met 'mathematical tool-box' used as a metaphor many times over the decades. The main occurrence is in undergraduate mathematicians' accounts of solving a problem—when their accounts refer to looking into their mathematical tool-box and choosing the appropriate algorithm or theorem or whatever to solve the problem. I think/hope I'm very tolerant of students' metaphorical reference to a tool-box and, like many metaphors, there is a sense in which it 'rings true' but I think these students have overlooked such things as 'recognise' and 'formulate' actions which preface their use of a tool to solve their problem. Further to this, I think much of school/institutional mathematics encourages students to view mathematics as a toolbox. For example, a question like 'Factorise x^2+3x+2' calls for the 'factorise tool' a bit like 'Dig that garden' calls for a digging tool but both directives will be (one hopes) embedded in a wider activity, otherwise the actions following the directives are what Chevallard (2005) calls 'monumental' acts.

With regard to our general purpose, we are writing this book because we think tool use is an essential part of mathematics (it is impossible to do mathematics without tools) and there isn't a book devoted to the place of tools in mathematics but mathematics is more than just tool use.

Otherwise, Jon, you described yourself (Chap. 3, Sect. 1.2) as 'a computer-assisted quasi empiricist'. Reading this, I remembered the famous address of the Russian mathematician Arnold at the discussion on teaching of mathematics in Palais de la Découverte of Paris on 7 March 1997. His first sentence was: 'Mathematics is a part of physics. Physics is an experimental science, a part of natural science. Mathematics is the part of physics where experiments are cheap'.[3] Jon, as an experimental mathematician, what could you say about that?

---

[3] See the whole address at http://pauli.uni-muenster.de/~munsteg/arnold.html (retrieved in 12th February 2015).

Mathematics, Experience, Experiment and Cheapness

Jon: This is a great aphorism. Yet it becomes less true as scientific experiments have moved from in vivo to in silico. As I have discovered over the years, mathematical problems can be as computationally demanding as any. That said, governments and agencies are more likely to pay for global warming models than for an attack on the Riemann Hypothesis. Moreover, in particle physics, it is hard to see any experiments to validate string theory. Some see this as an approaching crisis in Physics. At any rate, as I hope I illustrated, there are indeed many cheap and insight-laden computer experiments available to the working mathematician.

6.3.2 Mathematics and Computer Sciences

The French mathematician Jean-Pierre Kahane, who was always interested in the questions of teaching,[4] chaired the French CREM (Commission of Reflection on Mathematics Teaching) from 1998 to 2002. This commission wrote a report (Kahane, 2002), trying to define mathematics (our translation):

Mathematics is the oldest of the sciences and of whose values are more permanent. However, the approach and means of study varied according to civilisations and eras. Printing, navigation and astronomy contribute to shape the usual functions and calculus.

Today, computer science creates both new ways and new areas of study, all the sciences improve using mathematical tools and help to forge new, the link to the physical strengthens, and mathematical research benefits from the intuition of physicists.

The vision of mathematics has changed considerably over the last 50 years. Mathematical then seemed to have regained its unity on the basis of a solid construction of its foundations and structures. But she was impoverished. Then applied mathematics have made a breakthrough. Currently, the movement of math reveals a multitude of sources and impacts, together with a considerable work in constituted mathematics.

Mathematics enriches themselves with problems, methods and concepts from other sciences and practices, creating new concepts and theories, and provides material to sometimes unforeseen applications. The mathematical models, allowing simulations are everywhere, and mathematics develop through interactions with other disciplines together by interactions within them. Thus mathematics is far from being the affair of the only mathematicians.

[4] He chaired the ICMI—International Commission on Mathematical Instruction—from 1983 to 1990.

In the pumping process, distillation and irrigation they represent today, we must take into account the mathematical activity engineers, physicists, computer scientists, engineers, biologists, economists, chemists at the same time as that of mathematicians in the strict sense. It is good to no longer think only in terms of 'mathematic', 'pure mathematics and applied mathematics', but to consider all 'Mathematics' in the variety of their actors and their users.

In this view of the 'mathematics sciences', it could be relevant, in the frame of this book, to question the particular relationships between mathematics 'themselves' and computer sciences. These two fields develop complex interactions. A number of curricula (for example: France, see Sect. 12.4.3) now integrate from the primary and middle school elements of teaching on programing. There is then an emerging debate (see Sect. 12.4.3) on the relationship between mathematics and computer science teaching. Who should teach programing and algorithmic? The mathematics teacher? And why, on the contrary, computer science teachers should not teach mathematics? Or could we conceive mathematics and computer sciences as a new scientific field giving matter to a new teaching in secondary mathematics? What could you say on that, John and Jon?

Teaching Mathematics vs. Teaching Computer Science

John: I am more interested in Jon's response to this than mine as he is a research mathematician and the use of computers (and ideas/techniques from computer science) is essential in his research but my experience in computer science (other than simply using computers) is limited to teaching processor architecture a long time ago. Further to this, computer experiments in school mathematics, though 'well intentioned', are very different to the computer experiments Jon reports on in Chap. 3. They are often of the form 'use software X to generate a pattern and use mathematics (possibly computer-aided mathematics) to find mathematical relationships in the pattern generated'. A 'didactical transposition' (cf. Sect. 10.3) of sorts, from research mathematics to school mathematics, has taken place.

But I see no reason why school mathematics could not evolve to include aspects of computer science. There is an historical precedence for such an evolution. School mathematics (at the senior level) in England (and in some other countries influenced by British culture) includes Newtonian mechanics as a part of mathematics. The reason for this appears to be simple, Newton was an English mathematician and his mechanics was an important part of his oeuvre.

Jon: There are various nascent curricular coding proposals in different countries. At the school level, I share John's concern that they are largely ill considered. But my nine years old grandson is keen to learn to build his own

(continued)

video games; any tools that ensure he learns some of the fundamentals of programing while 'designing' his games could only do good.

At the university level, I would love to see computation deeply embedded in the curriculum but to the degree that this is happening it is slow and haphazard. Some years ago Penn State mandated the use of Maple in all entry-level classes. Six weeks later the hottest selling item on campus was a 'F**K Maple' T-shirt. The administrators had forgotten to tell the teachers!

6.3.3 Mathematics and Tools, Serendipity, vs. Intentionality?

John (in Chap. 1) used the beautiful word of *serendipity*[5] for describing the history of tools. Thinking of serendipity, I have always in mind the history of the steam engine:

> A tool does not have a pre-assigned function. The 'logical of use' can deviate at any time its trajectory. The twists of this sartorial give the history of technology, material and intellectual, a baroque and poetic charm that closer, for our greatest benefit and pleasure, its polar opposite: an anthology of wonderful. The first steam engine was not designed by Savery in 1698 to drive a vehicle, but to draw water from the bottom of a well (Debray, 2001, p. 106, our translation).

In the same time, a tool is also intentionally developed to meet a given need, to support a given activity. Actually the design of tools cannot be thought independently of their usages. Designing trajectories and usages trajectories appear completely interrelated.[6] Intentionality aims to meet a given necessity. We could understand with this perspective the so-called Mohr–Mascheroni theorem (Maschieroni, 1797), stating that each construction with a ruler and a compass can be realised with a single compass: a compass is a more accurate tool than a ruler. And that, perhaps, is the reason why this theorem had a practical interest.

I wonder, Jon, if, and how, you could identify, and feel, serendipity, intentionality, and necessity in your own activity of mathematician, as you state in your chapter that '[mathematicians] produce so many unneeded results'...

---

[5] Following Wikipedia (2015, February 12th), serendipity means a 'fortunate happenstance' or 'pleasant surprise'. It was coined by Horace Walpole in 1754. In a letter he wrote to a friend Walpole explained an unexpected discovery he had made by reference to a Persian fairy tale, *The Three Princes of Serendip*. The princes, he told his correspondent, were 'always making discoveries, by accidents and sagacity, of things which they were not in quest of'.

[6] I described this dialectic (Chap. 10) as an interplay between instrumentation and instrumentalisation processes.

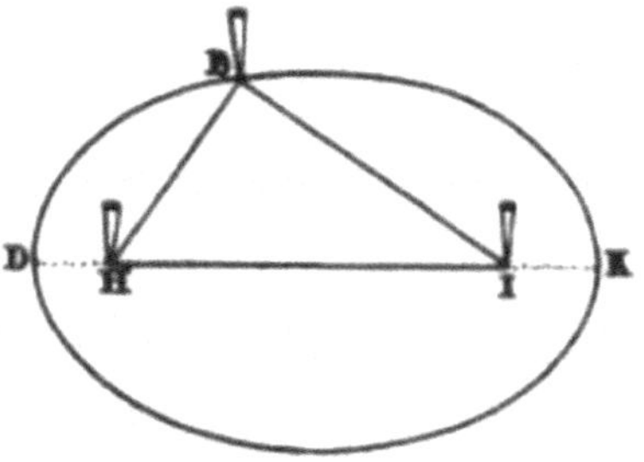

Fig. 6.1 The gardener's ellipse

Serendipity, Intentionality, Necessity and Mathematical Activity

Jon: I can do no better than quote Pasteur who said correctly that 'fate favours the prepared mind'. One attractive role for an intelligent agent in the future is to be aware of things that have been of interest in the past to a given researcher. Then metaphorically, the computer could tap the researcher on the shoulder when a potentially interesting phenomenon arose. It has often seemed to me that one difference between good and excellent researchers is in the ability to recall things in context.

I have also in mind the Descartes' description of the drawing of an ellipse (Fig. 6.1), linking one of its essential property to a practical way of drawing it:

> L'ellipse ou l'ovale est une ligne courbe que les mathématiciens ont accoutumé de nous exposer en coupant de travers un cône ou un cylindre, et que j'ai vu aussi quelquefois employer par des jardiniers dans les compartiments de leurs parterres, où ils la décrivent d'une façon qui est véritablement fort grossière et peu exacte, mais qui fait, ce me semble, mieux comprendre sa nature que la section du cylindre ni du cône. (Descartes, 1637)

The development of tools (here a nail and a string), using—implicitly—a mathematical property answers here to a practical (or aesthetic) necessity: drawing a given shape.

I wonder, John, how you feel about serendipity, intentionality and necessity? Do new tools really arise from serendipity? In Chap. 2, John, you describe the different achievements of one given task (bisection of one angle) by four tools. Could you describe this in terms of necessity? For what reason should we bisect an angle? Who needed to do this? A gardener? A painter? Which tools have been designed, and by whom, for achieving such a task?

Serendipity, Necessity and Tools

John: I said 'new tools often arise from serendipity'. Yes, I stick by what I said, though the word 'often' might be replaced by 'sometimes'. I suspect I was thinking of two things. The first is Wertsch's (1998) 9th thesis on mediational means: *Mediational means are often produced for reasons other than to facilitate mediated action*. I will cite myself from Sect. 13.2:

Sometimes they [*mediational means*] *are produced for the purpose for which they are used but this is not always the case, sometimes they are a spin-off. Wertsch cites fibreglass pole-vaulting poles. Fibreglass was developed by the military for reasons that had nothing to do with pole-vaulting. But once the material was produced it was available to be made into poles for pole-vaulting.*

We see spin-offs in mathematics classes at the moment in the form of calculators and spreadsheets. Small(ish) electronic calculators, as far as I have been able to ascertain, came about because advances in electronics enabled the production of such devices (cf. Sect. 7.1), not because of a need for such devices (though once they appeared they were put to use). Spreadsheets (cf. Sect. 7.1) were developed for financial use; their use in school mathematics could be said to be serendipitous (or not, depending on the value one attaches to the use of spreadsheets in school mathematics, the attribution of serendipity to something is a value judgement).

The second thing I was thinking was of the future, indeed, what I said in context was:

This book will also tentatively consider the future of mathematics and the role of new tools and new ways of using tools in this future. We say 'tentatively' because new tools often arise from serendipity and it is somewhat foolish to say that things will develop in this way.

I am wary about predicting the future and one reason for this is that new artefacts will appear and people will appropriate (some of) these and this appropriation of new artefacts will impinge on future practices in ways I, at least, cannot imagine prior to the new artefacts and practices appearing.

But the other side of spin-offs is 'need' (or, at least, the perception of need). In Sect. 4.4 I write of the need to calculate accurately in astronomy in the sixteenth century and the development of multi-digit trigonometrical tables was an artefactual design to satisfy this need. Needs arise in mathematics—to understand why something is so, to solve a problem, to prove a theorem, etc.—and new mathematical algorithms are developed specifically to satisfy these needs as Jon's 'top 10 twentieth century algorithms' (cf. Sect. 3.5) aptly show. But with regard to your question about who first needed to

(continued)

bisect an angle, I do not know the answer. The ancient Greeks could certainly do this using a compass and straight edge but whether this came from a need to understand or do something or whether it came about from serendipitous doodling with a compass, I do not know if this is in historical records.

Tools as well as mathematics arrive as answers to humans needs or questions. It has a practical consequence for teaching: as stated by Chevallard (2005), mathematics has to be taught and learnt as such, and not as monuments left by the great elders to the admiration of future generations.

6.3.4 Words, Images, Gestures and Proving

In Chap. 3, Jon underlines the power of visualisation for discovering new properties and proving them. We know well the power of images. Debray (2013) remarks that: 'magie et image ont même lettres'.[7] I wonder if this centration on *images* was not a reduced view of *experiencing/feeling* things. In French the deep dialectic between experiencing and proving appears more clearly through the duo éprouver/prouver.

I have in mind two examples for illustrating this idea:

- First example, about the question 'Is it possible to tile the plane with any quadrilateral?' The spontaneous students' answers are 'no!', and the using of a Dynamic Geometry Software appears not so easy for exploring such a question. But, if they use cardboard templates, and if they move them on a table (Fig. 6.2), they arrive, quite rapidly, to a solution, and this set of gestures convince them that the answer is 'yes'. In other words, the conviction is shaped by the activity, not the final image.
- Second example, about what is finally, for me, as my best memory of my years of teaching in a secondary school. My intention was to prove that the orthogonal projection of a right angle on a plane is a right angle if and only if one of the sides of the right angle is parallel to the plane. For this purpose (it was a quite long time ago), without announcing the targeted knowledge, I ask my students to come to school with four potatoes and a knife (I do not know if, today, it would be possible to enter a classroom with a such dangerous tool...). Then I ask them to cut their potatoes in order to have some cubes. And then comes the question: is it possible for a plane section of a cube to be a right triangle? The students try and try, and they felt the impossibility: 'to obtain such a triangle, I am obliged to curve the knife when cutting the potatoes'...

[7] Such a playing on words cannot happen in English with the corresponding words "magic" and "image".

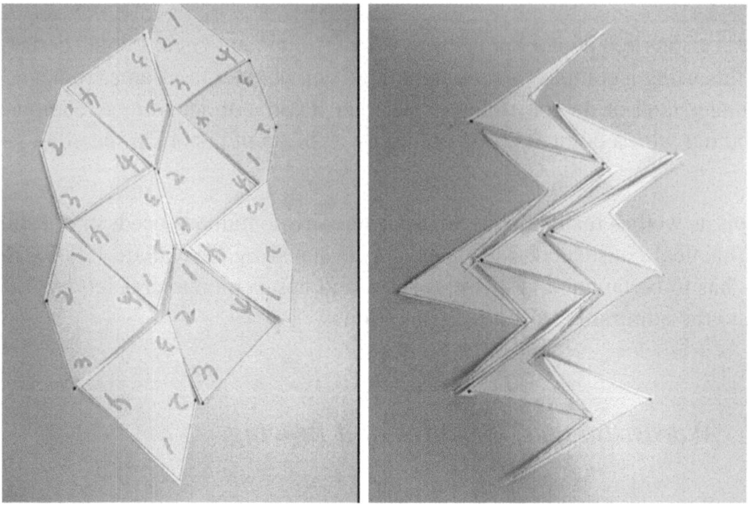

Fig. 6.2 Experiencing the tiling of a plan with given quadrilaterals

In these two situations (proving a possibility vs. proving an impossibility), the necessity of the result emerges from the acting, encompassing the seeing, but not reduced to it.

What comments, John and Jon, do these examples inspire you?

Experiencing, Seeing, Proving

Jon: In the Collin's Cobuild Dictionary (http://www.collinsdictionary.com/dictionary/english-cobuild-learners), the verb to see has many meanings and the optical is far from the most common usage. So image for me is really subordinate to visualisation in a much broader sense. In that spirit we have not exploited either haptics or movement nearly enough, nor indeed auditory data. Moreover, there are times when a movie shows much more than a still picture and others when it distracts. Sometimes the best visualisation is a potato.

John: Manipulatives can be good for fun and for getting mathematical ideas across to learners but DGS need not be that hard. Figure 6.3 shows a *GeoGebra* screenshot of the mid-stage of the construction of a tessellation (a stage that shows all the essential construction points, lines and vectors).

I've modelled my construction on Fig. 6.3 (think of C and E_1' being vertices labelled '1' in Fig. 6.2). E is the mid-point of AB and I've translated the point E by the vector CE to get E_1'. What you give the learner (the bare problem or a partially constructed tessellation as in Fig. 6.4) depends on their DGS and mathematics proficiencies.

(continued)

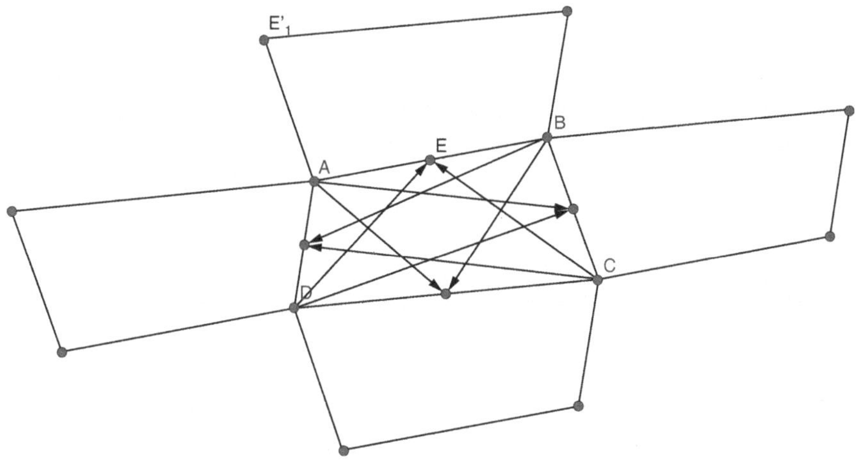

Fig. 6.3 GeoGebra screenshot of the mid-stage of the construction of a tessellation

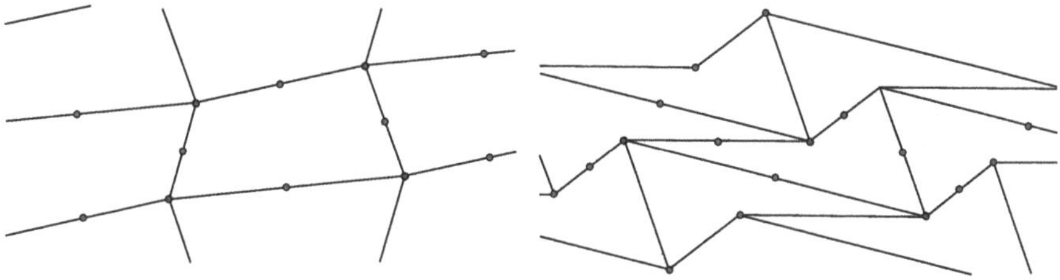

Fig. 6.4 The final stage of the construction and one result of dragging

But whatever one gives the learner as a starting point, the final results, a tessellation that 'sways' as you drag any of the points *A*, *B*, *C* or *D*, is virtually guaranteed to inspire 'awe' in the learner. There can be a kind of 'experimental proof' in this dragging/swaying: try to drag a point so that the quadrilateral does not tessellate. Trying to do this (which, of course, cannot be done) can also link the two sets of manipulatives in Fig. 6.2. Figure 6.4 shows the final stage of my construction and (on the left) one result of dragging.

I now move on to 'seeing'/visualisation. The philosophy and the psychology of perception are specialist and complex fields. They are important for this book on tools and mathematics (and not only for Jon's 'visual theorems'). A problem for us is that we three are not experts in these fields. My own views on perception have been influenced by the work of the philosopher Marx Wartofsky and the psychologists Eleanor and James Gibson. I came to reading works of these scholars from my interest in tools and mathematics and I report on the importance of their work for understanding tool use in Sect. 7.2.1, so I do not repeat it here.

(continued)

Work on visualisation has featured in mathematics education research and scholarship for many decades. An early and, in my opinion, an important paper is Eisenberg and Dreyfus (1991) which shows how diagrams can structure students' epistemic processes. The authors argue, however, that this is only possible for students who are able to perceive the abstract structure permeating the diagram. Visualisation thus benefits high achieving students more than low achieving students.

To return to 'seeing' the theorem that any quadrilateral can produce a tessellation, when one drags a construction like mine above in a DGS, visual images can be 'seen' but mathematical relationships may not be seen; 'seeing in a mathematical way' is, I hold, usually (always?) mediated by an artefact or a teacher. An advantage of a DGS for 'mathematical seeing' is that physical actions (what the hand drags) and what is seen are co-ordinated. The DGS thus affords (in Gibson's sense of the word, see Sect. 7.2.1) seeing mathematical relationships.

6.4 Finally, How Mathematics Teaching Could Develop Interest in Proving?

Following the previous discussions with John and Jon, I would like to evoke, among other possible perspectives, two answers that support my own teaching.

6.4.1 Understanding Being and Reasons of Being

The first answer is: Provoking students' curiosity *for a result* (is it true?) and *its reasons of being* (why is it true?)

Jon evokes such an example, in the rubric 'Proofs without words' (Chap. 3, Fig. 3.5). The objective is to find a synthetic formula for the sum of the first odd numbers. The result emerge from a number pattern: $1+3=4$; $1+3+5=9$; $1+3+5+7=16$... The sum of the n first odd numbers seems to be n^2. The questions are both: is it always true, and what is the reason of such an amazing result? There is of course a lot of proofs, but the geometrical one is very interesting because it evidences the law of building this result. To build a new border of a square (Fig. 6.5), I need at each step two more tiles, i.e. I need to add the following odd number (after 9, it will be 11). It may be a proof without any *written* words, but certainly a proof combining a lot of mental words and written images (i.e. artefacts) leading to this certainty: the result is true, because I have discovered its reason of being.

I would like to confront this first example with a second one that I had found in a book of a man fascinated by numbers (Le Lyonnais, 1983). The geometrical context is: given n distinct points on a circle, draw all the possible chords. These chords determine a certain numbers A_n of regions on the circle. For example (see Fig. 6.6):

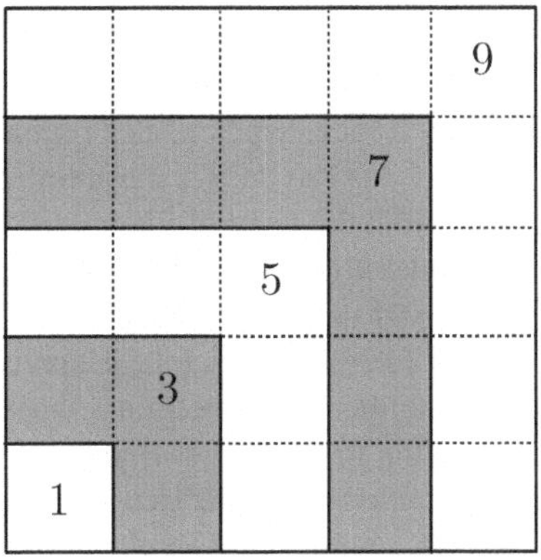

Fig. 6.5 The sum of the first odd numbers (see also Fig. 3.5)

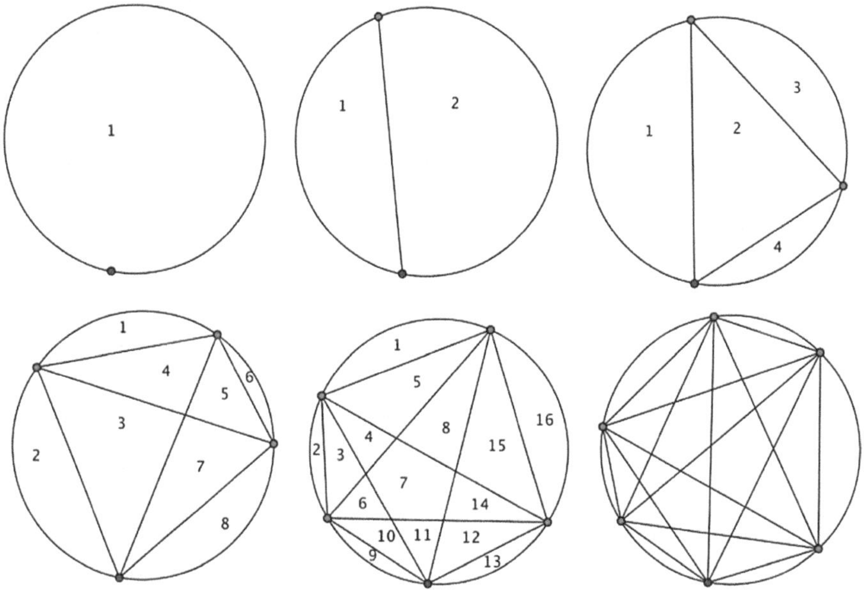

Fig. 6.6 Counting the number of sectors determined by chords in a disc

$A_1 = 1, A_2 = 2, A_3 = 4, A_4 = 8\ldots$ A conjecture quite 'natural' is that, at each step, A_n is multiplied by two. In other terms, (A_1) should be a geometrical sequence whose reason should be 2.

Number of points on the circle	Number A_n of regions in the circle
1	1
2	2
3	4
4	8
5	16
6	?
7	?

Then the two interrelated questions are: is it always true? And why, if I add a point, is the number of regions multiplied by two? These two questions feed one another. If the curious reader counts the number of sectors for 6 points, s/he would find that, instead of 32, the number is 31, leading to a reformulation of the two questions at the heart of the proving process: what is the general formula, and what is the rule of building such a sentence, its reason of being?

We will not answer here these questions, letting the reader thinking alone, or looking for the book of Le Lyonnais… But what is important, at this step, is to underline the complementary aspect of the two previous examples (Figs. 6.5 and 6.6): in the first case, what appears at once (the general rule) is true; in the second case, it is wrong. In the two cases, looking for the reasons of being of the potential rule is the motor of the mathematical activity. Meeting these two kinds of situation, for a student, is certainly necessary to avoid the feeling that 'what is true a sufficient number of times is probably always true'. As long as the reasons of being of a result have not been elucidated… the reasonable doubt remains.

6.4.2 *Extending the Domain of the Validity of a Given Result*

The second answer is: *engaging students in a path* allowing them *to enlarge, at each step, their view on a given mathematical landscape.*

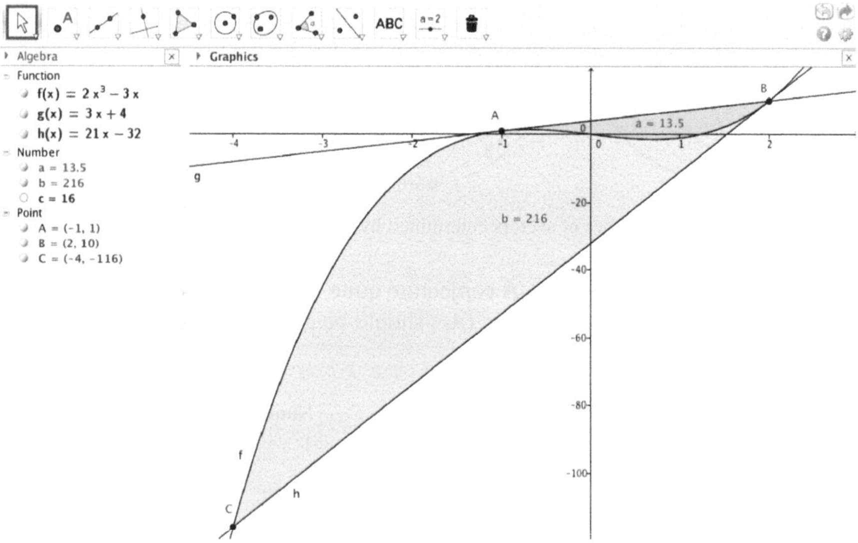

Fig. 6.7 Comparing two areas, and looking for invariants (Trouche, 1998b)

I draw the following example from the work I did, during a whole year, with a 12th grade class experiencing an environment of symbolic calculators (Trouche, 1998a, 1998b). The question at stake was (Fig. 6.7):

One considers the function $f : x \mapsto f(x) = 2x^3 - 3x$, its curve F, and the tangent T to its curve at the point A $(-1, 1)$. T crosses again the curve F at a point B. The tangent T′ to the curve F crosses again the curve F at a point C (we do not examine here the existence and uniqueness of the points B and C, which are far from being obvious). The students were asked to calculate two areas:

- The area a of the surface comprised between the straight lines $x = -1$, $x = 2$, the tangent T and the curve C.
- The area b of the surface comprised between the straight lines $x = -4$, $x = 2$, the tangent T′ and the curve C.

There is a certain relationship between a and b. Roughly speaking, b seems much bigger than a. We could model this relationship with the quotient b/a. Is this quotient a feature of the departure point A? The amazing thing is that the quotient does not depend on A… Then, is it a feature of the third degree polynomial f? The amazing thing is that the quotient does not depend on the coefficients of the third degree polynomial f… Then, is it a feature of a polynomial function?… I let the reader to engage herself in the exploring.

Advancing in the way of studying the property, extending its domain of validity, the students are led to move from calculating to reasoning again on the shape and property of mathematical objects.

6.4.3 Interacting with Objects and People

The third answer is: *creating conditions for fruitful interactions between students themselves* as well as *between students and mathematical objects*.

I draw the following example from the work I did, in 2015, during a school of high studies[8] in Recife (Brazil, reference to come), with a group of master students. During two months, a productive atmosphere develops, evidencing the potential of social interactions for proving. The mathematical question was formulated in a metaphorical way: 'Find the mother inside the daughter'… In other terms: the triangle 'mother' ABC (see Fig. 6.8) generates the triangle 'daughter' A′B′C′ by three reflections about the points A, B and C (A′ being the symmetrical of A through the reflection about B, etc.). Once removed ABC, is it possible to design the reverse process, constructing ABC from A′B′C′, i.e. to find the mother inside the daughter?

This geometrical situation is quite well known, and there are many ways for developing a given solution. The environment where the resolution took place, the

[8] Escola de Altos Estudos: Dos artefatos aos instrumentos do trabalho matemático: a dualidade essencial instrumentação-instrumentalização (http://lematec.net/EAE/).

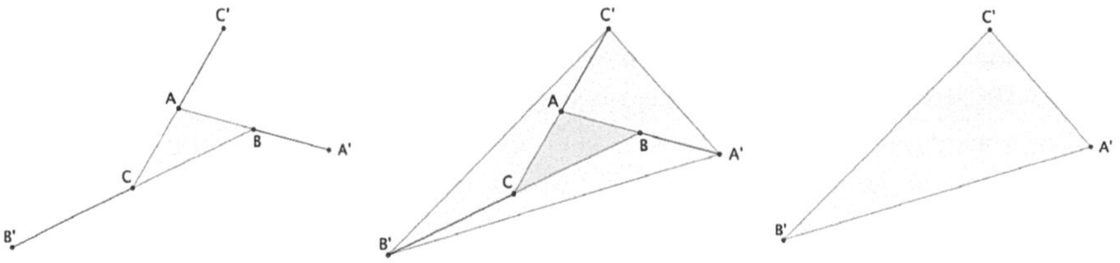

Fig. 6.8 From the mother triangle to the daughter triangle

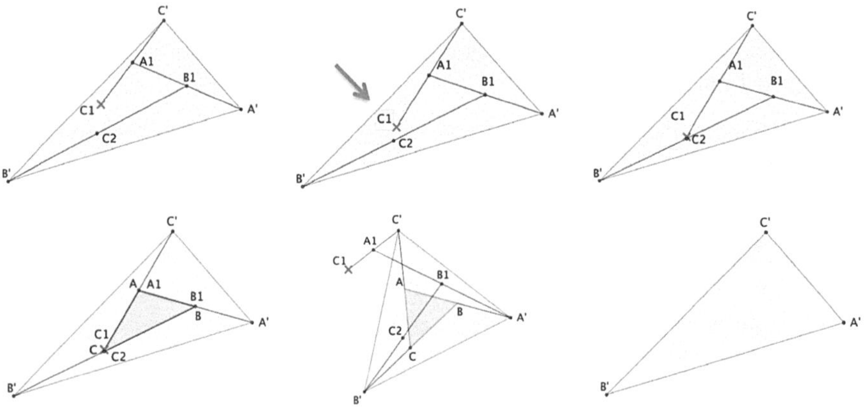

Fig. 6.9 From the mother triangle to the daughter triangle

dynamic geometrical software Geogebra, offers a lot of opportunities for interacting with the mathematical objects, for example (see Fig. 6.9): after hiding ABC, choosing randomly a point C1, one can construct A1, the mid-point of C1C′, then B1, the mid-point of A1A′, then C2, the mid-point of B1B′. Of course, C2 is not on C1. But it is possible to drag C1 to get it on C2. If we make the original mother triangle ABC appear again, we can check that ABC and A1B1C1 are alike. Obviously, the result is not stable: if we move the initiating point A, then A1 will break away from A. Therefore, if the objective was to find a solution that resists movement, the problem is not yet solved. Besides, this first construction opens perspectives for going further (using successive reflections for example…)

The discussions among students lead to the emergence of number of solutions. The Rodrigo's solution came from an extension of the initial metaphor: after the daughter triangle (see Fig. 6.10, what happens if one draws the grand-daughter triangle? It appears (and the property resists if we move the point A) that the sides of the mother triangle are respectively parallel to the sides of the grand-daughter triangle. If it is true, it opens a way for constructing from A1B1C1 the triangle A2B2C2, then the triangle ABC.

This new point of view gave students the idea of drawing parallel straight lines in the original figure, splitting each side of the mother triangle in three segments of the same length (Fig. 6.11). Extending this lattice outside the triangle gives the whole

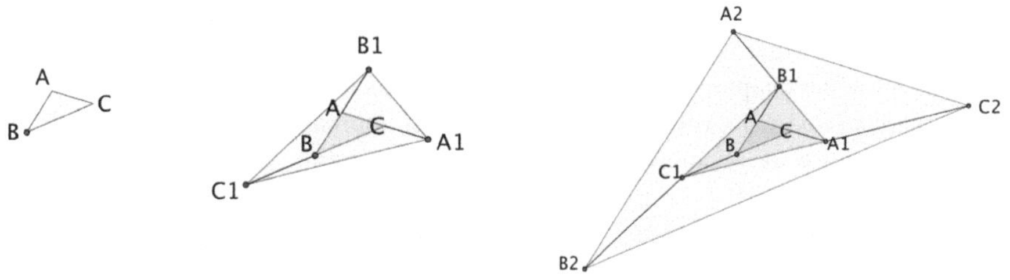

Fig. 6.10 From the mother triangle to the grand-daughter triangle

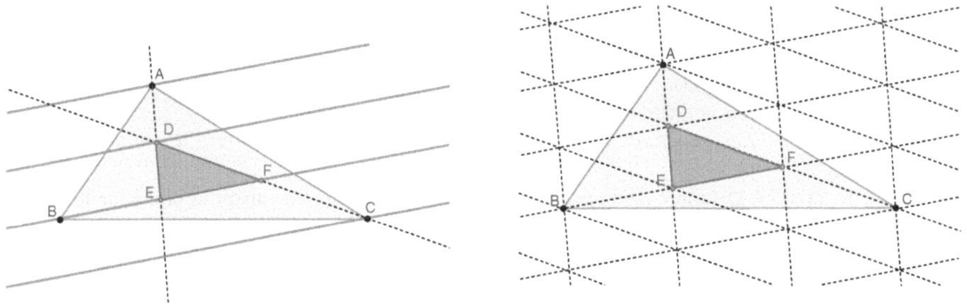

Fig. 6.11 *Parallel straight lines*, inside and outside the initial figure, structuring a set of geometrical properties

plan a structure allowing to understand the links mother–daughter triangle (and recalling a tessellation point of view (see Figs. 6.2, 6.3 and 6.4).

We have proposed in this section some examples and guidelines for designing a possible 'interesting route' in re-thinking mathematics teaching: provoking students' curiosity *for a result* and *its reasons of being*, *engaging students in a path* allowing them *to enlarge, at each step, their view on a given mathematical landscape*; *creating conditions for fruitful interactions between students themselves* as well as *between students and mathematical objects*. Situating 'Constructing, computing, proving, and understanding' at the heart of the mathematics curriculum seems to be nowadays an object of attention and research, as noticed by Hanna and de Villiers (2012): 'there has been an upsurge in research on the teaching and learning of proof at all grade levels, leading to a re-examination of the role of proof in the curriculum and of its relation to other forms of explanation, illustration and justification'.

We will focus, in the following chapters, on some crucial aspects of this research, regarding the integration of tools in mathematics education, tools seen as critical elements of *mathematics laboratories* (Maschietto & Trouche, 2010). New technological environments lead indeed to *new constraints*, *new opportunities* for the teacher (Trouche, 2000). It is particularly clear, in the previous examples, for the dynamic geometry software (DGS). But. . . we have to keep in mind, as stated by John (Sect. 6.2.4), that 'when one drags a construction [. . .] in a DGS, *visual images* can be seen, but *mathematical relationships* may not be seen; "seeing in a mathematical way" is [. . .] usually (always?) mediated by an artefact or a teacher.

An advantage of a DGS for 'mathematical seeing' is that physical actions (what the hand drags) and what is seen are co-ordinated'.

The following chapters will provide developments relevant to the use of tools in mathematics (Chap. 7), offering some theoretical approaches allowing to analyse students' activity in advanced technological environments, and teachers' role for *orchestrating* (Sect. 10.4) situations of research.

References

Chevallard, Y. (2005). Steps towards a new epistemology in mathematics education. In M. Bosch (ed.), *Proceedings of the Fourth congress of the European Society for Research in Mathematics Education* (pp. 21–44).

Debray, R. (2001). *Dieu, un itinéraire*. Paris: Odile Jacob.

Debray, R. (2013). *Le stupéfiant image: De la grotte Chauvet au centre Pompidou*. Paris: Gallimard.

Descartes, R. (1637). Discours de la méthode pour bien conduire sa raison et chercher la vérité dans les sciences. Plus la Dioptrique. Les Meteores. & la Geometrie qui sont des essais de cette Methode. Leyde : I. Maire. Retrieved from http://gallica.bnf.fr/ark:/12148/btv1b86069594/f5.image.r=.langEN

Eisenberg, T., & Dreyfus, T. (1991). On the reluctance to visualize in mathematics. In W. Zimmermann & S. Cunningham (Eds.), *Visualization in teaching and learning mathematic* (Notes Series, Vol. 19, pp. 25–37). Washington, DC: Mathematical Association of America.

Englund, R.K. (1998). Texts from the Late Uruk Period. In J. Bauer, R.K. Englund, & M. Krebernik (Eds.), *Mesopotamien. Späturuk-Zeit und Frühdynastische Zeit*, (Vol. 160, pp. 15–233). Freibourg, Göttingen: OBO.

Hanna, G., & de Villiers, M. (Eds.). (2012). *Proof and proving in mathematics education*. Springer: The 19th ICMI Study.

Kahane, J.-P. (Dir.). (2002). *L'enseignement des sciences mathématiques*. Paris: Odile Jacob.

Lakatos, I. (1976). *Proofs and refutations. The logic of mathematical discovery*. Cambridge: Cambridge University Press.

Le Lionnais, F. (avec la collaboration de Jean Brette). (1983). *Les nombres remarquables*. Paris: Editions Hermann.

Maschieroni, L. (1797). *La Geometria del Compasso*. Pavia: Pietro Galeazzi.

Maschietto, M., & Trouche, L. (2010). Mathematics learning and tools from theoretical, historical and practical points of view: The productive notion of mathematics laboratories, *ZDM. The International Journal on Mathematics Education, 42*(1), 33–47.

Netz, R. (1999). *The shaping of deduction in Greek mathematics*. Cambridge: Cambridge University Press.

Nissen, H. J., Damerow, P., & Englund, R. (1993). *Archaic bookkeeping. Writing and techniques of economic administration in the ancient near east*. Chicago: University of Chicago Press.

Singh, S. (2000). *The code book: The secret history of codes and code-breaking*. London: Fourth Estate.

Trouche, L. (1998a). *Faire des mathématiques avec des calculatrices symboliques, conjecturer et prouver. 37 variations sur un thème imposé*. Montpellier: IREM, Université Montpellier 2.

Trouche, L. (1998b). *Regards croisés sur quelques aspects de l'environnement TI-92*. Montpellier: IREM, Université Montpellier 2.

Trouche, L. (2000). New technological environments: new constraints, new opportunities for the teacher. *The International Journal of Computer Algebra in Mathematics Education, 7*(3), 165–180.

Vygotsky, L.S. (1978). *Mind in Society*. Cambridge, MA: Harvard University Press.

Wertsch, J.V. (1998). *Mind as action*. Oxford: Oxford University Press.

Part II
The Development and Interpretation of Tools

Part I sets out a view on tools and tool use in mathematics at both the school level and in mathematics research. It also presented a case for the importance of tools in the development of our species and in the development of mathematics. Part II jumps forward in time to the recent past (1960 to the early twenty-first century), and the main focus is on how academics in the field of mathematics education have interpreted tool use in mathematics. The decade from 1960 is an important period with regard to tools. It witnessed tremendous advances in digital technology and is also the period which ushered in mathematics education as an academic discipline. Chapter 7 sets the scene and considers developments in mathematics, computing, mathematics education, and scholarship relevant to understanding tools. Chapters 8–10 consider three schools of thought that we consider to be especially relevant to tool use in mathematics education: constructionism, activity theory, and developments originating in France (not really a school of thought but, rather, several schools of thought). Part II closes with Chap. 11 in which we discuss themes arising in Chaps. 7–10.

Chapter 7
Developments Relevant to the Use of Tools in Mathematics

John Monaghan

7.1 Introduction

This chapter explores developments in mathematics, computing, mathematics education and scholarship relevant to understanding tools from 1960 to the time of writing. This exploration is biased in accentuating influences relevant to tools and mathematics education. I am European and there is also a bias towards that with which I am most familiar, Western influences. This opening chapter could be a book in itself. To avoid this I describe a broad landscape and focus on selected technological advances, ideas and people that I consider important. The chapter begins with a section charting developments in mathematics, computing and education followed by a section on intellectual trends relevant to understanding tools and tool use. The final section focuses on the development of ideas in mathematics education regarding tools and tool use.

7.2 Developments in Mathematics, Computing and Education

The 1960s is interesting with regard to the joint development of mathematics and computing. In mathematics Paul Cohen solved the *continuum hypothesis* (Cohen, 1963/1964) and Abraham Robinson introduced *non-standard analysis* (Robinson, 1966). Both of these advances were due to developments in mathematical logic. Mathematical logic, coupled with advances in physics/electronics, was also behind advances in technology. Mathematicians such as Alan Turing and Johnny von Neumann were instrumental in the development of the computer pre-1960. Jon's top 10 algorithms (see Sect. 3.5) illustrate the co-development of mathematics and computing and the quicksort algorithm (#7 in Jon's list), developed in the early 1960s, remain the most used sorting method in databases. As we will see in Chap. 8,

J. Monaghan et al., *Tools and Mathematics*, Mathematics Education Library,
DOI 10.1007/978-3-319-02396-0_7

mathematicians also developed the first high level computer languages. The mathematics-computing influence was two way and in 1961 Shanks and Wrench computed π to 100,000 d.p. using an inverse-tangent identity and a computer. By the 1970s the computer, as a tool, was an active agent in mathematics. In 1975 Benoit Mandelbrot introduced the world to fractals and, soon after, was using computers to plot images of Julia sets. In the following year a proof of the Four Colour Theorem was published; this was significant because it was the first major proof in which a computer was essential (for parts of the proof). Soon after the developments in experimental mathematics with computers, which Jon describes in Chap. 2, started. Borwein and Devlin (2008, p. 7) write of this period:

> At the same time that the increasing availability of ever cheaper, faster, and more powerful computers proved irresistible for some mathematicians, there was a significant, though gradual, shift in the way mathematicians viewed their discipline. The Platonistic philosophy that abstract mathematical objects have a definite existence in some realm outside of humankind, with the task of the mathematician being to uncover or discover eternal, immutable truths about those objects, gave way to an acceptance that the subject is the product of humankind, the result of a particular kind of human thinking.

1967 heralded the first compact electronic calculator, Texas Instruments' Cal-Tech, though it used transistors and required mains power. Two years later Sharp's QT-8D appeared with semi-conductor technology replacing transistors. In 1972 Hewlett Packard introduced the first scientific calculator (with trigonometric and algebraic functions) followed in 1974 by the first programmable calculator. Calculators were not designed for education purposes (their design was largely forced by the technology available) and the early target for sales were business and scientific workers. In 1985 the first graphic calculator, the Casio fx-7000G, appeared. The *mathematicians' touch* is evident in this, and some other early scientific calculators, in that it used Reverse Polish Notation, i.e. 2×3 is input as $2\ 3\ \times$.

The development of semi-conductor technology also paved the way for small computers, which came to be known as 'micro computers' in these early years. These were, initially, often sold as kits to be assembled for amateur and professional 'computer boffs' to learn about microprocessors but the home and education markets were soon targeted. The ability to link micro computers, equipped with high level languages (rather than machine code), to a TV screen came around 1977 with three computers targeted at the home market: Radio Shack's TRS-80, the Apple II and the Commodore PET. Soon after this some mathematicians, such as David Tall, started writing BASIC programmes for educational graphing with sub-routines to provide such things as an 'intelligent scale' to the axes and the ability to recognise asymptotes; Tall's programmes were later marketed as *Supergraph* (Tall, 1985). Simultaneously with this (though the origins pre-date the advent of the micro computer) other mathematicians were developing educational ideas using the languages *Lisp* and *Logo*—this is discussed in some detail in the next chapter.

The 1980s witnessed a burgeoning of computer applications for doing and learning mathematics. The first interactive geometry system, the *Geometric Supposer*, appeared in the early 1980s. The *Geometric Supposer* did not allow geometric objects to be dragged but *Cabri*, which first appeared in 1986, did allow objects to be dragged and soon after the mathematical implications of 'dragging' for learning geometry became (and remains) a focus for mathematics educators. *Cinderella*, which Jon refers to in Chap. 2, first appeared in the late 1990s. The first computer algebra systems (CAS), *Axiom* and *Reduce*, were developed in the 1960s; *Maple* and *Mathematica* (referred to in Chap. 3) were developed in the 1980s. The first CAS aimed at student use (*muMATH*, later to become *Derive*) first appeared in the late 1970s; by 1994 a version of *Derive* was available on a graphic calculator, the TI-92 (which also included a version of *Cabri*). Statistics has been radically transformed by advances in computing, not least because of the ability of computers to handle large datasets. Recognition of the importance of computers for statistics is evident in the establishment, in 1972, of a Statistical Computing Section of the American Statistical Association. Two years later Generalized Linear Interactive Modelling (GLIM) software appeared and was, thereafter, a tool for university students. In 1993 the programming language *R* appeared which is used by professional statisticians today (2014) and is particularly suited to data mining. Statistical functions were quickly introduced into early scientific and graphic calculators but work on large data sets is more appropriate for computers than it is for calculators. The first appearance of what would now be recognised as a spreadsheet was again the early 1960s. It was, of course, developed for financial work, not mathematics, but it has been appropriated by mathematics teachers at all levels of education. The appropriation, by mathematics teachers, of a tool designed for finance may be viewed as strange from within the mathematics community but is understandable from the perspective of education as a leading contributor to a nation's economic future.

Developments in technology during the 1980s which would impact on education in the 1990s and beyond were the internet, interactive whiteboards and touchscreen technology. These developments affect all subject areas of education, not just mathematics. With regard to mathematics it appears they do not so much affect the subject matter itself (as a CAS might do) but they affect the means through which students and teachers can access and present mathematical explanations and ideas. It is clear that advances in digital technology, starting in the 1960s, ushered in a period of tremendous growth in tools for education and for doing mathematics. The development of technology and mathematical tools is ongoing but there is a real sense in which mathematicians and mathematics educators are struggling to understand the revolution (of sorts) which began in the 1960s. I draw a close to this opening section with a summary of developments in mathematics education since the 1960s.

There was considerable mathematics curriculum innovation in the 1960s originating in the birth of new mathematics in the previous decade. The 1960s was the period of the *cold war*, which negatively influenced the exchange of ideas between the two power blocs including Soviet work on activity theory (which we consider in Chap. 7). In 1957 the Soviets put the first artificial satellite, Sputnik, into space. Bybee (1997, p. 3) comments, 'Sputnik made clear to the American public that it was in the national interest to change education, in particular the curriculum in mathematics and science'. Further curriculum reform started in the 1980s due to advances in digital technology. In the UK, for example, an influential government report, wrote, 'We devote a separate chapter to electronic calculators and computers because we believe that their increasing availability at low cost is of the greatest significance for the teaching of mathematics' (Cockcroft, 1982, p. 109). Developments in mathematics education, however, were not just at the curriculum/policy level, they were also at the organisational and academic levels.

At the organisational level the International Congress on Mathematical Education, which is now held every 4 years under the auspices of the International Commission on Mathematical Instruction of the International Mathematical Union, held its first conference in 1969. This reflected the rise of national mathematics education organisations. In the UK,[1] for example, the *Association of Teachers of Mathematics* was founded in 1962 from the *Association for Teaching Aids in Mathematics* (a group who valued tool use!) and was effectively a breakaway group from the more 'traditional' *Mathematical Association* (MA) and, in 1971, the first issue of the MA's professional journal, *Mathematics in School*, appeared.

The 1960s was, effectively, the birth of academic mathematics education as a research field, 'In the 1960s and 1970s, research studies in mathematics education grew not only in number but in scope as researchers increasingly moved across the boundaries of disciplines and countries'. (Kilpatrick, 1992, p. 29), and the graphs Kilpatrick produces for research studies and for these are approximately exponential ($k > 0$) for the 1960s. Further to this, the current two academic mathematics education journals which are in the Social Science Index stem from this period: *Educational Studies in Mathematics* first appeared in 1968 and the *Journal for Research in Mathematics Education* first appeared in 1971. International academic exchange led to the first conference of the *International Group for the Psychology of Mathematics Education* in 1976 which continues as the main annual academic mathematics education conference.

I now look at considerations of tools within the wider intellectual climate of the times (1960 to the early twenty-first century).

7.3 Intellectual Trends Relevant to Understanding Tools and Tool Use

I divide this section into three subsections. In the first I look at what behaviourism had to say about tools and note its decline. I also introduce a construct—affordances—that is now regarded as important for consideration of tools and which rose as behaviourism declined as an influence. The second and longest subsection looks at the work of two scholars whose work should not be ignored

[1] Luc describes similar developments in France in Chap. 10.

in any serious consideration of tools, Marx Wartofsky and Lev Vygotsky. The section closes with a summary of theoretical approaches in which tools are regarded as agents in activity.

7.3.1 The Decline of Behaviourism

I am sure that we can look at any period and claim that various established and emerging ideas/approaches/paradigms competed in intellectual discourse, but the period around 1970 in the West is the period I focus on and this was a period of intellectual warfare. The century had been dominated in Western universities by 'positivism', founded on empiricism, and the psychological form of this was behaviourism. A feature of all forms of positivism was breaking phenomena of study into discrete parts which could be studied in isolation. Developments in mathematics, specifically in meta-mathematics, went hand-in-hand with this penchant for discrete study; Bertrand Russell's 'logical atomism' being a case in point, e.g. that non-logical expressions 'have meaning if, and only if, either they or the expressions that appear in their analyses (if any) signify existent things' (Pears, 1972, p. 9). Behaviourism was interested in external actions and how these could be initiated and channelled, *stimulus–response*. Tools were a means to initiate and channel external responses. The work of Patrick Suppes is interesting in this respect and I consider the legacy of behaviourism through a consideration of one of his papers.

Suppes was a mathematician, a philosopher and a psychologist, who applied his talents to issues in mathematics education. He was clearly interested in tools, as his paper 'Computer technology and the future of education' (Suppes, 1968) shows. I select him as an intelligent and well-meaning example of the late behaviourist school of thought in mathematics education; and as someone who spoke of tool use in education and used tools in his research in mathematics education without entering into the niceties of the agent–tool dialectic.

In Suppes (1968) he speaks with enthusiasm that 'individualised instruction once possible only for a few members of the aristocracy can be made available to all students' (Suppes, 1968, p. 41). This is possible 'because of its great speed of operation, a computer can handle simultaneously a large number of students' (Suppes, 1968, p. 41). Suppes does consider student–computer interaction, of which there are three possible levels 'individualised drill-and-practice systems ... tutorial systems and dialogue systems' (Suppes, 1968, pp. 42–44). The last level is viewed as speculative because of speech recognition 'technical problems must first be solved' (Suppes, 1968, p. 44) but Suppes conducted experiments at the second level where the 'intention is to approximate the interaction of a patient tutor ... As soon as the student manifests a clear understanding of a concept on the basis of his handling of a number of exercises, he is moved on to a new concept and new exercises' (Suppes, 1968, p. 43). I summarise this approach to tool use as the tool, the computer, imitates a human instructor in techniques and understanding is judged by 'correct' responses to specific stimuli. Similar tutorial level systems are common today (2014) and go under the name of 'individualised learning systems'. It is interesting to note that there is very little research, apart from reports from parent companies, on student learning through interaction with such systems and what there is, e.g. Baturo, Cooper, and Mc Robbie (1999), suggests that students merely learn strategies which generate responses which produce correct answers.

A consideration of one paper by one author does not prove a claim about an approach but I have faith in the claim that the legacy of behaviourism with regard to tool use in mathematics education could be summed up as: useful as a means to simulate or speed up the work of humans. By 1970, however, behaviourism was on the decline as an intellectual force, even if 'popular' forms were thriving in mathematics classrooms. I now consider two non-behaviourist academics who were working in the 1970s on ideas relevant to tool use.

In a series of papers and books over three decades, from the 1950s to the 1980s, the psychologists J.J. and E.J. Gibson developed a theory of visual perception that was not tied to stimulus–response theory. Visual stimuli featured in their approach but perceptual learning concerned 'differentiating qualities of stimuli in the environment rather than acquiring associated responses that cause greater differentiation by enrichment of stimuli as a result of past experience' (Greeno, 1994, p. 336). The environment is central to their approach; environments afford some actions and constrain others:

> The *affordances* of the environment are what it *offers* the animal, what it *provides* or *furnishes* ... If a terrestrial surface is nearly horizontal ... nearly flat ... and sufficiently extended (relative to the size of the animal) and if its substance is rigid (relative to the weight of the animal), then the surface *affords support*. (Gibson, 1979, p. 127).

Depending on the computer, the computer screen as an environment, affords clicking (anywhere on the screen) with a mouse or touching the screen with a finger (or other object). To the Gibsons, an affordance (or constraint) is a feature of agent–environment relation; it does not need to be perceived by the agent. An icon on a computer screen is not an affordance, as Norman (1999, p. 40) notes:

> The affordance exists independently of what is visible on the screen. Those displays are not affordances; they are visual feedback that advertise the affordances; they are perceived affordances.

A constraint of an environment is related to affordance in as much as it specifies what the environment does not afford: a large dog cannot lie down in a small broom closet; we cannot click with a mouse outside of the computer screen. Norman (1999) distinguishes between physical, logical and cultural constraints. The last example (mouse outside of the screen) is a physical constraint. Logical constraints require reasoning to determine alternatives (clicking on three points in a specific order to obtain the angle bisector in our dynamic geometry system (DGS) example in chapter P1). Cultural constraints are conventions shared by a community of users, e.g. dragging an object in a DGS. The constructs 'affordances' and 'constraints' have wide application in mathematics education, from investigations into the extent to which tasks and questions afford participation in mathematics classrooms (Watson, 2007) to valuations of software for doing mathematics (Monaghan & Mason, 2013). We pick up the thread of affordances and constraints in later chapters but now, in this overview, turn to two people who made significant contributions to an understanding of tools, Wartofsky and Vygotsky.

7.3.2 Two Approaches Which Take Tools Seriously

Western analytic philosophy, a key creator of positivism, was, by 1970, well into a period of questioning its assumptions. A thriving twentieth century branch of philosophy was the philosophy of science. A focal mid-twentieth century publishing outlet for this branch was the series *Boston Studies in the Philosophy of Science*. In this series every 'ism', from anarchism to positivism, was debated. Volume 48 was a collection of papers by the philosopher Marx Wartofsky. Chapter 11 (Wartofsky, 1979) is an essay on perception, representation and forms of action and advances an interesting perspective on artefacts/tools. Wartofsky's position with regard to perception is anti-empiricist and views perceptions as 'an *historically* evolved faculty, and therefore based on the development of historical human practice' (Wartofsky, 1979, p. 189), i.e. humans, in different epochs, actually perceive in different ways. Historical human practice or, to use a word favoured by Marxists, 'praxis', is firstly 'the fundamental activity of producing and reproducing the conditions of species existence ... human beings do this by means of the creation of artefacts ... the 'tool' may be *any* artefact created for the purpose' (Wartofsky, 1979, p. 200).

Wartofsky extends the artefact–tool idea to 'the acquisition of skills, in the process of production (...hunting ... agriculture ...) creates such skills as themselves "artifacts"' (Wartofsky, 1979, p. 201). Wartofsky created a new ontology of artefacts:

> *Primary* artefacts are those directly used in this production; *secondary* artifacts are those used in the preservation and transmission of the acquired skills or modes of action or praxis by which this production is carried out. Secondary artefacts are therefore *representations* of such modes of action (Wartofsky, 1979, p. 202)

Wartofsky goes on to describe a third kind of artefact, 'artefacts of the imaginative construction of "off-line" worlds' (Wartofsky, 1979, p. 208), where 'online' is 'in praxis', but the imagination is constrained by our experiences in our 'online' world. With regard to tool-design this is basically a statement that tool-design is bounded by the designer's past experience.

It was, unfortunately, many years before mathematics educators took note of Wartofsky's work but around the same time a book with a similar, but quite independent, emphasis on the development of historical human practice was published that would deeply influence mathematics educational thought on tool use. This was the Soviet Lev Vygotsky's *Mind in Society*. Vygotsky did not write this as a book. He died 40 years before it was published. Four American academics

Fig. 7.1 A representation of Vygotsky's position on mediation

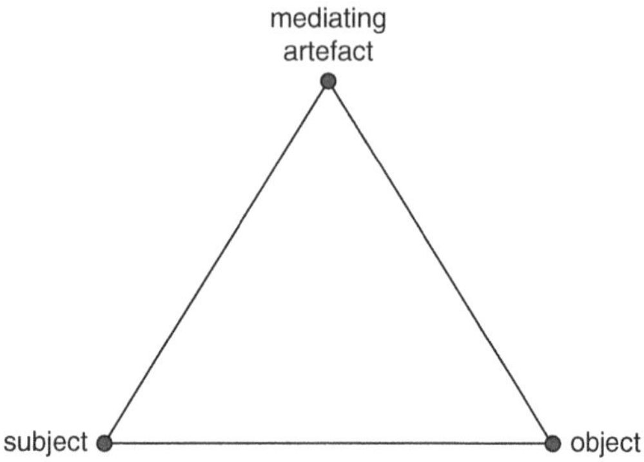

compiled it as a collection of Vygotsky's essays. Vygotsky, with colleagues, had a tremendous influence on Soviet psychology but the cold war effectively meant that Western and Soviet psychology largely developed without reference to the other. I explore the details of the influence of Vygotsky and colleagues on mathematics education in Chap. 9 but take the opportunity at this point to sketch the origins of this school of thought in the 1920/1930s.

Today (2014) the founder of activity theory is considered to be Vygotsky but he was not alone. His interests initially centred on literary studies and he was particularly interested in language, signs and 'mediation'. Physical tools were not, unlike the authors of this book, of interest in themselves, any interest was due to their mediating qualities, 'the basic analogy between sign and tool rests on their mediating function that characterises each of them' (Vygotsky, 1978, p. 54). The difference between signs and tools rests on:

> The tool's function is to serve as the conductor of human influence on the object of activity; it is *externally* oriented; it must lead to a change in objects . . . The sign, on the other hand, changes nothing in the object of a psychological operation. It is a means of internal activity aimed at mastering oneself; the sign is *internally* oriented. (Vygotsky, 1978, p. 55)

It is not unusual to see Vygotsky's position on mediation to be represented by a triangle (see Fig. 7.1):

There are at least three ways to misinterpret this diagram. The first is to regard the object as a thing; it is not, the object is the *raison d'etre* of activity. The second is as a form of behaviourism that takes account of 'artefact mediation'. This is not the case: the subject–object pair simply represents unmediated activity, e.g. drawing a round shape in the sand with your finger; the subject–artefact–object represents mediated activity, e.g. drawing a circle with a compass. The third is to regard the subject as going through the mediating artefact to the object; this is simplistic to the extent that it is wrong, as Cole (1998, p. 119) points out:

> the incorporation of tools into the activity creates a new structural relation in which the cultural (mediated) and the natural (unmediated) routes operate synergistically; through active attempts to appropriate their surroundings to their own goals, people incorporate auxiliary means (including, very significantly, other people) into their actions, giving rise to the distinctive, triadic relationship of subject-medium-object.

Vygotsky's interest in mediated activity centred on the shift from external processes to internal (mental) processes, which he calls 'internalisation'. He provides an enlightening example of the genesis of meaning in the act that becomes pointing. A baby initially attempts to grasp something out of his/her reach. This may look like s/he is pointing but s/he is not, s/he is trying to get the object. Then along comes an adult who sees the attempt and brings the object within reach of the baby. What is crucial here is that another human has come into the process of trying to grasp the object. Over time, and with repetition, the act of trying to grasp becomes a gesture of pointing. Vygotsky comments, 'At this juncture there occurs a change in that movement's function: from an object-orientated movement it becomes a movement aimed at another person, a means of establishing relations' (Cole, 1998, p. 56). This leads him to claim:

> Every function in the child's cultural development appears twice: first, on the social level, and later, on the individual level; first *between* people (*interpsychological*), and then *inside* the child (*intrapsychological*). (Cole, 1998, p. 57)

This is a powerful and persuasive claim but, as Bereiter (1994, p. 21) states, these lines

> make an empirical claim, and one that is almost certainly too strong. There is ample evidence ... that young children work out a substantial knowledge of the physical world, well before they could have gained much of it from the surrounding culture.

The word 'culture' in these quotes represents the phylogenetic accumulation of knowledge and this is important in the Vygotskian distinction between 'everyday' and 'scientific' concepts. Vygotskians regard everyday concepts as those acquired through our senses—the sun disappears from the sky and it is night—but scientific concepts have a theory behind them, such as the axial rotation of the earth in an orbit around the sun to explain night. A theory does not need to be correct for a concept to be scientific, as Scott, Mortimer, and Ametller (2011, p. 6) point out:

> ... scientific concepts are taken to be the products of specific scientific communities and constitute part of the disciplinary knowledge of that community. The term 'scientific' as used by Vygotsky is not restricted to the natural sciences, but covers all comparable communities such as those of history, philosophy, art and so on. As the agreed upon products of specific communities, scientific concepts are not open to 'discovery' by the individual but can only be learned through some form of tuition.

In Chap. 11 we shall see another criticism of Vygotsky's account of internalisation. My view is that Vygotsky pointed to something important but he was engaged in early work. From the point of view of tool use in mathematics, the opposite of internalisation, 'externalisation', the shift from internal processes to external processes, is also important and this is something I mentioned, in so many words, in my Sect. 1.3 definition of tools.

I take a brief pause at this point to make a disclaimer (of sorts) and move into the field of semiotics. The chapter I am writing is about tools and this focus has biased my presentation of Vygotsky's thought for, as I say above, he was particularly interested in language, signs—semiotic mediation—and his conceptualisation of internalisation was fundamentally as a semiotic process. Given the mediational similarity (there is, indeed, an overlap) between signs and tools, it is appropriate to consider their relationship.

In mathematics we tend to think of signs as 'our symbols': $\times$, +, $\int$, etc. In linguistics the focus is on language(s) and the signs of interest are words, speech, text, etc. But a sign in semiotics is just an arbitrary thing (a *signifier*) that signifies something (the *signified*). Mathematics lessons are rife with signs as teachers and learners attempt to communicate with each other not just through mathematical symbols and words but also through pointing, gestures, intensity of speech, etc. The signs used in this communication must have some common meanings to each individual involved or communication would not be possible. This is particularly tricky in mathematics instruction because teachers introduce new signs and tools that initially have no meaning for the learner. Consider, for instance, the long process (Vygotsky's internalisation) of turning calculus into tools for doing mathematics, e.g. $\int x dx = \frac{x^2}{2} + c$ considered as an algorithm. The individual signs, e.g. dx, in this equation have a meaning that has been passed down by the culture of mathematics but they are arbitrary and have no reference to the learner as s/he starts learning calculus. Learning about this sign-tool takes place over a considerable period of time during which teacher and learner draw on formal signs which have meaning to the learner including a great deal of pointing and gestures. Vygotsky was not, as far as I know, interested in algorithms and mathematics educators may not be interested in the writings of Vygotsky but semiotic mediation is something that is central to mathematics teaching whatever one's belief.

7.3.3 Tools as Agents

The final intellectual trend of this period I consider arises from scholarship into the history, philosophy and sociology of science. The publication of Thomas Kuhn's *The structure of scientific revolutions* in 1962 (alongside work by other scholars), in the words of Pickering (1995, p. 2), 'opened the way for new waves of scholarship ...work on the sociology of scientific knowledge (SSK) has increasingly documented the importance of the human and the social in the production and use of scientific knowledge'. One of these waves of scholarship, originating from the late 1970s is now called actor network theory (ANT); Latour (2005) is a fairly recent exposition. ANT is a theory about how to study social phenomena—by following the actors, where an actor is '*anything* that does modify a state of affairs by making a difference' (Latour, 2005, p. 71); ANT theorists would not look at a mathematics lesson as a given social structure but would describe the structure in terms of actors. It views social life as being in a state of flux and looks to the performance of the actors in situations. Objects (artefacts/tools) can make a difference in activity and so can be actors, exerting agency, in the playing out of social situations. Pickering (1995), who is 'almost ANT' in my opinion, examines the practices of late twentieth century physicists using machines to trace elementary particles. He accepts ANT's human and material agencies and adds 'disciplinary agency' (in our discipline $a + a = 2a$ regardless of what we might want it to be). He proposes a 'dance of agency where, in the performance of scientific inquiry, human, material and disciplinary agencies 'emerge in the temporality of practice and are definitional of and sustain one another' (Latour, 2005, p. 21). What ANT and Pickering bring to a consideration of tool use in mathematics is that tools have agency. This is a controversial claim but an interesting one that cannot be ignored in a serious study of tool use.

I now move on to consider the development of ideas in mathematics education regarding tools and tool use.

7.4 The Development of Ideas in Mathematics Education Regarding Tools and Tool Use

I sketch the development of ideas in academic mathematics education regarding tools (including, in some cases, an absence of regard to the role of tools) from the early 1960s.[2] The dominant approach in the West to what would now (2014) be called 'mathematics education' or the 'didactics of mathematics' was behaviourism. I have discussed this in the previous section and do not discuss it further here. But there were notable others around. Zoltan Dienes conducted learning experiments which centred on young children making and finding patterns in play-like activity with manipulatives and concrete representations. Many schools purchased 'Dienes blocks', wooden cubes (units), 1 by 1 by 10 cuboids (tens), 1 by 10 by 10 cuboids (hundreds) and 10 by 10 by 10 cubes (thousands). These wooden artefacts are tools (by my Sect. 1.3.1 definition of tools) when their overt purpose is to instill conceptual understanding (of place value)—they are tools for learning mathematics rather than as tools for doing mathematics. Dienes (1963) writes with attention to detail and respect for learners about children's mathematical activity with manipulatives. His interest is on the learners' appreciation of mathematical structure (e.g. arranging blocks to appreciate that $12^2 = 10 \times 10 + 4 \times 10 + 4 \times 1$) and their formation of abstraction and generalisation. His model of the abstraction–generalisation process (see Dienes, 1963, p. 67) focuses on the mind and on mathematics—the manipulatives do not enter the model and appear, to me, to be mere props for mental operations (interpret, symbolise) that are part of his model. It was several decades before the usefulness of manipulatives in the long-term learning of mathematics was questioned (see Hart, 1989). Another notable figure in the 1960s is Guy Brousseau whose classroom experiments included the use of manipulatives. I do not discuss this here as this work is considered in Chap. 10. But

[2] This sketch of the development of ideas cannot be comprehensive and is bounded by my knowledge of the field.

the most important figure, in terms of Western mathematics education, in the 1960s was Jean Piaget.

Piaget, who was initially a biologist, was a contemporary of Vygotsky but unlike Vygotsky he did not die young; his academic opus spanned many decades during which time his influence increased. His main interest was children's conceptions and he explored, amongst other things, their conceptions of geometry, logic, movement, number, space and the world. His interests centred on the *development* of thought from birth to adolescence. He posited that humans develop through stages and that growth within stages depended on schemes, 'the structure or organisation of actions as they are transferred or generalised by repetition in similar or analogous circumstances' (Piaget & Inhelder, 1969, p. 4). At any stage in the development of the child 'reality data are treated or modified in such a way as to become incorporated into the structure of the subject' (Piaget & Inhelder, 1969, p. 5), this is *assimilation*. By repetition schemes are modified to fit with the child's new interpretation of reality, this is *accommodation*. This is a far cry from behaviourism. If behaviourism can be represented as 'stimulus → response', then Piaget's version can be represented as 'stimulus ↔ response'. An interesting thing about Piaget's work is his neglect of the role of tools in cognitive development. This is obviously noteworthy in a book on tools but it also appears a little strange as he did pay specific attention to related things, signs (and semiotics in general) and objects. Respect for Piaget's work increased to an extent that he had 'guru status'. A prominent English mathematics educator of 1960s and 1970s was Kenneth Lovell and if you take any of his academic papers, then you will find Piaget's theory being expanded on with regard to new mathematical ideas or experiments but with virtually no criticism (to my knowledge) of Piaget's assumptions or theory. The nearest I have found to questioning Piaget in Lovell's work is:

> I believe that his position regarding the acquisition of certain kinds of new knowledge is of more value to the mathematics teacher than any other position at the moment, although I affirm with equal conviction that his theory does not cover all the facts and that one day it will be replaced or subsumed by a more all-embracing one. (Lovell, 1972, p. 165)

I make this point on 'guru status' because it may help to explain the virtual neglect on tools in pre-twenty-first century scholarship in mathematics education that paid homage to Piaget. The most influential (epistemological) theory in mathematics education in the 1980s was constructivism. Constructivists pay regular homage to Piaget's theory as the inspiration for their ideas, e.g. Von Glasersfeld (1991, p. xiv) in speaking of constructivists, 'They have taken seriously the revolutionary attitude pioneered in the 1930s by Jean Piaget'. Leslie Steffe was an early constructivist researcher. I consider a paper of his (Steffe, 1983) which considers children's algorithms as schemes. I select this paper because, in Chap. 1, I argue that algorithms may be regarded as tools. Steffe pays homage to Piaget, 'children's methods can be viewed as schemes. This premise has justification in the primordial seriation scheme studies by Inhelder and Piaget (1969)' (Piaget & Inhelder, 1969, p. 110). He considers operative and figurative schemes in the algorithms developed by two young children. Educational resources (strips with ten squares and blocks) are used by the children but the paper makes no mention of artefacts or tools.

During the 1980s constructivism divided into what are now called 'radical' and 'social' constructivism. The radical branch, of whom Steffe is an example, was concerned with ontogenic development of the individual child. The social branch was interested in microgenetic, i.e. child–environment, as well as ontogenic development. During the 1980s Paul Cobb moved from a radical to the social branch and early paper that could be called 'social constructivist' is Cobb (1987).[3] The paper investigates the sense young children make of statements such as $3 + 6 = 9$ and $9 = 3 + 6$ and notes conflicting models of early number development in the literature. Cobb uses clinical interviews (a research tool developed by Piaget) in worksheet tasks and tasks which employ felt squares to hide numbers. Cobb concludes that the 'academic context' of the task is crucial and influence children's goals, they tend 'to use primitive finger patterns in worksheet situations ... [as] ... these methods ... constituted viable ways of operating in their classrooms' (Cobb, 1987, p. 121) I have no criticisms of the conclusions reached by Cobb but, as with Steffe (1983), I find it curious that no mention is made to tools (as finger patterns may be regarded as arithmetic tools). Cobb, with various co-researchers, went on to develop a specific form of social constructivism, sometimes referred to as the 'emergent perspective'. An important paper from this perspective is Yackel and Cobb (1996). This paper examines teacher–young children discussions and argumentation in a classroom context and introduced a new construct to the field of mathematics education, 'sociomathematical norms', 'normative aspects of mathematical discussions that are specific to students' mathematical activity' (Yackel & Cobb, 1996, p. 458). The classroom is provided with various resources (centicubes and an overhead projector) but the paper does not mention tools. This neglect has been noticed by others, e.g. Hershkowitz and Schwarz (1999, p. 149) refer directly to Yackel and Cobb (1996) when they write '... sociomathematical norms do not arise from verbal actions only, but also from computer manipulations as communicative non-verbal actions'. In fairness to Cobb, however, by 2002 his output did explicitly consider tool use.[4] Cobb (2002) presents an analysis of seventh grade statistical data analysis; students were given data sets in which 'it would be essential that they actually begin to analyse data in order to address a significant question' (Cobb, 2002, p. 176). The instructional strategy behind students' data analysis was supported with two 'computer minitools' developed to fit with the instructional sequence. The paper explores how symbolising, modelling and tool use interrelate in students' data analysis.

[3] This paragraph could be taken as an attack on Cobb's work. This is not my intention and I hold his opus on high regard. I focus on Cobb partly because he is a 'major player' in the mathematics education community and because he developed as a researcher from a Piagetian base and, I believe, this base led him to overlook the role of tools in his twentieth century publications.

[4] Explicit reference by Cobb to tool use may have occurred prior to 2002 but I am not aware that this is the case.

During the 1980/1990s, however, influences of a non-Piagetian nature were stirring which Lerman (2000, p. 23) calls 'the social turn':

> the emergence into the mathematical education community of theories that see meaning, thinking, and reasoning as products of social activity. This goes beyond the idea that social interactions provide a spark that generates or stimulates an individual's internal meaning making activity. A major challenge for theories from the social turn is to account for individual cognition and difference, and to incorporate the substantial body of research on mathematical cognition, as products of social activity.

This was followed by what has been called 'the sociopolitical turn', which Gutiérrez (2013, p. 40) describes as 'theoretical [and methodological] perspectives that see knowledge, power, and identity as interwoven and arising from (and constituted within) social discourses'. There is, to my mind, overlap in these 'turns' which I illustrate via Lerman. The 'substantial body of research' which Lerman goes on to cite includes situated cognition (e.g. Lave, 1988), Foucauldian analyses (e.g. Walkerdine, 1988) and cross-cultural studies (e.g. Bishop, 1988); Foucauldian analyses are clearly sociopolitical in essence.

I do not disagree with what Lerman says but I think another turn, 'the technological turn', was rotating in the same period and the moment of this turn was the technological revolution that I describe in Sect. 7.2. Calculators and micro computer applications became objects of great interest to mathematics educators and, to paraphrase Lerman above, they began to 'see meaning, thinking, and reasoning as products of tool-based activity'. Further to this, many of these tool-focused researchers were also influenced by the social turn (e.g. the first two authors of this book). A recent (at the time of writing) paper (Morgan & Kynigos, 2014) illustrates social and technological foci. The paper concerns digital artefacts and external representations and I preface a consideration of the paper with some comments on representations.

I use the term 'external representation' (ER) to refer to 'a configuration of symbols, images or concrete objects standing for some other entity' (Fagnant & Vlassis, 2013, p. 149). Some authors use the term 'schematic representations' or just 'representations', but I like to make the external aspect explicit. An ER may be presented to the learner (e.g. a Cartesian graph or a number line) or it may be generated by the learner. The ER may or may not correspond to something mathematical (e.g. a smiley face) and when two ERs which correspond to something mathematical they may do so in different ways. For example, the two images in Fig. 7.2 could both represent '3 × 4' but the rectangular array may be presented by a teacher to focus on the idea of a Cartesian product whereas the 'dots in circles' may be learner generated and focus on repeated addition.

Both ERs in Fig. 7.2 are artefacts and, if they are used in doing mathematics, function as mathematical tools (by Monaghan's Sect. 1.3 definition of a tool). Van Dooren, Vancraenenbroeck, and Verschaffel (2013, p. 322) point out:

> an external representation of a problem situation can be a beneficial heuristic ... may lead to a reduction of working memory load ... to be effective, the external representation needs to be a correct display of the problem situation

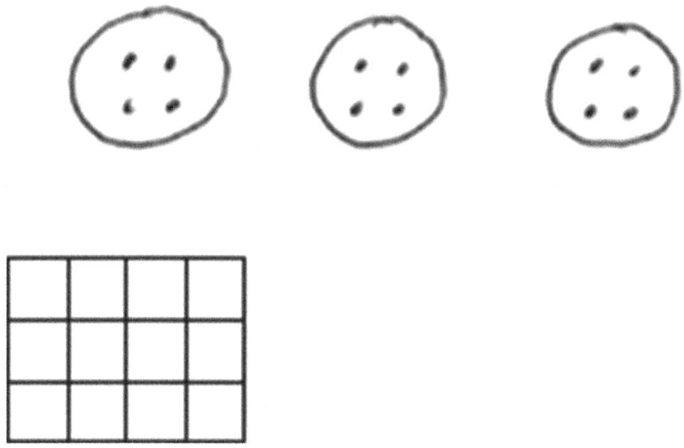

Fig. 7.2 Two external representations for '3 × 4'

Fagnant and Vlassis (2013) claim that ERs are central elements in expert problem solving and this certainly seems to be the case in Jon's Chap. 3, which is awash with ERs and his claims about the visual theorems are essentially claims about the central roles of ERs in expert problem solving. The history of mathematics is, in part, a history of external representations as Sect. 4.3 and Chap. 5 (on ancient Greek and on Babylonian mathematics) attest. Digital technology has opened the way for new ERs, for example 'sliders' in dynamic graphing and geometry packages. I now return to a consideration of Morgan and Kynigos (2014) to illustrate differences in social and technological foci.

Morgan and Kynigos (2014) appear in a special issue of *Educational Studies in Mathematics* on the ReMaths project (*Re*presenting *Math*ematics with digital media: Working across theoretical and contextual boundaries). Basically what ReMath did was get pairs (or groups) of researchers (and, often, mathematical software designers) from different countries and using different theoretical frameworks to try out each other's software and interpret the other's data (if possible) from the perspective of a different framework. Morgan and Kynigos are one such pair. The paper focuses on a digital artefact, MoPiX, designed for:

> constructing animated models using the principles of Newtonian motion. The objects of the MoPiX microworld were designed to behave in mathematically coherent ways providing an environment … intended to allow students to construct orientations to concepts such as velocity and acceleration consistent with conventional mathematical and physical principles (Morgan & Kynigos, 2014, p. 362)

Morgan represents the social turn:

> a multimodal social semiotic perspective on representation … From this perspective, the elements of spoken, written, diagrammatic or other forms of communication are not taken to have a fixed relationship to specific objects or concepts … Rather, the resources offered by language, diagrams, gestures and other modes are considered to provide a potential for

> meaning making. The term representation cannot therefore be taken to have an internal reference to some individual mental image or structure. Nor can it be taken to refer to a determined relationship between signifier (word, picture, symbol, etc.) and signified (represented object or concept). As the elements of communication acquire meaning in interactions within social practices, the notion of representation must also be understood relative to specific social interactions and practices. (Morgan & Kynigos, 2014, p. 359)

Kynigos represents the technological turn via a constructionist framework (see Chap. 8):

> The role of representations is important in the sense that they are perceived as integral components of artefacts-under-change and as a means for expressing, generating and communicating meaning. The nature of representations and the kinds of use to which they are put are at the centre of attention ... artefacts can embody a wide range of complexity and have been perceived and analysed as representations themselves ... Unlike the social semiotic perspective, digital artefacts are seen as representations designed by pedagogues to embed one or more powerful ideas ... representations are not seen simply as objects to which some kind of meaning may be attached but as artefacts for tinkering with. (Morgan & Kynigos, 2014, p. 360)

Morgan and Kynigos (2014) proceeds to analyse student work from perspective of each framework. I do not detail these analyses and you do not get a prize for guessing that the analyses differ. The two authors, however, agree that the two approaches 'are not on the whole incompatible and yield interpretations of the data that have some similarities' (Morgan & Kynigos, 2014, p. 375). There are also instances where the two authors use a construct in a similar way, for instance, the term 'meaning'. But even here the constructionist approach sees meaning residing in the individual and is often linked to 'tinkering' with artefacts, whereas, in the multimodal social semiotic approach, 'meaning is conceptualised as the establishment of shared orientations through communication in interaction between individuals' (Morgan & Kynigos, 2014, p. 377).

7.5 End Note and Anticipation of the Remaining Part II Chapters

Morgan and Kynigos (2014) illustrate social and technological foci in twenty-first century mathematics education research. It also illustrates two of the large number of theoretical frameworks used twenty-first century mathematics education research. This book on tools and mathematics is not the place for considering all of these frameworks but we consider frameworks that appear, to us, to be particularly important with regard to conceptualising tool use in mathematics in Chaps. 8–10. Chapter 8 focuses on constructionism, the framework that Kynigos, above, uses. Chapter 9 concerns activity theory which is a continuation of the work of Vygotsky. The flow of ideas in Chap. 7 did not provide a sensible place to outline Radford (2014) but it is a paper that readers may like to follow up if they are interested in representations because Radford (2014) is an activity theoretic critique of Morgan and Kynigos (2014) and has a lot of interesting things to say on ERs and on artefacts. Chapter 10 outlines several frameworks that emerged in France around the time of social and technological turns (late 1980s). In the 1990s researchers centred around Michèle Artigue following the approaches of Chevallard and of Rabardel were engaged in ground breaking research on the use of technology in mathematics classrooms that challenged the hegemony of constructivist interpretations of the role of technology in the learning of mathematics.

References

Baturo, A. R., Cooper, T. J., & Mc Robbie, C. J. (1999). Karen and Benny: Deja vu in research. In *Proceedings of the Conference of the International Group for the Psychology of Mathematics Education, 2*, 73–80.

Bereiter, C. (1994). Constructivism, socioculturalism, and Popper's world 3. *Educational Researcher, 23*, 21–23.

Bishop, A. (1988). *Mathematical enculturation: A cultural perspective on mathematics education.* Dordrecht, The Netherlands: Kluwer.

Borwein, J. M., & Devlin, K. (2008). *The computer as crucible*. Natick, MA: AK Peters.

Bybee, R. W. (1997). *The Sputnik era: Why is this educational reform different from all other reforms*. Prepared for the Symposium "Reflecting on Sputnik: Linking the Past, Present, and Future of Educational Reform". Washington, DC: National Academy of Sciences.

Cobb, P. (1987). An investigation of young children's academic arithmetic contexts. *Educational Studies in Mathematics, 18*(2), 109–124.

Cobb, P. (2002). Modeling, symbolizing, and tool use in statistical data analysis. In K. Gravemeijer, R. Lehrer, B. van Oers, & L. Verschaffel (Eds.), *Symbolizing, modelling and tool use in mathematics education* (pp. 171–195). Dordrecht, The Netherlands: Kluwer.

Cockcroft, W. H. (1982). Report of the committee of inquiry into the teaching of mathematics in schools.

Cohen, P. J. (1963/1964). The independence of the continuum hypothesis. *Proceedings of the National Academy of Sciences of the United States of America, 50* (1963, 1143–1148) and *51* (1964, 105–110).

Cole, M. (1998). *Cultural psychology: A once and future discipline*. Harvard: Harvard University Press.

Dienes, Z. P. (1963). *An experimental study of mathematics learning*. London: Hutchinson.

Fagnant, A., & Vlassis, J. (2013). Schematic representations in arithmetical problem solving: Analysis of their impact on grade 4 students. *Educational Studies in Mathematics, 84*(1), 149–168.

Gibson, J. J. (1979). *The ecological approach to visual perception*. Boston: Houghton Mifflin.

Greeno, J. G. (1994). Gibson's affordances. *Psychological Review, 101*(2), 336–342.

Gutiérrez, R. (2013). The sociopolitical turn in mathematics education. *Journal for Research in Mathematics Education, 44*, 37–68.

Hart, K. M. (1989). *Children's mathematical frameworks 8-13: A study of classroom teaching*. Windsor, England: Nfer-Nelson.

Hershkowitz, R., & Schwarz, B. (1999). The emergent perspective in rich learning environments: Some roles of tools and activities in the construction of sociomathematical norms. *Educational Studies in Mathematics, 39*(1–3), 149–166.

Kilpatrick, J. (1992). A history of research in mathematics education. In D. A. Grouws (Ed.), *Handbook of research on mathematics teaching and learning* (pp. 3–38). New York: Macmillan.

Kuhn, T. S. (1962). *The structure of scientific revolutions*. Chicago: University of Chicago Press.

Latour, B. (2005). *Reassembling the social—An introduction to actor-network-theory*. Oxford, England: Oxford University Press.

Lave, J. (1988). *Cognition in practice: Mind, mathematics and culture in everyday life*. Cambridge, MA: Cambridge University Press.

Lerman, S. (2000). The social turn in mathematics education research. In J. Boaler (Ed.), *Multiple perspectives on mathematics teaching and learning* (pp. 19–44). Westport, CT: Ablex.

Lovell, K. R. (1972). Intellectual growth and understanding mathematics. *Journal for Research in Mathematics Education, 3*, 164–182.

Monaghan, J., & Mason, J. (2013). Exploring the notion 'cultural affordance' with regard to mathematics software. In *Proceedings of the British Society for Research into Learning Mathematics*. Retrieved from http://www.bsrlm.org.uk/IPs/ip32-3/BSRLM-IP-32-3-23.pdf

Morgan, C., & Kynigos, C. (2014). Digital artefacts as representations: Forging connections between a constructionist and a social semiotic perspective. *Educational Studies in Mathematics, 85*(3), 357–379.

Norman, D. A. (1999). Affordances, conventions, and design. *Interactions, 6*, 38–42.

Pears, D. F. (1972). Introduction to Russell's logical atomism. London: Fontana.

Piaget, J., & Inhelder, B. (1969). *The psychology of the child*. New York: Basic Books.

Pickering, A. (1995). *The mangle of practice: Time, agency, and science*. Chicago: University of Chicago Press.

Radford, L. (2014). On the role of representations and artefacts in knowing and learning. *Educational Studies in Mathematics, 85*(3), 405–422.

Robinson, A. (1966). *Non-standard analysis: Studies in logic and the foundations of mathematics*. Amsterdam: North Holland.

Scott, P., Mortimer, E., & Ametller, J. (2011). Pedagogical link-making: A fundamental aspect of teaching and learning scientific conceptual knowledge. *Studies in Science Education, 47*(1), 3–36.

Steffe, L. P. (1983). Children's algorithms as schemes. *Educational Studies in Mathematics, 14*(2), 109–125.

Suppes, P. (1968). Computer technology and the future of education. In R. Atkinson & H. Wilson (Eds.), *Computer-assisted instruction: A book of readings* (pp. 41–47). New York: Academic Press.

Tall, D. O. (1985). *Supergraph*. Barnet, England: Glentop.

Van Dooren, W., Vancraenenbroeck, G., & Verschaffel, L. (2013). The role of external representations in solving multiplicative problems. In *Proceedings of the 37th Conference of the International Group for the Psychology of Mathematics Education* (Vol. 4, pp. 321–328).

Von Glasersfeld, E. (1991). *An exposition of constructivism: Why some like it radical* (pp. 229–238). New York: Springer.

Vygotsky, L. S. (1978). *Mind in society*. Cambridge, MA: Harvard University Press.

Walkerdine, V. (1988). *The mastery of reason*. London: Routledge.

Wartofsky, M. (1979). *Models: Representation and the scientific understanding*. Dordrecht, The Netherlands: Reidel.

Watson, A. (2007). The nature of participation afforded by tasks, questions and prompts in mathematics classrooms. *Research in Mathematics Education, 9*, 111–126.

Yackel, E., & Cobb, P. (1996). Sociomathematical norms, argumentation, and autonomy in mathematics. *Journal for Research in Mathematics Education, 27*, 458–477.

Chapter 8
Constructionism

John Monaghan

8.1 Introduction

Constructionism evolved through the work of Seymour Papert and his co-workers. The first definition of constructionism appears to have been in Papert (1987):

> The word constructionism is a mnemonic for two aspects of the theory of science education underlying this project. From constructivist theories of psychology we take a view of learning as a reconstruction rather than a transmission of knowledge. Then we extend the idea of manipulative materials to the idea that learning is most effective when part of an activity the learner experiences as constructing is a meaningful product.

Constructionism is an apt place to start our case studies of tool use in mathematics education because of its longevity (relative to the life of mathematics education as a field of inquiry), its influence in education software development and its influence on design. From a scholarly perspective constructionism is interesting because it is problematic: there is a sense in which it both is and is not a learning theory; it began life adhering to Piagetian principles but leading constructionist scholars, at the time of writing, can be located in the sociocultural field of thought; tool use is central to its philosophy but constructionist scholars are often not forthcoming about what a tool is. My aim in this chapter is to attempt to distill the/a constructionist view(s) of tools in learning mathematics[1] but this distillation must be accompanied by an understanding of what constructionism is and this, I hope, will emerge via a critical consideration of its history, achievements and the claims of constructionists. To this end I structure this chapter as follows. Section 8.1 presents a brief history of constructionism. Section 8.2 positions the programming language *Logo* (which is a very important player in the history of constructionism) in the history of computer developments. Sections 8.3 and 8.4 respectively consider

[1] I hesitated about inserting the word 'mathematics' after 'learning' because there is a sense (which I hope will become apparent in the course of this chapter) in which constructionism is more than just about learning mathematics.

J. Monaghan et al., *Tools and Mathematics*, Mathematics Education Library,
DOI 10.1007/978-3-319-02396-0_8

two influential constructionist books, Papert's, 1980 book *Mindstorms* and Noss & Hoyles' 1996 book *Windows on mathematical meanings*. Section 8.5 attempts to distill the/a constructionist view(s) of tools in learning mathematics.

8.2 A Brief History of Constructionism

Papert began academic life as a mathematician; he obtained two PhDs in mathematics in the 1950s. In the late 1950s and early 1960s he worked with Piaget. In 1963 he moved to MIT and, a few years later, became director of the MIT Artificial Intelligence Laboratory. In 1967 he designed, with Wally Fuerzeig, the first version of the programming language *Logo* (from the Greek 'logos'). *Logo* was to serve a vision of education Papert developed whilst he was with Piaget. A small robot, the floor turtle, was developed and Papert used this in his work with children. This was the start of what came to be known as 'turtle geometry' which has a slightly different orientation than school geometry. For example, an equilateral triangle is typically constructed by programming the turtle to repeat three times the move 'go forward by a fixed amount and then turn 120° right' (or left). This construction, with this tool in this language, focuses on linear movement, rotation and external angles of a triangle rather than the equal arc lengths and internal angle of a triangle that a ruler and compass construction focuses on. I recall, as a school teacher in the 1980s, that some of my colleagues did not like this alternative focus of turtle geometry, though I did not see this as a comparative evaluation of new and old tools for mathematics at the time. In 1980 Papert wrote a book, *Mindstorms: Children, computers, and powerful ideas*, on this work with children. Healy and Kynigos (2010, p. 63) aptly characterise *Mindstorms* as realising Papert's 'desire to use computers as mathematically expressive media with which to design an appropriable mathematics fitted to the learner.' *Mindstorms* probably remains (at the time of writing, 34 years later) the most widely read and influential text on computers in mathematics education. It could be said that constructionism publicly started with the publication of this book.

Papert went on to write and edit other books, Harel and Papert (1991) and Papert (1993, 1996), but it is fair to say that, after *Mindstorms*, his main influence was through projects, papers and through the work of others that he inspired. The project he was working on at the time of his serious accident, shortly after giving a plenary address at a mathematics education conference in Hanoi in 2006, was *One Laptop per Child*, with the aim to manufacture and distribute an inexpensive (≈$100 at the time) computer for children in developing nations. In the field of mathematics education he inspired the work of Celia Hoyles, Richard Noss, Andrea diSessa and Uri Wilensky, amongst many others.

My guess that *Mindstorms* is the most widely read and influential text on computers in mathematics education, could be followed by a guess that W*indows on mathematical meanings: Learning cultures and computers*, Noss and Hoyles (1996), is the second most widely read and influential text on computers in mathematics education. *Windows on mathematical meanings* established links between constructionism and other theories of learning, design and practice. I now flesh out this brief history with sections as described above.

8.3 Positioning *Logo*

My aim in this section is fourfold: (1) to position *Logo* with regard to other programming languages; (2) to consider the aesthetic appeal of *Logo* to some mathematicians; (3) to consider its appeal to the artificial intelligence community as a medium for problem solving; (4) to relate all three issues above to a past debate within the mathematics education community. I will assume that you, the readers have some familiarity with at least one of the many versions of *Logo*; you can download a version and get a self-help starter guide if this is not the case.

The first version of *Logo* was written in another programming language called *Lisp*, standing for '**list** **p**rocessing'. The first two programming languages were FORTRAN (1957) and *Lisp* (1958). FORTRAN (for FORmula TRANslating) was designed for numerical analysis and is an *imperative* language, the fundamental role of the code of FORTRAN programs consists of sequences of 'do this' commands. *Lisp* is a *functional* language, the fundamental role of the code of *Lisp* programs is to evaluate a function, i.e. result = f(input values).

Lisp has a rather illustrious mathematical background going back to Alan Turing's Ph.D. supervisor Alonzo Church. Church developed a formal logical system, lambda calculus, as a prelude to formalising the concept of computability (Church, 1932). In lambda calculus functions are represented via lists (of a sort):

> {F} (A) represents the value taken on by the function F when the independent variable takes on the value A. The usual notation is F(A). ... we treat a function of two variables as a function of one variable whose values are functions of one variable, and a function of three or more variables similarly. Thus, what is usually written F (A,B) we write {{F} (A)} (B), and what is usually written F(A, B, C) we write {{{F} (A)} (B)} (C), and so on. (ibid., 352)

Further to this lambda calculus writes expressions in prefix form, i.e. $x + 1$ is written $+x\ 1$ and is able to differentiate between defining and calling a function. The latter is done via its valid expressions, lambda terms: a variable, x, is a lambda term; if t is a lambda term and x is a variable, then $\lambda x.t$ is a lambda term; if t and s are lambda terms, then ts is a lambda term. $\lambda x.t$ is called a 'lambda abstraction' and defines a function that takes x as its input and substitutes it into the lambda term t. ts, however, calls on the function t with input s to produce what is commonly written as $t(s)$. Further to this, functions can operate on functions; for example $\lambda x.x$ represent the identity function $x \rightarrow x$ and $(\lambda x.x)y$ represents the identity function applied to y.

The notation of the lambda calculus was the inspiration for *Lisp*. This is evident in a *Lisp* code that defines the square function:

$$(\text{lambda}(x)(*xx))$$

Lists are fundamental data structures for *Lisp* and a *Lisp* list is written inside a pair of brackets with its elements separated by spaces; for example, $1+2$ is written (+1 2). *Lisp* had an appeal to mathematicians and early versions of the computer algebra system *Derive* were written in a version of *Lisp*. It also had an appeal, as a medium for problem solving, to the nascent artificial intelligence community in the 1960s. To appreciate this appeal I consider how *Lisp* and FORTRAN may be used to obtain a solution that would count as 'problem solving' if a human did it. I select a classic (origins in the eighth century) old problem, *The wolf, the cabbage and the goat*:

> A man needs to bring a wolf, a goat, and a cabbage across the river. The boat is small and can only carry one passenger at a time. If he leaves the wolf and the goat alone together, the wolf will eat the goat. If he leaves the goat and the cabbage alone together, the goat will eat the cabbage. How can he bring all three safely across the river?

It is hard to imagine using FORTRAN to solve this problem. The original FORTRAN had 32 reserved words and 15 of these were for input/output statements in the days of punched cards, tapes and drums. However, an expert early FORTRAN programmer could possibly find a devious way to configure a matrix and use IF statements and GOTO commands within DO loops to list all possible combinations of a representation of the objects in the problem and, from this list, to eliminate illegal (in terms of the problem) combinations. But FORTRAN was designed for numerical analysis, not to solve problems such as this one.

In *Lisp*, however, the problem can be represented in terms of the elements on each bank at given states: man (M), wolf (W), cabbage (C), goat (G), and boat (B). These can be programmed in terms of lists with the initial conditions:

Leftbank = [M W C G B][2]
Rightbank = [ ] (the empty list)

We would then want to create a procedure for possible moves, which could be a list of lists, something like:

Possiblemoves = [[M W B] [M C B] [M G B] [M B]]

And set up procedures for movement from bank to bank (two procedures for each direction of travel) which would update the Leftbank and Rightbank lists by deleting/augmenting elements from their lists depending on the moves. We would also have to set up procedures for checking that the Leftbank and Rightbank lists after any moves did not contain illegal elements such as [W G] and also set up procedures to try all possible moves and keep a record of moves. I will stop this account of the old problem here and note that a full explanation (with codes in both *Lisp* and *Logo*) can be found in Bundy (1980).

---

[2] *Lisp* code would actually be (SET LEFTBANK '(M W C G B)) but I will not present code in my description.

The above focus on machine problem solving, I hope, suggests why a functional language such as *Lisp* was preferred over an imperative language such as FORTRAN by early workers in the field of artificial intelligence. Before leaving this consideration of machine problem solving I would like to make a point about tools and what are the tools. Returning to my definition of a tool in Chap. 1 (that an artefact becomes a tool when it is used to do something) a programming language is not a tool but is simply an artefact but code written in a programming language and executed by a computer is a tool. Is *Lisp* better than FORTRAN for problem solving? The answer depends on the problem. *Lisp* is an appropriate medium for generating code to solve problems such as *the wolf, the cabbage and the goat* problem but FORTRAN is designed so that its code can function as a tool to provide numeric solutions to problems in numerical analysis.

I now move to the educational arena. Logo was designed as an educational programming language to teach concepts related to *Lisp* programming. The educational focus was extended to children when Papert introduced a *floor turtle* (a robot with a retractable pen that could be programmed to draw mathematical shapes on paper on the floor). The mathematical legacy of the lambda calculus in such drawing is open for all ages: for the young, in realising a square repeat two actions four times, as in

```
REPEAT 4[FORWARD 100 RIGHT 90]
```

And, for older children, in differentiating between defining and calling a function, as in

```
MAKE “SIDE 1
REPEAT 30[FD :SIDE RT 20 MAKE “SIDE :SIDE + 1]
```

In the early 1980s some school children were being introduced to code like that above as well as BASIC code like that below:

```
For X = 1 to 10
Print X*X
Next X
```

School children were taught ‘programming’ and the main computer languages were the imperative language BASIC and the functional language *Logo*. Programming was taught in some mathematics lessons and also in the new school subject Computer Studies. To many teachers, I believe, BASIC and *Logo* were just two languages but to others there were ‘values’ at stake—the kind of problems students engage in technomathematical tasks mattered. This was the focus in the BASIC vs. *Logo* debate that went on in the late 1980s and early 1990s in various countries. I illustrate this debate by considering two articles published in a professional journal for mathematics teachers, MacKernan (1992) and Noss (1992). These two articles represented the tail end of a debate that had been conducted through the journal *Micromath*.

MacKernan’s (1992) defends BASIC with the following arguments: ‘It is a valuable tool for doing awkward or long-winded or repetitive calculations’ (ibid.,

p. 17); a FOR-NEXT loop which prints the square numbers from 1 to 10 is cited in the National Curriculum algebra strand; it is easy to print hard copy in BASIC; and *Logo* is good for shapes but BASIC is good for numbers. Noss (1992) is a response to MacKernan (1992) and instead of arguing against MacKernan's points he offers 'a vision of what mathematics could be like using computers' (ibid., p. 18). He contrasts 'school maths' (with the implication that it is *humdrum*) with the power and potential fasciation of 'mathematics', 'the power of mathematics for constructing explanations of how the physical and social world works is unparalled. And I think that the computer can help with this task.' (ibid., p. 19). Noss sees the choice of programming language as central to this vision and sees the debate as centring on:

> a tool for solving problems or a means of expression . . . to incorporate far larger sections of society into mathematical culture, we had better look . . . towards the computer as a medium for expression rather than simply a convenient tool to 'deliver' . . . the National Curriculum (ibid.).

This debate is useful for my purpose in this chapter (to distill a constructionist view of tools) but a problem with debates (other than mathematical debates) is that the law of the excluded middle rarely operates in the lived-in-world; only two views are presented when, in fact, there are a myriad of phenomenologically valid points of view. The MacKernan–Noss debate is not so much about tools per se but how tools are and can be used (to preserve a past with a fixed number of mathematical problems for which to use the tools or a creative future). Tools can be used for problem posing as well as problem solving and this, in a nutshell, is what *Mindstorms* is about.

My final words about *Logo* are, first, to note that virtually every article/book by a *Logo*-enthusiast cited in this chapter has a note somewhere in it that says, in so many words, '*Logo* isn't perfect but it seems to be the best we have at the moment to realise our vision'. Second, to note that there are many versions of *Logo* and it has influenced other languages but a consideration of these is not necessary for this chapter and I now move on to consider *Mindstorms*.

8.4 Mindstorms

Rather than present a linear description of Papert's book I start this section by jumping about (in the book and in time) to present my interpretation of Papert's mindset. I start with the final chapter, *Images of the learning society*, which begins, 'THE VISION I HAVE PRESENTED is of a particular computer culture, a mathetic one, that is, one that helps us not only to learn but to learn about learning' (ibid., p. 177, capitals in the original). *Mindstorms* is a scholarly book in the 'learned' rather than the 'scientific' sense of the word 'scholarly'. Papert developed *Logo*, he spent years working with children using *Logo*, he was inspired by what these children did and he wanted to present the reader with the vision that came from this inspiration. I now present Papert's contextualisation of the events leading up to *Mindstorms* via of a summary of Chap. 8 of Papert (1993) because I feel it is important to attempt to communicate the genesis of Papert's vision.

Chapter 8, *Computerists*, is a personal account of developments in computing, artificial intelligence, educational computing and Papert's place in these developments; I focus on the last two items in this list. 'In the 1960s we were a small handful ... Patrick Suppes from philosophy and psychology, John Kemeny (who invented BASIC) ...' (ibid., p. 160). Papert notes a classification by Taylor of the modes of use of computers in education: tutor, tutee and tool. Papert notes that that calculators, simulation programs and word processors are tools and that 'tutor' refers to 'the most common image of the computer in education' (ibid., p. 161). But it is clear that 'tutee' is the term that captured Papert's imagination, that we (or, rather, a child) could teach a computer.

Papert then positions Suppes, Kemeny and himself with regard to Taylor's classification. Suppes (who was considered in Sect. 7.2), as the intellectual father of computer aided instruction (CAI), focused on the tutor mode of use whereas Kemeny focused on students programming computers and, in doing so, making the computer program, in his words, a tool that aids learning. In terms of 'modes of use' Suppes and Kemeny are different but Papert notes a similarity, 'They shared an acceptance of School' (ibid., p. 163), where 'School' refers to the institutionalised transmission of knowledge[3] (my paraphrase of pp. 1–21 of Papert, 1993). In School learning is viewed with regard to facts and skills; in School learning was *cold* but Papert's vision was *hot*.

Papert returns to Debbie, a fourth grade student considered earlier in Papert (1993), who overcame considerable problems with School fractions through *Logo* programming:

> I pose the educational goal not as giving her factlets but as encouraging her to make connections between different elements of what she already knows: for example, intuitive knowledge about fractions, knowledge about the "real world", and knowledge about strategies of learning. Making connections is something only Debbie can do. They have to be *her* connections. (ibid., p. 165)

Papert goes on to reflect on his time with Piaget and 'playful speculation about what would happen if children could play at building little artificial minds' (ibid., p. 169) and his subsequent (mid 1960s) realisation that children could do what workers in artificial intelligence do, write programs that simulate what people do and discuss differences between what machines and human can do. But he needed a programming language that matched the needs and capabilities of children and this was the impetus that led to the design of *Logo* and, later, a robot (eventually a floor turtle) that could enact their programs. I now briefly present a linear description of Papert's book followed by a consideration of one of its critics.

---

[3] This view of school mathematics as 'institutionalised mathematics' has links with Chevallard's anthropological theory of didactics which is considered in Chap. 10.

The person and thought of Piaget permeates *Mindstorms* and his/its influence is stated at the outset, ‘The powerful image of child as epistemologist caught my imagination while I was working with Piaget’ (ibid., p. 19). Papert, however, was less concerned than Piaget on the order/stages of development and more concerned with ‘the influence of the materials a particular culture provides in determining that order . . . the computer presence might have more fundamental effects on intellectual development than did other new technologies’ (ibid., p. 20). Further to this Piaget contrasted concrete and formal thinking but, with the computer, ‘Knowledge that was accessible only through formal processes can now be approached concretely’ (ibid., p. 21). Papert’s vision of liberation is a materialist one but one founded on computers rather than tools per se, ‘computers of the very near future **will be the private property of individuals** . . . Education will become more of a private act . . . There will be new opportunities for imagination and originality’ (ibid., p. 37, my emphasis). The highlighted part of this vision statement has been realised and I think Papert may have been the first person to publicly anticipate this evolution of computers. Noss (2001, p. 22) makes a related point about Papert’s vision, the computer ‘can be changed (even change itself) into any number of forms’. So, maybe Papert did not comment on tools because the computer, as an artefact, is capable of transformative change into many tools. The statement following the highlighted text in Papert’s vision, however, has turned out different to that which Papert anticipated, connectivity (considered in Chap. 18) has led to education to become a more public act.

Papert goes on to consider what he calls *Mathophobia*, fear of learning, and he believes that the computer can play a fundamental role in their liberation from this fear and ‘our culture’s hard-edged separation between the verbal and the mathematical’ (ibid., p. 45). He outlines the case of Jenny who did not understand English grammar until she taught (programmed) it to *Logo* and, thereafter, went from an average to high performing school student. He states that ‘every educated person vaguely remembers that $y = x^2$ is the equation of a parabola’ (ibid., p. 52) and goes on to suggest:

> The reason for what is included and what is not included in school math might be as crudely technological as the ease of production of parabolas with pencils! This is what could change most profoundly in a computer-rich world: The range of easily produced mathematical constructs will be vastly expanded. (ibid., p. 52)

In turtle geometry Papert sees the potential for *body syntonic* learning, that is learning ‘firmly related to children’s sense and knowledge about their own bodies’ (ibid., p. 63); and there is a sense in which he anticipated current (*circa* 2014) thought about embodied cognition (e.g. Lakoff & Núñez, 2000) by several decades. He presents many examples to illustrate this. An interesting example with regard to tools is, ‘Let us imagine, then, as I have seen a hundred times, a child who demands: How can I make the Turtle draw a circle?’ (ibid., p. 58). The instructor says ‘play Turtle’ and the child repeats ‘move forward a little and turn a little’. Through discussion, description and initial coding this eventually becomes:

TO CIRCLE REPEAT [FORWARD 1 RIGHT 1]

Papert contrasts this way of creating a circle (equal curvature; differential geometry) with the geometry of Euclid (constant distance between a centre point and points on the circumference) and of Descartes, $(x-a)^2+(y-b)^2=R^2$. A matter of interest to me, in writing this book on tools and mathematics, is that he does not explicitly focus on the tools used in these three methods of creating a circle (as we have done in a not too dissimilar example in Chap. 1 of this book). I shall return to this in the last section of this chapter.

Papert goes on to discuss structured programming: breaking a problem down into parts, describing these parts and programming these descriptions (akin to my sketch of a *Lisp* solution to the 'wolf, cabbage and goat' problem). He presents the case of a child, Keith, who wrote linear code (25 lines) to produce a stick man. It didn't work and 'Keith was unable to figure out what had happened' (ibid., p. 102). Papert shows us how we can write a shorter program using a procedure for the legs and arms and another procedure for the head. He contrasts Keith with Robert who breaks problems/programs down, 'I used to get mixed up by my programs. Now I don't bite off more than I can chew' (ibid., p. 103). Papert argues that, over time children get used to a structured way of thinking and debugging procedures rather than long linear programs and in this work, unlike classroom work, 'the teacher and the learner can be engaged in real intellectual collaboration' (ibid., p. 115). The aim of this collaboration is to produce a 'specific something' and this leads on to the idea of *microworlds*:

> The Turtle World was a microworld, a "place," a "province of Mathland," where certain kinds of mathematical thinking could hatch and grow with particular ease. (ibid., p. 125)

The learning theory behind the microworld concept in *Mindstorms* is, of course, Piaget's genetic epistemology and, in particular, the concept of 'assimilation', of absorbing new ideas into existing mental schemes. Hoyles (1993) traces the history of the concept[4] and cites Weir, that it 'was first used by artificial intelligence workers to describe a *small, coherent domain of objects and activities implemented in the form of a computer program*[5]' (ibid., p. 1). From this start they evolved, through the work of Papert and colleagues, 'Microworlds did not simplify and trivialise structural features of the knowledge domain: rather, they aimed to facilitate the building of conceptual and strategic foundations' (ibid., p. 3). Thereafter they were taken up by (mainly) university mathematics educators who appropriated 'Papert's vision but with the specific intention of provoking change through computer use *within* the practice of school mathematics' (ibid., p. 4). This last stage is, effectively, the narrative behind the next section of this chapter so I do not pursue it further here but return to Papert.

[4] See Healy and Kynigos (2010) for a more recent history of the concept.

[5] *The wolf, the cabbage and the goat* problem (see Sect. 8.2) is an example of this kind of microworld.

Papert describes a specific microworld with turtles which can be linked turtles and, in which, the turtles obey turtle forms of Newton's laws of motion (e.g. 'Every Turtle remains in its state of rest until compelled by a TURTLE COMMAND to change that state,' ibid., p. 127). His discussion of the Monkey problem is interesting with regard to tools:

> A monkey and a rock are attached to opposite ends of a rope that is hung over a pulley. The monkey and the rock are of equal weight and balance one another. The monkey begins to climb the rope. What happens to the rock? (ibid., p. 131)

Papert states that most MIT Physics undergraduates in their experience either cannot address this problem or give an incorrect answer (I suspect that this remains true in most universities):

> ... students ask themselves: "Is this a 'conservation-of-energy' problem?" "Is this a 'lever-arm' problem?" and so on. They do not ask themselves: "Is this a 'law-of-motion' problem?" ... conservation, energy, lever-arm, and so on, have become tools to think with. They are powerful ideas that organise thinking and problem solving ... For a student who has had experience in a "laws-of-motion" microworld this is true of "law of motion". This student will not be blocked by asking the right question ... but a student who sees laws of motion only in algebraic terms will not even ask the question ... (ibid., pp. 131–132)

The question 'What happens to the rock?' is an interesting one. It does not provide a hint that the rock goes up symmetrically with the monkey and by this deliberate omission the question can lead to some strange answers: it is a hard question from an institutional mathematics perspective but it is, somehow, an easy question if one is immersed in the laws of motion microworld. But for all its interest I am not sure how the institutional "tools to think with" that Papert refers to are tools.

There is more to *Mindstorms* than I have room to summarise but I feel I have covered the overall philosophy behind the book and some salient constructs so I will move on to reactions to the book. The first reaction I'd like to mention is my own. I read the book shortly after it was published and I embraced its vision—it provided a warrant for my work in mathematics classrooms at the time. Some years later I was critical of its 'scientific base' but a few years later these criticisms were relaxed as I realised it was a 'vision statement' and not a scientific text. Although I am really not clear what 'tool' means to Papert there is a sense in which this is unimportant. My aim in writing this book is to clarify (and communicate) my understanding of the meaning and use of tools in mathematics and a part of my quest is to look to at what others regard tools to be and their role in this thing called 'mathematics'. This was not Papert's aim.

Not surprisingly, there was no shortage of critiques of *Mindstorms* after its publication. The most noted of these were by a pair of authors, Roy Pea and Midian Kurland. Although comments on Papert (1980) are frequent in Pea and Kurland (1984), it is not a review of Papert (1980) but a discursive consideration of academic papers (circa 1984) concerned "about whether learning computer programming promotes the development of higher mental functions" (ibid., p. 137). Pea and Kurland published (together, alone and with others) a series of technical reports on the cognitive impact of computer programming (including *Logo*) on children; Pea and Kurland (1984) is a *grand overview* of sorts and does not include new primary data. The Conclusion calls for further research on: claims for the cognitive effects of learning to program; the developmental role of contexts in learning to program; what skilled programming is; levels of programming skill development; and cognitive constraints on learning to program. I present an extended extract from the Conclusion that summarises their findings:

> First, there are no substantial studies to support the claim that programming promotes mathematical rigor. . . .Secondly, there are no reports demonstrating that programming aids children's mathematical exploration. . . .Third, although Feurzeig et al. (1969) suggest that the twelve 7- to 9-year-old children to whom they taught Logo came to 'acquire a meaningful understanding of concepts like variable, function and general procedure,' they provide no evidence for the claim that programming helped the children gain insight into these mathematical concepts. Finally, we ask whether programming has been shown to provide a context and language that promotes problem solving beyond programming . . . Planning in advance of problem solving, and evaluating and checking progress in terms of goals, are important aspects of a reflective attitude to one's own mental activities . . . Results indicated that Logo programming experiences had *no* significant effects on planning performances, on any of the plan efficiency or planning process measures. (ibid., pp. 159–160)

This is pretty damning stuff. I do not intend to enter this debate and I move on to Noss and Hoyles (1996).

8.5 Windows on Mathematical Meanings

As we saw in Chap. 5, the period, 1980–1996, between *Mindstorms* and W*indows on mathematical meanings* (referred to by 'WMM' in the remainder of this chapter) saw a significant rise in mathematics education as an academic field. As befits a developing academic field, MWW has a methodological base that is far less open to Pea and Kurland-like criticisms than *Mindstorms* was. Nevertheless, WMM can be seen as the constructionist successor to *Mindstorms*. Indeed, the Foreword to WMM includes, 'We owe our greatest intellectual debt to Seymour Papert, whose ideas have provided continual inspiration'.

Like *Mindstorms*, WMM (1) synthesises years of work (including theorising) in mathematics education with computers (very often with *Logo*) and (2) presents a vision of learning mathematics. Chapter 5, *Webs and situated abstractions*, is, for me, the theoretical heart of the book. 'Webs' and 'situated abstractions' are deep constructs which cannot be easily summarised but elsewhere I have (with Fatih Ozmantar) linked them with contextual (sociocultural) views on abstraction (see Ozmantar & Monaghan, 2008). Nevertheless, a *quick fix* definition of these constructs may be useful to orientate the reader to their general nature. The idea of a web goes back to Papert's idea of making connections:

> We want to put forward a case for learning as the construction of a *web* of connections—between classes of problems, mathematical objects and relationships, 'real' entities and person-specific experiences. (WMM, p. 105)

The idea 'situated abstraction' stems from the tension between situated (e.g. Lave, 1988) and decontextualised (e.g. Piagetian) approaches to knowledge development and also a Papert-like belief that mathematics is not 'school maths' but 'Mathematics ... is activity-with-relationships' (WMM, p. 124).

> We intend by the term *situated abstraction* to describe how learners construct mathematical ideas by drawing on the webbing of a particular setting which, in turn, shapes the way the ideas are expressed. (WMM, p. 122, my emphasis)

The words 'which, in turn', to me, underline the dialectical materialist philosophy behind WMM, the full title of which is *Windows on mathematical meanings: Learning cultures and computers*. Dialectical materialism takes a view of the world in a constant state of flux where unidirectional 'cause and effect' arguments are replaced by interrelationships; the setting (culture) and the technology shape each other. These introductory descriptions allow me to present an extract that, as near as anything in WMM, encapsulates the place of tools in the vision of Noss and Hoyles:

> Webbing and abstracting are complimentary. Situated abstraction describes how learners construct mathematical ideas by breathing life into the web using the tools at hand, a process which, in turn, shapes the ideas. Tools are not passive: in a microworld, for example, the designer's intentions are constituted in the software tools. These tools wrap up some of the mathematical ontology of the environment and form part of the web of ideas and actions embedded in it. Yet it is students who shape these ideas ... A microworld comprises tools to construct objects. But these tools are themselves objects which encapsulate relationships. This process/object duality is at the root of mathematical activity ... (WNN, p. 227)

In the remainder of this section I further explore this view of tools.

By 1996 the influence of Piaget on constructionism was waning and the influence of Vygotsky was waxing. WMM acknowledges this influence but not without noting the special significance of mathematics:

> For Vygotsky, learning is mediated by language and signs are representational tools developed by social interactions ... since the mathematics comprises a duality of objects and relationships: there is a need to take into account how objects characterised and represented, the tools by which we act on them as well as the language which describes how they interrelate. This duality is evident in the origins of mathematics as a tool and as an object of study communicated in particular linguistic forms. (WMM, p. 42)

Chapter 3 of WMM is called *Tools and technologies* and provides further clarification of Noss and Hoyles' view of tools. Due, I posit, to their view that *mathematics is activity-with-relationships* (a view I share) they are interested in languages and tools that allow mathematical relationships to be *expressed*. This seems pretty important to me for the computer-strand theme of this book (the one you are reading) as, amongst other things, a computer can be used, at one extreme, to 'present' mathematics to a learner and, at another extreme, as a medium for the learner to 'engage in' mathematics by expressing mathematical relationships.[6]

> Software which fails to provide the learner with a means of expressing mathematical ideas also fails to open any window on the processes of mathematical learning. A student working with even the very best simulation, is intent on grasping what the simulation is demonstrating rather than attempting to articulate the relationships involved. It is the *articulation* which offers some purchase on what the learner is thinking and it is in the process of articulation that a learner can create mathematics. (WMM, p. 54)

Unlike *Mindstorms*, WMM considers systems (software and applications) other than *Logo* as suitable media for articulating/expressing mathematical relationships. They state a preference for *Logo* but clearly regard dynamic geometry systems as expressive media in which, like *Logo*, mathematical relationships can be seen and, further to this, there is a connection between gestures (through dragging) and mathematical relationships. But any talk of 'preferences' with regard to WMM clearly does not just concern the media but the manner of using the media: a learner using a simulation package that allowed considerable tinkering with the parameters in, say, simulating traffic flow, would be 'better' than a learner attending to the *Logo* task 'draw a square using the commands FORWARD, RIGHT and REPEAT'. I feel that an emphasis on the manner of using the media is evidenced by the reference, in WMM, to Illich's (1973) construct of a *convivial tool*:

> To the degree that he masters his tools, he can invest the world with his meaning; to the degree that he is mastered by his tools, the shape of the tool determines his own self-image. Convivial tools are those which give each person who uses them the greatest opportunity to enrich the environment with the fruits of his or her vision . . . They allow the user to express his meaning in action. (ibid., pp. 10–11)

Like *Mindstorms*, WMM provides numerous examples to illustrate learning in expressive computational environments and I close this section by considering one example which sheds light on WMM's view of tools within microworlds to that of two 14-year-old girls working on a dynamic geometry system (DGS). The DGS had two flag-shaped figures on it and the girls were told that one flag was the reflection of the other in a hidden line of symmetry. Their task was to find the 'mirror line', the line of symmetry. The girls did not know how to construct the line and they dragged one of the flags and observed the movement of the two flags. Within a short period of time they had located the line of symmetry informally in as much as they could indicate where the line of symmetry was with their fingers. Then one of the girls dragged the extreme points of the flags together and they both exclaimed 'that's it' and went on to explain, 'the mirror line is what you see on the screen if you drag points and their reflections together'. This statement evidenced a situated abstraction. But the girls did not stop here and went on to focus on the three objects on the screen (the two flags and the mirror line) and the mathematical relationship between

[6] Curiously, in the BASIC vs. *Logo* debate presented earlier in this chapter, both MacKernan and Noss argue within the 'expressing mathematics' pole, suggesting, of course, that the debate on how to use digital tools in mathematics education is not unidimensional.

these objects; this eventually led them to see that the mirror line could be constructed by constructing the line formed by midpoints of lines connecting corresponding points in the two flags.

This example/story of a mathematical construction is pleasant in itself (and it is told with greater attention to detail in WMM than in my summary paragraph) but what interests me with regard to tools is the discussion of this story, which I again summarise.

A midpoint command in a DGS is an abstraction that:

> wraps up much of the process of construction into a set of actions . . . which become part of the web of ideas and action embedded in the medium . . . it is a tool to construct objects and *at the same time* an object which encapsulates a relationship (WMM, p. 116)

Noss and Hoyles see 'using the webbing of the medium' in the girl's solution and ask if the midpoint strategy extends beyond the medium. They consider compass and measurement constructions of the midpoint and state that the actions with these different tools "are only 'the same' as the *Cabri* construction from a particular perspective—one in which the midpoint is already understood". Noss and Hoyles suggest that generalising from one setting to another might be 'to become aware explicitly of the relationships wrapped up in the setting, to notice precisely what elements of the computational web are interacting with one's current state of understanding'. They conclude, 'Using the web as a tool is a necessary but not a sufficient condition for this awareness to emerge' (WMM, p. 117).

I see a continuation of the ideas sketched in *Mindstorms* in this discussion of the story but with constructs and attention to the niceties of tool use that extend the frame of *Mindstorms*. The consideration of the midpoint command as a tool which can be used to construct objects whilst being an object which encapsulates a relationship illustrates an important aspect of 'expressive tools'. The statement that the different tools are only the same if the midpoint is already understood is interesting; I feel 'the same' implicitly refers to a sort quasi mathematical equivalence class with regard to understanding a concept (the actions are certainly not the same). Finally, 'using the web as a tool' appears to be an extension of Papert's notion of 'tools to think with' in microworlds (discussed in the previous section in the *Monkey problem*).

8.6 A Constructionist View of Tools?

In this closing section I attempt to characterise the view of tools in *Mindstorms* and in WMM.

In his essay on the internalisation of higher psychological functions Vygotsky takes pains to emphasise similarities and differences between signs and tools. He then writes:

> Here we want to be as precise as possible. Leaning for support on the term's figurative meaning, some psychologists have used the word "tool" when referring to the indirect

> function of an object as the means for accomplishing some activity. Expressions such as "the tongue is the tool of thought" or "aides de memoire" are usually bereft of any definite content and hardly mean more that than what they really are: simple metaphors or more colourful ways of expressing the fact that certain objects or operations play an auxiliary role in psychological activity. (Vygotsky, 1978, p. 53)

Following Vygotsky I characterise phrases such as 'the computer is a tool to think with' and 'the midpoint command on a DGS is a tool to construct objects' as, respectively, 'metaphorical' and 'precise' use of the word 'tool'.

In, *Mindstorms*, and thereafter, Papert is clearly very interested in programming languages, computers, children, learning and the intersection of these four areas. I do not think that he is particularly interested in tools per se. The word 'tool' is rarely used in *Mindstorms* and, when it is, it is used in a metaphorical sense, e.g. 'conservation, energy, lever-arm, and so on, have become tools to think with. They are powerful ideas that organise thinking and problem solving' (Papert, 1980, p. 132).

In WMM Noss and Hoyles retain the vision and foci of interest of *Mindstorms* but, while there is occasional metaphorical use of the word 'tool', e.g. 'programming in its widest sense might be thought of as a tool for expressing and articulating ideas' (WMM, p. 57), the majority of references to the word 'tool' are precise. I summarise their view of tools as:

> Their framework respects post-Vygotskian thought (learning is activity mediated by language, signs and tools) but mathematical learning is special (and was not addressed by Vygotsky) as it is essentially activity with relationships and tool are important in learners' construction of mathematical relationships. Tools are not passive actors in learning in as much as they can transform existing instructional regimes. Digital tools (for mathematics) are especially important in such transformations because they display symbols. An important division within such digital tools is between tools which allow the expression/articulation of mathematical relationships and those which do not. Tools which are different with respect to physical actions may be 'the same' to someone who already understands the mathematics at hand in a task but are (or can be) different to the learner who is developing an understanding with the use of a tool. Finally, tool use enters a dialectical relationship with task, social relationships and context.

WMM certainly raised the level of consideration of tool use in mathematics. Given the respect given to its authors in constructionist circles (and beyond), it seems reasonable to assume the view of tools in WMM influenced many constructionists but the appreciation of the subtleties of consideration of tool use in WMM is likely to be uneven within the constructionist community and individuals in this communities will have their own agendas. It would be foolish for me to assume that some sort of *telos* operates in this community with regard to a developing understanding of tool use in mathematics education research. I think it is safe to say that a constructionist view of tools remains, for the time being, that is described in WMM.

References

Bundy, A. (Ed.). (1980). *Artificial intelligence: An introductory course*. Edinburgh, Scotland: University of Edinburgh.

Church, A. (1932). A set of postulates for the foundation of logic. *Annals of Mathematics, 33*, 346–366.

Harel, I., & Papert, S. (Eds.). (1991). *Constructionism: Research reports and essays, 1985-1990*. Norwood, NJ: Alex.

Healy, L., & Kynigos, C. (2010). Charting the microworld territory over time: Design and construction in mathematics education. *Zentralblatt für Didaktik der Mathematik, 42*(1), 63–76.

Hoyles, C. (1993). Microworlds/schoolworlds: The transformation of an innovation. In C. Keitel & K. Ruthven (Eds.), *Learning from computers: Mathematics education and technology* (pp. 1–17). Berlin, Germany: Springer.

Illich, I. (1973). *Tools for conviviality*. London: Calder and Boyars.

Lakoff, G., & Núñez, R. (2000). *Where mathematics comes from*. New York: Basic Books.

Lave, J. (1988). *Cognition in practice: Mind, mathematics and culture in everyday life*. Cambridge University Press.

MacKernan, J. (1992). In defense of basic. *MicroMath, 8*(1), 17–18.

Noss, R. (1992). In defense of mathematics. *MicroMath, 8*(1), 18–19.

Noss, R. (2001). For a learnable mathematics in the digital culture. *Educational Studies in Mathematics, 48*, 21–46.

Noss, R., & Hoyles, C. (1996). *Windows on mathematical meanings: Learning cultures and computers*. Dordrecht, The Netherlands: Springer.

Ozmantar, M. F., & Monaghan, J. (2008). Are mathematical abstractions situated? In A. Watson & P. Winbourne (Eds.), *New directions for situated cognition in mathematics education* (pp. 103–127). New York: Springer.

Papert, S. (1980). *Mindstorms: Children, computers, and powerful ideas*. Brighton, England: The Harvester Press.

Papert, S. (1987). *Constructionism: A new opportunity for elementary science education*. DRL Division of Research on Learning in Formal and Informal Settings. Retrieved from http://nsf.gov/awardsearch/showAward.do?AwardNumber=8751190

Papert, S. (1993). *The children's machine: Rethinking school in the age of the computer*. New York: Basic Books.

Papert, S. (1996). *The connected family: Bridging the generation gap*. Atlanta, GA: Longstreet Press.

Pea, R. D., & Kurland, D. M. (1984). On the cognitive effects of learning computer programming. *New Ideas in Psychology, 2*(2), 137–168.

Vygotsky, L. S. (1978). *Mind in society*. Cambridge, MA: Harvard University Press.

Chapter 9
Activity Theoretic Approaches

John Monaghan

9.1 Introduction

Activity theory[1] (AT) is an approach to the study of human practices—any human practice and human practice in itself. It warrants a chapter in this book on tools and mathematics because artefacts/tools[2] are intrinsic to its approach and many mathematics educators use theoretic approaches to study mathematical practices. I first consider this approach in general but then focus on the practices of doing, learning and teaching mathematics, and the light that activity theoretic approaches shed on tool use in these mathematical practices. The roots of AT go back to early Soviet approaches and the section on Vygotsky in Sect. 7.2 serves as an introduction to these roots. This chapter has four sections. Section 9.1 provides an overview of AT. Section 9.2 traces early influences of AT in mathematics education research. Section 9.3 considers foci of a set of mathematics education papers recent at the time of writing. Section 9.4 explores emphases and tensions in papers considered in Sects. 9.2 and 9.3.

---

[1] As will soon become apparent, there are a number of schools of thought within what is called 'activity theory' and I use the term 'activity theoretic approaches' as a collective noun for these different approaches.

[2] A note for readers who are reading this as a 'stand alone chapter'. In Sect. 1.3.1 I stated my distinction between and artefact and a tool as, an artefact becomes a tool when it is used by an agent to do something. I use this distinction in this chapter. For example, a compass as a metal thing which holds a pencil and rests on a desk is an artefact but when it is picked up by someone to draw a circle it is a tool. When its status is ambiguous I use the term 'artefact/tool'.

J. Monaghan et al., *Tools and Mathematics*, Mathematics Education Library,
DOI 10.1007/978-3-319-02396-0_9

9.2 The Development of Activity Theory

It is important to start with a clarification of the word 'activity' as 'activity' is an everyday word for 'doing something' and it is not always the same as the word 'activity' in AT. *Activity* became a focus for Vygotsky in the 1920s in his consideration of consciousness as a *problem* for psychology. Kozulin (1986, pp. xxiii–xxiv) explains:

> The major objection Vygotsky had to the mentalist tradition was that it confined itself to a vicious circle in which states of consciousness are "explained" by the concept of consciousness. Vygotsky argued that if one is to take consciousness as a *subject* of study, then the *explanatory principle* must be sought in some other layer of reality. Vygotsky suggested that socially meaningful activity (Tätigkeit) may play this role and serve as a generator of consciousness.

Activity, going way back into our ancestors' prehistory, can be conceived as that which continues the species. Hunting, gathering, cooking and schooling are such activities *writ large*. In AT 'object orientated activity' is the *unit of analysis*, that which preserves the essence of concrete practice. The 'object' here is not the object-thing but the object-*raison d'etre*; indeed if two individuals perform similar actions but have different objects, then it can be said that they are involved in different activities. Although activity theorists all agree that object orientated activity is the unit of analysis, they argue amongst themselves about what constitutes this 'essence'. The unit of analysis is a means to understand the *Piaget vs Vygotsky* debate (see Monaghan, 2007). The cognitive activity (that Piaget was interested in) of a student engaged in a mathematical activity is, to an activity theorist, only a part of the unit of analysis which includes why the student is doing this mathematics, who s/he is doing it with and what tools s/he is doing it with—and the why/who/what cannot, to an activity theorist, be separated and analysed in themselves.

Such thinking was, though not through this example, present in the original work of Vygotsky and this was continued after his death by Leont'ev who considered individual and collective *actions* (usually with tools) and *operations* (things to be performed or modes of using tools) involved in socially organized *activity* (Leont'ev, 1978). Tool use here can be considered to include the primary, secondary and tertiary tools of Wartofsky (considered in Sect. 7.2.2); tool use is not, by this thinking, an activity in itself though tool use and activity are dialectically related (the activity shapes the tool use and the tool use shapes the activity). Leont'ev emphasised that all activity is motivated (though the motive may not be explicit) and transforming the object into an outcome is essential to the existence of an activity; this has immediate implications for considerations of the role of the mathematics teachers who may be mere facilitators to post-Piagetians but who are central, to activity theorists, in ensuring that students realise the object of learning. The upshot of learning is 'change', the student and the object are involved in a dialectical transformation: the object transforms the activity of the student and at the same time the object is transformed by the psychological reflective activity of the student. Parallel with the work of Leont'ev was activity theoretic work in

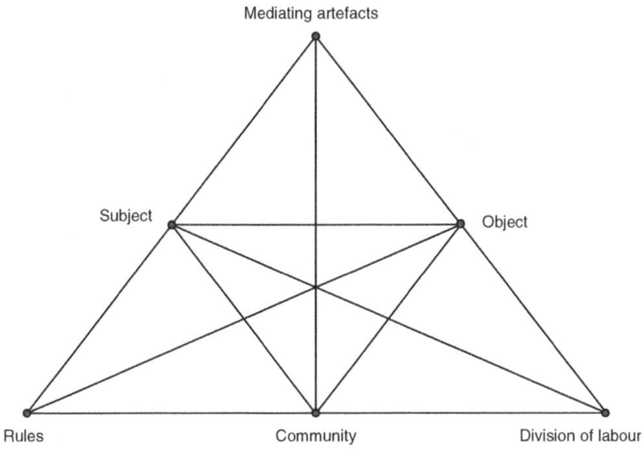

Fig. 9.1 Engeström's expanded mediational triangle

neurology but I do not consider this as it appears, to me, to have had little impact on mathematics education or tool use to date.

The ideas of Leont'ev were barely known outside the Soviet Union until the 1980s. Around the same time Scandinavian work in AT, and that of Yrjö Engeström in particular, began to attract the attention of education researchers. Engeström (1987) extends Vygotsky's focus on mediation through signs and tools to multiple forms of mediation and extends Leont'ev's frame to 'activity systems' to include the community and social rules underlying activity. These ideas are commonly schematised as in Fig. 9.1 below.

Figure 9.1 is designed to show multiple forms of mediation, for example: the top triangle (subject—mediating artefact—object) is the mediational triangle considered in Sect. 7.2; in the lower left triangle (subject—rules—community), social rules (norms and conventions) are mediational means; in the lower right triangle (division of labour—community—object) the division of labour mediates the object-oriented actions of the community. Figure 9.1 as a whole is used, in specific cases, to represent *activity systems* and the subsystems considered in this paragraph should be considered only in relation to the activity system. Activity systems research often examines interactive activity systems such as a hospital and an outpatient clinic with a focus on the objects of activity in the two systems. Engeström (2001) presents five principles for his form of AT: the activity system as a whole as the unit of analysis; *multi-voicedness*, 'multiple points of view, traditions and interests' (Engeström, 2001, p. 136); *historicity*, 'Activity systems take shape and get transformed over lengthy periods of time' (Engeström, 2001); *contradictions*, 'as sources of change and development' (Engeström, 2001, p. 137); and 'the possibility of expansive transformations ... A full cycle of expansive transformation may be understood as a collective journey through the *zone of proximal development* of the activity' (Engeström, 2001). It might be thought that tool mediation attracts less attention in activity systems research than in those of

Vygotsky and Leont'ev (and there is some truth in this) but activity systems research emphasises tool use in the context of the whole system; and it is appropriate to take a paragraph to emphasise 'tool use in context' in all the above forms of AT.

AT in Vygotsky, Leont'ev and Engeström's forms is often referred to as cultural–historical activity theory (CHAT). Cole (1996, p. 108) is an eloquent proponent of CHAT and states that the central thesis is that 'the structure and development of human psychological processes emerge through culturally mediated, historically developing, practical activity' and these three components are interrelated. I have addressed practical activity above but I feel a few words on culture and on history are appropriate. 'Culturally mediated' includes tool mediation. The book you are reading is focused on tools and mathematics but CHAT is focused on all tools in activity and language is 'an integral part of the process of cultural mediation' (Cole, 1996). We downplay language as a tool in this book to address our focus but we do not deny its place as the 'tool of tools' (Cole, 1996). There is a sense in which the interrelated set of tools (with no special status given to mathematical tools) used in collective activity is, to a CHAT researcher, the basis for the culture of that collective. With regard to history, we are each born into a culture based on a set of interrelated artefacts/tools, and our immersion in this culture continues in our (mathematical) development/education. We attend school where we have a teacher who was born into a prior form of our culture (who had a teacher . . . who had a teacher . . .). Our teacher looks to our future (including what tools she/he/society feels we need to master) but this vision of our future needs is grounded on valuations of what should be preserved from the past. So, tool use is of fundamental importance to CHAT researchers but this is tool use in the context of cultural–historical activity.

Apart from scholars, such as Michael Cole, who could read Russian texts, AT came to be known by Western scholars after the appearance of Vygotsky (1978), Leont'ev (1978) and Wertsch (1981).[3] The third book has not been mentioned until now. It is a primer on AT edited by James Wertsch which contains a preface by Cole, an introduction by Wertsch and translations of key Soviet AT texts grouped under the headings: theoretical foundations; Vygotsky's influence; the role of sign systems; and empirical studies. AT (in its various forms) is used as a framework in many fields of study. Three fields of study relevant to this book are human–computer interaction (HCI; see Nardi, 1996), ergonomics (see Daniellou & Rabardel, 2005) and education (see Daniels, 2002). Wertsch (1991, 1998) has attracted the attention of mathematics educators interested in tool use because it focuses, amongst other things, on the person-tool dialectic or, as he puts it, 'the *irreducible bond* between agent and mediational means' (1997, p. 27); the bond in, say, a person using a calculator, is irreducible because the act of calculating with a

[3] I will focus on English language texts due to (1) the dominance of the English language in Western academic writing, (2) English is my first language and (3) to keep this chapter to a reasonable length.

calculator cannot be reduced to what the human alone can do or to what the calculator can do, the calculation is done by a human-with-calculator. Wertsch emphasises that "the relationship between action and mediational means is so fundamental that it is more appropriate, when referring to the agent involved, to speak of 'individual(s)-acting-with-mediational-means' than to speak simply of 'individual(s)'" (1991, p. 12). I now consider the genesis of activity theoretic influences in mathematics education (mainly in English language writing).

9.3 The Genesis of the Influence of Activity Theory in Mathematics Education Research

In this subsection I trace, to the best of my ability, the early influence of activity theory in Western[4] mathematics education research. I do this via two subsections. In the first I consider two books from around 1990. I then consider the influence of AT in academic journal papers.

9.3.1 Two Activity Theoretic Mathematics Education Books

I believe that the first English language text by a Western mathematics educator was a book on the politics of mathematics by the Norwegian Stieg Mellin-Olsen (1987). This book focuses on the alienation of learners of mathematics and he employs the approaches and constructs of Vygotsky and Leont'ev. Mellin-Olsen considers tools in a broad sense, 'both thinking-tools and communicative tools ... Their functionality is dependent on whether they are experienced in the process of Activity or not' (Mellin-Olsen, 1987, p. 48) and that language is the basic human thinking tool. Three years after Mellin-Olsen (1987) the National Council of Teachers of Mathematics published a translation of a 1972 book by the Soviet educator Davydov. The aspect of Davydov's work that attracted most attention was his consideration of abstraction and generalisation, as Jeremy Kilpatrick, in the Introduction to Davydov (1990, pp. xv–xvi), wrote:

> Much work on the learning of concepts and principles has assumed that such learning occurs "from the ground up." Students need to see many examples so that they can use induction to form a generalization. The generalization reduces the diversity in the specific examples. Davydov argues that we ought to conceive of learning differently. The specific examples should be seen as carrying the generalizations within them; the generalization process ought to be one of enrichment rather than impoverishment.

---

[4] This caveat is important as the influence of AT in mathematics education research in (what was known as) 'communist bloc' countries was long standing at the time AT started to influence Western mathematics education research.

Davydov's views on abstraction, his *ascent to the concrete*, which refers to the development of an idea via a dialectical *to and fro* between the concrete and the abstract, was to become the basis for a well-respected framework of 'abstraction in context' which stemmed from Hershkowitz et al. (2001). Fascinating as Davydov's work in this area is and despite his references to tools and social interactions, he says virtually nothing of the place of tools in the formation of abstractions and generalisations.

9.3.2 The Genesis of the Influence of AT in Academic Journal Papers

The remainder of this section considers the early influence of AT in Western mathematics education research journals.[5] In planning this subsection I encountered two problems which I relate for the sake of intellectual transparency. First, how do I overcome the bias of simply considering papers with which I am familiar? My solution to this problem was to adopt a systematic means of considering paper. The second problem is, how do I do this in a reasonably short word length? My solution was to choose one primary source, the highly respected international journal *Educational Studies in Mathematics* (ESM). I searched the ESM web site using the keywords 'activity', 'Vygotsky', 'Leont'ev' and 'Engeström'. I choose the three names to ensure that I considered all the dominant approaches to AT. I stopped my search when I had papers that I considered represented all current approaches to AT employed by scholars in mathematics education research (at the time of writing). The remainder of this subsection provides a summary (with specific regard to tool use) of six papers from the period 1996 to 2003.

Two AT papers appeared in ESM in 1996, Bartolini Bussi (1996) and Crawford (1996). Although Crawford (1996) does report research it is largely an exposition of Vygotskian AT. It asks question such as 'What difference does the use of tools such as computers and calculators make to the quality of human activity?' and states that these, to Vygotsky, 'are cultural artefacts' (Crawford, 1996, p. 57) but the paper does not explore the nature of tools further. Three aspects of this paper in a leading academic journal suggest that AT was not, in 1996, widely known: the fact that the paper has an expository style (such styles are often used when a subject matter is new); there is no reference to Leont'ev or Engeström; the paper references only three works (all books not journal papers) from the field of mathematics education and these three books are only loosely associated with AT in having a social/ practice orientation (Lave, 1988; Papert, 1994; Walkerdine, 1988).

Bartolini Bussi (1996) also has an expository style (with regard to the AT of Vygotsky and of Leont'ev) but its main focus is a report on a 3-year primary

---

[5] I focus on academic journals as I regard them as a dominant media through which ideas are circulated in academia.

mathematics teaching experiment on geometric perspective which was part of a wider project on mathematical discussion. The paper analyses the teaching experiment 'by means of the theoretical construct *semiotic mediation* (Vygotsky, 1978) in an attempt to substantiate its crucial effect on pupils' learning and metalearning' (Vygotsky, 1978, p. 13) and 'The *theory of activity, actions and operations* developed by Leont'ev (1978) is supposed to offer a suitable tool to either differentiate or coordinate the analysis of long term and short term processes' (Leont'ev, 1978, p. 15). The design of the teaching experiment includes tasks, mathematical discussions and 'appropriation of existing cultural artefacts (e.g. devices, texts and so forth' (Leont'ev, 1978, p. 22). The word 'tool' has two uses in the paper: Leont'ev's theory as a tool for analysis (see the quote two above); 'semiotic tools'. The term 'semiotic tool' is actually not defined in the paper but examples of semiotic tools are provided. One such example is a 'two column scheme', 'In the left column there was *reality*, in the right column *representation*' (Leont'ev, 1978, p. 26), which was 'built collectively in a discussion orchestrated by the teacher' (Leont'ev, 1978, p. 33). The two column scheme was created to highlight invariant and non-invariant properties of 3D objects in 2D representations; one column was for 'reality', the other for 'representation'. The two column scheme served to focus students' attention not only on what has changed but on what has not changed (the cultural–mathematical idea of invariance). Once the scheme had been created, it 'acted as a *semiotic tool* in perspective drawing for either producing or reading an image' (Leont'ev, 1978, p. 33).

Bartolini Bussi (1996) and Crawford (1996) show that 'AT had arrived' in mathematics education research in 1996 and Bartolini Bussi (1996) reveals a very specific appropriation of the word 'tool'. In 1998 there were two ESM papers that considered tool use in very different ways to Bartolini Bussi (1996), Chassapis (1998) and Pozzi, Noss, and Hoyles (1998), which I now consider.

Chassapis (1998) focuses on the processes by which children develop a formal mathematical concept of the circle by using various instruments to draw circles. It considers drawing circles: by hand; using circle tracers and templates; and using the compass. The primary theoretical influences are Vygotsky (the similarities and differences between signs and tools in activity; the difference between spontaneous/everyday and cultured/scientific concepts) and Soviet and Western interpreters of Vygotsky, e.g.: Zinchenko, 'tool-mediated action must be considered as the primary unit of analysis for a Vygotskian account of human mental functioning' (Chassapis, 1998, p. 276); and Wertsch, that tools "have been developed in a culture over extended periods of time and have become an integral part of human activity, being 'the 'carriers' of socio-cultural patterns and knowledge'" (Chassapis, 1998, pp. 275–276). Chassapis (1998) stresses that:

> The process of learning to use a tool, for example, an abacus, involves the construction of an experiential reality that is consensual with that of others who know how to use an abacus. As a consequence, when we use an abacus individually or while interacting with others, we participate in a continual regeneration of a consensual reality which both constrains and enables our individual ways of thinking and calculating. (Chassapis, 1998, p. 276)

Chassapis (1998) concludes that children's everyday concepts of a circle are global and static curvature concepts, not that of a set of points equidistance from a fixed point. These everyday concepts are in the realm of perceptual thinking and the use of freehand circle drawing and of circle tracers and templates does not radically change these everyday concepts. The use of the compass, however, 'structures the circle-drawing operation ... may give rise to concepts constructed in the realm of action-bound practical thinking ... constituting a potential ground for the development of analytical, more formal mathematical concepts of the circle' (Chassapis, 1998, p. 292).

Pozzi, Noss, and Hoyles (1998) results from research on nursing. As the chapter you are reading is on activity theoretic approaches, it is appropriate to mention that this research is one of several studies by this team where the object of the research activity is to understand mathematical practices in workplaces. The goal of this paper is to address the question 'how do resources enter into professional situations, and how do they mediate the relationship between mathematical tools and professional know–how?' (Pozzi et al., 1998, p. 110).The paper focuses on nurses administering drugs and monitoring fluid balance. The opening paragraph of the paper includes an unambiguous homage to the value of activity theory in such work:

> the entire corpus of work on activity theory, offers compelling evidence that individual and social acts of problem solving are contingent upon structuring resources, including a range of artefacts such as notational systems, physical and computational tools, and work protocols (Gagliardi, 1990). These artefacts are 'crystallised operations' (Leont'ev, 1978), borne out of needs within a given set of social practices, and in turn playing their part in shaping and restructuring future practices: artefacts exhibit an ongoing dialectic of producing and being produced by activity. (Leont'ev, 1978, p. 105)

The paper's conclusion is also framed in activity theoretic terms. Mathematics is bound into nurse's action, especially when there are concerns, for example that the wrong dose of a drug may have been given, but nursing activity is not arithmetic activity. Nursing activity includes mathematical artefacts/tools, such as rules for drug dosages and fluid balance charts, but the use of these tools is but a part of the activity of nursing.

Two years after the papers by Chassapis and Pozzi et al. ESM published an AT paper, Radford (2000), that signalled a new emphasis in mathematics education research, semiotic-cultural analysis. Radford's focus is on the early algebraic thinking (generalisation) which "is considered as a sign-mediated cognitive *praxis*" (Radford, 2000, p. 237) where the term 'sign' includes symbols, words, gestures, indeed anything that signifies. He grounds this conception in the work of Vygotsky, Leont'ev[6] and also Bakhtin (but a consideration of the later would be inappropriate in my brief exposition). I select a long extract which gives a flavour of the radical action/activity regard that Radford has towards signs in mathematics education:

[6] Leont'evs actually, father and son.

> instead of seeing signs as the reflecting mirrors of internal cognitive processes, we consider them as tools or prostheses of the mind to accomplish actions as required by the contextual activities in which the individuals engage. As a result, there is a theoretical shift from what signs *represent* to what they *enable* us to do ... the signs *with* which the individual acts and *in* which the individual thinks belong to cultural symbolic systems which transcend the individual *qua* individual. Signs hence have a double life. On the one hand, they function as tools allowing the individuals to engage in cognitive praxis. On the other hand, they are part of those systems transcending the individual and through which a social reality is objectified. The sign-tools with which the individual thinks appear then as framed by social meanings and rules of use and provide the individual with social means of semiotic objectification ... the conceptual and the signifying aspects of signs need to be studied in the activity that the signs mediate in accordance to specific semiotic configurations resulting from, and interwoven with, social meaning-making practices and cultural forms of signification (Radford, 2000, p. 241).

Radford focuses on small groups of Grade 8 students engaged in tasks in which they are to use circular counters ('chips') to generalise from visually presented sequences representing linear algebraic expressions (e.g. $2n - 1$). The role of the teacher is not only for the students to get the answer but to see for themselves the kind of answer they are to get. The analysis of the activity includes discourse analysis as students struggle to express the general through the particular, 'you always add 1 to the bottom, right?'. I cannot summarise the paper in this paragraph but I can point to Radford's focus on the intersection of semiotic means which allow the students to appropriate cultural forms (the use of letters):

Student: How many chips to have vertically ... you would subtract 1 from how many chips
Teacher: But now you have to say it without using words! Use letters! OK?
Student: You have to do 1 n minus ...

There is a sense in which Radford both continues and breaks with the traditions of cognitive studies in mathematics education, AT and semiotics: cognition is reconceived as social and cultural sign-mediated cognitive *praxis*; Vygotsky's distinction between signs and tools is blurred; the classic semiotic approach where sign, object and signified are regarded in isolation is replaced by an approach which focuses on joint acts of symbolising in context.

I close this section on the genesis of the influence of activity theory in mathematics education research by bringing the work of Engeström (and co-workers) into the picture. The first mention of Engeström in ESM appears in Jaworski (2003). This paper outlines a framework for 'both *insider* and *outsider* research and *co-learning* between teachers and educators in promoting classroom inquiry' (Jaworski, 2003, p. 249). It is not essentially concerned with tool use in doing mathematics though it does consider classroom inquiry as a 'developmental tool' (which may, in AT terms, be taken as 'a mediational means to assist the development of teachers and researchers in pursuit of their educational objectives'). The paper has a brief afterword:

> It seems important to mention the suggestion of one reviewer that discussion of knowledge and learning relating to social and societal significance might be recast in terms of an

> *activity theory* perspective. Subsequent work on these ideas led to my development of a mapping between the framework here and Engeström's "mediational triangle". (Jaworski, 2003, p. 276)

Barbara Jaworski ends the paper with 'I plan to work further on these ideas' which, as we shall see in the next section, she did. There are a number of issues worthy of discussion arising from the papers considered above but I leave this until the final section of this chapter and I now move on to consider activity theoretic approaches in mathematics education in the early part of the twenty-first century with, of course, particular regard to tool use.

9.4 Activity Theoretic Approaches in Mathematics Education in the Twenty-First Century

Since the Jaworski paper, 2003, AT has exerted a strong influence on research in mathematics education and it would be rather foolish of me to attempt a summary of this research. Further to this, in selecting research reports to review I wished, as I stated in the preamble to Sect. 9.2.2, to avoid bias by simply considering work with which I am familiar, so I once again looked for a source. ESM would be a suitable source but 2012–2013 saw the publication of a two-volume special edition of *The International Journal for Technology in Mathematics Education* devoted to *Activity Theoretical Approaches to Mathematics Classroom Practices with the Use of Technology* and this seemed a closed but appropriate set of papers in which to examine current AT research practices as it is likely, with its focus on technology, to raise issues related to tool use.

I first describe the corpus of papers in this Special Issue. Of the 11 papers 5 are more or less 'straight AT': Abboud-Blanchard and Cazes (2012), Chiappini (2012), Jaworski, Robinson, Matthews, and Croft (2012), Ladel and Kortenkamp (2013), and Maracci and Mariotti (2013). Another two, Robert (2012) and Abboud-Blanchard and Vandebrouck (2012), are AT with a specific French interpretation. A further three jointly consider several approaches (called 'networking theories'): Fuglestad (2013) networks AT and the instrumental approach; Kynigos and Psycharis (2013) networks constructionism and the instrumental approach; Lagrange (2013) networks AT and the anthropological theory of didactics. Monaghan (2013) considers a socio-cultural theory, Valsiner's 'zone theory', that shares Vygotskian roots with AT. The diversity of papers illustrate that AT is open to national variation and networking with related theories. I now summarise the purportedly 'straight AT' papers with specific regard to artefacts/tools.

Chiappini (2012) focuses on the teaching and learning of mid-level algebra (equations, functions, inequalities and equivalence associated with expressions such as $x^2 - 2x - 4$) with software, called *Alnuset*, with a visual 'algebraic line' and conventional algebraic notation, to draw students' attention to the culture of mathematics (see Fig. 9.1). Chiappini is a software designer as well as a

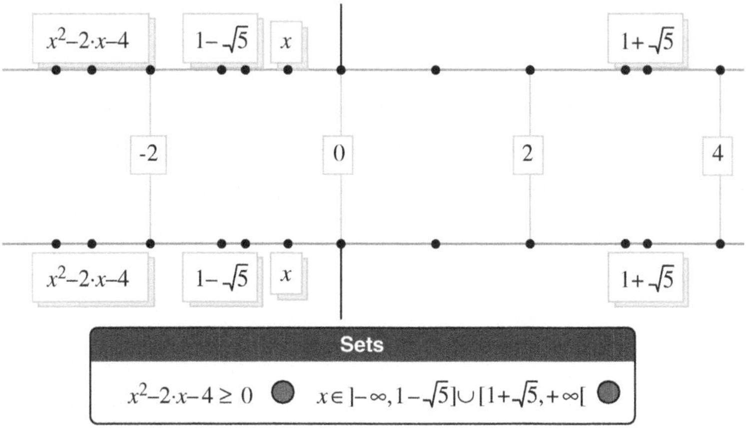

Fig. 9.2 A screen shot from A*lnuset*

mathematics educator and draws on work from Fig. 9.2, *Alnuset*'s algebraic line the HCI strand of AT which employs the Gibsons' construct of *affordances* (considered in Sect. 7.2) with regard to 'usability research both as an heuristic and an ad hoc design principle to describe the potential of a (computer) system with regard to its effectiveness' (Chiappini, 2012, p. 135). There are two interpretations of affordances (of a system for a user) in the HCI community which hinge on whether the affordances are perceived or not; the significance of this difference for HCI work lies in the potential for user actions. This difference leads to a distinction between the usability of a system (how a task can be completed) and its usefulness (how a system responds to user actions). Chiappini regards this distinction and, in particular, the construct 'usefulness' as

> important in educational contexts where students may not focus clearly on the objectives of the task at hand and teachers' goals do not necessarily coincide with those of their students in a didactical activity mediated by a digital artefact. In particular, the notion of usefulness makes it possible to evaluate the affordance provided by the system software: to promote in students the emergence of the objectives for the solution of the task they are engaged in; to support the development of the teacher's cultural goals (development of knowledge, meaning, principles and values of the discipline) that may also transcend those of the task in which students are involved. (Chiappini, 2012, p. 135)

Chiappini seeks to employ this distinction to evaluate student–teacher use of *Alnuset* and, to this end employs an HCI breakdown of affordances: perceived affordances; ergonomic affordances, which allow 'embodied actions involved in solutions of tasks and sub-tasks peculiar to the context' (Chiappini, 2012); and cultural affordances, which concern

> the cultural teaching/learning objectives underlying the system being used. Evaluation of cultural affordances can be carried out through the analysis of how meanings, values and principles underlying the action mediated by the use of the embodied actions, get to be known through the artefact-mediated activity. (Chiappini, 2012).

Chiappini's framework can be viewed as an elaboration of Leont'ev's activity–action–operation triple specifically designed for the evaluation of mathematics education with software systems. The final section of Chiappini (2012) develops a framework to evaluate the cultural affordances of *Alnuset* which uses Engeström's expansive learning cycle in four phases. In terms of tools these are as follows.

1. The students are given a task (an open algebraic problem) and the artefacts/tools embedded into the software provide output to student input, some of this output surprises the student and produces cognitive conflict.
2. Tasks are then:

   > designed in order to exploit the visuo-spatial and deictic ergonomic affordance of the algebraic line to allow students to explore the conditions, causes and explicative mechanisms of conflicts ... the teacher's crucial task consists in the introduction of terms and algebraic notions found in the visuo-spatial and deictic narration of the various problematic situations (Chiappini, 2012, p. 139)

3. The teacher encourages the student to recast their work using *Alnuset*'s algebraic manipulation facilities in order for them to mathematicise surprises encountered in earlier work:

   > In this phase the teacher encourages both the establishment of the algebraic axiomatic model in the student's practice and the development of meta-cognitive processes involved in the re-configuration in symbolic terms of the algebraic meanings expressed beforehand in visuo-spatial and deictic terms. (Chiappini, 2012)

4. Students, with a transformed understanding of algebraic activity, and teachers, with a transformed understanding of their students' understandings, engage in teacher-led whole class consolidation of their understandings.

There are strong parallels in this paper to Radford's (2000) paper considered above. Both emphasise the cultural–historical (one might say 'unnatural') objectification of mathematical knowledge. But there are differences too: Chiappini employs Engeström's expansive learning cycle; Radford places greater emphasis on semiotics and does not employ digital technology.

Jaworski et al. (2012) focuses on an undergraduate mathematics module for engineering students that employs inquiry-based tasks and a computer system *GeoGebra*. The teachers had put a lot of effort into designing a module to enhance student engagement in mathematics and the object of their research was to evaluate this design from a learner and a teacher perspective; the paper focuses on the aspect of this evaluation related to the use of *GeoGebra*. An initial evaluation using student surveys revealed some positive comments in terms of better understanding but also 'Just because I understand maths better doesn't mean I'll do better in the exam'. The AT analysis conducted put the *GeoGebra* aspect of the work in perspective, 'It is the whole with which we work and in which we participate'. They analyse the whole from the perspective of both the students and the teachers using separately both Leont'ev's activity–action–operation triple and Engeström's expanded mediational triangle. Both analysis reveal differences between students and teachers:

> Perhaps the most important difference is the *object* of activity (Engeström) or *the motivating force* (Leont'ev) for the two systems. Both are valid, but the fact that they are different means that along with other factors—values placed on forms of understanding (the *rules* of the enterprise) or whether GeoGebra is positively helpful in promoting learning (mediating artefacts)—they result in the tensions observed. (Jaworski et al., 2012, p. 151)

This paper says virtually nothing with regard to tool use. There is no explicit mention of tools in the paper and two instances where the word 'artefact' is used (one in the quote above and one in relation to Engeström's expanded mediational triangle). At one level this is surprising in a paper considering the use of *GeoGebra* in a mathematics module but the paper does take a holistic view of the module (we shall return to this via a consideration of the unit of analysis later in this chapter).

Ladel and Kortenkamp (2013) focuses on the design and use of a multi-touch-table (a large touch-screen artefact that registers input from fingers, not just a finger) to engage young children (5–7 years of age) in meaningful work with whole number operations. The paper notes a feature of the child-technology environment which has similarities to the Gibsons' construct of *affordances*, they note that 'such technology ... enables children to work with virtual manipulatives directly instead of being mediated through another input device' (Ladel & Kortenkamp, 2013, p. 3); and they also note that 'We want to restrict the students' externalizing actions to support the internalization of specific properties of the objects[7] in consideration ... Thus the mediation through the artefact is characterized by restriction and focussing.'

Ladel and Kortenkamp (2013) adapt Engeström's expanded mediational triangle in what they call an 'artefact-centric activity theory' model (ACAT, see Fig. 9.3). They note that:

> the artefact itself does not have agency and is only mediating ...[but] the artefact changes the way children act drastically and in non-obvious ways ...we use Activity Theory not only for analyzing the interaction between subject and object, but in addition for designing the artefact. We adapted the activity system diagram of Engeström ...We believe that *Rules* ... should also affect the design of the artefact, thus we need a new relation between these two nodes. For clarity we omit the division of labor from the diagram. Because our focus lies on the artefact, we are not considering the relations between the rules and subject, object and community in this article, though they are important for a full activity system ... (Ladel & Kortenkamp, 2013, p. 3)

Ladel and Kortenkamp (2013) view students' arithmetic work in the light of Leont'ev's activity–action–operation triple and conclude that 'Through the lens of ACAT that places the artefact in the center of attention we can locate the various areas of didactic and pedagogic design that have to be taken into account' (Ladel & Kortenkamp, 2013, p. 7). In contrast to Ladel and Kortenkamp's account of artefact mediation Maracci & Mariotti (2013) present a very human-centred view of mediation.

[7] Ladel and Kortenkamp (2013) consistently use the word 'object' to mean a 'thing'. This occurs elsewhere in papers in this Special Issue. We consider this interpretation later in the chapter.

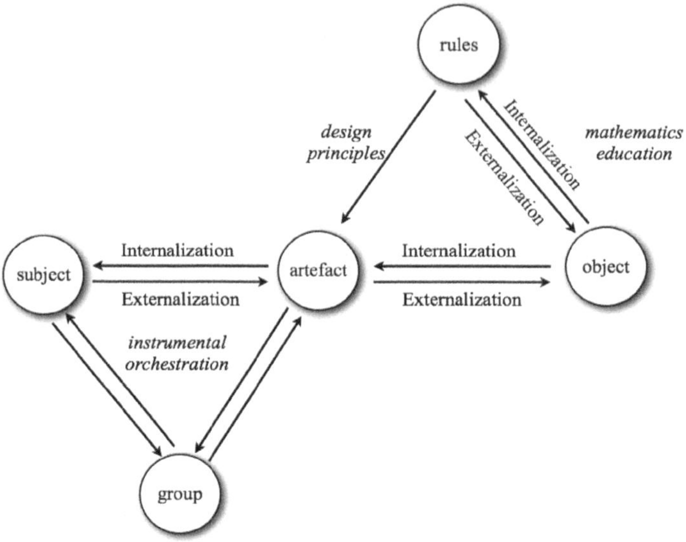

Fig. 9.3 A representation of the ACAT theory

Maracci and Mariotti (2013) outline the Theory of Semiotic Mediation (TSM) with regard to 'the use of artefacts to enhance mathematics teaching and learning, with a particular focus on technological artefacts' (Maracci & Mariotti, 2013, p. 21). This paper continues a long line of papers on semiotic mediation originating from Bartolini Bussi (1996) considered in the previous section. TSM draws on the AT of Vygotsky and of Leont'ev but they are critical of research where 'the mediating function of the artefact is often limited to the study of its role in relation to the accomplishment of tasks' (Bartolini Bussi, 1996). TSM is essentially semiotic in that:

> teaching-learning ... originates from an intricate interplay of signs... individuals have to be involved in semiotic processes leading to the explicit formulation of the meaning they have developed in relation to an activity, in order to become conscious of such meanings ... mathematical meanings can be crystallized, embedded in artefacts and signs ... (Bartolini Bussi, 1996)

TSM also draws on the work of the socio-linguist Hasan who distinguishes between the mediator, the thing (which may be a concept) that is mediated, the mediatee and the *circumstances* for mediation. Following Hasan, Maracci and Mariotti (2013), claim that:

> The mediator is not the artefact itself but it is the person who takes the initiative and the responsibility for the use of the artefact to mediate a specific content... artefacts are among the constitutive elements of the "circumstances for mediation". In fact, the modalities of use of the artefacts, the tasks to be accomplished, the whole organization of the classroom work, the classroom interactions among students and between them and the teachers are constituents of the "circumstances for mediation". (Bartolini Bussi, 1996, p. 22)

Leont'ev's activity–action–operation triple provides a frame for studying *circumstances* for realising the desired semiotic mediation. To mediate the learning of mathematics the teacher has to design specific circumstances, a didactical cycle, aimed at fostering specific semiotic mediation processes: accomplishing a task with the artefact; producing signs related to the artefact use; and classroom discussion. A central aim of the didactical cycle is the '*unfolding of the semiotic potential* of the artefact' which I interpret as having similarities to what Radford (above) calls 'objectification': students sitting in a mathematics classroom at the beginning of a (sequence of) lesson(s) are there to learn mathematics and do not know what they are to learn; the central aim of the teacher is that the students appropriate cultural (scientific) meanings. It is crucial that teachers design tasks which 'lead students to develop personal meanings related to the artefact use having the potential to evolve towards mathematical meanings' (Bartolini Bussi, 1996, p. 23). All three parts of the didactical cycle are essential for personal meanings to become shared meaning and for the teacher to shape these shared meanings into public scientific meanings. Artefacts are an essential part of this cycle but they are not mediating agents in the TSM.

Abboud-Blanchard and Cazes (2012) interprets research on Electronic-Exercise-Bases (EEB), digitised mathematical exercises. The research was carried out over 3 years with 30 teachers with a focus on three phases of teachers' use of EEBs, 'the preparation of the lesson, its progress and the reflexive return that the teacher makes on this lesson' (Abboud-Blanchard & Cazes, 2012, p. 142). The research questions that the paper addresses are, 'Why and how do teachers use EEB? What effect does this use have on their teaching activity?' (Abboud-Blanchard & Cazes, 2012, p. 141). The paper uses Engeström's expanded mediational triangle (unamended) to interpret the data (teacher interviews and classroom observations). Like Ladel and Kortenkamp (2013), the authors sometimes appear to use the word 'object' to mean a 'thing', e.g. AT 'studies a subject acting on an object to produce a result' (Abboud-Blanchard & Cazes, 2012, p. 142). The paper does not use the word 'artefact' but does use the words 'tool' and 'instrument', for example in explaining the terms of Engeström's expanded triangle they write 'The tool allows the subject to exercise her/his activity. It is a set of tools or of instruments. The essential instrument in this study is the EEB' (Abboud-Blanchard & Cazes, 2012).

Abboud-Blanchard and Cazes are French researchers but, apart from the use of the word 'instrument', there is nothing particularly French about Abboud-Blanchard and Cazes (2012). In contrast, the final two papers I consider (Abboud-Blanchard & Vandebrouck, 2012; Robert, 2012) do present 'a French take' on AT. I would like to add here that I do not regard one's nationality as determining one's theoretical framework: Chiappini, Mariotti and Maracci are all Italian but their papers present differing foci within AT. But there is a specific line of inquiry within French mathematics education research that takes its cue from Leplat (1997). Leplat focused on the psychology of the workplace, viewing the characteristics of the tasks and the characteristics of the workers in two dialectical feedback loops with 'activity' in the middle, the 'double approach': a production loop in which activity is object-oriented to the task(s) at hand; a construction loop in which activity is subject-oriented to the development or well-being of the workers. Rogalski (2013, p.7) summarises this thus:

> The situation is a determining factor of the activity, and is simultaneously itself modified by the activity. This modification primarily affects the object of the activity, but can also include modification of resources and constraints. Subjects, too, both determine the activity and are modified in turn by their own activity.

Some (not all) double approach researchers also 'network' this approach with Rabardel's *instrumentation* theory and Chevallard's *anthropological theory of didactics* (considered in Chap. 10). I now consider the papers of Robert (2012) and Abboud-Blanchard and Vandebrouck (2012).

Robert (2012) outlines the 'double approach' with regard to 10 years of research on students' and teachers' activities in and out of mathematics classrooms. She stresses that this work addresses AT 'from a cognitive individual perspective, not as a whole system ... [not] the socio-cultural contexts of students and teachers' (Robert, 2012, p. 151). The main foci of this work has been on students solving exercises and teachers' monitoring of student work (this is consistent with Leplat's characteristics of the tasks and of the workers). The focus on student work continues a French cognitive strand and Robert references Douady and Vergnaud and stresses 'knowledge': old knowledge, new knowledge, knowledge to be used, states of knowledge, reorganisation of knowledge, recognition of knowledge, lack of knowledge, knowledge to be adapted, ... Further to this:

> student learning is tied to the quality of the so-called "scenario," but it is also tied to the precise way the students work on the corresponding tasks. So, the better we describe the offered (proposed) tasks, the better we succeed in understanding students' actual activities (Robert, 2012, p. 155)

The focus on the teacher in this body of research is at a local and a global level. At the local level this involves studying the ways that teachers interact with students and their mathematical work and (with an implicit references to Leplat) distinguishes between procedural help (directed at the task completion) and constructive help (with a focus on the students' interpretations of the task). The global level considers the management of student activities with respect to the craft knowledge of the teacher-in-context. The local and global level are interrelated. Robert also stresses (again with an implicit reference to Leplat) the interrelated productive (students completing tasks) and affective dimensions of teaching. Robert (2012) is a general introduction to the double approach and does not focus on the artefact/tools used in mathematics classroom.

Abboud-Blanchard and Vandebrouck (2012) is written as a follow up to Robert (2012) with a focus on teachers' practices in technology-based lessons with particular regard to Leplat's production and construction loops. The genesis of teachers' practices with technology are considered to have 'external aspects which correspond to the evolutions of the teachers' productive activity throughout technology-based lessons, but also have internal aspects related to the constructive activity which accompanies these evolutions' (Abboud-Blanchard & Vandebrouck, 2012, p. 159). It is assumed that teachers' practices are stable and the evolution of these practices involves three levels: the micro level, which has similarities to Leont'ev's 'operations'; the local level, which refers teachers' goals and actions; and the global level, which refers teachers' motives. Abboud-Blanchard and Vandebrouck (2012) explanation of teachers' technological geneses puts forward a two-stage process in which these levels interact.

In the first stage the local level is regulated by the micro level. When a teacher first uses a new tool in the classroom the 'the automatic regulation of teaching practice at the micro level allows the teacher to cope with difficulties emerging during the technology session at the local level' (Abboud-Blanchard & Vandebrouck, 2012, p. 160) but this is usually short-lived and 'some teachers feel the need to build new specific practices with technology, while others will tend to reduce the role of technology within their teaching' (Abboud-Blanchard & Vandebrouck, 2012). This stage concerns Leplat's 'production loop' but 'it generates constructive activity at the medium and long-term' (Abboud-Blanchard & Vandebrouck, 2012, p. 161) and this leads to the second stage.

The second stage has two parts. The first part concerns the movement from the local to the global level and includes an evolution of the production loop and the development of the construction loop, 'There is a new balance between traditional sessions and technology sessions, between collective work and individual phases of students' activity or between old and new mathematical knowledge in students' activity' (Abboud-Blanchard & Vandebrouck, 2012). The second part concerns the movement from the local level to the micro level (the refinement of the teachers' understanding of the artefacts/tools they are using in their classrooms). This part develops over time as a teacher goes from 'tinkering' with an artefact, to using it as a tool for personal mathematics, to 'tinkering' with an artefact in the classroom, to assisting the technomathematical development of their students' use of a new tool for doing mathematics.

I now consider emphases and tensions in these papers together with the approaches considered in the previous section.

9.5 Emphases and Tensions in Mathematics Education Activity Theoretic Approaches

There are similarities in the approaches in the mathematics education papers considered in Sects. 9.2 and 9.3. Every paper: pays homage to Vygotsky by mentioning his works directly or indirectly (via Leont'ev or Engeström); places a positive valuation on considering 'practice' (though what 'practice' involves varies); attempts to describe (rather than prescribe) a practice (bar, possibly, Chiappini, 2012). But there are also differences which I shall consider briefly under the following interrelated categories: sign and tool; unit of analysis; cognition; the cultural–historical dimension; mediation.

I have mentioned (in Sects. 7.2 and 9.1) Vygotsky's observation on the similarities and differences between signs and tools. Vygotsky's view of these similarities and differences is present in the early papers (Chassapis, 1998; Crawford, 1996; Pozzi et al., 1998), is implicit in Bartolini Bussi (1996) and, as mentioned above, Vygotsky's view is extended and, to some extent blurred, in Radford (2000). But when we consider the more recent IJTME papers there are differences and omissions. Neither Jaworski et al. (2012) or Ladel and Kortenkamp (2013) consider signs or tools in the body of the text, though the latter does place emphasis on artefacts. Maracci and Mariotti (2013) make much of signs and artefacts but only mentions tools once in a quote. In the three papers by French authors 'sign' is only mentioned (twice) in Robert (2012). These differences, I feel, go beyond terminology used and reflect differences in the basic fabric of scholastic mathematical activity.

I shall consider differences with regard to the unit of analysis and cognition together as it seems important, to me, for mathematics education research, whether cognition is an explicit part of the unit of analysis. Despite the importance of the unit of analysis for AT research, not all papers explicitly consider the unit of analysis. Considering the early ESM papers it is explicitly mentioned by Chassapis (1998, p. 276), 'tool-mediated action must be considered as the appropriate primary unit of analysis' and Radford (2000, p. 244) who used it to guide his data analysis, '*situated discourse analysis* whose elementary unit (i.e. the unit of analysis) was constituted by the refined (i.e. contextualised and cadenced) identified salient segments'. Cognition, mathematical thinking with signs/tools, is central in both of these papers. Neither Bartolini Bussi (1996), Crawford (1996) or Pozzi et al. (1998) explicitly mention the 'unit of analysis' but (1) Bartolini Bussi (1996) clearly considers the long-term teaching and learning process as the unit of analysis, and (2) in Crawford (1996) and in Pozzi et al. (1998) cognition is viewed in a wider context where there is bi-directional 'shaping': 'mathematical knowledge increasingly shapes and is shaped by human activity' (Crawford, 1996, p. 46);

> In the past, the issue tended to be seen in purely cognitive terms . . . Now investigations tend to focus on how activities are shaped by the social practices and goals of the working culture, and to examine how this shaping informs our understanding of mathematical behaviour and learning. Pozzi et al. (1998, p. 105).

There is no mention of 'unit of analysis' in any of the IJTME papers but my reading of the seven papers puts them into four camps with regard to what this unit might be and the place of cognition in this unit. The first is a 'systems approach' (Engeström's model with reference to Leont'ev's triple) where the implicit unit of analysis is the activity system and, in which, cognition is an implicit part of this system; I put the papers by Abboud-Blanchard and Cazes (2012) and Jaworski et al. (2012) in this camp. The second camp is reflected in the papers by Robert (2012) and Abboud-Blanchard and Vandebrouck (2012) which consider 'Activity Theory from a cognitive individual perspective, not as a whole system. It does not address a more general point of view, involving the socio-cultural context of students and teachers' (Robert, 2012, p. 153). The third is the papers by Chiappini (2012) and Ladel and Kortenkamp (2013). Both of these papers use the Engeström model (adapted in the case of the second paper) where the model is the implicit unit of analysis but, in which, individual cognition (with artefacts) is an intrinsic component. The fourth camp is a singleton, the paper by Maracci and Mariotti

(2013) where the implicit unit of analysis is the teacher-mediated didactical cycle and cognition is an intrinsic component:

> A didactical cycle, or an iteration of didactical cycles, can be seen as an activity whose motive is to promote the generation of students' personal signs related to the accomplishment of a task through an artefact and their evolution towards desired mathematical signs. (Maracci & Mariotti, 2013, p. 23)

With regard to the cultural–historical dimension, the papers considered, in my opinion, fall into two camps: those that embrace this dimension (Bartolini Bussi, 1996; Chassapis, 1998; Chiappini, 2012; Crawford, 1996; Ladel & Kortenkamp, 2013; Maracci & Mariotti, 2013; Pozzi et al., 1998; Radford, 2000); and those that appear to ignore this dimension (Abboud-Blanchard & Cazes, 2012; Abboud-Blanchard & Vandebrouck, 2012; Jaworski et al., 2012; Robert, 2012). Those in the first camp do not view mathematical activity as a 'natural' unfolding of psychological development. Mathematics has a culture steeped in a history and, in workplace mathematics:

> Fluid balance charts, like many informational resources in the workplace, are not products designed for the benefit of individuals; they are cultural products, in constant use by members of a working community. (Pozzi et al., 1998, p. 115)

Radford's (2000, p. 240) provides a non-ambiguous statement of the importance of the cultural–historical dimension:

> as long as the relation subject/object is seen as a non-culturally-mediated, direct one, meaning construction appears to be the result of the relation that the isolated subject entertains with the *a*historical object

I do not claim that those in the second camp view mathematical activity as a 'natural' unfolding of psychological development but they do not say that it is not this.

My final consideration of differences in the approaches in the mathematics education papers considered in Sects. 9.2 and 9.3 concerns mediation. Mediation, by people and/or language and/or sign/artefacts/tools, is a central concept in the majority of papers considered except in Robert (2012), which does not mention 'mediation', and in Abboud-Blanchard and Cazes (2012) and Jaworski et al. (2012), where consideration of 'mediation' is mainly restricted to mentioning its importance in the theoretical frameworks of Leont'ev and Engeström. But behind the 'and/or's in the previous sentence are different emphases with regard to mediator. These emphases are most clearly marked in the papers by Ladel and Kortenkamp (2013) and Maracci and Mariotti (2013). To Ladel and Kortenkamp (2013) artefact mediation is in the centre of their model, which goes by the name of 'artefact-centric activity theory' but to Maracci and Mariotti (2013, p. 22), 'The mediator is not the artefact itself but it is the person who takes the initiative and the responsibility for the use of the artefact to mediate a specific content'. I suspect that behind 'theoretic statements' on mediation, there are the phenomena that interest us as researchers. Ladel and Kortenkamp are clearly interested in the potential of their artefacts, multi-touch-tables, to improve learning. Maracci and Mariotti, as noted above, continues a line of papers on semiotic mediation that can be traced back to Bartolini Bussi (1996) who focused on mathematical discussion, where human mediation is a central consideration. Further to this, they state that they follow Hasan in formulating their interpretation of mediation. Hasan, as noted above, is a socio-linguist who has based her academic career on the study of everyday discourse, for example, between mothers and daughters. Hasan is interested in such things as daughters' appropriation of the language of their mothers and Maracci and Mariotti appear to have appropriated Hasan's focus for mediation.

I feel that the differences considered above show that AT is a loose collection of approaches (at least in mathematics education) and is not a unified theory. In closing this section I would like to bring in my interests in tools and mathematics and consider tensions in activity theoretic approaches with regard to Leont'ev's activity–action–operation triple. Artefacts/tools are important in each element of the triple but the focus on artefacts/tools is different in each element. In the *operation* element we may focus on the details of manipulating an artefact/tool. Such a focus is likely to interest a mathematics educator who is convinced that a calculator or a dynamic geometry system (DGS) or whatever can help students do/learn mathematics by establishing relationships between mathematical objects but that certain configurations (e.g. modes of dragging in a DGS) of the artefact/tool are important to optimise learning. In the *action* element the mediating qualities of an artefact/tool become paramount and mathematics educators may focus on the transformation of actions by different artefacts/tools; the differences, for example, in drawing the graph of a specific function using pencil, ruler and graph paper compared to drawing the graph of the same function using *GeoGebra*. The human part of this focus on action may be an individual or a group of individuals but when we consider the *activity* element it is always a group. The analysis *activity* element includes the *operation* and *action* elements but its consideration of mediation goes beyond artefact or person mediation to 'include the institutional contexts and history of the systems of activities' (Cole, 1996, p. 333) under investigation.

Although Leont'ev's activity–action–operation triple is not supposed to be ripped asunder, it can be difficult to combine the elements. Cole (1996, pp. 332–334) considers similar matters in relation to Wertsch's focus on mediated action and Engeström's focus on activity systems, he concludes:

> Mediated action and its activity context are two moments of a single process, and whatever we want to specify as psychological processes is but a moment of their combined properties. It is possible to argue how best to parse their contributions in individual cases, *in practice*, but attempting such a parsing "in general" results in empty abstractions, unconstrained by the circumstances to which they are appropriate. (Cole, 1996, p. 334)

LaCroix (2009) goes further than Cole (though not with regard to Wertsch) and argues, in the context of an individual case, that Engeström's approach and Radford's approach 'do not sit well together'. The case concerns adult students (pre-apprentices in the pipe-trades) learning to read fractions-of-an-inch on a measuring tape (an essential trade-skill) in a course. LaCroix was a participant observer and collected data from multiple sources over the 8 week course. He analysed the data in two separate stages using Engeström's approach and then Radford's approach and notes that the analysis from the point of view of Radford's approach sometimes required 'frame-by-frame analysis of videotape to assess the role and co-ordination of spoken language with the use of artifacts and gestures' (LaCroix, 2009, p. 856). Both analyses produced interesting results but:

> [Engeström's] foci, while useful for research in many contexts, run counter to mathematics educators' practical interests in teaching and learning activity, that is, individual students' mathematical enculturation on a day-to-day, if not minute-to-minute basis ... [Radford] provides a way of defining and positioning mathematics as a cultural practice within particular forms of activity ... [Engeström] theorizes learning within activity theory as change in the activity itself, Radford focuses on the learning of individuals as they come to be part of an existing historical-cultural activity. (LaCroix, 2009, p. 859)

This statement is similar to my statement above on the phenomena that interest researchers and is linked to my statement, early in this chapter, that activity theorists argue amongst themselves about the appropriate unit of analysis. I think that AT has contributed much to our understanding of tool use in mathematics but it offers us nuanced understandings.

References

Abboud-Blanchard, M., & Cazes, C. (2012). How does an activity theory model help to know better about teaching with electronic-exercise-bases? *International Journal for Technology in Mathematics Education, 19*(4), 141–146.

Abboud-Blanchard, M., & Vandebrouck, F. (2012). Analysing teachers' practices in technology environments from an activity theoretical approach. *International Journal for Technology in Mathematics Education, 19*(4), 159–164.

Bartolini Bussi, M. G. (1996). Mathematical discussion and perspective drawing in primary school. *Educational Studies in Mathematics, 31*, 11–41.

Chassapis, D. (1998). The mediation of tools in the development of formal mathematical concepts: The compass and the circle as an example. *Educational Studies in Mathematics, 37*(3), 275–293.

Chiappini, G. (2012). The transformation of ergonomic affordances into cultural affordances: The case of the Alnuset system. *International Journal for Technology in Mathematics Education, 19*(4), 135–140.

Cole, M. (1996). *Cultural psychology: A once and future discipline*. Cambridge, MA: Harvard University Press.

Crawford, K. (1996). Vygotskian approaches in human development in the information era. *Educational Studies in Mathematics, 31*(1–2), 43–62.

Daniellou, F., & Rabardel, P. (2005). Activity-oriented approaches to ergonomics: Some traditions and communities. *Theoretic Issues in Ergonomics Science, 6*(5), 353–359.

Daniels, H. (2002). *Vygotsky and pedagogy*. London: Routledge.

Davydov, V. V. (1990). *Soviet studies in mathematics education: Vol. 2. Types of generalization in instruction: logical and psychological problems in the structuring of school curricula* (J. Kilpatrick, Ed. & J. Teller, Trans.). Reston, VA: National Council of Teachers of Mathematics (Original work published in 1972).

Engeström, Y. (1987). *Learning by expanding: An activity-theoretical approach to developmental research*. Helsinki, Finland: Orienta-Konsultit.

Engeström, Y. (2001). Expansive learning at work: Toward an activity theoretical reconceptualization. *Journal of Education and Work, 14*(1), 133–156.

Fuglestad, A. B. (2013). Mediation of a Teacher's Development of Spreadsheets as an Instrument to Support Pupils' Inquiry in Mathematics. *International Journal for Technology in Mathematics Education, 20*(1), 9–14.

Hershkowitz, R., Schwarz, B. B., & Dreyfus, T. (2001). Abstraction in context: Epistemic actions. *Journal for Research in Mathematics Education, 32*, 195–222.

Jaworski, B. (2003). Research practice into/influencing mathematics teaching and learning development: Towards a theoretical framework based on co-learning partnerships. *Educational Studies in Mathematics, 54*(2–3), 249–282.

Jaworski, B., Robinson, C., Matthews, J., & Croft, T. (2012). An activity theory analysis of teaching goals versus student epistemological positions. *International Journal for Technology in Mathematics Education, 19*(4), 147–152.

Kozulin, A. (1986). Vygotsky in context. Pp.*xi-lxi* in Vygotsky, L. (1934/1986). *Thought and language*. Cambridge, MA: The MIT Press.

Kynigos, C., & Psycharis, G. (2013). Designing for instrumentalisation: Constructionist perspectives on instrumental theory. *International Journal for Technology in Mathematics Education, 20*(1), 15–20.

LaCroix, L. (2009). Iconicity, objectification, and the math behind the measuring tape: An example from pipe-trades training. In *Sixth Congress of the European Society for Research in Mathematics Education, Working Group* (Vol. 6, pp. 852–861).

Lagrange, J. B. (2013). Anthropological approach and activity theory: culture, communities and institutions. *International Journal for Technology in Mathematics Education, 20*(1), 33–38.

Ladel, S., & Kortenkamp, U. (2013). An activity-theoretic approach to multi-touch tools in early mathematics learning. *International Journal for Technology in Mathematics Education, 20*(1), 3–8.

Lave, J. (1988). *Cognition in practice: Mind, mathematics and culture in everyday life*. Cambridge, England: Cambridge University Press.

Leont'ev, A. N. (1978). *Activity, consciousness, and personality*. Englewood Cliffs, NJ: Prentice-Hall.

Leplat, J. (1997). *Regards sur l'activité en situation de travail*. Paris: PUF.

Maracci, M., & Mariotti, M. A. (2013). Semiotic mediation within an at frame. *International Journal for Technology in Mathematics Education, 20*(1), 21–26.

Mellin-Olsen, S. (1987). *The politics of mathematics education*. Dordrecht, The Netherlands: Reidel.

Monaghan, J. (2007). Computer algebra, instrumentation and the anthropological approach. *International Journal for Technology in Mathematics Education, 14*(2), 63–71.

Monaghan, J. (2013). Towards a socio-cultural framework for the analysis of joint student-teacher development over technology-based mathematics lessons. *International Journal for Technology in Mathematics Education, 20*(1), 27–32.

Nardi, B. A. (Ed.). (1996). *Context and consciousness: Activity theory and human-computer interaction*. Cambridge, MA: The MIT Press.

Papert, S. (1994). *The children's machine: Rethinking school in tech information age of the computer machine*. Hertfordshire, England: Harvester Wheatsheaf.

Rogalski, J. (2013). Theory of activity and developmental frameworks for an analysis of teachers' practices and Students' learning. In F. Vandebrouck (Ed.), *Mathematics classrooms: Students' activities and teachers' practices* (pp. 3–22). Rotterdam, The Netherlands: Sense Publishers.

Pozzi, S., Noss, R., & Hoyles, C. (1998). Tools in practice, mathematics in use. *Educational Studies in Mathematics, 36*(2), 105–122.

Radford, L. (2000). Signs and meanings in students' emergent algebraic thinking: A semiotic analysis. *Educational Studies in Mathematics, 42*(3), 237–268.

Robert, A. (2012). A didactical framework for studying students' and teachers' activities when learning and teaching mathematics. *International Journal for Technology in Mathematics Education, 19*(4), 153–157.

Vygotsky, L. S. (1978). *Mind in society*. Cambridge, MA: Harvard University Press.

Walkerdine, V. (1988). *The mastery of reason: Cognitive development and the production of rationality*. London: Routledge.

Wertsch, J. V. (Ed.). (1981). *The concept of activity in Soviet psychology*. New York: M. E. Sharpe Inc.

Wertsch, J. V. (1991). *Voices of the mind*. Cambridge, MA: Harvard University Press.

Wertsch, J. V. (1998). *Mind as action*. Oxford, England: Oxford University Press.

Chapter 10
Didactics of Mathematics: Concepts, Roots, Interactions and Dynamics from France

Luc Trouche

Abstract This chapter analyses specificities of the French field of 'didactics of mathematics', questioning its reasons, tracing the geneses of concepts related to artefacts and following influences on, and interactions with the international communities of research. This complex genesis is traced in four sections: a first section on the roots of the didactics of mathematics in France, a second section on two founding theoretical frameworks (the theory of didactical situations of Brousseau, and the theory of conceptual fields of Vergnaud), a third section on the anthropological approach of Chevallard, a fourth focusing on specific approaches dedicated to artefacts and resources in mathematics education. Beyond historical and cultural specificities, the chapter aims to evidence the potential of interactions between teachers and researchers, as well as interactions between researchers in mathematics and mathematics education for improving our understanding of learning and teaching issues in mathematics.

10.1 Introduction

In this chapter, I analyse specificities of the French field of 'didactics of mathematics', questioning its reasons, tracing the geneses of concepts related to artefacts, and following influences on and interactions with the international communities of research. Questioning the dynamics of the theoretical frameworks, that we bear and that leads us, is complex, as each theory is a result of individual and collective pathways (Trouche, 2009), which meet a set of sometimes critical facts and are subject to multiple influences. I have organised this chapter in four sections, giving voice to some main actors[1] involved in these complex geneses: a first section on the roots of the didactics of mathematics in France, a second section on two founding theoretical frameworks (the theory of didactical situations of Brousseau, and the

[1] I have chosen to give sometimes long quotations, keeping the words and the language—sometimes in French—of these authors, in order to allow the reader to have a direct contact with their works.

J. Monaghan et al., *Tools and Mathematics*, Mathematics Education Library,
DOI 10.1007/978-3-319-02396-0_10

theory of conceptual fields of Vergnaud), a third section on the anthropological approach of Chevallard, a fourth focusing on specific approaches dedicated to artefacts and resources in mathematics education.[2]

10.2 The Emergence of Didactics of Mathematics in France as the Result of an Exceptional Conjunction of Phenomena

It is impossible to summarise in a few lines a tumultuous history, made of contrary motions. We will try to underline here some major facts and trends: the position of mathematics teaching in the French educational debate, the role that mathematicians took in this debate, the joint action of psychologists, mathematicians, and teachers themselves in it, and finally, the creation of the IREM as a 'total social fact'.

10.2.1 A Strong and Questioned Position of Mathematics as a Subject of Teaching

The position of mathematics in French curricula appears to be getting stronger over time, if we consider, for setting the scene, three key moments: 1802, 1902 and 2002 (change of centuries of course, but also of economic, social and political conditions). This position appears, however, largely questioned if we consider, beyond these benchmarks, the discontinuity of the curricular development.

1802: Napoleon's ordinance of 19 *frimaire* of Year XI (December 10, 1802) stated: 'The *lycées* will essentially teach Latin and mathematics'. For Gispert (2014, p. 230), 'In placing mathematics at the same level as Latin in the male secondary curriculum, [this ordinance] took into account the new situation following the French Revolution, in which mathematics had become *a core aspect* of an intellectual education *combining theory and practice'*.

1902: a new reform, following a great survey launched by the French Parliament, reasserted these two structuring aspects of mathematics education (Gispert, 2014, p. 233):

- The educational importance of mathematics and science: 'It was, for a time, the end of the monopoly on classical humanities by the *lycées*, through the creation of a modern curriculum that was on par—at least in theory—with the classical curriculum. It also furthered the development of new disciplines such as the living languages, sciences, and mathematics'.

[2] I would like to thank Janine Rogalski and Rudolf Straesser for their comments and advices.

- The importance of the *experience* for learning mathematics and connecting them to sciences: '[In the first cycle], it was recommended to use the concrete experience and induction as the first step necessary before the transition to deductive reasoning. In the second cycle, it was necessary to introduce the new and connected notions of functions and their variations: teaching has to be now linked to physics and its need'.

2002: a report of the CREM[3] Commission (appointed in 1999 by the French Ministry of education for rethinking the teaching of mathematics for the new century) stated 'La mathématique est la plus ancienne des sciences et celle dont les valeurs sont les plus permanentes' (Kahane, 2002). It situates the mathematics among the other sciences and underlines the necessity of connecting their teaching in combining *rigor* and *imagination.*

Beyond this apparent continuity, the situation is more contrasted. First of all, in two centuries, the school system underwent a true metamorphosis, from a school for males and elite to a school for (almost) all, with compulsory education until 16 years of age. Secondly, there is often a large gap between the prescribed and the real curriculum. For example, after the ordinance of 1802, Gispert (2014, p. 230) notices that 'actually the real teaching, after this ordinance, continues to favour Latin and the classical humanities until the end of the nineteenth century, and to separate theory and practice': it appears that two kinds of mathematics teaching existed, according to the social class and schooling structure (lycées vs. primary schools): formation of the mind on one side, training for the practice on the other side. Thirdly, questions at the heart of mathematics teaching appeared very sensitive to social and political events (and the twentieth century was fertile in such major events). Gispert (2014, p. 235) indicates, for example, 'that the reform of the beginning of the twentieth century was accused of being inspired by the German model of the *Realschule* to the detriment of the specificity of a 'French spirit' based on Latin and the classical humanities. In 1923, the Chamber, strongly dominated by conservatives, voted for a new reform that revoked the 1902 programs and principles. Secondary instruction, including mathematics, was again dominated for decades by a theoretical and abstract conception'. Contrary evolutions happened at the end of the 1930s, under the left-wing regime of the Popular Front.

Last but not least, mathematics teaching appears very sensitive to intellectual and scientific pressure. A major event was the constitution, after the Second World War, of the Bourbaki group of mathematicians, who wrote the manifesto *The Architecture of Mathematics*, characterising mathematics as follows: 'In the axiomatic conception, mathematics appears all in all as a reservoir of abstract forms—the structures of mathematics; and one finds—without knowing well why—that certain aspects of experimental reality mold themselves in some of the forms, as a sort of pre-adaption' (Bourbaki, 1962, p. 46, our translation). This theoretical

[3] CREM: Commission de réflexion sur l'enseignement des mathématiques.

construct resonates with the structuralist movement: for example, Weil, a member of Bourbaki, closely collaborates with the anthropologist Lévy-Strauss on the structure of parenthood (Lévy-Strauss, 1949). As Brousseau (2012, p. 104) states:

> Avec les espérances d'une après-guerre et l'aisance financière des trente glorieuses, des propositions d'origines diverses, concernant entre autres, l'éducation (Langevin-Wallon), la psychologie (Piaget), « la » mathématique (Bourbaki), la linguistique (Chomsky), etc. se rassemblent sous une même bannière épistémologique : le structuralisme.

This synergy puts forward the position of mathematics in society 'The new mathematics and its structures were recognised not only by mathematicians but even by scholars in other fields, in particular in the humanities, as a language and scientific tool that were essential for having access to any knowledge' (Gispert, 2014, p. 236) and led to a deep reform of mathematics teaching, the so-called 'modern mathematics'. This reform concentrates on all the objectives of the society: to be modern, to be in line with the development of science and to be democratic:

> In December 1966, in this context of profound institutional changes, the National Education Ministry gave in to the demands of mathematics teachers and created a commission for the study of teaching mathematics, led by André Lichnerowicz [...] The program of the Commission was clear. It had to first work out new guidelines for teaching mathematics in primary and secondary school and assess their viability by pilot experiments. [...] In force as of the 1969 school year, the reform based itself on a critique of traditional mathematics teaching (symbolized by classical geometry), considered too far removed from living mathematics, that is to say mathematics as taught and done in universities since the mid-1950s, with algebra of sets, probability theory, and statistics. Euclidian geometry and calculus were no longer taught as such.

'Convinced that mathematics has to act as a driving force in the development of hard sciences and of human and social sciences as well, in citizens' daily lives, and, beyond that, in the modernisation of society, the proponents of the reform saw in mathematics above all a new language that allowed all citizens to understand its functioning. One of the principal challenges of these reformers was to offer to all children, no matter what their academic future, the most modern mathematics' (Gispert, 2014, p. 238).

This led to a very abstract teaching of mathematics and, in primary schools, to the use of manipulatives (e.g. the famous Dienes blocks, cf. Dienes, 1970), and also put less emphasis on classical mathematical instruments. Ten years later, after considerable discussion, enlisting a large part of the society, this ambitious reform was abandoned but mathematics remains as 'the decisive discipline discriminating between student academic orientations [...], a true subject of selection' (Gispert, 2014, p. 237).

Despite, or perhaps because of these upheavals, mathematics teaching remains, till this time, at the heart of the educational debate in France. One of the characteristics of this debate was the place that mathematicians took in its animation that is the purpose of the following section.

10.2.2 An Important Role of the Mathematicians in the Society, and Their Interest for Mathematics Teaching

To take the same span than the previous section, we could underline that from the time of Napoleon to the current period, the interest and influence of French mathematicians for education was important: Monge, in the first case, creating new institutions (as Ecole Normale and Ecole Polytechnique in 1794) and programmes of teaching; Villani, in the second case, Fields medal 2010, frequently advocating a renewing of mathematics teaching in the French media. Between these two examples, we could follow a real continuity of mathematicians' interventions in France, on three complementary aspects: their international engagement for discussing the issues of mathematics teaching, their engagement in the national educational institutions for designing new curricula and their interventions related to the use of tools.

The birth and the development of ICMI (International Commission on Mathematical Instruction) evidences the engagement of—not only French—mathematicians on teaching questions: ICMI was created in Roma in 1908 by the IV International Congress of Mathematicians. 'Its first president was Felix Klein, an eminent mathematician and promoter of an important reform for teaching of mathematics in Germany. A substantial role in establishing the commission was played by David Eugene Smith, a professor at Teachers College of New York, who was deeply interested in education and the history of mathematics. Thus the commission was born of the closest collaboration between mathematicians and educators' (Menghini et al., 2008, p. 1). To be noticed: Henri Fehr (a Swiss mathematician) and Charles Laisant (a French politician and mathematician) had created the international research journal written in French *L'Enseignement Mathé matique* in 1899, and from early on this journal became the official organ of ICMI in 1908. The use of the French language in this journal indicates the importance of this language, at this time, as an international means for scientific communication. The 11th edition of this journal[4] gave the composition of the commission, including 3 delegates from France among 34 members. Four French mathematicians were elected president during the history of ICMI: Jacques Hadamard (1932–1936), André Lichnerowicz (1963–1966), Jean-Pierre Kahane (1983–1999) and Michèle Artigue (2007–2009), the first woman to occupy such a position. The first three were mathematicians deeply interested in questions of education; the last one, Artigue, is a didactician of mathematics, with a strong mathematics background[5]; we will meet these names again below. We could consider such an evolution as

[4] The digital copies of the journal can be found at http://www.unige.ch/math/EnsMath/EM_fr/welcome.html.

[5] Informations from the ICMI website http://www.mathunion.org/icmi/icmi/executive-committee/past-executive-committees/.

symbolic of the emergence of didactics of mathematics as a new recognised field of research, we will go back to this point further.

We retain from this short evocation of the ICMI birth the interest of mathematicians, not only for discussing general questions of teaching, but also for implementing new curricula in their country. It was the case in Germany with Klein, and the case in France for three essential moments (see Sect. 10.2.1): the reform of 1902, the reform of the 'Modern Math' in 1967, and the reflection for a new curriculum in 2002:

- Regarding the reform of 1902, the commission for designing the curricula in sciences was chaired by a mathematician (Gaston Darboux). Poincaré, Borel and Hadamard made lectures in the 'Musée pédagogique'[6] for supporting its implementation (Belhoste, 1990)
- The second moment was the reform of 'Modern Math' in 1967, led by a commission chaired by André Lichnerowicz. We have seen in the previous section that the implementation of such a reform was a true catastrophe. In 1967, as in 1902, the driven idea was to bring closer real mathematics and mathematics taught,[7] with the illusion that 'closer to the real mathematics, closer to the real need of their teaching'
- The third moment was the reflection of the CREM, chaired by Jean-Pierre Kahane (Sect. 10.2.1). This commission gathered mathematicians, teachers, but also didacticians (among them Michèle Artigue). Retaining the lesson of the 'Modern math', this commission did not want to reform the whole curricula at once, but offer some perspectives for thinking the teaching of mathematics on the long term, conceiving its work in relation to the experience of the teachers on the field:

  > La réflexion sur l'enseignement des mathématiques est donc, par nature, une réflexion à long terme [. . .]. Elle prend point d'appui sur ce que nous savons du mouvement des sciences, et sur une vision implicite de l'avenir à long terme: des possibilités sans nombre, des dangers déjà identifiés, et une multitude de problèmes auxquels l'humanité ne pourra faire face qu'en mobilisant toutes les ressources de l'imagination, de la curiosité, de la créativité, des capacités d'analyse critique et de raisonnement, et des connaissances engrangées par les générations précédentes. La réflexion doit prendre en compte le mouvement actuel de la science comme son histoire et tout ce qui doit être revisité de son passé. Elle doit être ambitieuse, audacieuse, et en même temps tenir compte des contraintes de terrain. Elle doit marier les analyses épistémologiques et didactiques. Au

[6] *The Musée pédagogique* is the forerunner of the National Institute for Pedagogical Research, which became in 2010 the French Institute of Education.

[7] Note that this did not come, for the 1967 reform, from the whole Bourbaki group, but only from its members interested in changing the teaching at a secondary level. Houzel (2004, p. 57) wrote, on this particular question: "In the late 60s, a reform movement in secondary mathematics education has been launched in most countries and this movement has unfortunately claimed Bourbaki. From it, came what was called the 'new math', whose harmfulness is no longer in doubt. But it is unfair to shift the burden to Bourbaki, whose only fault was to ignore the problem of Dieudonné propaganda rather dangerous to teachers (our translation)"

> sein de la commission elle a bénéficié d'une grande variété d'expériences et de sensibilités. Elle doit se poursuivre à l'extérieur, et de façon permanente (Kahane, 2002).

This last moment differs from the two previous ones, in the way of thinking the distance between the mathematics currently developed by the mathematicians and the mathematics to be taught. This distance has been conceptualised further as the *didactical transposition*, a major concept, developed by the didactician Chevallard (see Sect. 10.4). To be noticed: another difference between the reforms of 1902 and 1967 on one side, and the reform of 2002 on the other side is that the propositions of the CREM... had not been really applied, probably due to their financial as well as didactical cost.

A last aspect of the mathematicians' interventions concerns the use of tools and the context of their use, reflecting a constructive point of view on mathematics and its teaching.[8] Poincaré (1904, p. 275) insisted on this aspect, supporting *the incessant use of mobile instruments* in geometry teaching: 'J'ai dit que la plupart des définitions mathématiques étaient de véritables constructions. Dès lors, ne convient-il pas de faire la construction d'abord, de l'exécuter devant les élèves, ou, mieux, de la leur faire exécuter de façon à préparer la définition ?'. Maschietto and Trouche (2010, p. 34) evidence how this issue of tool use runs through most of the issues of *L'enseignement mathématique*. They underline also (p. 39) the productive notion of *mathematics laboratory*, as places to learn mathematics from experiments (see Chap. 3), proposed by French mathematicians at the beginning of the twentieth century, particularly Borel (1904) and rediscovered one century later: for Kahane (2006), 'The main feature of math laboratories is that they are places for experiments. Experiments in mathematics need time and freedom. The pupils should be provided with subjects to explore, they should not have a task to stick to. They should feel free, not under pressure'. This proposition is clearly in line with the pedagogical and philosophical tradition of active methods for learning (cf. Dewey in USA, Freinet in France, Montessori or Pestalozzi in Italia), but also with the spirit of Klein's propositions in the first international movement for reforming mathematics teaching, supporting it by the use of *geometrical models* and artefacts (Schubring, 2010). In doing this, the mathematicians wished to bring closer the mathematicians' practices and the mathematics learning and teaching practices. We will see further how this notion of mathematics laboratory meets the notion of *a-didactical situation* of Brousseau (Sect. 10.3).

We have evidenced in this section the engagement of mathematicians to reform mathematics teaching, bringing them closer to the contemporary mathematics and the conditions of their production, with a growing awareness of the necessary distance between 'mathematics for mathematicians' and 'mathematics for teachers and students'. This awareness is probably due to the lessons of history, particularly

[8] It is indeed significant that the first ICMI study, 1985) was dedicated to the Influence of Computers and Informatics on the Mathematics and its Teaching (Cornu & Ralston, 1992).

the lesson of 'Modern math', and also to the interactions of a range of actors in the field of mathematics teaching. This is the purpose of the following subsection.

10.2.3 A Joint Action of Teachers, Psychologists and Mathematicians for Rethinking Mathematics Education

We will evidence, in this section, the growing and convergent views of teachers, psychologists and pedagogues, particularly in the francophone area, in the debate on mathematics to be taught. Among scholars, a double movement took place in the second half of the twentieth century.

The first movement was a convergence of interest between some mathematicians, psychologists and philosophers, leading in 1950 to creation of a new organisation, CIEAEM (*Commission Internationale pour l'Etude et l'Amélioration de l'Enseignement des Mathématiques* [International Commission for the Study of and Improvement of Teaching Mathematics]), 'in which French mathematicians played an important role' (Gispert, p. 236), due to the Bourbaki influence. The first book coming from this new organisation is in French: *L'Enseignement des mathé matiques* (Piaget et al., 1955). It is a collection of chapters written by a psychologist (Piaget, first author, from Geneva), a logician (Beth, from Amsterdam), three—French-mathematicians (Choquet, Dieudonné et Lichnerowicz) and a pedagogue (Gategno, from London). The introduction makes clear the goal of the book, aiming to enlighten what is possible to teach (the psychologist point of view), what has to be taught (the mathematicians' point of views) and how to teach it (the pedagogue point of view). It should be pointed out that the book is a succession of chapters, not really articulated; this reflection is driven without the inputs of the teachers, but wishes 'to attract the interest of high school teachers supposed to be thought-provoking to them in a way that could renovate their teaching' (Piaget et al., 1955, p. 8). Teachers, however, will soon invade the scene...

The second movement was a specialisation of some mathematics scholars towards the issues of education, taking them as a main interest of research (cf. Sect. 7.2). Hans Freudenthal, chairing ICMI from 1967 to 1970, was the first president considering mathematics education as a field of research in its own right. It corresponded to his view of mathematics, seen 'not primarily as a body of knowledge, but as a human activity' (Bass, 2008, p. 12), his view on mathematics education, seen essentially as a development from the concrete to the general, and his view on research on mathematics education to be developed as a new field, not restricted to a statistical or psychological point of view. He drew all the consequences of such a position in creating a journal, a conference, and an institute: he founded in 1968 a journal dedicated to this question (*Educational studies in mathematics*); he launched the first ICME (International Congress on mathematical education) in 1969 in Lyon, as a manifestation of independence from IMU:

> Freudenthal's bold and adventurous launching of ICME 1 (in Lyon) was essentially, and characteristically, a unilateral action, for which he sought approval and authorization from no one, not the Executive Committee of ICMI, nor that of IMU. And this provoked some anger and concern about ICMI inside ILU. But ICME 1 was a great success... (Bass, 2008, p. 12).

Finally, he founded in 1971 the Institute for the Development of Mathematical Education (IOWO) at Utrecht University, which, after his death, was renamed the Freudenthal Institute. Such creativity cannot be fully understood independently of the social and political context of 1968, as we will see soon.

The intervention of teachers in the debate was certainly decisive. The French APMEP (*Association des Professeurs de Mathématiques de l'Enseignement Public)* [The Association of Mathematics Public School Teachers], was created in 1910, in a moment of academic, social and political upheaval (marked by the creation of trade unions and associations[9]). A former president of this association, Eric Barbazo, dedicated his Ph.D. (Barbazo, 2010) to its history, and evidenced the engagement of this association, from its beginning, in the public debate, reflecting the position of its members and the organisation of schooling. For example, in 1912 (the teaching of mathematics concerned at this moment an elite), in response to a survey initiated in 1912 by the Chamber on the implementation of the reform of 1902 (see Sect. 10.2.1), 'the Association expressed reservations concerning the method of relying on the concrete in mathematics. It highlighted the potential dangers of such a method and the harm that it could do if it was used to substitute experience for proof more generally. Students should not be deprived of the advantages they could obtain from the study of mathematics, which had always been a school of logic' (Gispert, 2014, p. 234).

Convinced that the Bourbaki's ideas were a means for promoting 'mathematics for all', The APMEP created a commission named 'axiomatic and re-discovery', evidencing the double teaching necessity of 'learning logic and structures' and 'favouring students activity'. Convinced that teacher education were a critical issue, 'APMEP, together with the *Société Mathématique de France*, organised between 1955 and 1963, lectures for secondary school teachers on the notions of structures that, for the most part, they had not seen in their studies' (Gispert, 2014, p. 236). The conjunction of the influence of the group Bourbaki, of the structuralist intellectual spirit and of the pressure of APMEP towards the ministry of education leads en 1967 to the constitution of the official commission Lichnerowicz, giving birth, in 1971, to the Modern Math reform (Sect. 10.2.1) and launching the ideas of constituting new institutes, the IREM (Institute for research on mathematics teaching) for supporting this radical reform. Actually the creation of the IREM is a more complex story, concentrating on all the features of this period, as we will show in the following section.

---

[9] In 1910 was also created the association « L'école émancipée » [The emancipated school], gathering pedagogical activist teachers and revolutionary syndicalists.

10.2.4 The IREM as a Total Social Fact, and the Incubator of the French Didactics of Mathematics

The creation of the IREM can be considered indeed as a total social fact (Mauss, 1966), i.e. a fact that informs and organises seemingly quite distinct practices and institutions.

A first institution (and associated practices) is the APMEP. In 1958, the president of this association, Gilbert Walusinski, already proposed the creation of *Institutes for training and pedagogical research*, dedicated to the development of theoretical as well as practical pedagogical research, and to the development of interactions between teachers of different levels (from schools to university) for improving teacher training, in the perspective of the 'Ecole unique'.[10]

A second (emerging) institution is the collective (still not a community) of research on mathematics education. In 1964 a young primary teacher of Bordeaux, having a bachelor in mathematics, Guy Brousseau (who will be the 'hero' of the following section of this chapter) asks Lichnerowicz for a question to be studied on mathematics teaching (cf. Sect. 10.3). In order to gather the conditions of an answer, he created, with the support of Lichnerowicz, a *Center for research on mathematics teaching,* becoming later the COREM (*Center of Observation and Research on Mathematics Teaching*). In the text describing the organisation of this centre, established in Bordeaux, directed by two mathematicians and involving himself, Brousseau writes:

> Avant de chercher à influencer l'enseignement, il convient d'abord de l'observer et de le comprendre en n'agissant que de façon limitée, contrôlée a priori par des connaissance scientifiques et a posteriori par des expériences reproductibles. L'important et le difficile était de rendre possible l'établissement de rapports *appropriés* entre des chercheurs mathématiciens et un système d'enseignement.[11]

The link between research and teacher training, the link between research and experimentation, as well as the link between teachers of different levels and between researchers from different scientific fields appear as the major features of these centres that the Commission Lichnerowicz will retain later in its recommendations:

> It is necessary to progressively create, in each university, Institutes for Research on mathematics teaching with the double objective of performing teacher training at all the levels, to organize necessary experiments, in order to implement their conclusions as facts,

[10] The democratic goal of the French republic was to move from a schooling system founded on 'orders' (schools for people vs. schools for upper classes) to a schooling system founded on 'levels' (primary level vs. secondary level), i.e. a same school for each student: 'l'école unique'. Several successive laws (1959, 1963, 1966) constituted a progress towards this objective, never fully achieved.

[11] http://faculty.washington.edu/warfield/guy-brousseau.com

> in a progressive way. The commission estimates that the IREM have to facilitate or provoke the teamwork and to weave a network of teams in each academic region.[12]

It is well known that there is a long way from an institutional proposition, to its implementation. What makes this implementation possible was certainly the social and political pressure of 1968: the third institution involved in the creation of IREM could be considered as the social movement, including teachers and trade unions. Bass (2008, p. 13) confirms the essential role of the social pressure, of the APMEP and its president at this time, Maurice Glaymann:

> In 1968, at the time of the student demonstrations, Glaymann asked, and received an audience with the new Minister of National Education, Edgar Faure. Faure has decided to move things and Glaymann reiterates the APM proposition of creating IREMs. Faure, after one week of reflection, to evaluate the cost of such an operation, proposed to create an IREM in Paris and says that he has evaluated the coast to be 3 million francs. Glaymann answers him that with the same budget, the APM thinks that three IREMs could be created. This came to pass, Glaymann was named first director of the IREM at Lyon, and this, just in time, put him in a position to offer to host the international congress on mathematics education [in 1969] proposed by Frendenthal.

That is a point where a set of actors already evoked met.

Once created, the IREM developed as a network of Institutes in each university in France. The centre of Brousseau, became the Centre of observation and research on mathematics teaching, associated with primary schools, depending on the IREM of Bordeaux. The IREM became then an incubator for a new field of research. The dynamic of the research in IREM is well summarised by Artigue and Douady:

> This evolution is due to the running of these institutes. IREM gather indeed teachers of several levels and, thus they forced the research which was born in them to not be isolated in an academic ghetto, but on the contrary to keep in touch with the schooling institution, the classes, the teachers. The difficulties created by the new curricula [...] evidenced the inadequacy of the points of view leading to the reform. They evidenced also the limits of the research centred on action and innovation, the necessity for the didactics of mathematics, taking into account the neighbouring fields (psychology, epistemology, sociology, linguistic, sciences of education) to constitute a theoretical field specifically fitted to its problematic and to the methods of research that it developed (Artigue & Douady, 1986, p. 70).

It is thus an exceptional conjunction of phenomenon (mathematical, pedagogical, scientific, intellectual, social and political) that leads, in this country and this period, to the emergence of the French didactics of mathematics, as 'a fragment of the history of the IREM' (Rouchier, 1978, p. 153). This emergence of didactics is not a French *exception*: Biehler et al. (1994) evidence the richness of the interactions between several national communities, facilitated by a network of international scientific conferences and commissions, steering, at an international level, Didactics of Mathematics as a Scientific Discipline.[13] The interaction with the

---

[12] Extract of the third part of the preliminary report of the commission Lichnerowicz, published in the « Bulletin of APMEP », no. 258.

[13] Artigue and Douady (p. 85) underline that "the expression *didactics of mathematics* has been introduced by Klein in 1910".

German community has been particularly intense (Strässer, 1994). But there is no doubt that there is a French *specificity*: the context of IREM, the number and diversity of persons involved in this network, the particular status of mathematics in France (where such institutes, exist only for mathematics) and the mathematical Bourbaki context where the field was born, leads probably to a more theoretical structured field. I propose, in the following parts of this chapter, a visit of this field (keeping in mind the place of artefacts), aiming to evidence both its diversity and its unity, confirming Kilpatrick's point of view (1994, p. 90): « Aux yeux des américains, la didactique des mathématiques française possède une remarquable unité ».

10.3 Two Founding Theoretical Frameworks: Brousseau's Theory of Didactical Situations; Vergnaud' Theory of the Conceptual Fields

From the 1970s, the growth of the French community of didactics of mathematics was very rapid (Artigue & Douady, 1986, p. 71): first master teaching in 1975, a national seminar in 1978, a new journal (Recherches en didactique des mathématiques) and the first summer school in 1980, a first group of research recognised by the CNRS[14] in 1981.[15] Referring to Brousseau and Vergnaud as the founders of the field does not come only from a personal choice. The French community of didactics of mathematics, 20 years after its birth, acknowledged this role in a collective book: *Vingt ans de didactique des mathématiques en France. Hommage à Guy Brousseau et Gérard Vergnaud* (Artigue et al., 1994).[16] It is, indeed, impossible, in the frame of this chapter section, to summarise the scientific contributions of these two preeminent researchers. I will only try to enlighten some major aspects of their works in line with the tool focus of this book.

10.3.1 Guy Brousseau and the Critical Notions of 'Situation' and 'Milieu'

The question Lichnerowicz asked to Brousseau was: 'You ought to study the limiting conditions for an experiment in the pedagogy of mathematics' (Brousseau et al., 2014, p. 172). Brousseau described his reaction: 'my questions were not of the type of 'how many experimental and model classes should the administration set up, and what would be the budget for that?' But rather how to reconcile the

[14] French national centre for scientific research.

[15] This growth mirrors the growth of mathematics education as a field noted in Sect. 7.2.

[16] I have chosen to give, in this section, recent references to the work of Brousseau and Vergnaud, offering a more synthetic view on their work, but it has to be clear that the foundations of their theories come from the 1970s.

flexibility necessary in order to adapt the project to a class with a respect for conventional conditions common to a whole cohort of schools—which notions were indispensible and how to make them accessible' (Brousseau et al., 2014, p. 173). The decision resulting from this reflection was the creation of the COREM (Sect. 10.2.4), which was linked to a primary school, 'during 25 years the most advanced laboratory of experimental didactics of mathematics'.[17] Brousseau's theoretical framework draws its substance from the work in this laboratory.[18] His major work, *The theory of didactical situations*, has been translated into English in 1997 (Brousseau, 1997), and its fundamental idea, *Teaching through situations*, regarding the theme of fractions, gave matter to a recent book (Brousseau et al., 2014). The fundamental idea of 'situations' is defined by Warfield (2014), in a short book 'inviting to didactics': 'A Situation describes the relevant conditions in which a student uses and learns a piece of mathematical knowledge. At the basic level, these conditions deal with three components: a topic to be taught, a problem in the classical sense and a variety of characteristics of the material and didactical environment of the action'.[19]

In this section, I would like to deepen this idea, focusing on three structuring notions of Brousseau's theory: *a-didactical situations*, *didactical situations*, and the *milieu*, and I will do that through an example, then, a quotation from Brousseau. The example is the emblematic situation of the puzzle, described by Brousseau et al. (2014, p. 51). (Fig. 10.1)

Let me now introduce Brousseau's main concepts (Brousseau et al. 2014):

> *A-didactical situations* occur in the classroom, and have the goal of reproducing the conditions of a real mathematical activity dealing with a determined concept: i.e. a mathematical situation. In the course of an a-didactical situation, the students are supposed to produce a correct and adequate action or mathematical text without receiving any supplementary information or influence.
>
> With this definition in hand, a *didactical situation* can be defined as the actions taken by a teacher to set-up and maintain an a-didactical situation designed to allow students to develop some goal concept(s). In particular, the teacher sets up the *milieu*, which includes the physical surroundings, the instructions, carefully chosen information, etc. The milieu may, or may not include a material element (for example Cabri geometry), and other cooperating or concurrent students, etc. but it does at the least include the *savoirs*[20] of the subject, and certain of her current *connaissances*.[21] It is essential that the mi: lieu is

[17] http://guy-brousseau.com/le-corem/

[18] In a recent paper, Brousseau (2012) came back to the "psychological and didactical roots" of his theory, acknowledging the importance, among other researchers, of Greco and Piaget. He has developed a website http://faculty.washington.edu/warfield/guy-brousseau.com where could be found various elements grounding his approach.

[19] For distinguishing Brousseau's notion from the non-specific, standard uses of the word 'situation', she chose to capitalize this as soon as it is used in the frame of the Theory of didactical situations. I have not retained here such a convention: it is enough to consider that, in this section related to Brousseau's theory, the word 'situation' is used with respect to this theory.

[20] In French in the text/

[21] In French in the text.

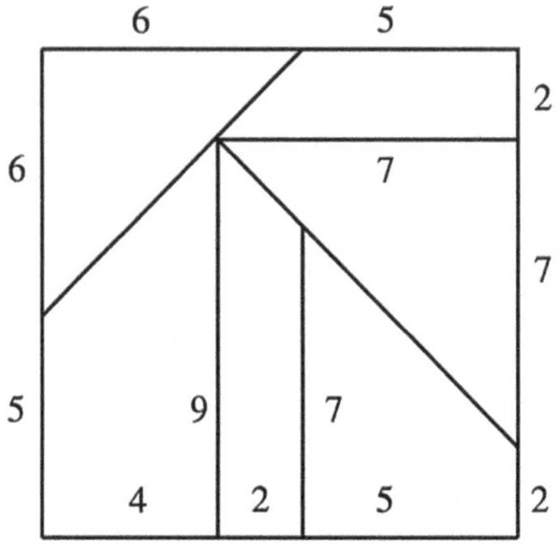

Fig. 10.1 One example of the Brousseau's puzzles. '*Instruction:* Here are some puzzles. You are going to make some similar ones, larger than the ones I am giving you, according to the following rules: (a) The segment that measures 4 cm on the model must measure 7 cm on your reproduction. (b) When you have finished, you must be able to take any figure made up from pieces from the original puzzle and make the exact same figure with the corresponding pieces of the new puzzle. (c) I will give a puzzle to each group of four or five students, and every student must either do at least one piece or else join up with a partner and do at least two. *Development:* After a brief planning phase in each group, the students separate to produce their pieces. The teacher puts (or draws) an enlarged representation of the complete puzzle on the chalkboard'

> designed in a way that it only obeys "objective" necessities, and that the student be convinced of that fact. Once that design is in place, the teacher's mandate is limited to making sure the students focus on the milieu and not on the teacher" (p. 203).

The previous example allows me to illustrate the three fundamental notions proposed here by Brousseau:

- The *a-didactical situation* is constituted in the classroom by the problem of enlarging a puzzle according to a given constraint (adding 3 cm at a given dimension); the determined concept the students are dealing with is the concept of proportionality
- The *didactical situation* is constituted by the actions taken by the teacher to set up and maintain this a-didactical situation. This setting up lies on very subtle adjustments (the initial dimensions of the puzzle pieces, the organisation of students' collective work, the cutting of time in successive phases...) are not randomly chosen, but the result of a very careful analysis of numbers of experiments in the frame of the COREM
- The *milieu* of the situation is all the 'things' the students are acting on, and all the 'things' which are providing feedbacks to the students. In the situation at stake, the puzzle, the ruler, the 'savoir' of the students on numbers, their 'connaissance' of the additive model... are part of the milieu

The feedbacks of the milieu allow the development of the a-didactical situation: a student does not need a validation from the teacher, as the feedback from the

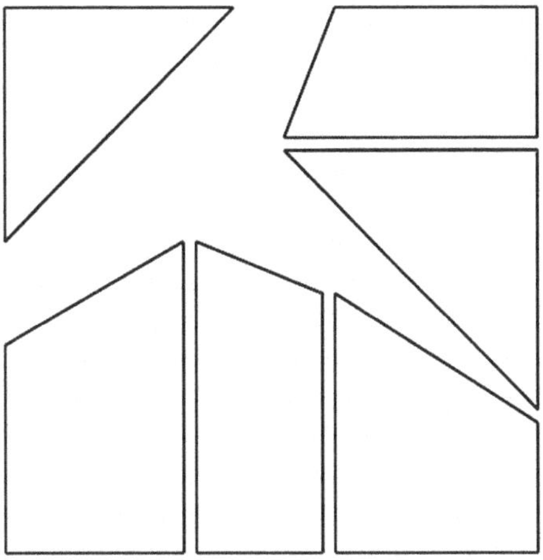

Fig. 10.2 The result of students' action. *One of the strategies and behaviours observed.* 'Almost all the students think that the thing to do is to add 3 cm to every dimension. Even if a few doubt this plan, they rarely succeed in explaining themselves to their partners and never succeed in convincing them at this point. The result, obviously, is that the pieces are not compatible'. [The authors detailed some other students' strategies: adding 3 cm to each segment on the outside square, leading to obtain a rectangle measuring 17 cm × 20 cm; multiplying each measurement by 2 and subtracting 1, as they observe that $4 \times 2 - 1 = 7$...]. *The results:* 'All the children have tried out at least one strategy, and most have tried two. By the end of the class, they are all convinced that their plan of action was at fault, and they are ready to change it so they can make the puzzle work. But no one group is bored or discouraged. At the end of the session, they all are eager to find the right way' (Brousseau et al., 2014, p. 53)

milieu (see Fig. 10.2) is enough to convince her that her method is wrong. And finally the targeted knowledge—the proportionality—is the necessary way for solving the problem.

It is not possible, in the frame of this section, to further develop the other essential concepts constituting this theory (situation of devolution, of institutionalisation, didactical contract...). But what I have presented here is, to me, the heart of Brousseau theory, modelling the learning of mathematics as a social game, with specific rules, the targeted knowledge constituting the optimal way for winning, individually, and with the other students, the game.

The game develops through the interaction with a milieu. This notion of 'milieu', that Brousseau did not translate in English, is very interesting to be analysed: a milieu seems to be always, and it will not be a surprise for the reader of this book, 'full of artefacts'. But it contains more, and is permanently fed by interactions with other students. For me, a possible English translation of this term, in line with further conceptualisations (Sect. 10.5.1), should be 'the student's resources in the situation'.[22]

[22] This proposition results also of interactions with a French didactician, Alain Mercier I want to thank here.

10.3.2 Gérard Vergnaud, and the Conceptualisation Seen as a Cognitive Mediated Process

Brousseau came from the community of mathematicians, Vergnaud came from the community of developmental psychologists. But there were a number of connections. Brousseau mentions that 'the experimental designs imagined by Piaget [were] directly inspired by his exchanges with the mathematician Gonseth' (Brousseau et al., 2014, p. 192), and Piaget, situated at the borders of several fields (epistemology, biology, logic) was, together with prestigious mathematicians, a founding member of the CIEAEM (Sect. 10.2.3). But this original difference between the two researchers could explain some major differences in point of view, Vergnaud saw conceptualisation as a developmental process, and according to a crucial importance to the connections between the *operational* form of knowledge and the *predicative* one. I will present his theory focusing on these two aspects.[23]

First of all, Vergnaud shared with Piaget the idea that even highly structured concepts develop from the most elementary actions of a subject. These actions apply in *situations* (opening a door, solving an equation, climbing stairs...), and facing these situations lead the subject to develop *schemes* (the more or less flexible ways of opening a variety of doors, etc.). Studying the processes of mathematics learning leads Vergnaud to specify this notion of scheme:

> The function of schemes, in the present theory, is both to describe ordinary ways of doing, for situations already mastered, and give hints on how to tackle new situations. Schemes are adaptable resources: they assimilate new situations by accommodating to them. Therefore the definition of schemes must contain ready-made rules, tricks and procedures that have been shaped by already mastered situations; but these components should also offer the possibility to adapt to new situations. On the one hand, a scheme is *the invariant organisation of activity for a certain class of situations;* on the other hand, its analytic definition must contain open concepts and possibilities of inference. From these considerations, it becomes clear that schemes comprise several aspects defined as follows:
>
> - The intentional aspect of schemes involves a goal or several goals that can be developed in sub-goals and anticipations.
> - The generative aspect of schemes involves rules to generate activity, namely the sequences of actions, information gathering, and controls.
> - The epistemic aspect of schemes involves operational invariants, namely concepts-in-action and theorems-in-action. Their main function is to pick up and select the relevant information and infer from it goals and rules.
> - The computational aspect involves possibilities of inference. They are essential to understand that thinking is made up of an intense activity of computation, even in apparently simple situations; even more in new situations. We need to generate goals, sub-goals and rules, also properties and relationships that are not observable.

[23] More on Vergnaud theory and publications can be found on the French mathematics didactics website: http://www.ardm.eu/contenu/gérard-vergnaud-english.

> The main points I needed to stress in this definition are the generative property of schemes, and the fact that they contain conceptual components, without which they would be unable to adapt activity to the variety of cases a subject usually meets. I also feel the need to add several comments in what follows. The dialectical relationship between situations and schemes is so intricate that one sometimes uses an expression concerning situations to refer to a scheme, for in-stance *high jumping,* or *solving equations with two unknowns,* as well as an expression concerning schemes to refer to a situation, for instance '*rule of three' situations* (the rule of three is a scheme, not a situation) (Vergnaud, 2009, p. 88).

What is certainly crucial for mathematical learning, is the *conceptual component* of schemes, namely the *operational invariants*: concepts-in-action and theorems-in-action, that is implicit properties, that are not necessarily true, but appear as relevant in a certain domain. For example, when learning to multiply two integers, students used to develop a strong theorem-in-action as 'the product of two numbers is a number bigger than the two initial numbers'; and a strong concept-in-action as 'the multiplication is a machine for increasing numbers'. Other examples are given in the case of symmetrical figures (Fig. 10.3), and further for the use of graphing calculators (Sect. 10.5.1). Such operational invariants are relevant in a certain domain (that is a reason for their resistance), and turn into obstacles as soon as the mathematical context exceeds this domain.

A concept is, for Vergnaud (2009), related to a given subject and to a moment of her conceptualisation, and it is defined by a triplet: a set of situations, a set of operational invariants and a set of representatives. For a given student, the concept of symmetry exists as soon as she is able to associate to this word a set of *situations* (in or out of school), a set of operational invariants (for example, 'the figure and its image are separated by an axis') and a set of representatives (figures drawn on paper, objects that can be moved from either side of a rule, sentences for describing such situations...). A concept is, in this frame, always associated to a set of artefacts allowing it to be set in different situations. Noticeably, Vergnaud uses the same word of 'concept', for designating something well recognised by a scientific community, and for designating a subject's temporary construction: Vergnaud justified this ambiguity in arguing that a concept is always a living entity, engaged in a genesis, personal or collective. A concept never lives in isolation, but takes sense in the frame of a *conceptual field*, that Vergnaud (2009, p. 86), giving the example of the conceptual field of the additive structures, defined as 'a set of situations and a set of concepts tied together'.

The second main idea of Vergnaud addresses the different forms of knowledge. For him, there is a gulf between the operational form of knowledge, which allows one *to do something*, and the predicative form of knowledge, which allows one *to state/justify what has been done, or what is to be done*, as explained in the following extract.

> Some researchers even consider that the difficulty of mathematics is mainly a linguistic difficulty. This view is wrong, because mathematics is not a language, but knowledge. Still, understanding and wording mathematical sentences play a significant role in the difficulties students encounter. To illustrate this point, let us take two situations (Fig. 10.3) in which students have to draw the symmetrical shape of a given figure. These situations contrast with each other, both from the point of view of the schemes that are necessary for the

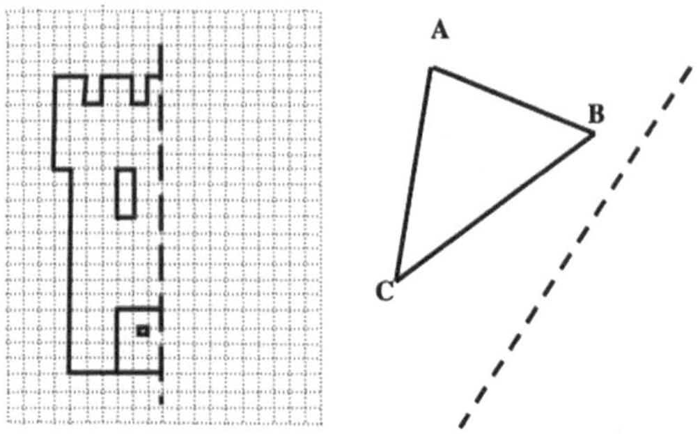

Fig. 10.3 Two figures associated to symmetry

construction and from the point of view of the sentences that one may have to understand or produce on these occasions. The first figure corresponds to a situation likely to be presented to 8- or 10-year-old students, in which they have to complete the drawing of the fortress symmetrically from the vertical axis

The second one could be typically given to 12- or 14-year-olds in France: construct a triangle symmetrical to triangle ABC in relation to d ('d' here refers to the dotted line).

In the first case, there are some coordination difficulties because the child needs to draw a straight line just above the dotted line, neither too high nor too low, and everybody knows that it is not that easy with a ruler; there is the same kind of awkwardness for the departure point and the arrival point. There are also conditional rules. For example, 'one square to the left on the part already drawn, one square on the right on the part to be drawn,' or else 'two squares down on the figure on the left, two squares down on the right,' or else 'one square to the right on the left figure, one square to the left on the one on the right,' starting from a reference point homologous to the point of departure on the left. These rules are not very complex. Nevertheless they rely upon several concepts-in-action and theorems-in-action concerning symmetry and conservation of lengths and angles. As all angles are right angles and lengths are expressed as discrete units (squares), the difficulty is minimal.

In the second case, drawing the triangle A′B′C′, symmetrical to triangle ABC in relation to line d, is much more complex, with the instruments usual in the classroom (ruler, compass, set square). Even the reduction of the triangle to its vertices as sufficient elements to complete the task is an abstraction that some students do not accept easily because they see the figure as a non-decomposable whole. One step further, using d as the axis of symmetry for segments AA′, BB′, CC′, is far from trivial. Why draw a circle with its centre in A, and why should we be interested in the inter-sections of that circle with line d? One can also use a set square and draw a perpendicular line from A to d, measure the distance from A to d, go across line d to construct A′ at the same distance of A to d. But how can I think of the distance to be the same when there is no line yet?

The epistemological jump from the first to the second situation is obvious. But there are also big jumps between different sentences that are likely to be articulated on these occasions. I will use French rather than English because of the syntax of definite articles in French:

1. *La forteresse est symétrique* ('The fortress is symmetrical')
2. *Le triangle A′B′C′ est symétrique du triangle ABC par rapport à la droite d* ('Triangle A′B′C′ is symmetrical to triangle ABC in relation to line d').
3. *La symétrie conserve les longueurs et les angles* ('Symmetry conserves lengths and angles').
4. *La symétrie est une isométrie* ('Symmetry is an isometry').

> Between sentence 1 and sentence 2, there is already a qualitative jump: the adjective *symétrique* moves from the status of a one-element predicate to the status of a three-elements predicate (A is symmetrical to B in relation to C).
>
> Between sentence 2 and sentence 3, the predicate *symétrique* is transformed into an object of thought, *la symétrie,* which has its own properties: it conserves lengths and angles. Nominalisation (i.e., to form a noun from another word class or a group of words) is the most common linguistic process used to transform predicates into objects. In sentences 1 and 2, the idea of symmetry is a predicate (propositional function); in sentence 3, it has become an object (argument). Lower-case 's' is the kind of symbol used by logicians for arguments, whereas upper-case 'S' is used for predicates. The two new predicates, conserving lengths and conserving angles, are thus properties of this new object 's'.
>
> When we move from sentence 3 to sentence 4, a new transformation takes place; the retention of lengths and angles then becomes an object of thought: isometry. This time the predicate is the inclusion relationship between two sets: the set of symmetries S and the set of isometries (Vergnaud, 2009, p. 90).

For Vergnaud, the predicative form of knowledge is a necessary means for building knowledge, but not the main one: the operative form is more subtle, richer than the predicative one. For him, solving a problem is the *source* and the *criterion* of knowledge. Schemes appear thus at the centre of the Vergnaud theoretical framework, as an essential link between *gestures* and *thought.* This importance given to gestures and artefacts situates the work of Vergnaud at crossroads of influences: Piaget of course, but also Vygotsky for the structuring place of mediations (Vergnaud, 2000) and Bourdieu for the social founding of psychology (Bronckart & Schurmans, 1999). Looking further, I could relate the notion of scheme to Eastern culture, considering the dialectic interaction between hand and mind, as in the following quote, which describes the gradual synthesis of 'proper gestures' to a very complex scheme:

> Entre force et douceur, la main trouve, l'esprit répond. Par approximations successives, la main trouve le geste juste. L'esprit enregistre les résultats et en tire peu à peu le schème du geste efficace, qui est d'une grande complexité physique et mathématique, mais simple pour celui qui le possède. Le geste est une synthèse (...). L'adulte ne se rend plus compte qu'il lui a fallu accomplir un travail de synthèse pour mettre au point chacun des gestes qui forment le soubassement de son activité consciente, y compris de son activité intellectuelle (Tchouang Tseu, in Billeter, 2002).

Brousseau/Vergnaud: two different views on mathematics teaching, the first one focusing on a micro-level (the interactions student-*milieu* through very finely tuned *didactical situations*), the second one on a macro-level (the process of conceptualisation, through the encountering of various situations, most of them at school, drawing attention on intermediate forms of knowledge). Both views share an understanding of learning mathematics through mathematical situations, from interactions with specific resources (*milieu* vs. *mediations*).

10.4 Chevallard and the Anthropological Theory of Didactics

Roughly speaking, as the main source of inspiration of Brousseau (resp. Vergnaud) was mathematical (resp. psychological), it could be said that the main source of inspiration of Chevallard[24] was anthropological. I will present in this section his theory, focusing on two essential points: the concept of praxeology, and the importance of tools. Then I will evidence some convergence and tensions, contrasting Chevallard's approach with Brousseau's and Vergnaud's ones.

10.4.1 A View on Mathematical Activity Through Artificial Praxeologies, Products of Human Cultures

Chevallard, in developing his theoretical framework, the so-called 'Anthropological approach of didactics' (ATD in the following), often evokes the work of the French anthropologist Marcel Mauss (1872–1950), who introduces the notion of *total social fact* for designating essential social phenomena:

> These phenomena are at once legal, economic, religious, aesthetic, morphological and so on. They are legal in that they concern individual and collective rights, organized and diffuse morality; they may be entirely obligatory, or subject simply to praise or disapproval. They are at once political and domestic, being of interest both to classes and to clans and families. They are religious; they concern true religion, animism, magic and diffuse religious mentality. They are economic, for the notions of value, utility, interest, luxury, wealth, acquisition, accumulation, consumption and liberal and sumptuous expenditure are all present... (1966, pp. 76–77).

For Chevallard, referring to the English anthropologist Douglas (1986), a given *institution* constitutes a total social fact. The word institution stands here for each social structure which allows—and impose to—its members, occupying various positions in this structure, different 'ways of doing': a classroom, in this sense, constitutes an institution, as well as a school, as well as the schooling system, in a given country and at a given period.

Chevallard defines, in a given institution, a *didactic fact*:

> What I shall henceforth call a didactic fact is any fact that can in some way be regarded as the effect of a socially situated wish to cause someone to learn something. Let me add—this is a more difficult point, on which I shall not dwell any longer—that a didactic fact is, considered to be so only to the extent that it is effective in influencing the learning process (2005, p. 22).

This definition of a-didactical fact is very powerful, and leads to a very general definition of didactics: 'Didactics should, in my view, be defined as the science of

[24] Chevallard offers, on his website, most of his publications: http://yves.chevallard.free.fr.

the diffusion of knowledge in any social group, such as a class of pupils, society at large, etc.' (Chevallard, 2005, p. 22).

Didactics, as a science, analyses didactical facts in a structured way, as elements of local or global *praxeologies*. Chevallard defines a praxeology in the following:

> What exactly is a praxeology? We can rely on etymology to guide us here—one can analyse any human doing into two main, interrelated components: praxis, i.e. the practical part, on the one hand, and logos, on the other hand. "Logos" is a Greek word which, from pre-Socratic times, has been used steadily to refer to human thinking and reasoning—particularly about the cosmos. Let me represent the "praxis" or practical part by P, and the "logos" or noetic or intellectual part by L, so that a praxeology can be represented by [P/L]. How are P and L interrelated within the praxeology [P/L], and how do they affect one another? The answer draws on one fundamental principle of ATD—the anthropological theory of the didactic—, according to which no human action can exist without being, at least partially, "explained", made "intelligible", "justified", "accounted for", in whatever style of "reasoning" such an explanation or justification may be cast. Praxis thus entails logos which in turn backs up praxis. For praxis needs support—just because, in the long run, no human doing goes unquestioned. Of course, a praxeology may be a bad one, with its "praxis" part being made of an inefficient technique—"technique" is here the official word for a 'way of doing'—, and its "logos" component consisting almost entirely of sheer nonsense—at least from the praxeologist's point of view! (2005, p. 23).

A praxeology is made of four components: a set of tasks, a set of techniques, as a way of accomplishing these tasks, a set of technologies, as discourses justifying the techniques, and a theory accounting for these technologies. Let me illustrate this with an example drawn from Chevallard (2005) (Fig. 10.4).

This deep idea of a socially and culturally built mathematics world is essential to understand Chevallard's frame, as evidenced in the following quotation:

> Why do mathematicians seem so attracted to triangles for example? Why does geometry tell us about angles, lines and rays, or about crossing lines and parallel lines? Why does geometry make room for the notions of acute angle, obtuse angle, and reflex angle? If you are tempted to answer: "Mathematicians are interested in all these entities simply because there do exist crossing lines, rays, acute angles, reflex angles, etc., that is, just because these 'things' are out there, in the natural world, waiting for us to study them", then you have been infected with the evil "monumentalistic" doctrine that pervades contemporary school epistemology. If indeed you accept such a poor, unspecific reply, it is more than likely that you have secretly espoused a naturalistic view of the human world—including the mathematical world—, forgetting that almost everything out there, as well as everything in our minds, is socially contrived. A straight line is a concept, not a reality outside us. It is something created in order to make sense of the outside world and to allow us to think and act more in tune with that reality (Chevallard, 2005, p. 26).

This anthropological point of view enlightens the seminal Chevallard's work on *didactical transposition* as a social construct:

> The transition from, knowledge regarded as a tool to be put to use, to knowledge as something to be taught and learnt, is precisely what I have termed the *didactic transposition* of knowledge [. . .] Although long-established, teaching, or the project to have someone *learn* some *knowledge* and *know* it, is therefore a peculiar undertaking. The very first predicament that faces this undertaking is related to its *definition* as a social reality. In defining itself, teaching must draw on culturally accepted concepts. Essentially it defines itself as a process by which people who *do not know* some *knowledge* will be made to *learn* it, and thereby come to *know* it. Such is the *social contract* by which the teaching institution, whatever its concrete institutional forms, binds itself to society (Chevallard, 1988, p. 6).

If I had to write the number... $A = \left(\frac{2}{1+\sqrt{3}}\right)^4$

in standard form (i.e. $a + b\sqrt{3}$, where a, b are rational numbers), I can know *that* $x = 1 + \sqrt{3}$ is a non-zero root of a quadratic equation, and I can know *how* to generate this equation, which is $(x-1)^2 = 3$, or $x^2 - 2x = 2$. It then follows that

$$\frac{2}{x^2} = 1 - \frac{2}{x} = 1 - (x-2) = 3 - x,$$

and therefore that

$$\frac{4}{x^4} = 9 - 6x + x^2 = 9 - 4x + 2 = 11 - 4(1+\sqrt{3}) = 7 - 4\sqrt{3},$$

so that $A = \frac{16}{x^4} = 4(7 - 4\sqrt{3}) = 28 - 16\sqrt{3}$.

Fig. 10.4 A task carried out in the frame of a given praxeology (Chevallard, 2005, p. 24). The task consists in writing a given number under the form $a + b\sqrt{3}$. The technique consists in seeing each number $x = a + b\sqrt{3}$ as a root of the equation $(x-a)^2 = 3b^2$. The technology consists in knowing that $Q + Q\sqrt{3}$ is a field. The theory is that of algebraic structures

This point of view leads also Chevallard, with Marianna Bosch, to pay attention to tools in/for mathematics doing, learning as well as teaching activity.

10.4.2 *The Tools at the Heart of Mathematical Activity*

In the following quotation, Bosch and Chevallard evidence, in contrast to the dominant 'western cultural axiology', the importance of tools, materia, ls, visual, audible or tactile, conditioning mathematics activity. Far from being isolated, they constitute a complex of working tools, at the heart of the mathematicians as well as of the students' activity:

> Writing, symbols, words, speech, gestures and graphic objects used in mathematical activity—or what we call, due to their material and perceptible characteristics, ostensive objects—are reflected in very different ways in mathematics education research work, according to the concept of mathematical activity that is implicitly or explicitly assumed by researchers. In the framework of the anthropological approach, ostensive objects, in dialectical interaction with non-ostensive objects (notions, concepts, ideas, etc.), appear as the raw material of tasks, techniques, technologies and theories of the different praxeological organisations (praxeologies) mathematical knowledge is made of. This conceptualisation, which highlights the instrumental value of ostensive objects side by side with their semiotic value, allows us to evidence how specific praxeologies may be affected by generic constraints concerning the ostensive dimension, for example the difficulties of writing or

the supposed transparency attributed to verbal discourse. Similarly, the problem of 'loss of meaning' which affects certain types of ostensive manipulations is easier to approach in this framework when it is considered with regard to the technological and theoretical needs of the corresponding praxeological organisations.

Our inquiry allows a presentation in more simple terms. It starts from the premise that Western culture establishes, in the range of human practices, a structural opposition between activities considered to be 'manual' and activities considered to be 'intellectual'. This opposition is not neutral. Western cultural axiology prioritises activities of 'the spirit' (in English 'mind', in Spanish 'mente') over the work 'of the hand', that is to say, the work that involves the body—with the exception of those 'body parts' that are located 'in the head' . . .

It goes without saying that what is regarded as 'mathematics' is considered to be of the first type of activities, that is, working 'with the head' with notional tools, reasoning, ideas, insights and very little material elements. In fact, the few material instruments used in school mathematics (paper and pencil, blackboard and chalk, ruler and compass, calculator, computer) are generally regarded as simple 'aids', sometimes as an indispensable aid but not actually a part of the activity itself. Other objects, if not material at less sensitive (writings, formalisms, graphics, words, speeches, etc.), activated by mathematicians can sometimes play a specific role in the activity, but they are assumed to play the role of 'signs', replacing other objects they are supposed to represent.

We now understand that mathematics does not spontaneously appear as an activity in the true (economic) sense of the word; an act or intervention which involves actors and material objects, as instruments which extend the human body to increase its capacity (in strength, accuracy, etc.), or as external objects against which the action is realised. The current conceptualisation of mathematical activity tends to repress the place of material tools which are part of the activity and, if it takes into account particular objects such as discourse, writing and graphs, the focus is not on these objects themselves (and how to handle them), but on what they are supposed to refer to, what they 'represent' or 'signify', in short, their meaning To do mathematics it is necessary to have words, writings, figures and symbols, but what is important would be beyond words and writing. The dominant point of view in this respect can be considered as idealistic in the sense that it seems to only take into account one aspect of the concrete observable mathematical activity: its signifying function, the production of concepts.

Detaching ourselves from this common vision we suggest considering how mathematical activity is conditioned by the material, visual, audible and tactile instruments it puts into play. It is known that the absence of a concept can block the development of mathematical 'thinking', at both the historical level of a community and at the individual levels of a researcher or a student. One may ask to what extent is this absence merely an absence of an idea, a way of 'thinking' or 'conceiving' the world, or the absence of a complex of work tools (which are, for the most part, material in nature), the availability or absence of which could change in a 'catastrophic' manner the performance of the activity. We believe that the didactic analysis of the development of mathematical knowledge—in history as well as in the life of a person or a class—cannot consider this dimension as secondary, assigning it to a purely instrumental function in the construction of concepts"[25] (Bosch & Chevallard, 1999, pp. 89–90).

What is a 'complex of working tools' is to be explained in more depth. Chevallard distinguishes two kinds of objects: ostensive objects (which can be concretely manipulated), and non-ostensive objects. For example the notation 'log',

[25] Translation resulting from interactions between Marianna Bosch and John Monaghan.

$$(x^3 + x + 1) + (x^2 + 4x) =$$
$$1 + 5x + x(x^2 + x)$$

Fig. 10.5 A co-activation of ostensive and non-objective objects for transforming an algebraic expression. The technique used for transforming the first expression into the second one needs to mobilise: (a) Several ostensive objects, belonging to different registers: *written* ones (parentheses, figures, letters. . .), *oral* ones (small discourses like 'x plus $4x$ equals $5x$'), *sign languages* (for gathering the terms of the same degree (The notion of ostensive objects is close to the notions of representatives introduced by Vergnaud (Sect. 10.3.1). For a discussion on this point, see Chevallard (2005, p. 31)). . .. (b) Non-ostensive objects guiding the usage of the ostensive ones: the notions of: 'terms having the same degree', 'factorisation', 'reminders of order 2'. . . Chevallard (1995)

the word 'logarithm', as well as the graphical representation of the function logarithm are ostensive objects; the *notion* of logarithm is a non-ostensive object.

Looking at an example (Fig. 10.5) to illustrate the difference of these objects and their joint mobilisation in the mathematical activity:

For Chevallard, ostensive and non-ostensive objects are intrinsically articulated, at each level of the mathematical activity (expressing a task, using a technique, explaining a technique, integrating a technology in the frame of a theory). In this approach, tools are not relegated to a lower level of mathematical activity: manipulative, concepts and theorems are permanently and jointly involved in this activity.

10.4.3 Some Convergence and Tensions, and a Productive Atmosphere

These three theoretical frameworks share some fundamental ideas, grounding the French community of mathematics didactics: each teaching and learning analysis starts from the *mathematical content* of what is to be learnt; student learning is viewed as an individual and a social activity; and interactions with objects (milieu, tools, instruments. . .) are viewed as crucial for developing this activity.

However some essential tensions can be distinguished:

- While Brousseau analyses what *should be done* for teaching mathematics, Chevallard analyses *what can be done, or cannot be done*, according to institutional constraints; I could illustrate this tension through the pair *didactical* contract (Brousseau)/*social* contract (Chevallard); a-didactical situation (Brousseau)/tasks and techniques (Chevallard).
- While Vergnaud situates the knowledge as an individual and social cognitive construct in progress, Chevallard relates to knowledge as a social and historical construct; I could illustrate this tension through the pair conceptual fields (Vergnaud)/praxeologies (Chevallard); schemes (Vergnaud)/techniques (Chevallard).

For the three approaches, addressing the interrelations between *tools* and *knowledge* appears as a node of complexity: they are caught by Brousseau with the notion of *milieu*, by Vergnaud with the notion of *schemes* (and the dialectic relationship between gestures and operational invariants), by Chevallard with the notion of *ostensive objects* (and the dialectic relationship between ostensive and non-ostensive objects).

Noticeably, these interrelations between tools and knowledge motivate several conceptualisations in this productive atmosphere of an emerging scientific emerging field:

- Douady (1987) defined the *tool-object dialectic* as 'a cyclic process organising the role of the teacher and the pupils, in which mathematical concepts appear successively as tools for the solution of a problem and as objects with a place in the construction of an organised knowledge'.
- Duval (2006) defines a *semiotic register* as a system of signs allowing a transformation of representations (Fig. 10.6).

Working with registers is especially important in mathematics as they are the only way for accessing mathematics objects. For Duval, internal representations come from a process of internalisation of external representations. The notions of semiotic registers seem to be close to the Chevallard's notion of ostensive objects (see Sect. 10.4.2), but where Duval analyses difficulties in terms of structure of semiotic registers, Chevallard relates to the structure of mathematics, i.e. praxeologies[26];

- Balacheff (1996), studying the new problems arising from the computational tools, evidences the necessity to take into account a new transposition (after the Chevallard's *didactical transposition*, see Sect. 10.4.1) the *computational transposition*: 'Let us call computational transposition the process which leads to the specification and then the implementation of a knowledge model. Computational transposition refers to the work needed to fit the requirements of symbolic representation and computation'.
- Finally, I would like to mention the work Robert and Rogalski (2002), the so-called double approach (ergonomics and didactics), taking into account both the teacher's didactical goals (aiming to organise students' activity towards the learning of mathematics) and professional constraints (linked to the Activity theory, see Sect. 9.4). This approach leads to determine five components of teachers' activity: cognitive, mediative, institutional, social and personal, and to analyse teachers' activity both in class and out of class (preparing lessons, discussing with colleagues).

These interconnected conceptualisations opened the way for further theoretical developments, both for these theories themselves, and for new frames to emerge:

[26] For a discussion on this point, see Bosch and Chevallard (1999, p. 29).

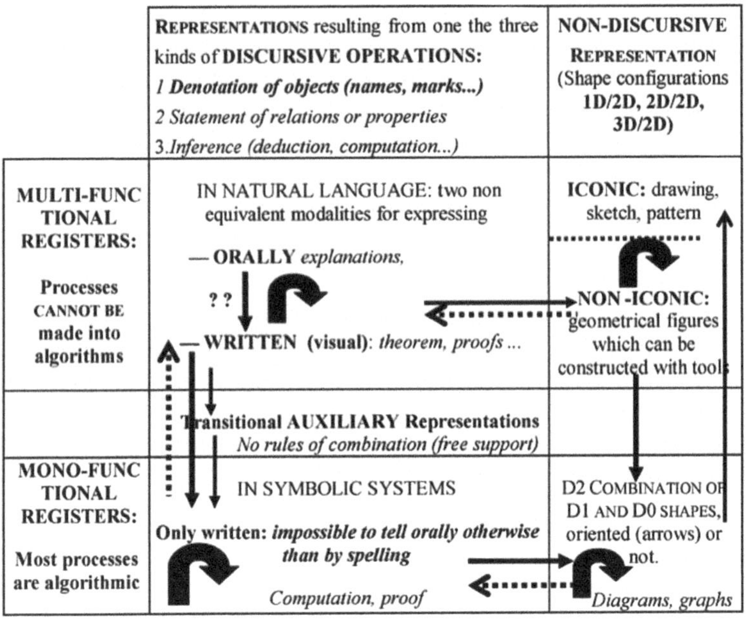

Fig. 10.6 Classification of four types of registers that can be mobilised in mathematical processes (Duval, 2006, p. 110). He distinguishes four types of semiotic registers (see Fig. 10.6) and analyses the problems of understanding in mathematics learning, from a cognitive point of view, through the difficulties of using these registers. Using semiotic registers means either doing a *treatment* (i.e. moving from one representation in a given register to another one in the same register, or a *conversion* (i.e. moving from one register to another one). Doing conversion is often a source of difficulties, because no two distinct semiotic registers will have the same structure

- Brousseau's description of a didactical situation as 'the actions taken by the teacher to set up and maintain an a-didactical situation' (Sect. 10.3.1) inspired me for introducing the notion of instrumental orchestration (Sect. 15.4).
- The variety of concepts introduced for taking into account teachers' resources suggest the need for developing a unified framework; this is discussed in the next chapter.

Intellectual geneses are also individual geneses: Vergnaud was the supervisor of the Ph.D. of Rabardel, who developed the notion of scheme in the context of instrumented activity, and proposed a new approach with regard to artefacts, as we will see in the following section.

10.5 The Instrumental Approach as a Search of New Theoretical Tools for Analysing Tools in Mathematics Education

In this section, I would like to evidence how the proliferation of very new tools in mathematics education motivated the emergence of a new theoretical framework in France, nourished by the soil of the existing interrelated frameworks we have just described. Then, zooming out, I will relate these ideas to an international survey to show how the French situation resonates with wider international trends.

10.5.1 The Emergence of the Instrumental Approach

The proliferation of graphic calculators imported in classrooms by students themselves aroused a lot of debates in society, and analyses in the community of mathematics education (cf. Sect. 13.2). The study of students' mathematical activity, using such material, brings up (what appears as) new phenomena, evidencing the influence of tools on conceptualisation. In my Ph.D. dedicated to this question (Trouche, 1997), I pick up several such phenomena, among these *the influence of images on conceptualisation* (see Fig. 10.7).

This phenomenon and other similar examples (Guin & Trouche, 1999) led me to look for new concepts, taking into account the potential of tools for mathematics education. In a living scientific community, a new approach rarely emerges from the initiative of a single researcher: it emerges for answering to practical needs. Noticeably, at a French summer school on the integration of complex calculators in

The computation of $\lim_{x\to+\infty} \ln(x) + 10\sin(x)$ in a graphic calculator environment

The students had to compute this limit. For the students working without graphic calculators, it was clear that the limit was $+\infty$, as the logarithm function grows towards $+\infty$, and the sin function oscillates between -1 and $+1$.

For the students using a graphing calculator (having the same knowledge, about the functions at stake, as the previous students), this limit cannot exist, due to the oscillations of the function (see screen copies below).

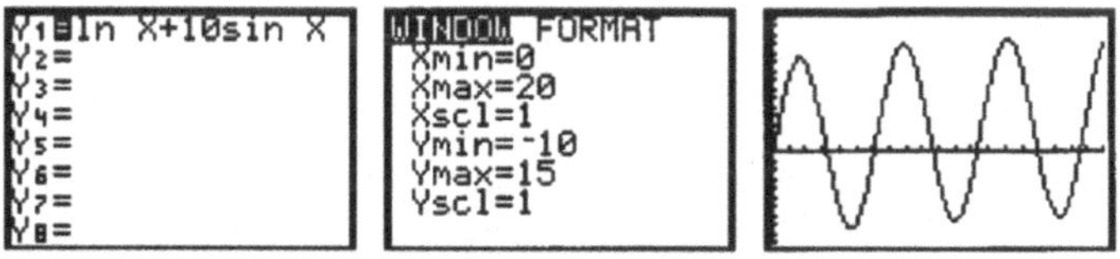

Asking the students for justifying their answers, the ones regularly working with a graphing calculator argued that:

- "the function tending towards $+\infty$ are functions strongly increasing from a given value of the variable (what could be interpreted in term of concept-in-action", § 10.2.1);
- "if a function has a limit, it is necessarily monotonous from a given value of the variable", what could be interpreted in term of theorem-in-action, § 10.2.1).

Such phenomena evidenced the deep influence of such tools on conceptualisation, and pleaded for considering them as essential components of the didactical milieu (§ 10.2.1).

Fig. 10.7 A standard view of a function on a graphic calculator screen (Guin & Trouche, 1999, p. 198)

1997, Michèle Artigue and myself (Artigue, 1997; Trouche, 1997), independently, borrowed the same essential concepts of *artefacts* and *instruments*, *instrumentation* and *instrumentalisation* (Vérillon & Rabardel, 1995), for our respective communications, installing the first milestones of what will become 'the instrumental approach of mathematics didactics'. Guin and Trouche (1999) describe this first attempt for developing a new frame:

> Verillon and Rabardel's studies focusing on learning processes involving instruments in the area of cognitive ergonomy are based on this idea. If cognition evolves through interaction with the environment, accommodating to artefacts may have an effect on cognitive development, knowledge construction and processing, and the nature itself of the knowledge generated (Vérillon & Rabardel, 1995, p. 77). They suggest models and concepts to analyse the instrumented activity of children confronted with tasks involving artefacts.
>
> Verillon and Rabardel stress the difference between the artefact (a material object) and the instrument as a psychological construct: "The instrument does not exist in itself, it becomes an instrument when the subject has been able to appropriate it for himself and has integrated it with his activity" (Vérillon & Rabardel, 1995, p. 84). The subject has to develop the instrumental genesis and efficient procedures in order to manipulate the artefact. During this interaction process, he or she acquires knowledge, which may lead to a different use of it. Similarly, the specific features of instrumented activity are specified: firstly, the constraints inherent to artefacts; secondly, the resources artefacts afford for action; and finally, the procedures linked to the use of artefacts. The subject is faced with constraints imposed by the artefact to identify, understand and manage in the course of this action: some constraints are relative to the transformations this action allows and to the way they are produced. Others imply, more or less explicitly, a prestructuration of the user's action.
>
> The reorganisation of the activity resulting from the introduction of instruments also affords new possibilities of action which are offered to the user; they may provide new conditions and new means for organising action. Thus, it can be argued that, because the instrument is not given but must be worked out by the subject, the educational objectives stated above require the analysis of the instrumented activity of artefacts involved in the learning processes.
>
> It seems quite natural that mathematics education has borrowed from cognitive ergonomics ways of thinking appropriation processes of artefacts. These concepts basically distinguish on one hand what was given to the subject (artefacts, historically and culturally situated) and on the other hand, what was built by the subject (the instruments) during its finalized activity finalized. To be taken into account the long and complex process (the genesis) supporting the construction, combination of two developments, instrumentation and instrumentalisation.

What appears as essential, in this preliminary construction:

- The distinction between an artefact (a product of human activity,[27] that a subject can appropriate for performing a given task) and an instrument (resulting from this appropriation process).
- The distinction of two processes, structuring the instrumental genesis, from an artefact to an instrument: a process of instrumentation, directed from the artefact

[27] Contrary to what is said in the previous quote of Guin and Trouche (1999), corresponding to a previous step of the genesis instrumental approach of didactics, an artefact is not necessarily material. It can be also symbolic, as an algorithm, or a language. Its structural characteristic is to be a result of human activity, and to be potentially engaged in new activity.

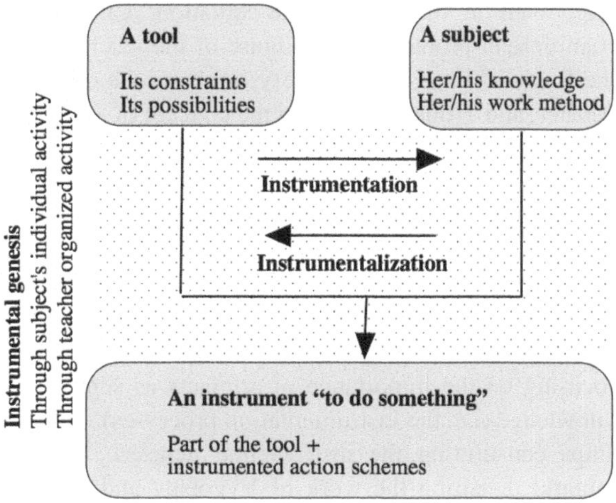

Fig. 10.8 A representation of an instrumental genesis (Trouche, 2004, p. 289)

towards the subject, and a process of instrumentalisation, directed from the subject towards the artefact.

This distinction leads to a careful analysis of artefacts, their constraints and potentialities (to be related to affordances, see Chap. 7): in which way does the computational transposition (Sect. 10.4.3) transform the knowledge involved? What are the gestures favoured by the artefacts and in which way do they influence the student's knowledge in progress? The discussion was deepened through discussions in the French community of mathematics education, particularly during the summer school of this community in 1999, through a lecture by Rabardel (2000). This allowed scholars to establish links between this approach and other existing approaches (Fig. 10.8).

This link was made easier, as Rabardel himself situates his work in the thread of Vergnaud (Sect. 10.3.2), framed by the couple scheme/situation. Defining an instrument as a *mixed entity* with two components (artefact and schemes) leads to describe the instrumented action schemes, and the operational invariants involved in them, and to think *the dialectics between technical and conceptual work* (Artigue, 2002). We had also to re-think the design of *situations*, particularly a-didactical ones (Sect. 10.3.1) according to the constraints of the artefact and the targeted knowledge. At least, we had to take into account the *institutions* in which the artefacts were integrated, and, beyond experimental classes, re-think the conditions for an integration of artefacts in ordinary settings. In ordinary settings, artefacts rarely live alone: several artefacts are engaged in students' activity, making necessary to study the *system of instruments* they develop.

Finally, the instrumental approach appears as a new frame fully integrated in a network of theoretical approaches, marked by the theory of didactical situations, the anthropological approach of didactics and the theory of the conceptual fields. These interrelations appeared also in the theoretical developments arising in the French community at the end of the last century, giving more importance to the teacher's role, even in the a-didactical situation: Chevallard (1997) elaborates on the 'familière et problématique' figure of the teacher, Margolinas (2002), working in the frame of Brousseau's theory, analyses the different *situations* and *milieus* of a teacher, and Trouche proposed the concept of *orchestration* (Sect. 15.3) for modelling the teacher's role in rich technological environments.

But a theoretical approach never develops in an isolated national frame. The instrumental approach was discussed in the Third Computer Algebra in Mathematics Education Symposium held in Reims in 2003, and this discussion gave birth to two papers (Hoyles et al., 2004; Trouche, 2004). This led a number of scholars to give more importance to the *instrumentalisation* process, i.e. to the creative power of students developing their own instruments from the available artefacts. After focusing on the importance of artefacts as supports of activity and mediators of knowledge (i.e. the instrumentation processes), it implies to rebalance the relationships constituting the instrumental geneses. It resonates with an Engeström's remark, revisiting the work of Vygotsky and his colleagues: 'it seems to be all but forgotten that the early studies led by Vygotsky, Leont'ev, and Luria not only examined the role of artifacts as mediators of cognition, but was also interested in how children *created* artifacts of their own to facilitate their performance' (Engeström et al., 1999, p. 26). This discussion was further developed in a journal issue dedicated to the work of Celia Hoyles (Trouche & Drijvers, 2014).

After having grounded the instrumental approach in the 'French field', I would like to link, beyond the local interaction during the CAME symposium, French and international trends in the field of mathematics education, and that is the purpose of the following section.

10.5.2 Zoom Out, Where the National Characteristics Join International Trends (and Vice Versa)

In this section I draw on a meta-study of a comprehensive corpus of publications, driven by a French team (Lagrange et al., 2003), supported by the French Ministry of research. This study addresses the field of research and innovation in the worldwide field of the integration of ICT in mathematics education from 1994 to 1998. In contrast with classical meta-studies,[28] this study did not focus only on the findings of publications, but considered also characteristics like the questions addressed, approaches, cognitive theoretical background, etc. The authors expected that 'analysing this material would help to identify as many as possible aspects of the complexity of the integration, some of them widely addressed and others less considered by the literature' (p. 241). I draw on this study in order to have a means for comparing the evolutions in the French and the international community

[28] For example, Hembree and Dessart (1986), see Sect. 13.2.

of mathematics education in the field of information and communication technology (ICT).

The study proceeded in four stages. The first stage consisted in gathering *all the references* related to ICT and mathematics education in a variety of international sources (the Zentralblatt für Didaktik der Mathematik database, four international journals on mathematical education, seven international journals on computers for mathematics learning, books with chapters on technology and mathematics education, etc.) as well as French works (professional and research journals, dissertations, research and official reports, etc.); this resulted in a corpus of 662 published works. The second stage consisted in a first sorting of this large corpus, distinguishing the *type of analysis* (presentation of a technological product, experimentation-innovation, research report or general reflection), the mathematical field, the type of technology, and the country of the first author. Figure 10.9 shows that: research publications were not in a majority. The literature about ICT includes classroom innovation and pure speculation as well as research studies; a number of papers did not specify a mathematical field, focusing on the support of technology in 'general' mathematical learning.

The third stage consists in a reduction of the corpus, removing papers which had insufficient substance (for example: technical descriptions, or simple description of an innovative classroom activity), whilst keeping a large enough selection to respect the diversity of approaches and to avoid biases, resulting in a corpus of 79 papers. The fourth stage was dedicated to an in-depth analysis of this corpus. For each of these papers, one participant of the project established a detailed review showing the following characteristics: problematic,[29] theoretical background, details of the questions addressed, methodology used, specific findings and an appreciation by the reviewer. This analysis led the authors to distinguish seven different orientations for characterising these papers, named 'dimensions'. For each of these dimensions, a set of indicators was designed, resulting into a grid (Fig. 10.10).

Using a statistical procedure based on a cluster analysis, we obtained clusters of papers sharing specific indicators. The procedure also selected one or two papers at the centre of each cluster, which were the publications statistically best represented. In each cluster, French and international papers were represented.

Globally, the 1994–1998 literature appeared to restrict its analysis to potentialities of ICT itself (easier and more varied representations, new aspects of mathematical knowledge, etc.) rather than questions raised by its insertion into the 'ordinary' mathematics teaching. The general picture of ICT in the teaching and learning of mathematics emerging from this analysis is that of a field where

[29] The notion of 'problematic' comes from the French 'problématique', well defined by Edward Said: 'The idea of beginning, indeed the act of beginning [a research], necessarily involves an act of delimitation by which something is cut out of a great mass of material, separated from the mass, and made to stand for, as well as be, a starting point, a beginning; [...] such notion of inaugural delimitation is Louis Althusser's idea of the problematic, [...] is something given rise to by analysis' (Said 1978, p. 24).

Presentation of product	20%
Experimentation, Innovation	20%
Research report	37%
General reflection	15%
Not specified	6%

Geometry	23%
Calculus (or Algebra and Calculus)	14%
Algebra	11%
Graph/functions	10%
Arithmetic	5%
Other fields	7%
No specific field	30%

Fig. 10.9 Elements of description of the first corpus (Lagrange et al., 2003, p. 242)

1. General approach of the integration	• Type of hypothesis (assumption of improvements, questions, etc.) • Methodology and validation processes (comparing experimental and control groups, comparing a priori analysis and expectations with an experiment, etc.)
2. The epistemological and semiotic dimension	Influence of ICT • on the mathematical knowledge and practices • on the representatives used in this activity
3. The cognitive dimension	• Cognitive frame (constructivist, socio-cultural, ...) Concepts used (schemes, webbing, etc.) • Cognitive role of ICT (visualisation, expression, connection, etc.)
4. The institutional dimension	• Interaction of ICT with tasks and techniques in the culture of a school institution • Role of instrumented techniques in conceptualisation of mathematics
5. The instrumental dimension	• The tool's possibilities and constraints • Instrumentation processes
6. The situational dimension	Influence of ICT on • the structure of the situation • students' solving strategies • the didactical contract
7. The teacher dimension	• Teacher's beliefs and representations of mathematics and of ICT • New teaching situations • Influence of research and pre/in service programs

Fig. 10.10 The dimensions of analysis and their indicators (Lagrange et al., 2003, p. 247)

publications about innovative use or new tools and applications dominated. Research studies differed with regard to the way they considered the potentialities of ICT, but they converged in a focus on the student, the cognitive role of ICT (third dimension) and in an emphasis on epistemological and semiotic aspects (first and second dimensions). The other dimensions appeared as emergent:

- The institutional dimension gathers papers focusing on the difficult viability of technology in schools. Papers with a pioneer spirit started from today's difficulties to motivate the use of tomorrow's technology, while others looked for reasons in more permanent characteristics of technology and of the educational institutions
- The instrumental dimension gathers papers analysing constraints, evidencing complexity of appropriation processes, conjecturing relationships between gestures and conceptualisation
- The situational dimension gathers a few papers taking into account new economy of problem solving, and new situations to be though for integrating technologies; changes to be made in the curriculum, and the need for thinking about the use of technology together with other learning situations
- Very few papers considered the teacher dimension

It is amazing to realise that the 'dimensions' distinguished by the French team were close to the theoretical frameworks they were close to: the cognitive dimension with Vergnaud, the institutional dimension with Chevallard, the instrumental dimension (the 'French instrumental approach') and the situational dimension with Brousseau. But this shows that the French papers were not isolated: they share in each of these dimensions elements of analyses, and through the cross references, they feed other theoretical frames as well as being fed by them.

It is also interesting to analyse evolutions along the period covered by the survey. The authors 'discerned a long-term motion towards awareness of a more complex integration and the subsequent necessity of new dimensions of analysis. It is confirmed by what we know of the institutional and instrumental dimensions in today's research studies and of the emerging reflections on the teacher' (p. 260). They noticed a growing interest for the Brousseau theory of situations (Sect. 10.3.1) which could 'help when looking in depth into changes in the learning situations and when showing precisely what is at stake in these new situations' (Sutherland & Balacheff, 1999, p. 259). They note that 'elements of evolution appear in the convergence towards dialectical approaches to issues like visualisation and contextualisation. These approaches contribute to the development of new dimensions by helping to better consider the institutional contextualisation of knowledge as well as the schemes attached to the use of a technological tool in their instrumental dimension' (p. 259). Evoking Mariotti (2002), the authors provide evidence that, just after this period, the instrumental dimension experienced a strong development. Regarding the complexity of teaching and learning situations with ICT, researchers became more cautious. The authors note that 'Interesting research studies start from the observation of teachers struggling to integrate ICT into the real teaching' (Monaghan, 2001, p. 259).

Thus the period 1992–1998 appeared as a period of transition, from a naive point of view on integration to a more balanced point of view, to which the instrumental approach gives means of expression: for preparing the ICMI study on mathematics education and technology, the introductory document (Hoyles & Lagrange, 2006) proposed as one of the key publications, the book by (Guin et al., 2005) which constitutes precisely both a presentation and a discussion of this approach at an international level.

10.6 Elements for Discussion

In this chapter, I have tried to explain the exceptional conjunction of phenomena leading, from 1970, to a strong development of theoretical frameworks in mathematics didactics in France. This productivity had been acknowledged by international distinctions (the Felix Klein medal to Guy Brousseau and Michèle Artigue, and the Hans Freudenthal medal to Yves Chevallard). Beyond this national context, we have shown that these constructs draw their sap from various theoretical traditions, and resonate with international trends.

The instrumental approach appears as a theoretical construction, starting from taking into account of 'artefacts for doing mathematics', and developing thanks to the theoretical ground provided by the other frames. In the field of ICT in mathematics education, interaction between the French community and the international one have strongly developed, due probably to the novelty of phenomena arising with the rapid evolution of technologies.

The instrumental approach, in the thread of these interactions, has deeply evolved:

- The evolution can be related to a concept, for example the concept of *orchestration*, enriched by Drijvers et al. (2010), see also Chap. 15.
- The evolution can also touch the relationships between concepts: the upheavals of digital resources and of Internet led in 2007 to expand the vision beyond technology, to the set of resources that sustain the activities of teachers, including textbooks (Sträßer, 2009). It leads Gueudet and Trouche (2009, see also chap. 15) to substitute the dialectic resources-documents to the dialectic artefact-instrument. This new model has resulted in the *documentational approach* that also induces other openings: considering the documentary work of the teacher in a variety of places and long time involving new methodological developments; new concepts are emerging, for example the notion of the *teacher resource system*, paving the way for new theoretical fruitful interactions (for example, for analysing the structure of the resource system, a successful track seems to be the study in terms of praxeologies, Sect. 10.4.1).

The interactions between frameworks allow more generally to deepen the concepts at stake. It was the case for the interactions between: the instrumental approach and the semiotic approach (Maschietto & Trouche, 2010); the instrumental approach and the ontological semiotics approach (Drijvers, Godino, Font, & Trouche, 2012); the documentational approach and the double approach (Sect. 10.4.3, Gueudet & Vandebrouck, 2011). This work was, at an international level, theorised as a set of possible strategies to grow the theories themselves (Prediger, Arzarello, Bosch, & Lenfant, 2008), see Chap. 11.

The documentational approach has developed a methodology of reflective investigation (Chap. 15), involving the gaze of the teacher on its own resources, and providing new tools to the researcher for analysing teachers' work. We could all struggle through this self-confrontation with our own resources in research. The representation that I have, at this point, of my own resource system is pretty close to that of the worker Demarcy, looking at his working place at the factory, as 'heterogeneous supports, improvised vices for stalling pieces…' (Linhart, 1978). What is true for a researcher is also probably for a community. Research communities in mathematics education, especially the part thereof that look to technology have long been committed to this cross questioning of theoretical tools, far beyond the French research communities (Trouche & Drijvers, 2014). A history far from being finished…

References

Artigue, L. (1997). Rapports entre dimensions technique et conceptuelle dans l'activité mathématique avec des systèmes de mathématiques symboliques. In *Actes de l'Université d'été « Des outils informatiques dans la classe aux calculatrices symboliques et géométriques : quelles perspectives pour l'enseignement des mathématiques ? »* (pp. 19–40). Rennes, France: IREM.

Artigue, M. (2002). Learning mathematics in a CAS environment: The genesis of a reflection about instrumentation and the dialectics between technical and conceptual work. *International Journal of Computers for Mathematical Learning, 7*(3), 245–274. doi:10.1016/s0277-5395-00-00166-7.

Artigue, M., & Douady, R. (1986). La didactique des mathématiques en France—Emergence d'un champ scientifique. *Revue française de pédagogie, 76*, 69–88. Retrieved September 15, 2014, from http://www.persee.fr/web/revues/home/prescript/article/rfp_0556-7807_1986_num_76_1_1503

Artigue, M., Gras, R., Laborde, C., & Tavignot, P. (Eds.). (1994). *Vingt ans de didactique des mathématiques en France. Hommage à Guy Brousseau et Gérard Vergnaud*. Grenoble, France: La pensée sauvage éditions.

Balacheff, N. (1996). Advanced educational technology: Knowledge revisited. In T. T. Liao (Ed.), *Advanced educational technology: Research issues and future potential* (pp. 1–20). Berlin: Springer.

Barbazo, E. (2010). *L'APMEP, un acteur politique, scientifique, pédagogique de l'enseignement secondaire mathématique du 20e siècle en France* [*APMEP, a political, scientific and pedagogic actor in 20th century French secondary mathematics teaching*]. PhD. Paris, France: EHESS.

Bass, H. (2008). Moments in the life of ICMI. In M. Menghini, F. Furinghetti, L. Giacardi, & F. Arzarello (Eds.), *The first century of the international commission on mathematical instruction (1908-2008). Reflecting and shaping the world of mathematics education* (pp. 9–24). Rome: Instituto della Enciclopedia Italiana.

Belhoste, B. (1990). L'enseignement secondaire français et les sciences au début du XXe siècle. La réforme de 1902 des plans d'études et des programmes. *Revue d'histoire des sciences, 43(4)*, 371–400. Retrieved from http://www.persee.fr/web/revues/home/prescript/article/rhs_0151-4105_1990_num_43_4_4502

Biehler, R., Scholz, R. W., Sträßer, R., & Winkelmann, B. (Eds.). (1994). *Didactics of mathematics as a scientific discipline*. Dordrecht, The Netherlands: Springer.

Billeter, J.-F. (2002). *Leçons sur Tchouang-Tseu*. Paris: Editions Allia.

Borel, E. (1904). *Les exercices pratiques de mathématiques dans l'enseignement secondaire* [Practical exercises in mathematics for secondary instruction]. *Revue générale des sciences pures et appliquées, 14*, 431–440.

Bosch, M., & Chevallard, Y. (1999). La sensibilité de l'activité mathématique aux ostensifs. Objet d'étude et problématique. *Recherches en didactique des mathématiques, 19*(1), 79–124.

Bourbaki, (1962). *L'architecture des mathématiques* [The architecture of mathematics]. In François Le Lionnais (Ed.), *Les grands courants de la pensée mathématique* (2nd ed., pp. 35–47). Paris: Albert Blanchard.

Bronckart, J.-P., & Schurmans, M.-N. (1999). Pierre Bourdieu—Jean Piaget : habitus, schèmes et construction du psychologique. In B. Lahire (Ed.), *Le travail sociologique de Pierre Bourdieu. Dettes et critiques*. Paris: La Découverte.

Brousseau, G. (1997). *Theory of didactical situations in mathematics*. Dordrecht, The Netherlands: Springer.

Brousseau, G. (2012). *Des dispositifs Piagétiens... aux situations didactiques* [From Piagetian experimental designs ... to didactical situations]. *Education et didactique, 6(2)*, 103–129. Retrieved from http://educationdidactique.revues.org/1475

Brousseau, G., Brousseau, N., & Warfield, G. (2014). *Teaching fractions through situations: A fundamental experiment*. Dordrecht, The Netherlands: Springer.

Chevallard, Y. (1988, August). On didactic transposition theory: Some introductory notes. In *International Symposium on Research and Development in Mathematics, Bratislava, Czechoslavakia*. Retrieved November 14, from http://yves.chevallard.free.fr/spip/spip/IMG/pdf/On_Didactic_Transposition_Theory.pdf

Chevallard, Y. (1995). Les outils sémiotiques du travail mathématique. « *petit x* », *42*, 33–57.

Chevallard, Y. (1997). Familière et problématique, la figure du professeur. *Recherches en didactique des mathématiques, 17*(3), 17–54.

Chevallard, Y. (2005). Steps towards a new epistemology in mathematics education. In M. Bosch (ed.), *Proceedings of the Fourth European Conference on Research on Mathematics Education* (pp. 21–30). FUNDEMI IQS—Universitat Ramon Llull. Retrieved from http://ermeweb.free.fr/CERME4

Cornu, B., & Ralston, A. (Eds.). (1992). The influence of computers and informatics on mathematics and its teaching. *ICMI Study Conference Held in Strasbourg, France, March 1985*. Second edition published by UNESCO. Retrieved from http://unesdoc.unesco.org/images/0009/000937/093772eo.pdf

Dienes, Z. P. (1970). *Les six étapes du processus d'apprentissage en mathématiques*. Paris: OCDL.

Douady, R. (1987). Jeux de cadres et dialectique outil objet. *Recherches en didactique des mathé matiques, 7*(2), 5–31.

Douglas, M. (1986). *How institutions think?* Syracuse, NY: Syracuse University Press.

Drijvers, P., Doorman, M., Boon, P., Reed, H., & Gravemeijer, K. (2010). The teacher and the tool: Instrumental orchestrations in the technology-rich mathematics classroom. *Educational Studies in Mathematics, 75*, 213–234.

Drijvers, P., Godino, J. D., Font, V., & Trouche, L. (2012). One episode, two lenses. A reflective analysis of student learning with computer algebra from instrumental and onto-semiotic perspectives. *Educational Studies in Mathematics, 82*, 23–49.

Duval, R. (2006). A cognitive analysis of problems of comprehension in a learning of mathematics. *Educational Studies in Mathematics, 61*, 103–131. Retrieved from http://www.cimm.ucr.ac.cr/ojs/index.php/eudoxus/article/viewFile/162/297

Engeström, Y., Miettinen, R., & Punamaki, R.-L. (1999). *Perspectives on activity theory*. Cambridge, England: Cambridge University Press.

Gispert, H. (2014). Mathematics education in France, 1900-1980. In A. Karp & G. Schubring (Eds.), *Handbook on the history of mathematics education* (pp. 229–240). New York: Springer.

Gueudet, G., & Trouche, L. (2009). Towards new documentation systems for mathematics teachers? *Educational Studies in Mathematics, 71*(3), 199–218.

Gueudet, G., & Vandebrouck, F. (2011). Technologies et évolution des pratiques enseignantes : études de cas et éclairages théoriques. *Recherches en didactique des mathématiques, 31*(3), 271–313.

Guin, D., Ruthven, K., & Trouche, L. (Eds.). (2005). *The didactical challenge of symbolic calculators: Turning a computational device into a mathematical instrument.* New York: Springer.

Guin, D., & Trouche, L. (1999). The complex process of converting tools into mathematical instruments. The case of calculators. *The International Journal of Computers for Mathematical Learning, 3(3),* 195–227. Retrieved from http://link.springer.com/article/10.1023/A:1009892720043#page-1

Hembree, R., & Dessart, D. J. (1986). Effects of hand-held calculators in precollege mathematics education: A meta-analysis. *Journal for Research in Mathematics Education, 17*(2), 83–99.

Houzel, C. (2004). Le rôle de Bourbaki dans les mathématiques du vingtième siècle. *Gazette des Mathématiciens, 100*, 53–63.

Hoyles, C., & Lagrange, J.-B. (2006). The seventeenth ICMI study. Technology revisited. Discussion document. Retrieved December 1, from http://www.mathunion.org/fileadmin/ICMI/files/Digital_Library/DiscussionDocs/DD_icmiStudy17_02.pdf

Hoyles, C., Noss, R., & Kent, P. (2004). On the integration of digital technologies into mathematical classrooms. *International Journal of Computers for Mathematical Learning, 9*, 309–326.

Kahane, J.-P. (dir.). (2002). *L'enseignement des sciences mathématiques. Rapport au ministre.* Commission de réflexion sur l'enseignement des mathématiques. Odile Jacob.

Kahane, J.-P. (2006). Cooperation and competition as a challenge in and beyond the classroom. In *ICMI study no. 16 conference.* Retrieved from http://www.mathunion.org/fileadmin/ICMI/files/Conferences/ICMI_studies/Study16/icmis16pkahane_03.pdf

Kilpatrick, J. (1994). Vingt ans de didactique française depuis les USA. In M. Artigue, R. Gras, C. Laborde, & P. Tavignot (Eds.), *Vingt ans de didactique des mathématiques en France. Hommage à Guy Brousseau et Gérard Vergnaud* (pp. 84–96). Grenoble, France: La pensée sauvage éditions.

Lagrange, J.-B., Artigue, M., Laborde, C., & Trouche, L. (2003). Technology and mathematics education: A multidimensional study of the evolution of research and innovation. In A. J. Bishop, M. A. Clements, C. Keitel, J. Kilpatrick, & F. K. S. Leung (Eds.), *Second international handbook of mathematics education* (pp. 239–271). Dordrecht, The Netherlands: Kluwer Academic Publishers.

Lévy-Strauss, C. (1949). *Les structures élémentaires de la parenté.* Paris: PUF.

Linhart, R. (1978). *L'établi.* Paris: Éditions de minuit.

Margolinas, C. (2002). Situations, milieux, connaissances : analyse de l'activité du professeur. In J.-L. Dorier, M. Artaud, M. Artigue, R. Berthelot, & R. Floris (Eds.), *Actes de la 11ème Ecole d'Eté de Didactique des Mathématiques* (pp. 141–156). Grenoble, France: La pensée sauvage.

Mariotti, M. (2002). The influence of technological advances on students' mathematics learning. In L. English, M. G. Bartolini Bussi, G. Jones, R. Lesh, & D. Tirosh (Eds.), *Handbook of international research in mathematics education* (pp. 695–723). Hillsdale, NJ: Lawrence Erbaum Associates.

Maschietto, M., & Trouche, L. (2010). Mathematics learning and tools from theoretical, historical and practical points of view: The productive notion of mathematics laboratories. *ZDM, International Journal on Mathematics Education, 42*(1), 33–47.

Mauss, M. (1966). *The gift: Forms and functions of exchange in archaic societies.* London: Cohen & West (original edition).

Menghini, M., Furinghetti, F., Giacardi, L., & Arzarello, F. (Eds.). (2008). *The first century of the international commission on mathematical instruction (1908-2008). Reflecting and shaping the world of mathematics education* (pp. 9–24). Rome: Instituto della Enciclopedia Italiana.

Monaghan, J. (2001). Teachers' classroom interactions in ICT-based mathematics lessons. In M. van den Heuvel (Ed.), *Proceedings of the 25th International Conference for the Psychology of Mathematics Education* (Vol. 3, pp. 383–390). Utrecht, The Netherlands.

Piaget, J., Beth, E. W., Dieudonné, J., Lichnerowicz, A., Choquet, G., & Gattegno, C. (1955). *L'enseignement des mathématiques*. Neuchâtel, Switzerland: Delachaux & Niestlé.

Poincaré, H. (1904). Les définitions générales en mathématiques. *L'Enseignement mathématique, 6*, 257–283.

Prediger, S., Arzarello, F., Bosch, M., & Lenfant, A. (Eds.). (2008). Comparing, combining, coordinating—Networking strategies for connecting theoretical approaches. *ZDM, The International Journal on Mathematics Education, 40*(2), pp. 163–327.

Rabardel, P. (2000). Eléments pour une approche instrumentale en didactique des mathématiques. In M. Bailleul (Ed.), *Actes de l'Ecole d'été de didactique des mathématiques*, IUFM de Caen, pp. 203–213.

Robert, A., & Rogalski, J. (2002). Le système complexe et cohérent des pratiques des enseignants de mathématiques : une double approche. *Canadian Journal of Science, Mathematics and Technology Education (La revue canadienne de l'enseignement des sciences, des mathé matiques et des technologies), 2*(4), 505–528.

Rouchier, A. (1994). Naissance et développement de la didactique des mathématiques. In M. Artigue, R. Gras, C. Laborde, & P. Tavignot (Eds.), *Vingt ans de didactique des mathé matiques en France. Hommage à Guy Brousseau et Gérard Vergnaud* (pp. 148–160). Grenoble, France: La pensée sauvage editions.

Said, E. W. (1978). *Orientalism*. London: Routledge & Kegan.

Schubring, G. (2010). Historical comments on the use of technology and devices in ICMEs and ICMI. *ZDM, The International Journal on Mathematics Education, 42*(1), 5–9.

Strässer, R. (1994). A propos de la transposition franco-allemande en didactique des mathématiques. In M. Artigue, R. Gras, C. Laborde, & P. Tavignot (Eds.), *Vingt ans de didactique des mathématiques en France. Hommage à Guy Brousseau et Gérard Vergnaud* (pp. 161–176). Grenoble, France: La pensée sauvage editions.

Sträßer, R. (2009). Instruments for learning and teaching mathematics an attempt to theorize about the role of textbooks, computers and other artefacts to teach and learn mathematics. In *Proceedings of the 33rd Conference of the International Group for the Psychology of Mathematics Education* (Vol. 1, pp. 67–81).

Sutherland, R., & Balacheff, N. (1999). Didactical complexity of computational environments for the learning of mathematics. *International Journal of Computers for Mathematical Learning, 4* (1), 1–26.

Trouche, L. (1997). A la recherche d'une méthode d'étude de "l'action instrumentée". In *Actes de l'Université d'été « Des outils informatiques dans la classe aux calculatrices symboliques et gé ométriques : quelles perspectives pour l'enseignement des mathématiques »* (pp. 113–148). Rennes, France: IREM.

Trouche, L. (2009). Penser la gestion didactique des artefacts pour faire et faire faire des mathématiques : histoire d'un cheminement intellectuel. *L'Educateur, 0309*, 35–38.

Trouche, L. (2004). Managing the complexity of human/machine interactions in computerized learning environments: Guiding students' command process through instrumental orchestrations. *International Journal of Computers for Mathematical Learning, 9*, 281–307.

Trouche, L., & Drijvers, P. (2014). Webbing and orchestration. Two interrelated views on digital tools in mathematics education. *Teaching Mathematics and Its Applications, 33*, 193–209. doi:10.1093/teamat/hru014.

Vergnaud, G. (2000). *Lev Vygotski. Pédagogue et penseur de notre temps*. Paris: Hachette.

Vergnaud, G. (2009). The theory of conceptual fields. *Human Development, 52*, 83–94.

Vérillon, P., & Rabardel, P. (1995). Cognition and artifacts: A contribution to the study of thought in relation to instrumented activity. *European Journal of Psychology of Education, 10*(1), 77–101.

Warfield, V. M. (2014). *Introduction to didactics*. SpringerBriefs in education 30. New York: Springer.

Chapter 11
Discussion of Issues in Chapters in Part II

John Monaghan, Luc Trouche, Jonathan M. Borwein, and Richard Noss

11.1 Introduction

This chapter is the second 'space for reflection' in this book; an opportunity for Jon and Luc to comment on John's three chapters and Luc's chapter. Richard Noss, a noted scholar and designer in the constructionist tradition in the area of mathematics and digital tools, has kindly agreed to join the discussion that led to this reflective chapter.

To structure the discussion John designed seven questions under four headings and Jon, Luc and Richard responded as they saw fit. The bulk of the text below presents the questions and the responses. 'I/my' refers to John and the questions follow the sequence of chapters in Part II.

11.2 Space for Alternative Conceptions on the Development of Tools

Chapter 7 provides a 'potted history' of the development: of tools; in understanding of the place of tools in activity; in mathematics education (of tool use in this field). Chapter 7 is my interpretation and, as such, is open to bias from my experiences, understandings and interests. In Sect. 7.2 I focus on the period from the 1960s to the present as a period which witnessed a flowering of ideas and technological developments relevant to tool use in mathematics education. My temporal focus here may simply reflect my own development as it was the period when I grew up.

Question 1

Is the period from the 1960s to the present a period which witnessed a flowering of ideas and technological developments relevant to tool use in mathematics and mathematics education?

J. Monaghan et al., *Tools and Mathematics*, Mathematics Education Library,
DOI 10.1007/978-3-319-02396-0_11

Responses

Richard: Well it's certainly true that the sixties represented the beginning of time: the first point at which anyone could reasonably claim that computer-use might mediate learning of anything, let alone maths. I think one can reasonably make the case that although tools were a topic in what there was in maths education (of course much less than now—and mercifully so perhaps), it was the computer—its expressive power and now its ubiquity—that has disrupted mathematical learning design and teaching practice to the point where 'tool' is hardly broad enough to characterise it.

Luc: To answer this question, we have to distinguish between mathematics and mathematics education. "For me, there are four elements leading to 'the flowering of ideas and technological developments relevant to tool use in mathematics education': the evolution of tools as supports of thinking, the evolution of schooling, the evolution of 'who is using tools'? and the evolution of perspectives in mathematics education". For example:

- Evolution of tools as supports of thinking: see the creation of writing and the developments of tools for mathematics learning in the scribal schools or the invention of printing
- Evolution of schooling: the necessity of addressing a large audience of heterogeneous students leads to the introduction of blackboards in school (and subsequent discussions on their legitimacy, as they replace oral interaction by written interactions)
- Evolution of 'who is using tools': the discussion is all the more important that the first users of tools are far from the math teachers (see the discussion on the abacus or on calculators in classroom)
- Evolution of perspectives in mathematics education: see the beginning of the twentieth century, where the mathematicians pleaded for a more active way of teaching mathematics

The feature of the period 'from the 1960' is that it meets these four evolutions: digital metamorphosis, generalisation of instruction, 'digital natives', inquiry-based mathematics teaching. Probably the first time in history where these four conditions meet with such an intensity.

Jon: I think the current cascade of new technological resources has much to offer and the ride has just begun. I hope I have illustrated this in my Chapter on homo habilitation mathematicus. That said, as I have responded in question 3, I suspect the long-term consequences remain to be identified. Moreover larger sociotechnological issues dominate which technologies flourish—if Facebook or Google sees the merit in a current tool then it will be developed but if not it is very hard for the community to find the level of resources needed to ensure successful robust and accessible implementation.

Remaining in Sect. 7.2, I am rather scathing about Piaget on tools, that he said nothing about them. Maybe Piaget had so many other important things to say that he simply didn't have time to focus on tools. Further to this, I ascribe a form of 'tool blindness' to researchers who continued work along Piagetian lines (starting from radical constructivists)—maybe I am simply unaware of post-Piagetian research on tool use in ontogenetic development.

Question 2
Am I being unfair on Piaget and post-Piagetians?

Responses
Richard: Well yes, you're being a bit unfair, although the failure to conceptualise tools (or contexts) limits the generalizability of Piaget's findings. This is the key contribution of Papert's work.
Luc: Yes, a bit unfair. Actually, for Piaget, learning comes from interactions with objects in various contexts. He probably underestimates the importance of mediations (of tools as well of institutions, mainly schools). This is the key contribution of Vygotsky's work.

11.3 On Theory and Theories

In mapping the content of this book, Luc and I took an early decision that there were three 'movements' in mathematics education that were particularly interesting with regard to tool use: constructionism; activity theory (AT); and work originating in twentieth century French didactics. I shall come on to questions specific to each of these movements in the next section but here I would like to consider the place and importance of 'theories' (constructionism and activity theory could be called 'theories' and the chapter on French didactics outlines several approaches that could be called 'theories'). Before framing my question I'd like to note that I think 'theoretical considerations' (including stating epistemological and ontological assumptions and principles regarding what it is to do and to learn mathematics) are important but (1) the theories used in mathematics education are quite different things to theories in the physical sciences and (2) theories do not exist without people to interpret them and different people may interpret a theory in different ways. I mention this simply to note that I do not see theories in mathematics education as being without problems.

Question 3
To what extent is a theory needed to understand tool use in mathematical activity?

Responses

Richard: I don't know the answer to this question but I'm sure that anyone attempting to answer it ought first to have read diSessa and Cobb (2004).

Luc: A complex question, that I could subdivide into different issues: is a theory needed to understand a specific aspect of human activity? Is a theory needed to understand this *specific* aspect of human activity that is 'tool use in mathematical activity'? Is a theory needed to understand this *more specific* activity that is 'tool use in mathematics education'? Is a *specific* theory needed to understand this specific aspect of human activity that is 'tool use in mathematical activity'? And, at least, what does 'to understand' mean?

Some elements of a personal point of view:

- Each human, aiming to accomplish a given activity needs to *understand* it. No need for a theory, but effective need for developing a reflective point of view on this activity (supported by social practices, in school, in a community of practice, etc.).
- The purpose of a science is not only to understand a given phenomenon, but to make it socially understandable.
- In this perspective, different theories could allow one to understand what is at stake in 'tool use in mathematical activity' (as in this book)...
- I do not think that a single theory is able to grasp the whole complexity of tool use in mathematical activity: personally, when I think 'didactical situations', I have in mind the theory of didactical situations; when I think 'institutions', I have in mind the anthropological theory of didactics (ATD), when I think 'mediation', I have in mind Vygotskian theory...
- This kind of theoretical ubiquity is viable only if, for the particular topic I am working on—as it happens, the interaction between teachers and resources, I am able to build a kind of theoretical ecosystem, combining diverse theoretical approaches, being aware that this combination is relevant only within the perimeter of the topic I am working on.

Jon: To 'understand' I suppose one must have a theory. But as with the logical foundations of mathematics which are central to the interests of mathematical philosophers and logicians, explicit theorising has little direct impact on either mathematics teachers or researchers.

As long as education faculties function largely independently from mathematics departments and as long as evidence-based educational theory remains unusual, I do not expect things to change. There will be

(continued)

periodic edicts from ministries of education and changes to curricula in university, none of which will have the intended impact. Eventually, various tools will become embedded in educational practice whether or not their impact is understood and whether or not the teachers are sufficiently expert to properly exploit their use. Since the media and technologies are still in rapid evolution, it may well be impracticable to expect more. Have we as educators yet properly integrated the Gutenberg revolution into our teaching style?

I also suspect that for profound cultural reasons the answer to this question looks quite different in each of, say, Hong Kong, France and Australia.

11.4 Constructionism and Activity Theory

I had an interesting experience in writing the chapters on constructionism and on activity theory. I felt I knew them (and the mathematics education literature related to tool use they stimulated) quite well before I started each chapter and I expected them both to say a great deal about tool use in mathematics. But in my reading, and the subsequent synthesis of this reading for each chapter, I was a little disappointed with what they had to say about tools. I summarise these little disappointments as follows:

11.4.1 Constructionism

Mindstorms is a fascinating book but it says very little about tools per se. *Windows on mathematical meanings* (WMM) gives greater insight into tool use in mathematical activity than anything that went before and, oddly, after—that is, the constructionist community (of which Richard is a part) post WMM (1996) did not take 'the place of tools in learner meaning making in mathematical activity' beyond anything done in WMM.

11.4.2 Activity Theory

I did not find an AT view on tools in mathematical activity but, instead, found multiple AT views on tools in mathematical activity. In retrospect I should not have been surprised because the 'unit of analysis' matters a great deal in consideration of tool use (basically that AT provides and insight on tool use when the unit of analysis has mediated action tools but when the unit of analysis is the activity system itself, AT does not provide great insight on tool use).

Question 4 (for Richard)
Is my 'little disappointment' with constructionism (from a tool use perspective) justified?

Richard's Response
Yes it is justified. Constructionism isn't really a 'theory' in the sense of, say, constructivism or 'evolution' or 'string theory' (yes I know science and social science have different criteria and meaning for the word). But it is true that the constructionist community has so far manifestly failed to situate the idea into the broader theoretical culture—a great failing of 'Windows' too: one day we will finally say more!

I also had a question on activity theory but there was only a short comment from Richard, 'I alternate between thinking it's mainly obvious and that it's used too formulaically to be useful (triangles!!)'.

11.5 On French Didactics

I found the chapter on French schools of thought fascinating in terms of the context provided. I have followed these schools of thought for several decades, so there was little new for me in terms of what theoretical frameworks say, but in terms of contextualising these frameworks within wider mathematical and educational movements I learnt a great deal. Of the many questions I could ask I have selected two. These questions are rather specialised and so I do not really expect anyone except Luc to answer them.

My first question relates to similarities and differences constructionism (as advanced in *Windows on mathematical meanings*—WMM), the theory of didactical situations (TDS) and the ATD with regard to the place of tools in learner meaning making in mathematical activity. My interpretation of the similarities and differences in these three frameworks is briefly summarised as follows. All three frameworks are centrally interested in learners' mathematical actions. In WMM the focus is on the joint design of tasks and tools, which allow learners to make connections/mathematical relationships. In TDS the teacher designs the milieu (which includes tools) to facilitate learners formulating and validating a pre-determined mathematical understanding. ATD is also interested in the milieu (which includes ostensives) but individual meaning making is viewed via institutional practices which overshadows individual meaning making.

Question 5
What is your reaction to my summary?

Luc's Response
I certainly agree with John's implied point that constructionism and didactiques are fully compatible as theories, although this might be because they belong to different forms of theory (see the diSessa and Cobb paper mentioned above).

- Two nuances: For the TSD, the *milieu* is not done once for ever. Students interact with the milieu, and, in this measure, contribute to its design. In the tread of this theory, Sensevy (2009) and his colleagues developed a theory of the *joint action* of a teacher and her students, all of them having a responsibility to the progression of the knowledge in a given classroom.
- For the ATD, instead of 'institutional practices', I would speak of 'institutional constraints', that influence the relationships of the individuals to knowledge and the way they accomplish tasks, using various ostensives.

My second question on French didactics concerns the instrumental approach and its relation to Leont'ev's approach to activity theory (activity–actions–operations). A strength of the instrumental approach is that it makes few assumptions but has a wide field of application. Leont'ev's approach to activity theory can be used, as was seen in Chap. 9, to shed light on the relationship between learners and their environments including the process by which an artefact becomes a tool for learners. The instrumental approach has the potential to enhance our understandings of the action and operation aspects of Leont'ev's approach without compromising any of its basic assumptions.

Question 6
What is your reaction Luc? Can the instrumental approach and Leont'ev's approach to activity theory be 'networked'?

Luc's Response
Before answering to your question, I would like to be sure to correctly understand what do you mean by 'the instrumental approach makes few assumptions'. For me, precisely, it can be used 'to shed light on the relationship between learners and their environments including the process by which an artefact becomes a tool for learners'. Perhaps we need to distinguish the Rabardel's approach, and the result of its appropriation by some French didacticians?

(continued)

For my own experience, the limitation of the instrumental approach is in its consideration of social processes (even if Rabardel himself, in his seminal book in 1995, evokes social schemes, or a social part of schemes). This 'missing resource' leads my doctoral students, who aimed to capture social processes, to use other theoretical frameworks, as communities of practice (Sabra, 2011) or activity theory (Hammoud, 2012). The communities of practice framework were useful for its concepts of participation and reification whilst the activity theory framework was useful for its notion of rules and division of labour. In a recent paper Gueudet et al. (2015), we have used both the documentational approach of didactics and CHAT to study the collective design of an e-textbook, analysing both the activity system of the community of designers, and the documentational genesis of the designed resources.

References

diSessa, A. A., & Cobb, P. (2004). Ontological innovation and the role of theory in design experiments. *The Journal of the Learning Sciences, 13*(1), 77–103.

Gueudet, G., Pepin, B., Sabra, H., & Trouche, L. (2015, online). Collective design of an e-textbook: teachers' collective documentation. *Journal of Mathematics Teacher Education.*

Hammoud, R. (2012). *Le travail collectif des professeurs en chimie comme levier pour la mise en œuvre de démarches d'investigation et le développement des connaissances professionnelles. Contribution au développement de l'approche documentaire du didactique.* PhD, University Lyon 1.

Rabardel, P. (1995). *Les Hommes et les Technologies, une Approche Cognitive des Instruments Contemporains.* Paris: Armand Colin.

Sabra, S. (2011). *Contribution à l'étude du travail documentaire des enseignants de mathé matiques : les incidents comme révélateurs des rapports entre documentations individuelle et communautaire.* PhD, University Lyon 1.

Sensevy, G. (2009). Outline of a joint action theory in didactics. In V. Durand-Guerrier, S. Soury-Lavergne, & F. Arzarello (Eds.), *Proceedings of the Sixth Congress of the European Society for Research in Mathematics Education*, January 28th–February 1st 2009, Lyon (pp. 1645–1655). France: INRP.

Part III
Selected Issues with Regard to Tool Use in Mathematics

The chapters in this part consider selected issues in tool use in mathematics education. There are many issues regarding tool use in mathematics education which could be considered. The issues selected (the curriculum, the calculator debate, mathematics in the real world, and the mathematics teacher) were debated (thrashed out!) by John and Luc over 2 days in the initial planning of this book. There is, then, a sense in which they reflect personal bias and also compromise. But these issues are, we maintain, important issues.

Chapter 12 considers curriculum and assessment policies with regard to the integration of digital technologies into the learning and teaching of mathematics. With regard to curricula, it focuses on interrelationships between tools and mathematics curricula and argues that: school mathematics develops from 'really used' tools; the development of tools is related to the implemented as well as the intended curricula. Assessment is viewed as a 'problem area', and the final section of the chapter uses French policy on assessment as a case study to examine tensions that can arise.

Chapter 13 considers the longstanding debate on the place and value of the calculator in the learning, teaching, and assessment of mathematics. The calculator has inspired one of (if not the most) the controversial debates regarding tool use in mathematics education. After an introduction to the issues, actors, and charges, this debate will be viewed by a framework informed by Wertsch's ten theses on mediational means. The chapter ends with a consideration of the future of this debate.

Chapter 14 focuses on mathematics in the real world and the problem of linking this mathematics with school mathematics. This leads us to address questions concerning the nature of mathematics. We consider the use of tools in leisure and in working practices. Tool use is omnipresent in out-of-school mathematics, but school mathematics privileges specific tools. The chapter considers the problem of 'suspension of sense making' in school mathematics and opportunities for using real-life artefacts to link in-school mathematics to out-of-school mathematical activities. The increasing presence of digital technology in everyday life and work opens up new opportunities (and problems) for linking in-school to out-of-school mathematical activities.

Chapter 15 considers the teacher with regard to mathematical knowledge and the use of technology. The teacher, once jokingly referred to as something that could be replaced by teaching machines, is arguably more important in classrooms where digital technology is a central feature than those without. But mathematics teachers, en masse, are often reluctant to enact deep integration of digital technology in their classrooms—why is this? A consideration of this question will include a critical review of attempts to categorise forms of teacher knowledge and what teacher should do.

Chapter 12
Integrating Tools as an Ordinary Component of the Curriculum in Mathematics Education

Luc Trouche

12.1 Introduction

We have evidenced the importance of tools for meeting the needs of computation in society (and for the development of culture itself); the development of computation tools and the development of writing appearing deeply interrelated (Chaps. 4 and 5). The crucial importance of tools for mathematicians themselves has been illustrated in Chap. 3. I would like, in this chapter, to focus on the use of tools for doing mathematics in the part of a society dedicated to learning: schools.

The use of tools in school mathematics can be seen through the curriculum, understood as the *vision* that a given society has of/for its school. This vision is the result of interrelated pressures of multiple agents (politicians, tools manufacturers, pedagogues, mathematicians, parents. . .), what Chevallard (1987) names the '*noosphère*'. The curriculum designers have indeed—theoretically—to take into account many factors: school equipment, equity, teachers training. . .

This use of tools depends also on the actors directly involved in school mathematics, mainly teachers and students, who can import proscribed tools into mathematics classes, or, on the contrary, be reluctant towards tools prescribed by the curriculum (see Chap. 13, the 'calculator debate'). We have thus to distinguish between the *intended, implemented, achieved* and even the *hidden* curriculum (Kelly, 1977).

I would like to show, in this chapter, how tools condition the development of these different curricula. Mathematics learning indeed develops 'under the umbrella' of the 'really used' tools: the fact that the computation clay tablets, in Mesopotamian tribal schools (Sect. 5.2), were used as a construction material, 'trapped in walls, floors and foundations' of schools, gives us a nice metaphoric illustration of this structuring effect of tools.

Each curriculum develops as a complex structure with some 'tools-sensitive' points. Among them, the question of *assessment* constitutes certainly a keystone; I will try to justify this.

J. Monaghan et al., *Tools and Mathematics*, Mathematics Education Library,
DOI 10.1007/978-3-319-02396-0_12

This chapter is organised in three sections setting the scene: the first section proposes a vertical (historical) point of view, aiming to evidence the continuity of some issues over the time; the second gives an horizontal (international comparative) point of view, aiming to evidence, beyond the national peculiarities, some common features; the third section proposes a case study, the French policy on assessment, seen as paradigmatic. The conclusion addresses some questions and draws some perspectives for further studies.

12.2 An Ancient Story

I wish to analyse, in this section, interactions, over the time,[1] between the development of curricula in mathematics and the evolution of tools. I analyse these interactions, firstly, as a 'mechanical' effect of news tools on the curriculum in mathematics; deepening this analysis, I evidence secondly a tension between the evolution of curricula and the integration of tools; finally, these tensions led to rebalance the needs for experiencing tools *at the fringe of curriculum*, and the needs for integrating them as an ordinary, even if it is crucial, *component of this curriculum.*

12.2.1 A 'Mechanical' Effect of Tools on Curriculum in Mathematics

Tools, being historically and culturally situated artefacts, emerge in a given society, and vanish, with the associated usages. Naturally, what happens in a given society, after a time of transition, happens in its school. The effects of the change of a tool on the mathematics curriculum may be very strong when this tool occupies a central place in the process of learning. It is the case for the tools dedicated to writing as clay tablet and calame in scribal schools (see Chap. 5). Lavoie (1994) shows also the consequences of the introduction of the *iron* quill (instead of the *goose* quill) for the learning of arithmetic in the nineteenth century in Québec: the easier writing allowed younger students to do computations by hand (instead of mentally), allowed them then to do longer computations, and at last, led to an earlier introduction of arithmetic in the curriculum.[2]

---

[1] For this analysis 'over the time', I have chosen, in this section, some particular moments appearing as critical: moments of transition for tools or/and curricula. This choice leads to jump over time, giving perhaps to the reader the impression of surfing over history: references are given for having means to analyse in more depth continuities and breaks.

[2] Such evolutions, linked to the writing tools, could also be analysed in the case or the transition from 'pen and pencil' to 'keyboard and screen', and then from 'keyboard and screen' to 'touch screen' (see Chap. 17).

Table 12.1 Evolution of curriculum (12th grade, scientific class), following the evolution of tools (Trouche, 2005, p. 25)

Year	Prescriptions for tools	Prescriptions for numerical computation	Prescriptions for graphic representations
1966	Use of numerical tables of standard functions and slide rules	Notably a specific chapter on numerical computation	**No reference to representative curve** except for exponential and logarithm functions
1971	Use of numerical tables, slide rules and computing machines	Numerical calculations are included in the chapter Real numbers, numerical computation, complex numbers	**The expression graphic representation appears in connection with general study of functions**
1982	Calculators will be **widely** used	No chapter specific to numerical computation: it is integrated into other chapters	**Usual use of graphic** representation will be promoted, because it plays a significant role in the behaviour of functions
1986	Calculators will be systematically used (a **basic model** is sufficient)	Numerical problems and methods play an essential role in the understanding of mathematical notions	**Graphic representations must hold a very important place** in the curriculum
1991	Calculators with statistical functions are recommended	Idem.	Idem.
	On the other hand, graphical screens are **not required**		
1998	Graphic calculators are **prescribed**	Numerical topics are introduced as an additional specialist option in mathematics	**Favouring argumentation supported by graphs**
2002	The power for investigation of computer tools and the existence of high-performance calculators, frequently at students' disposal, represent welcome progress and their impact on mathematical education is significant. This evolution has to be supported by using these tools, particularly in the phases of discovery and observation by students	Numerical topics form a domain with which informatics strongly interacts; use of various means of computation will be balanced: by hand, with the help of a spreadsheet or a calculator	

For a more recent period, I have evidenced (Table 12.1) the gradual effect of the integration of simple calculators, then of graphic calculators, on the prescription for numerical computation and graphic representation (see Sect. 12.4.1 for the taken-into-account of tool during the examinations).

Considering this evolution, I concluded:

> The introduction of graphic calculators leads to graphical representations being taken more and more into account: "A geometrical vision of problems will be developed in calculus, because geometry supports intuition with its language and its representation procedures". Consequently, we should notice that the *graphic frame* is privileged in the calculus part of curriculum, as for example in the following comment: "Deeper work is suggested on the limit of sequences, easier to tackle than the function limit at a point: the objective is ambitious, so it is advisable to remain reasonable in implementing it and to favour arguments supported by graphs". One may conjecture that the type of tool favoured has an influence on the frame of work: with scientific calculators, the numerical frame is favoured, whereas it is the graphical one with a graphic calculator (Trouche, 2005, p. 26).

I have entitled this sub-section 'mechanical effects of tools on curriculum' for enlightening the importance of these effects, but we have to keep in mind that:

- As I state in Chap. 5 (Sect. 5.6) 'new artefacts do not necessarily make old ones redundant', and phases of *transition* are often phases of *cohabitation.*
- The evolutions of the *intended* curriculum do not lead 'mechanically' to evolutions of the achieved *curriculum* (see Chap. 15).

Actually, even regarding the intended curriculum, the relationships between tools and curriculum are far for being a one-way relationship, as I will underline in the following section.

12.2.2 Critical Tensions Between Evolution of Curriculum and Integration of Tools

It is difficult sometimes to know if tools move from society to school (tools conditioning then mathematics teaching), or if the needs of mathematics teaching call for new tools to be introduced for learning and teaching purpose (mathematics teaching conditioning the development of tools). Again, I will illustrate this with regard to Mesopotamian scribal schools: I have explained (Sect. 5.3) how the incisions of signs, with a 'calame' (a stiletto), on a malleable media—a clay tablet—result in an embossed writing, and that signs should be read with lighting that allows the reader to identify all incisions in order to avoid misinterpretation. Then one can hypothesis that the scribal schools took place in open air, outside any house: 'It is now agreed that much teaching, reading and writing was necessarily done outside in the courtyard rather than indoors; this can be inferred not only from the need for bright light that most Assyriologists recognise from their own experience with tablets but also from telling passages of the E-dubba literature' (George, 2005, pp. 130–131). It is difficult to decide if this kind of writing imposes school to stand 'out of the walls', or if the way of meeting people and discussing in open air, in this geographic area, allows such a writing to be invented... There is clearly a dialectical relationship between the birth of a form of schooling (seen at large, including a form of curriculum) and the birth of a form of writing.

Four thousand years after the scribal schools, if we look at schools in Africa, standing outside, using slates for writing, we could consider them as a remote echo of the scribal schools. Four thousand years after also, looking, in the developed countries, at the use of tablets by students out of school, we can remark on some main differences: light has to come *from outside* in the clay tablets case, light comes *from inside* in the digital tablet case; the teacher is 'in the circle' in the Mesopotamian case, this is not necessarily the case today ...

Dialectical relationships between tools and curriculum are particularly visible at each moment of renewing mathematics teaching. It was the case at the beginning of the twentieth century, in France and Germany, following the reflections of prominent mathematicians (Poincaré and Klein), leading to deep reforms of science and mathematics teaching. The French reform (1902 in France, see Sect. 10.2.1) underlined the importance of *experience* for learning mathematics and science. Maschietto and Trouche (2010) relate, in an article written in the journal 'L'enseignement des mathématiques'[3] just before this reform:

> Integrating tools [geometrical instruments, striped and squared papers...] into mathematics teaching appears as interesting both from a practical and pedagogical points of view:
>
> These devices could become a precious help from a practical point of view with some improvements, which will probably happen. Moreover, the explanations of what underlies these devices enable the sticking into the mind, the fixing in memory, the concentration of students' action on some theories, which become in this way more visible. Here there is, on a pedagogical point of view, a set of questions, which are of the highest interest for all the teachers (Maschietto & Trouche, 2010, p. 35).

The integration of (sometimes ancient) tools into mathematics teaching and the evolution towards a more active way of learning mathematics appear thus closely linked. And the questions related to these two evolutions are interrelated. It appears clearly at the occasion of the foundation of ICMI, in 1908, in their proceedings:

> In elementary teaching, one can mention by, for instance, the folding of paper, the open air activities, the usage of simple instruments for measuring the geometry of observation, etc., practical and approximate computations (degree of precision, logarithms with different numbers of decimals, the usage of slide rules, etc.), the general question of graphics in algebra, the more widespread use of squared paper. Mathematics laboratories have been recently evoked. What has been done in this direction? With what results? (Maschietto & Trouche, 2010, p. 35).

Ferdinand Buisson was the director of the French primary teaching from 1879 to 1896. He coordinated, with more than 350 collaborators, an impressive editorial project, 'Le dictionnaire de pédagogie et d'instruction primaire', which had two editions, 1887 and 1911 (Buisson, 1911), setting the scene, from the noosphère's point of view, of what was to be taught and how, and witnesses major controversies.

---

[3] Founded at the end of the nineteenth century, the journal 'L'enseignement mathématique' was the place where mathematicians exchanged ideas about teaching. It became the official journal of the International Commission for Mathematical Instruction from its creation in 1908 (http://www.unige.ch/math/EnsMath/).

One of them is about the use of abacus for learning mathematics, a controversy which was replayed in the twentieth century (see Chap. 13, the ‘calculator debate’):

> What matters, however, is to determine in what way and to what extent the use of the abacus is to be approved. It met serious opponents. One of them, Mr. Rambert, professor at the Institute of Technology in Zurich, said about abacuses contained in Vienna World’s Fair (1873): “The abacus corrupts the teaching of arithmetic. The main purpose of this teaching is to exercise, in early childhood, the faculties of abstraction, to teach her to see through the eyes of the mind. Putting things in front of the body it goes directly against the spirit of this teaching. Nature has given children their ten fingers as a natural abacus; instead to give them a second, she must learn to do without the former. It is said that the abacus gives much ease teachers for explanations. I think so. We quickly counted on the abacus as 10 and 10 are 20; but the child who relies only on the abacus lost his time, while the one who counted by head made the most useful exercise. We need a complement and corrective to teaching by sight; it is the mental calculation, which will give them. »
>
> The discerning and critical spirit has perhaps confused here abacuses with calculators. We have done elsewhere our express reservations about calculating machines, as ingenious they are. A judge of great authority, Mr. Sonnet, perfectly said: “Mental mathematics is the basis of any instruction regarding the calculation; any machine that pretends to supplement the mental calculation goes against the purpose of teaching.” But the abacus is not an arithmometer: it facilitates the work of the student, but it does not remove it; and indeed it is only intended for very young children;
>
> As well observed MA Lenient in a series of studies on the abacus, “pointing to the child, making him see the results of addition, subtraction, multiplication or division, the abacus reduces the effort and fatigue of the child; but by the testimony of his eyes, he graves in his mind and in his memory all these results that he has to memorise. The abacus prepares and initiates mental calculation: we never thought it could be replaced. »
>
> We want the child gets used to “see with her mind,” that’s fine; but it is still necessary that he first learns to learn with his eyes. Before the abstract, the concrete; before the formula, the image; before the pure idea, the sensible idea: it is the general law of a true pedagogy (Buisson 1911, entry ‘abacus’—*le boulier* in French, our translation).

Thus, the discussion about tools and curriculum cannot be summarised as ‘Are new tools to be integrated in mathematics teaching and how’. The relationship between tools and curriculum is a dialectical one (Sect. 10.5), the issue at stake is ‘how is it possible to profit from the available artefacts for achieving the goals assigned to mathematics teaching and learning in a given curricula’. In this perspective, Chevallard (1992) analysed ‘the integration and viability of digital tools’ in mathematics teaching; he stated that the question was not, following the evolution of tools, to change *the content* of mathematics to be taught, but to rethink *the way* they were taught, taking into account the conditions and constraints of the educational system. Thinking the way for renewing a given curriculum leads one to think the evolution of a whole system, and not only of its fringes.

12.2.3 From a Naïve Idea of ‘Tools Improving Mathematics Education’ to a More Balanced Point of View

Focusing on recent history—the digital area—is interesting for at least two reasons: the digital revolution affects simultaneously mathematics, tools for doing mathematics and tools for communicating; and this revolution rapidly develops over the time, offering more and more complex tools for mathematics doing and teaching—for human thinking. Therefore it leads to very intensive discussions in the mathematics education community, opening a window for analysing some dramatic evolutions over a short period of time: it clearly appears a trend from studies focusing on *natural influences* of tools on mathematics and its teaching to studies focusing on *complexity of curricular evolutions*.

Even if they have not the ambition to give official recommendations for the teaching of mathematics, the 'ICMI studies'[4] constitute certainly a precious means to know how the international community of mathematicians and mathematics educators understands a given problem, and think of the curriculum in mathematics. They aim to give a state of the art on a problem that is 'ripe for a serious international study' (Kahane, 2008, p. 19). Two ICMI studied addressed the issues of tools in mathematics teaching: the first one in 1985—and it was also the first ICMI study—(Cornu & Ralston, 1992), the second one in 2006 (Hoyles & Lagrange, 2010). Looking at them allows us to highlight the main problems at stake, and the evolutions.

The titles of these two studies are: '*The influence of computers and informatics on mathematics and its teaching*' on the one hand, '*Mathematical education and digital technologies: Rethinking the terrain*' on the other hand. These titles are significant of the evolutions happening in 21 years between the two conferences:

- Technological evolutions: the first title evokes computers and informatics, while the second title evokes digital technologies
- Didactical evolution: the first title focused at first on mathematics themselves, while the second title focuses only on mathematics education
- Epistemological evolution: the first title mentions a one-way *influence* of computers on mathematics teaching, while the second title situates at the same level of mathematics education and technologies

Looking at the content of the two studies itself, some major features appear:

- The first ICMI study focused on mathematics teaching at the university level, while the second study addresses the whole curriculum from primary schools onwards
- In the first study, technologies are considered as given, while in the second one technologies are considered as open to improvements through their usages, the questions of design being addressed as a central challenge
- Questions of teacher education are addressed in the two studies, but the second one seems to have a wider scope, taking into account the whole system, from pre-service education to professional development programmes, the final objective being to make teachers capable to master technology in their mathematics classrooms

[4] The ICMI studies are launched by the International Commission on Mathematical Instruction, see http://www.mathunion.org/icmi/conferences/icmi-studies/introduction/.

- In the first study, the teacher mainly appears alone in her classroom, while the ideas of communities (of learners, teachers), networks and collaboration strongly appear in the second study
- The issues of assessment are quite marginal in the first ICMI study (the word itself does not appear in its table of contents), whereas a whole section (of 4) is dedicated to this issue in the second ICMI study.

Following the ideas evoked at the creation of ICMI in 1908 (Sect. 12.2.2), the two ICMI studies underlined that technology—computers—had greatly increased the possibilities of observation and conjecturing in mathematics was changing the way of teaching mathematics and had the potential to bring school mathematics closer to the conditions of researching in mathematics. In this perspective, the learning of mathematics, in the context of laboratories activity, seems to be close to the work of a 'computer-assisted quasi-empiricist mathematician', as Jon describes himself in Chap. 3.

Besides, the second study had a more balanced point of view, distinguishing between *potential* and *actual* use, intended and achieved curriculum, and focusing on access, equity and social–cultural issues. In the same period, the necessary didactical transposition, from the experimentation approaches of a mathematician in a mathematics laboratory to the experimentation approaches of teachers and students in classroom are also carefully analysed (Lagrange, 2005).

These evolutions seen through the eyes of the two ICMI studies can be confirmed by the meta survey (Lagrange, Artigue, Laborde, & Trouche, 2003) of the research literature on mathematics and technology, taking into account the period 1992–1998 (just 'in the middle' of the two ICMI studies, cf. Sect. 10.5.2), evidencing, during this period, emergent dimensions: the *institutional* dimension (focusing on the difficult viability of technology in school), the *instrumental* dimension (analysing constraints, evidencing complexity of appropriation processes) and the *situational* dimension (taking into account the new economy of problem solving, and thinking the curriculum as a whole).

After this historical view, evidencing some major trends, I would like to propose now a cross-national view on the current curricular situations, researchers speaking to institutions.

12.3 International Situation, Specificities, Invariants and Challenges

Researchers speaking to institutions: I refer in this section to two reports written by researchers for enlightening educational policies. We have chosen recent reports (for the reasons evoked Sect. 12.2.3), covering the whole mathematics curriculum (basic education in the first case, upper secondary education in the second case), addressing a large range of curricula all over the world, and written by experts in the field of ICT in mathematics education. In the first part, I lean on a report realised in

2011 for UNESCO by Michèle Artigue, former president of ICMI, in the second part, I rest on a report realised in 2014 by four researchers, experts in this field (Drijvers, Monaghan, Thomas, & Trouche, 2014) for International Baccalaureate (http://www.ibo.org). This report raises the issue of assessment, which is important to consider as the assessment of a curriculum is an integral part of a curriculum; assessment will be analysed in the third part of this section.

12.3.1 Main Common Challenges for Technology in/for Basic Mathematics Education

The report coordinated by Artigue (2011) is in response to a request of UNESCO. It concerns basic mathematics education, and leans on the literature of research and the contribution of a range of experts, well aware of the curricular situation and challenges faced by their own country. The audience targeted by this report is, beyond the policy makers, the actors of mathematics education: 'It will be of use not only to decision-makers wanting to mainstream quality SME (Science and Mathematics Education) education into their systems, but also to stakeholders who wish to participate in the change process' (p. 3).

Technology appears, from the beginning, as a question of major interest, as the UNESCO Assistant Director-General for Education underlines:

> All the experts agreed that the last decade has witnessed the development of a substantial body of knowledge on SME and the production of valuable tools and resources, many of which are now widely accessible thanks to technological advances (p. 3)

The technological issues indeed appear at a number of places in the report, and one of its sections is dedicated to technology and mathematics, underlining two major shared challenges:

- Questioning the gap between the potential of technology for the teaching and learning of mathematics and their weak effective use in basic mathematics; trying to reduce this gap, particularly in developing new modes of teacher training:

> These technologies [calculators, spreadsheet, dynamic geometry software and micro worlds such as Logo] have undeniably enriched opportunities for experimentation, visualization and simulation and have modified relations with calculation and geometrical figures. They have brought school mathematics closer to the outside world by making it easier to process more complex data and to handle more realistic problems. However, in spite of their undeniable potential for enhancing the teaching and learning of mathematics and their many positive achievements, they have to date had little effect even in education systems that strongly encourage their use. Recent work on teachers' practices in computer environments is beginning to give insights into this situation, and forms of training properly adapted to teachers' needs are being considered. Nevertheless, the issue of widespread effective use of these technologies in basic mathematics education remains for the moment unresolved (Artigue, 2011, p. 35).

- Questioning the gap between the abundance of burgeoning Internet resources and the lack of resources well fitted to teachers' needs and capacity of use; trying to reduce this gap by supporting the development of collaborative tools and teachers communities[5]:

> Recent developments, such as those relating to the growth of collaborative learning tools, the Internet and mobile technologies, have given rise to different opportunities with differing impacts, as evidenced by the option of using technology to support forms of collaborative mathematics learning by students, free online access to a range of resources, new options for organising distance education and support for the collaborative production and sharing of resources, for the emergence of communities of teachers and researchers and for networking and remote exchanges between students and teachers (Artigue, 2011, p. 36).

The emergence of Internet as 'a potential active partner in the field of teaching mathematics' appears, in this period, in other reports, for example the report *Knowledge and Training of Secondary School Mathematics Teachers* supervised by the Israel Academy of Sciences and Humanities, written in the frame of the Initiative for Applied Education Research.

> Continuing and even increasing support for the existing Internet site at the National Center for Secondary School Math Teachers. The site is a rich resource that supports the professional development of teachers and the formation of a community of teachers. It is very important to continue to develop the Internet site and to use it to increase internal communication within the community of teachers and between the community of teachers and other professional communities. Today it is clear that the Internet is an active tool accompanying the teaching of mathematics in Israel and abroad, and that due to its great potential it should be transformed from a tool that accompanies the learning process into an active partner in the field of teaching mathematics in general (Gutfreund & Rosenber, 2012, p. 109).

Some years later, these two challenges pointed out by these reports seem to be still relevant, questioning national policies, that try to find solutions, not always the same ones, as I will analyse in the following part.

12.3.2 Diverse Answers for Common Challenges

I will draw in this part from the recent report *Use of Technology in Secondary Mathematics* written to the request of the International Baccalaureate by a team of four researchers coordinated by one of the author of this book, John Monaghan (Drijvers et al., 2014). The International Baccalaureate (IB) commissioned these researchers to write this report:

> to provide insights into the use and integration of technology into curriculum, classroom practice and impact on learning in secondary mathematics courses and inform possible direction and focus for the coming curriculum review of IBDP (international baccalaureate diploma program) mathematics (Drijvers et al., 2014, p. 3).

[5] To be underlined: two annexes of the report are dedicated to such communities: the IREM network (see Chap. 6), and the Sesamath association (see Chap. 15).

The scope is thus different of the UNESCO report, which was dedicated to the basic mathematics education. The IB report concerns an upper level, where the mathematics is more advanced. The IB proposed a set of questions to be addressed, some of them related to IB priorities, other of a more general scope, about curriculum, assessment and implementation strategies. These questions were answered by the authors regarding their own countries, or the countries they had close association with. Six curricula were thus examined: England, France, the Netherlands, New Zealand, Singapore and Australia. For each country, a wide range of both curricular documents and research papers[6] supported the team members' assessments. I will retain here, for reasons of space, only the results concerning the four countries of the authors, with the hypothesis that one knows better the curricular situation of his own country.

I will examine here the issues related to curriculum, and implementation strategies, and in the following part (Sect. 12.3.3) the issues related to assessment. About curriculum, the researchers interpreted these questions as such:

Q1. Is the use of ICT explicitly part of the mathematics curricula? If yes, how is this addressed and described?

Q2. Do the opportunities that ICT offers impact on curriculum choices (e.g., integration by parts no longer needed, approximate solutions rather than exact ones, . . .)?

Q3. Is ICT used in mathematics classes on a regular basis? If yes, what type of technology (IWB, GDC, laptop, desktop, . . .)? Who uses it, the teacher or the student? Are there specific computer labs in schools, or do regular classes have ICT facilities?

Q4. Is there any funding, e.g. by governmental institutions, for ICT integration? Or other kinds of resources?

Q5. Do textbooks anticipate the availability of ICT?

Q6. Are Internet resources used in mathematics courses? Are there any plans to extend the use of digital technology in mathematics classes in the nearby future? If yes, what kind of plans? What kind of technology? Are graphic display calculators (GDCs) being replaced by other hardware such as tablets or smartphones?

For each answer, official-intended curriculum and real-implemented curriculum were distinguished. The authors were aware that these countries were all developed ones. The objective was not to design a global map of the integration of ICT in school mathematics, but to analyse the level of development being more or less the same, the differences and communalities of curricular choices regarding the integration. I present (Table 12.2) what appear as the more significant authors' answers regarding the interaction ICT-curriculum.

[6] The selected research papers were written from the years 2000 to 2014 (exceptions were made for key papers) from journals rated as A* and A (and some rated as B) in a recent European rating of mathematics education journals (Drijvers et al., 2014, p. 8).

Table 12.2 Comparison between fur countries (extracts from Drijvers et al., 2014)

About mathematics, curriculum and technology		
Country	Answers regarding the official, intended curriculum	Answers regarding the real, implemented curriculum
England	Three private organisations (AQA, Edexcel and OCR) called Examination Boards (EB) publish GCE curricula and associated examinations. An aim of all AS/A-level specification includes 'acquire the skills to use technology such as calculators and computers effectively, to recognise when such use may be inappropriate and to be aware of limitations' and the assessment objectives includes 'use contemporary technology and other permitted resources . . . understand when not to use such technology, and its limitations'	There is great variation. All students have, at least, a scientific calculator and many have a GDC. Schools, and teachers within schools, vary to the extent in which they embrace the use of technology. The writer has anecdotal evidence to suggest where technology is used extensively it is often 'local', that is a particular teacher shows students how to use a GDC to solve a specific type of question
	[AS and A-level stand for Advanced Supplementary and Advanced level]	
France	The national curriculum is under the authority of the Ministry of Education	« At a large scale, teachers consider that the new curriculum supports ICT integration. More and more teachers see ICT as real pedagogical tools
	ICT is explicitly part of the curriculum at each level, for example, for grade 11: « using software, tools for visualisation or simulation, of computing (both CAS and scientific) and of programming, deeply changes the nature of teaching in favouring inquiry-based learning » (Inspectors report)	However ICT usages change according to the high school. DGS are more and more used in classes, but there is not really analysis of their effects
	In addition, the inspectors are very supportive for ICT integration in teaching: « a reasonable use of different kinds of software is particularly fitted to mathematics teaching: it is the case for calculators, spreadsheet, CAS and DGS » (Inspectors report)	Teachers are waiting for an assessment of such tools during the final official examination (baccalauréat) » (Inspectors report)
The Netherlands	The official curricula—or rather targeted learning objectives and goals, as there are no time schedules or other prescriptions—are described in so-called syllabi available at www.examenblad.nl. The syllabi address different domains. The first domain, called Skills, mentions the use of ICT in general terms, e.g. 'The candidate can, also through the use of ICT, gather, select, process, judge and	In the reality of the mathematics classroom, ICT seems to be used more and more Pisa 2012 findings and national studies show that ICT infrastructures are relatively good in Dutch schools However, exact data for mathematics teaching seem to be lacking

(continued)

Table 12.2 (continued)

About mathematics, curriculum and technology		
Country	Answers regarding the official, intended curriculum	Answers regarding the real, implemented curriculum
	present information' and, under the heading of Algebraic skills, 'The candidate can [...] perform operations with, but also without ICT means such as a graphing calculator'. The domain-specific descriptions also in some places refer to the use of ICT, for example in the domain Differential and integral calculus: 'In appropriate situations, the candidate can set up an integral, calculate its exact value and approximates it using ICT'	
New Zealand	The written curriculum (http://nzcurriculum.tki.org.nz/National-Standards/Mathematics-standards) is divided into eight levels (with eight the highest) for the subject Mathematics and Statistics, with Achievement Objectives for each level. It is very short, comprising four pages in total, and makes general statements such in calculus:	This depends on the individual school and teacher
	Level 7: Sketch the graphs of functions and their gradient functions and describe the relationship between these graphs	
	Level 8: Form differential equations and interpret the solutions	
	The only explicit mention of technology is in Level 7 Statistics: '...calculating probabilities, using such tools as two-way tables, tree diagrams, simulations, and technology'	

CAS Computer Algebra System, *GDC* Graphic Display Calculator, *ICT* Information and Communication Technologies, *IWB* Interactive White Board

The four responses reveal some differences:

- Difference of nature between the different curricula, sometimes designed by private organisation (the case of England), sometimes by the Ministry; sometimes delivered under the form of a description of the teacher's work, sometimes delivered under the form of general standards
- Some curricula (New Zealand) seem to evoke at large the use of 'technology' whereas other curricula (France) designate specific tools (DGS, CAS,

calculators...), the reason being, perhaps, that in the second case there are a larger number of available technologies for teachers
- Some curricula (The Netherlands) make explicit the usage of technology for performing a given task, whereas in other curricula (New Zealand) these usages are left implicit (one can conjecture that 'Sketch the graphs of functions and their gradient functions and describe the relationship between these graphs' is made by using a GDC)
- Some curricula (England) underline the possible *inappropriate use* of technologies, and their *limitations*, whereas other curricula (France) underline the *reasonable* use and the *potential* of technologies

Beyond these differences, linked to cultural and historical differences, there are also commonalities, signs of a period of transition: in the four countries, the general use of technology in mathematics classrooms is globally increasing, with huge differences between schools and teachers. The integration of technology seems to remain 'local', 'that is a particular teacher shows students how to use a GDC to solve a specific type of question' (the case of England), and the teachers are reported (France) to be not always able to analyse the effects of the technology used. The taking into account of the technology during the final examinations seems to be a key point for a larger integration in classroom (Sect. 12.3.3).

One may remark that, if the description of the official curriculum is quite easy, the description of the real one is much more difficult: the answers lean on institutional reports (inspectors), or personal experience. There is clearly a lack of research analysing the real integration of technology in mathematics classroom at a large scale.[7]

The second point I would like to examine is about *implementation strategies*. The researchers interpreted the IB questions as such:

Q1. Is there a debate going on concerning the use of ICT in mathematics classes? If yes, what are the main issues and opinions?
Q2. Is there support for teachers' professional development with respect to integrating ICT in their teaching? If yes, is this technically oriented, or also pedagogical?
Q3. Is ICT used for supporting ICT integration, for example blended teacher education (pre- and in-service), online courses for professional development, MOOCs?
Q4. Are there any future plans to implement new curricula with a different role for ICT than is the case at present? If yes, how would you describe this changing role?

[7] To be noticed: the French Ministry of education and research has launched a large study for analysing the use of educational resources in four disciplines (mathematics, english, physics/chemistry and technology): the ReVEA project (Ressources vivantes pour l'enseignement et l'apprentissage) will develop from 2014 to 2018.

Q5. Please add other comments and information that you consider relevant but that is not addressed in the questions

I present (Table 12.3) what appear as the more significant authors' answers regarding the implementation strategies, focusing on teacher training, appearing as crucial in a time of transition.

The four answers echo the challenges underlined by recent international reports (Sect. 12.3.1): the need, for profiting from the potential of ICT, to develop new modes of teacher training and to sustain teachers' collaborative work:

> In all countries there are some supported initiatives for teacher professional development concerning the use of ICT. These initiatives, however, seem to be incidental, local and small-scale, rather than structural and widespread. Exceptions seem to be the work done by NCETM and MEI (EN). In the frame of research projects, small-scale PD initiatives exist in most countries. Online resources for teacher support are available in all countries, but the initiative to their use is mostly left over to schools and teachers (Drijvers et al., 2014, p. 33).

There are, of course, clear differences between countries having a centralised national curriculum (France and the Netherlands) and countries with a partition of responsibilities between local and national authorities, as well as between public and private initiatives. In the second case, it appears more difficult to analyse the reality of teacher development programmes. Even if some phenomena appear more clearly in one country than in the other ones, I analyse these as the demonstration of *tensions*, which seem to be characteristics of a period of transition:

- The tensions between the top-down development of CPD programmes and the bottom-up development of teachers associations designing resources of their own
- The tensions between a back-to-basis trend, aiming to avoid the uncertainty of the digital area, and an innovative trend, searching in the ICT potential the seeds for improving both students' learning and teachers' competencies
- The tensions between CPD organised for better using of a given technology, and CPD organised for better teaching in using a range of technologies
- The tensions between testing advanced methods for teacher education and fitting to average competencies of teachers
- The tensions between developing distance learning (one educator speaking to a number of teachers) and developing collaborative teachers' work deeply linked to local experiences

The word 'blended' (for learning, or teaching, or training) seems to be a right word for describing these tensions, resulting in provisional balances.

I have tried, in this part, to compare curricula and technology implementation strategies in four countries, analysing the communalities and variation in terms of institutional differences and tensions characterising a period of transition. I would like now to compare the curricular situation regarding the way of assessing students' learning.

Table 12.3 Supporting teachers in integrating ICT and professional development (extracts from Drijvers et al., 2014, with some updating, regarding the most recent events, as the French MOOC)

About implementation strategies		
Country	Answers regarding the official, intended curriculum	Answers regarding the real, implemented curriculum
England	There are many professional development (PD) providers in England. In 2005 the National Centre for Excellence in the Teaching of Mathematics (NCETM) was established and one of its main briefs was to co-ordinate and validate the diverse provisions of PD. Some of this PD relates to the use of technology in mathematics but the writer is not aware of any relating to technology in 16–19 mathematics other than offers for bespoke PD and the MEI. MEI's provision includes a yearlong part-time course, Teaching Advanced Mathematics, for 11–16 teachers who are starting to teach 16–19 academic stream mathematics; technology is integrated into this course	Cornerstone Mathematics (www.cornerstonemaths.co.uk) concerns 11–14 rather than 16–19 mathematics but it is worthy of comment as it aims to integrate digital technology into mathematics lessons to present mathematical ideas using dynamic representations and simulations. A pilot study, involving 19 teachers and 490 students, is being 'scaled up' to 100 schools across England (see Hoyles, Noss, Vahey, & Roschelle, 2013)
	MEI *Mathematics in Education and Industry* (MEI) is a private foundation, which publishes textbooks and runs CPD	Two independent organisations have a strong interest in technology in 16–19 academic mathematics education are:
		– *Technology for Secondary/College Mathematics (see* http://www.tsm-resources.com/), centred on the software *Autograph* and provides training, resources and an annual residential workshop
		– *Wolfram Research*, the founders of *Mathematica* support the use of technology in mathematics education and organises events for teachers (see http://www.wolfram.com/events/cambridge-feb-2014/)
France	Specific CPD programmes had been developed by the Ministry of Education, aiming to train teachers with/for ICT: Pairform@nce first (2006–2013) http://national.pairformance.education.fr/, then on the basis of an appraisal of this programme, a new programme, more flexible, developed from 2014 M@gister, both based on the collaborative work of teachers, and	Potential and constraints of the Pairform@nce programme have been analysed by Gueudet and Trouche (2011b). Several features of this programme are underlined: the links between teacher education and classroom practices, teachers collaborative work as a necessary condition for ICT integration, interrelations between teachers resource system, collective

(continued)

Table 12.3 (continued)

About implementation strategies		
Country	Answers regarding the official, intended curriculum	Answers regarding the real, implemented curriculum
	on a combination of face-o-face and distant training	resource systems, and institutional resource systems
	The Ministry of research and higher education has created a new structure (FUN: France Université Numérique) dedicated to the design and implementation of MOOCs; among them a MOOC dedicated to 'Teaching and Training mathematics with ICT'	It remains quite marginal among teachers, and the new programme M@gister can be analysed as a tentative of scaling-up
	Innovations come also from the field. The emergence of large teachers online association, collaboratively designing resources, as Sesamath, appears as a sign of the digital metamorphosis	The first MOOC dedicated to ICT integration in mathematics has gathered about 3000 teachers and reveals the complexity for involving, during a 6-weeks session, distant teachers in a collaborative work
		Sesamath is considered as a new way of professional development, helping teachers to face digital challenges
The Netherlands	As the GCD is already around for many years, there is no PD focusing on that. For other ICT tools, it is very limited and only small scale. For using the IWB, there has been some training, rather button oriented. There are no national PD courses for using ICT in mathematics education. In research projects or pilot PD courses, a blended approach is sometimes used, with Moodle like environments combined with face-to-face meetings. Pre-service teacher education does make use of online content	The new curricula make a case for 'mathematical thinking activity'. The question is if ICT can be used for this, or if it is detrimental. Of course, the answer depends on the type of ICT, and above all on the type of task and the type of use
	The new curricula that will be implemented in 2015 are a compromise between the back-to-the-basic movement, which is not in favour of ICT, and more twenty-first century like ideas that match better with ICT integration	
New Zealand	There are a lot of curriculum resources online to assist teachers, but little on the use of ICT	Teachers' professional development with respect to integrating ICT is organised only on an ad hoc basis, often organised locally by the teacher organisations. No central government or Ministry of Education assistance

PD Professional Development, *CPD* Courses for Professional Development, *MOOC* Massive open Online Courses, *GCSE* General Certificate for Secondary Education

12.3.3 The Sensitive Point of Assessment

Assessment, particularly 'high-stakes' assessment, is a sensitive point for a given curricula because, even if a teacher is not concentrating only on 'teaching to the test', s/he is aware that the content as well as the shape of an examination reflect the expectation of a given institution, and of a given society. Assessments reflect also the values that this institution and this society bestows on the tools supporting learning and teaching. A final assessment reflects also the state of the current teaching: assessment designers aim to evaluate what is really taught. This assessment thus cannot be more advanced than what is estimated to be learnt through average teaching. Therefore one can say: 'such a final exam, such a teaching' and it is all the more true in the secondary level.

I will still rely on Drivers et al. report. The researchers interpreted the IB questions as such:

Q1. Are there national examinations for mathematics? If yes, how are they set up (duration, one or more parts, ...)

Q2. Is the use of ICT allowed during the national examination? If yes, which types of technology? What are criteria? Are specific types or brands allowed?

Q3. If GDCs are allowed, do they need to be reset before the start of the examination? Are additional applications and text files allowed? Is press-to-test mode used? How are all these regulations controlled in schools?

Q4. Are tasks phrased in such a way that the student knows if algebraic/exact answers are required, or if approximations found with the GCD will do? Are there 'magic words' to indicate this?

Q5. Is the use of ICT during examinations rewarded, in the sense that the student gets credits for appropriate use, or for answers that are found by just using ICT? Or are tasks designed in such way that technology just supports the solution process, or that is of no value at all?

I present (Table 12.4) what appear as the more significant authors' answers regarding the assessment, focusing on the requested use of ICT, conditioning the actual use in the courses preparing this examination.

Drijvers et al. analyse these results as follows (remember that the original report addressed the situation of six countries: England EN, France FR, The Netherlands NL, New Zealand, Singapore, and Australia state of Victoria VI):

> All six countries have high stakes national examinations. There are differences between the countries in session time, marking schemes and grading procedures. In VI, the examination of the Mathematical Methods CAS course takes place in two sessions, a 1-hour non-calculator session and a 2-hours CAS calculator session. In NL, the national examination grade only determines half of the final grade, the other half being the result of local school examinations. The latter provides opportunities for other assessment formats, including the integration of digital technology.
>
> All six countries allow the use of calculators during (some/most) examinations; in EN, NL, NZ and SG these are GDCs without CAS facilities In FR and VI, as well as in some EN MEI and some NZ level 3 courses, CAS calculators are allowed. In all countries, criteria are

Table 12.4 The roles assigned to ICT for mathematics assessment during the final examinations (from information provided by Drijvers et al., 2014)

About assessment		
Country	Answers regarding the official, intended curriculum	Answers regarding the real, implemented curriculum
England	All of the modules have timed examinations	Core 1 examinations appear to be a vehicle for the examination of paper and pencil techniques, e.g. simple co-ordinate geometry, simplification of surd forms and simple calculus techniques
	In all (except Core 1 Pure Mathematics) examinations, students may use a scientific calculator or a GDC (no CAS) without retrievable information stored (including databanks; dictionaries; formulas; text). Some words (as 'show') indicate that GDC is not appropriate: 'Show that the equation of the tangent at $A(8,0)$ is $y + 8x = 64$'	The writer has been in Examination Board meetings dedicated to finalising the wording of examination questions and mark schemes where specific questions are revised so as to not disadvantage students without GDCs
	MEI has a unit *Further Pure with Technology* where 'Students are expected to have access to software for the teaching, learning and assessment that features a graph-plotter, spreadsheet, CAS and programming language. For the examination, each student will need access to a computer with the software and no communication ability'	
France	There is a national examination at the end of grade 12 (baccalauréat), opening doors of the university	Some tasks need a calculator to be performed (performing an algorithm, or computing an approximate value), corresponding to a kind of mechanical use of a calculator
	All kinds of calculators are allowed (including CAS) on two conditions: their use has to be autonomous (no Internet connexion, no connexion with other calculators), and no printer. They do not need to be reset. So it is possible to have downloaded specific applications or text files	Some new tasks appear, in order to avoid the use of a GDC (for example: analysing a curve that is done), introducing a new spirit (more inquiring) in the examination
	NB: mathematics assessment has been subject to deep changes, and will change again in 2016, see Sect. 12.4	

(continued)

Table 12.4 (continued)

About assessment		
Country	Answers regarding the official, intended curriculum	Answers regarding the real, implemented curriculum
The Netherlands	There are national final examinations for the different mathematics courses (www.examenblad.nl)	If a student would bring a CAS calculator, this probably would not be noticed by the examination officers. So control is weak
	The use of a GCD is allowed (no CAS). Specific types are permitted. The list of these types is updated yearly. Main criteria are: no communication and no printing options	The increasing capacities of GCD apps now question the current policy (e.g. ZoomMath, see http://www.zoommath.com/). The assessment authorities have installed a committee to assess the situation and to advise on future strategies
	GCDs do not need to be reset, so additional apps or text files can be used during the examination. The ministry's argument for this is that resetting in school practice is hard to carry out, also because students programme reset simulation programmes and expert math teachers are not always around during the examination	This list of magic words always raises discussion, as, in spite of efforts to communicate this clearly, teachers seem to miss this, do not explain these conventions to their students and, as a result, want to grade the examinations against the guidelines. Debate...
	There is an official list of 'magic words'. For example, 'calculate the exact value' or 'prove' means that no GDC facilities may be used, whereas 'calculate' means that GDC facilities (including procedures such as calc intersect or zeros or nderiv) may be used	Now that GDCs are so common, there is a tendency to less reward their use than was the case shortly after their introduction. Also, the requirement for a student to describe the techniques used is not as tight as it used to be. All together, the role for the GCD in the examination is decreasing
	In application tasks, the focus is on modelling, or on mathematisation, and the resulting equations or other mathematical problem can be solved using the GDC. If by-hand techniques are to be assessed, the above-mentioned magic words are used, and there usually is less context or application in such tasks	
New Zealand	There are nation al examinations at Levels 1, 2 and 3. None of the standards speak about technology directly, the comments are in the additional notes	
	Level 1: calculators not allowed in Mathematics standards 91027, but all approved scientific or graphing calculators may be used for standards 91028, 9103, and 91037 (a GCD is an advantage in 91028)	
	For levels 2 and 3: Candidates must bring an approved calculator (preferably a graphing calculator). Candidates who do not have access to graphing calculators will be disadvantaged	
	They have to be reset. Any calculator used in NZQA examinations must be silent, hand-held, non-printing and must contain its own power source. It MUST	

(continued)

Table 12.4 (continued)

About assessment		
Country	Answers regarding the official, intended curriculum	Answers regarding the real, implemented curriculum
	NOT be able to wirelessly transmit or receive information to or from another source; be used to bring in stored information; be used as a dictionary	
	Some kind of intermediate working is expected rather than simply a final answer from the Calculator. For example: When graphing calculators are used to solve a problem, candidates must provide evidence of their differentiation and integration skills. Find the area enclosed between the graph of $y = \sin(2x)$, the x-axis, and the lines $x = \frac{\pi}{6}$ and $x = \frac{\pi}{3}$. *Give the result of any integration needed to solve this problem*	
	The central aim seems to be to set calculator neutral examinations where there is no advantage in the calculator. In practice this is not really accomplished	

that communication (including internet access) and printing facilities are not allowed. In the EN MEI Further Pure with technology courses, students are expected to have computer access during the examination.

Four countries (NL, NZ, SG, VI) provide a list of approved calculators, whereas the other two (EN, FR) do not. In three countries (EN, NZ, SG), the calculator's memory needs to be cleared. This is phrased in different ways. In the three other countries, there is no need to clear the memory (NL, FR, VI). An argument for this may be that a calculator reset in school practice is hard to check, also because students program reset simulation programs and expert math teachers are not always around during the examination. As a result, students can bring specific applications (including CAS capabilities, e.g., the ZoomMath app for TI devices) or text files (e.g., with examination papers from earlier years). This leads to debate on the calculator's memory size (FR) and CAS capabilities (NL). Where GDCs are allowed, some phrasing conventions are established to make clear to the student if exact by-hand results of GDC approximations are expected (Drijvers et al., p. 31).

The authors summarise the national ICT rewarding policies as such:

- EN: Calculator allowed rather than expected;
- FR: Calculator needed for some algorithmic work, but not rewarded. Type of tasks has changed;
- NL: Calculator needed for procedures such as finding numerical solutions of equations. Credits assigned to this seem to decrease over the years (Drijvers et al., p. 32).

I would like to add some extra reflections arising from this survey:

- It appears that the integration of ICT in mathematics assessment increases the divide (Sect. 12.3.1) between the technology potential and its actual use: instead of developing experimentation, visualisation, experimentation, conjecturing abilities, it seems to induce the development of mechanical behaviour (for example, in Table 12.4: automatic answer to 'magic' words)
- The objectives of the ICT regulation during examinations seem to radically oppose the epistemology of tools (see Chap. 4): whereas tools develop for giving new means to human activity, the institution seeks to limit as far as possible the power of tools (see in Table 12.4 'It MUST NOT be able to wirelessly transmit

or receive information to or from another source; be used to bring in stored information; be used as a dictionary'...)

- We assist thus to a kind of competition between the development of tools, more and more powerful, and the development of institutional regulation, more and more restrictive, that goes up to avoid the use of available tools (giving for example a curve printed on a paper, with the legend: 'a GCD has provided such a curve'...)

Such a situation is not viable over the time. A tool, deprived of its substance, loses its legitimacy, as has happened in The Netherlands 'now that GDCs are so common, there is a tendency to less reward their use than was the case shortly after their introduction. Also, the requirement for a student to describe the techniques used is not as tight as it used to be' (Table 12.4). Making viable tools in a given institution—here the final assessment—we need to rethink the assessment itself, and finally the whole curricula.

I have analysed, in this section, the educational policies, and the related didactical issues, for curricula, assessment, and ICT implementation, in four countries, at a given moment of the digital evolution. In the following part, I will focus on one of these countries, France, to examine in more depth how she faced these challenges from the beginning of this century.

12.4 Tools in Mathematics Examinations, a Tumultuous French History

I have, in the previous sections, analysed the interactions between tools and curriculum with two lenses: an historical one, examining major trends over the years, and a geographical one, comparing four countries. I would like to focus now on assessment, on a (quite) short period (since 2000) and a single country (France). France is, among the countries we have analysed (Sect. 12.3), the only one allowing all kinds of calculators in its national examinations. The analysis in more depth of the evolutions of these regulations over time could thus inform us on the opportunities and difficulties resulting of such on opening. Regarding France, I have already 'set the scene' (Sect. 10.2). I analyse now the issues of using tools in mathematics examinations, choosing three angles: the continuous evolution of the official regulation, trying to integrate tools without changing the spirit of the examination; a tentative for a radical change of the examination mode, aiming to take profit of the full potential of technology; and the current curricular projects, proposing successive steps with a perspective of profound evolution. In so doing, I will justify the title of the article (Gueudet & Trouche, 2011a) dedicated to ICT in French mathematics education: Development of Usages, Institutional Hesitations and Research Questions.

12.4.1 Calculators in French Baccalauréat, Three Successive Regulations

What follows has to be related to the calculator debate (Chap. 13), this debate condenses a number of issues regarding the vision of mathematics, of mathematics learning and tools. The calculator (named 'pocket calculators' in the regulation of 1986), imported into classes by students themselves, had, at the beginning, a very weak legitimacy and are, since this importation, regularly questioned. The three successive regulations were decided by the French Ministry, in 1986, 1999 and 2015 (about each 15 years), and these changes are linked, as I described it in Sect. 12.2.1, to curricular changes. The curricular policy is an incarnation of institutional prescriptions: I have chosen then to give the full text giving these requirements.

The first regulation (Table 12.5) was decided in 1986. It corresponds to a deep change in the French curriculum: from 1980 France progressively abandoned the 'modern math' reform (Sect. 10.2.1), linked to a very abstract view of mathematics. With this abandonment came a renewing of geometry teaching and use of graphical representations, clearly appearing in the curriculum (cf. below an extract of Table 12.1 related to the mathematics curriculum for 12th grade preparing to the baccalauréat):

Year	Prescriptions for tools	Prescriptions for numerical computation	Prescriptions for graphic representations
1982	Calculators will be **widely** used	No chapter specific to numerical computation: it is integrated into other chapters	**Usual use of graphic** representation will be promoted, playing a significant role in a function
1986	Calculators will be systematically used (a **basic model** is sufficient)	**Numerical problems and methods play an essential role** in the understanding of mathematical notions	**Graphic representations must hold a very important place** in the curriculum

The regulation for the calculator use (Table 12.5) in the baccalauréat are coherent with this curriculum, stating that 'numerical problems and methods play an essential role', and that 'graphic representations must hold a very important place in the curriculum'.

It may be strange, today, to discover that the limitation of the tool potentiality was seen through the limitation of it size (to be noticed: the height of the device was

Table 12.5 Calculator's regulation during the French examinations in 1986 (circulaire n° 66-228 du 28 juillet 1986)

All the pocket calculators, including programmable and alphanumeric calculators, are allowed, provided that their functioning is autonomous, and they do not use printing. In order to limit the devices to a reasonable size, their surface should not exceed 21 cm × 15 cm. In order to prevent any cheating risk, exchanging calculators during the examination is forbidden, as well as the reading of the instructions of use supplied by the manufacturers

Table 12.6 Calculator's regulation during the French examinations in 1999 (circulaire n° 99-186 du 16 novembre 1999)

Mastering the use of calculators is an important goal for the education of all students because it is an effective tool as part of their studies and in professional, economic and social life. This is why their use is provided in many educational programmes and their use should be widely permitted in examinations and competitions
The authorised equipment includes all pocket calculators including programmable calculators, alphanumeric or graphic display providing that their functioning is autonomous, and they do not use printing
As part of the regulation of examinations and competitions, it is up to those responsible for the design of the examination to decide, for each event, whether the use of all calculating instruments (calculators, numerical tables, abacus) is allowed or not. This needs to be clarified ahead of the text of the examination
The authors of subjects take all necessary steps to encourage owners not too sophisticated equipment, providing, for example, candidate documents with subjects
The candidate uses only one machine on the table. However, if it comes to experience failure, another can replace it
In order to prevent any cheating risk, exchanging calculators during the examination is forbidden, as well as the reading of the instructions of use supplied by the manufacturers as well as the exchange of information via the calculator's transmission functionalities [...]

not restricted, opening the way for using, during the examination a computing tower: that never happened, but this semi-limitation was, during this time, a subject of jokes among mathematics teacher)... It is significant of this period the power of computers was distinguished according to its size (the essential difference between a 'pocket calculator' and a computer).

This regulation was strongly criticised, as it appears that 'symbolic calculators', including CAS and DGS, developing from 1990, could respect this limitation of size. These discussions lead to a second regulation, in 1999 (Table 12.6), abandoning the size restriction, and allowing all types of calculators, including symbolic calculators. The evolution of this regulation comes with new evolution of the curriculum (cf. below an extract of Table 12.1 related to mathematics curriculum for 12th grade): graphic calculators are henceforth prescribed, and the very sensitive (for each mathematics curriculum) argumentation activity may now lean on the analysis of graphs.

Year	Prescriptions for tools	Prescriptions for numerical computation	Prescriptions for graphic representations
1991	Calculators with statistical functions are recommended One the other hand, graphical screens are **not required**	Numerical problems and methods play an essential role in the understanding of mathematical notions	**Graphic representations must hold a very important place** in the curriculum
1998	Graphic calculators are **prescribed**	Numerical topics are introduced as an additional specialist option in mathematics	**Favouring argumentation supported by graphs**

This 'spirit' clearly appears in the 1999 regulation (Table 12.6), that not only allows this type of material, but encourages its use: 'their use should be widely permitted in examinations and competitions'. Besides this, regulation appears also as a text of compromise, with respect to the pedagogical liberty of the examination designers: it was up to you to decide if they allow, or not, the students' calculators.

Compromise was present also regarding the power of the calculators: all types of calculators are allowed, but, at the same time, it is recommended to avoid too sophisticated equipment, and the examination designers can provide candidates with documents (meaning: documents displaying the results that a CAS could provide: the derivative or a limit of a function for example). The regulator encourages the use of calculator, and, at the same time, it provides material making pointless such calculator...

As for the changes in 1986, these new changes, both in the curricula and in the regulation of tools in examination, do not come from nowhere. They are linked to reflection in the communities of mathematicians and mathematics educators about what has to be taught, and under which forms. This reflection wanted to answer new questions: how mathematics should be taught in universities facing increasing number of students (increasing in the first year of the French universities mathematics courses, from 5000 students in 1986 to 30,000 in 1994)? How should the curriculum be adapted for the expansion of computer science, the new needs of a number of disciplines for mathematics, the development of very powerful means of computation, the taking into account of the 'mathematics for citizen'? A new commission was appointed, in 1999, by the French ministry, mainly composed of mathematicians and mathematics educators, and chaired by Jean-Pierre Kahane, member of the French Academy of sciences (Sect. 10.2.3). From its beginning, its members discussed interactions between mathematics and computer science, and of the integration of ICT in mathematics education (Kahane, 2002). During its third meeting (27th November 1999), Michèle Artigue proposed orientations for addressing the issues of computation in mathematics teaching:

- In order to structure the reflection, she proposes to choose a common thread: the distinction between exact and approximate computation, these two dimensions being present from the first contacts with the world of computing and of the notions of number, size, measure and dimension
- She wishes too to give a particular attention to the influence of tools both to the issues related to computing and the way to address them; to the diversity of forms taken by computing according to the different mathematical domains[8]

The 1999 changes in calculator regulations, as well the curricular changes, echo these reflections running in the mathematics education community.

[8] See the report of the history of this commission, the CREM (Commission de Recherche sur l'Enseignement des Mathématiques), written in 2006 by its secretary J.-C. Duperret: see http://educmath.ens-lyon.fr/Educmath/ressources/etudes/crem/.

From 1999, this regulation was applied (Sect. 12.3.3). It appears that calculators (graphic as symbolic) were allowed at each session of the baccalauréat. This normalisation had some positive effects: calculators appear as *usual* tools for learning and teaching mathematics (and, as such, were integrated into teacher training and teacher recruitment competition). The nature of the problems to be solved during the baccalauréat evolved slowly, integrating for example some reflective questions on the shape of curves, but... avoiding most of the time the use of tools. Thus the potential of tools was rarely used. At the same time the calculators developed, particularly regarding their *connectivity* potentialities (see Chap. 18), and it became more and more difficult to control the transmission of information from a candidate to another candidate (or to outside).

Therefore this regulation appears difficult to keep under this form. When I was writing this chapter, I received the information of a new regulation (Table 12.7), to be applied for the 2018 baccalauréat session.

This regulation represents a deep evolution: development of a new kind of calculators, including an examination mode; an examination mode erasing calculator memory and removing the communication potential; GCD, CAS and DGS allowed, as far as they are compatible with this examination mode. We could hypothesise that the availability of such a calculator dedicated to the examination

Table 12.7 *Calculator's regulation during the French examinations in 2015* (note n° 2015-056 du 17 mars 2015)

[...] The use of the calculator is allowed if the subject of the examination explicitly forecast it. The flyleaf of the subjects has to indicate whether the use of the calculator is permitted or prohibited.

Is considered 'calculator' any autonomous electronic device, without any distant communication functionality, whose essential function is performing mathematical or financial computation, achieving graphs, statistical studies, and all treatments of mathematical data through charts or diagrams.

Authorised materials are: non-programmable calculators without alphanumeric memory; calculators with alphanumeric memory and/or graphic display including a feature of 'examination mode' meeting the following specifications:

- Temporary disabling access to the calculator memory or permanent erasure of this memory
- The blocking of any data, whether through Wi-Fi, Bluetooth or any other remote communications device;
- The presence of a f lashing signal on the high side of the calculator, attesting the transition to 'examination mode'
- The non-reversibility of the 'examination mode' for the duration of the event. The output of the 'examination mode' requires a physical connection, via a cable, to a computer or a calculator.

The 'Examination Mode' should only be activated by the applicant for the duration of the test, under instruction of the room supervisor when the subject of the event allows the use of the calculator. The candidate uses only one machine on the table. However, if it comes to experience failure, another can replace it. To prevent the risk of fraud are barred exchanges between candidates machines, consulting manuals supplied by the manufacturers as well as the exchange of information via the calculators transmission functions. The use of improper calculator technical features leads to a disciplinary procedure. Is prohibited the use of any plug-in module or extension and any cable, whatever the length, and connectivity [...]

may lead the designers of the baccalauréat to really adapt their questions to one mathematical environment including, as a regular tool. A new survey could be done on the implementation of this new regulation!

I have presented the successive *official* regulations of tool use in the French baccalauréat, from 1986 to 2015. I have to mention now what happens at the fringe of the system, in the thread of the reflections arising at the beginning of the twenty-first century. The alternative, for the stakeholders, was to *adapt tools* (i.e. restricting their potential) to the context of the current baccalauréat (one individual working session, aiming to assess the mathematical knowledge of a given student), or *to change the mode of assessment* for profiting from the potential of tools. This second way was experimented In France, in some volunteer classes, from 2007 to 2009, as I will explain in the following part.

12.4.2 The 'Mathematics Practical Test' in the French Baccalauréat, a Promising Parenthesis

From 1996, the necessity of renewing the modes of mathematics assessment appears in the noosphère (Sect. 10.4). A commission chaired by an inspector, Paul Attali, addresses both the issues of introducing the calculators in baccalauréat and of opening the kind of problems proposed to students during this examination.[9] A pilot study (for 5000 students) was organised to check the feasibility of such new modes of assessment. The analysis of this pilot study reveals the interest of teachers and students, and the difficulties linked to the implementation of such an examination (questions of materials, space and time needed). The inspectors' reflection go on to improve the organisation of the examination.

In 2007, in the light of this reflection, a *mathematics practical test* ('épreuve pratique de mathématiques') is introduced, for a limited number of students (5000 students) in the French scientific baccalauréat, by the Inspectors of mathematics ('Inspection générale de l'Education nationale, groupe de mathématiques'). Its designers describe its motivation thus:

> Thinking about the introduction of a practical test in mathematics bachelor S is situated in a threefold context:
>
> - A still marginal and disparate use of ICT in the teaching of mathematics even if such use is a displayed program objective;
> - A calculator's use in the written examinations is increasingly problematic;
> - A will to develop and evaluate scientific skills that are not just of writing, such as the ability to conjecture, to take initiative and use ICT (Fort, 2007, p. 3).

The purpose of the test is to assess students' skills in the use of calculators and specific software in mathematics, i.e. to assess students' ability to mobilise ICT to

---

[9] See the evocation of this experiment in the History of the CREM: http://educmath.ens-lyon.fr/Educmath/ressources/etudes/crem/.

Table 12.8 Example of a student sheet for the mathematics practical test (Fort, 2007, p. 10)

Expression of the term of rank n of a recursive sequence Consider the recursive sequence (u_n) of first term $u_0 = 0$ and such that for any natural number n, $u_{n+1} = u_n + 2n - 11$ 1. Using a spreadsheet or a calculator, calculate and plot the first 20 terms of the sequence. Has the cloud of points obtained a particular feature? If so, which? *Call the examiner in order to check the conjectured feature* 2. n is given, one can calculate the value of u_n if one knows u_{n-1}. We would now be able to calculate, for any value of non-zero natural number n, the value of u_n without knowing the value of u_{n-1}. To do so would require a formula giving u_n as a function of n (a) Using the observations made in the first question, conjecture a formula giving, for any value of the integer n, u_n as a function of n *Call the examiner in order to check the conjectured formula* (b) Prove that formula *Production requested* – The cloud of points asked in question 1 and the conjectured feature – The proof of strategy chosen to question 2, and the steps of this proof

solve a mathematical problem. It needs to conceive problems where the use of ICT (programmable graphing calculator, computers and specific software, spreadsheet, spreadsheet grapher, DGS, CAS) intervenes significantly in solving the given problem.

The examination takes place in the school attended by the student, in a room equipped with computers and calculators (a list of material is nationally established). Students, in groups of 4 in each examination room, individually face the same problem, monitored by a teacher observer–examiner. This teacher has to note, for the four students, the relevance of their use of ICT, their capacity to take initiative, to conjecture properties from observed regularities, to search for proofs. The teacher can also assist a student who meets a technical barrier.

A bank of examination subjects is developed at a national level. Each subject includes:

1. A description for supplying the national public list of examination situations
2. A 'student sheet' giving the statement and stating what is expected of the candidate (see Table 12.8)
3. A 'teacher sheet' describing the author's intentions, considerations on the ICT environment, and comments on assessment
4. An 'assessment form' designed to be fulfilled by the teacher which is placed in the candidate's file (see Table 12.9)

These subjects must meet three constraints: consistency between the assessment objectives: skills developed correspond to those developed by mathematics curriculum (skills related to different parts of the programme or 'transversal skills'); consistency of assessment (the teacher monitoring four students, must note the moments of assessing students' work, and give a means for quick checking); consistency of notation (balancing between the assessment of ICT part and the more theoretical one, depending on the subject itself).

Table 12.9 Example of an assessment form for the mathematics practical test (Fort, 2007, p. 12)

General recommendations	
We will not try to note each competency. For the final mark, the overall performance of the candidate will be taken into account within the following frame of reference:	
– The ability to experiment (which takes into account dialectically performance in the use of tools and the faculty to propose conjectures) must represent three-quarters of the final mark	
– The ability to report results derived from this experiment (proof, argumentation...) represent the remaining quarter	
– The ability to take initiative and to take advantage of interaction with the examiner will be considered substantially	
It is not necessary for a competency to be fully mastered for being considered as granted	
Competencies assessed	*Elements allowing to situate the student (to be completed by the examiner)*
The student is able to experiment; to test ... she uses appropriately the calculator or computer tools... It is able of emitting a conjecture consistent with her trials	
The student takes advantage of indications, which may be provided orally	
The student is able to represent the situation (dynamic figure, spreadsheet, curve ...) using ICT	
The student takes advantage of indications, which may be provided orally	
Following a possible oral questioning, the student is able to refine the explorations in using ICT effectively. She reveals a reflective thinking, with a possible return to his conjecture	
The student takes advantage of indications, which may be provided orally	
The student evidences some mathematical knowledge and know-how on the subject	
The student provides a correct resolution of the exercise and is able to formulate a critical review of her observations	
Additional notes	

The student sheet (Table 12.8) is totally unusual: it solicits explicitly the use of ICT, asks for answers leaning only on the observation of the results provided by ICT. In addition, the students may ask, during the examination itself, the teacher if they are doing the problem in the right way. We are here close to the situation of a scientific laboratory, where the important thing is to organise the research well, to proceed according to this organisation, not to find a result at once. One essentially evaluates the process, not the result.

The assessment form (Table 12.9) confirms the 'experimentation spirit' (Sect. 12.2.3) of the practical test, where the following competencies are evaluated: interaction with the teacher, experimenting, taking advantage of the available tool, proving and critical thinking.

This mode of assessment is indeed in line with the objectives of its initiators: 'develop and evaluate scientific skills that are not just of writing, such as the ability to conjecture, to take initiative and use ICT'. What were its effects?

The report of the inspectors coordinating this experiment was quite enthusiastic (what is not often the case for an institutional report on an innovative pilot study):

> The practical test in mathematics as it was experienced this year introduced two innovations:
>
> - An effective integration of ICT in solving a mathematical question;
> - Original assessment methods compared to normal practice in the discipline.
>
> These innovations have had several consequences:
>
> - They induce a different relationship of students to mathematics, because: this event is a place which can be assimilated to an experimental activity where the student can do various tests using the ICT within the frame of the problem at stake [...];
> - They encourage different teaching practices, leaving the possibility of a greater role of the investigation processes;
> - They involve different assessment practices: assessing the candidate during activity, appreciate his efforts, his qualities to experiment, perseverance or taste to seek, to take initiatives.
>
> Furthermore,
>
> - This experiment has received a favourable opinion of the educational community;
> - It did not raise any particular problems in its organization, as well as having educational material;
> - It has generated interest on the part of high school mathematics teachers who participated in the experiment, where they saw there, among others, the opportunity to update their practices;
> - It has generated real enthusiasm from students who have discovered other approaches of mathematical activity [...]
>
> The generalization of this test, which remains under the programs, should push mathematics education towards greater consistency with its purpose: how mathematics, with the tools currently available to them, solve problems, develop experimentation, taste and practice of research? This modernization of mathematics education responds to the evolution of the professional practice of the discipline (Fort, 2007, p. 16).

The website EducMath opened a discussion,[10] in 2007, questioning different actors (researchers, teachers, responsible for the schooling system at different levels) about this new model of assessment. Michèle Artigue asked: 'this new practical test seeks to assess, via the use of ICT, students' competencies for conjecturing from a careful observation of processes, engaging in a manner of

[10] See http://educmath.ens-lyon.fr/Educmath/en-debat/epreuve-pratique/.

proving, communicating their results. Does this new practical test constitute a possible way for such an assessment?'

The answers raised some weaknesses of this new model: the preparation of such a test needs some material that the high schools do not always have; this preparation needs also time, that teachers do not always have; the notion of competencies remains too vague (do we assess ICT competencies, or mathematics competencies using ICT); the experimental aspect of mathematics seems to be quite vague, and the quality of the problems proposed in this frame appears quite unequal; a real assessment of 'solving problems with ICT' should require assessors to follow the students' work over a year; the evaluation, by the teacher, of the students' activity, 'on the fly', appears quite complex and need specific training; the difficulty, for the students, is to write down the uncertain phases of their research, often the most interesting part of the mathematical activity.

Despite these critiques, most of the answers to Artigue's questions converge to underlining the interest of this model, and the necessity for going on, making sure that some essential conditions are considered, in terms of schools equipments, teachers' training, time for preparing this test and balancing experimental and more formal aspects of mathematics teaching and learning.

On the basis of the positive institutional appraisal, and the support of the mathematics education community, the experience was reproduced during the two following years, involving more and more high schools: 25,000 in 2008, 50,000 in 2009 (the total number of students attending the French baccalauréat is about 150,000). At the time where the generalisation of this model appeared certain, the French Ministry announced its 'suspension': the main argument was that this generalisation would require the Ministry to revise the general equilibrium of the baccalauréat, leading to a modification of the conditions of assessment for all the disciplines. Therefore it was necessary to wait for a global reform of the baccalauréat, allowing the integration of the mathematical practical test as one major element of the evolution of the whole examination process. It was then the end of this promising parenthesis. Six years after, here we are: the baccalauréat has not really changed, and the mathematical practical test is still suspended.

Some lessons can be drawn from this experimentation: the deep interrelations between the evolution of tools and the evolution of curricula; the interest, for renewing the teaching, to plan to change the assessing modes; the difficulty, in a centralised system, to change a part of it without considering it as a whole. Above all, it reveals the possibility of engaging teachers in a profound evolution of their teaching, thenceforth this evolution lies on well-prepared experimentations, conducted in close relationships with all the actors.

Seven years on, the French actors of mathematics teaching are engaged in a new challenge, as I will show in the following part.

12.4.3 *French New Curricular Perspectives and 'Mathematics Strategy'*

In 2014 the French Ministry of Education made two decisions for addressing important issues impacting on mathematics teaching: the first one concerns the creation of a Higher Council for Teaching Programmes (CSP, for 'Conseil supérieur des programmes'), the second one concerns the development of a 'mathematics strategy'. New opportunities for re-opening the parenthesis of new modes of assessment?

The first decision, the creation of CSP, came with a new way of conceiving teaching programmes. The CSP was created in the thread of the French law for rebuilding the school of the Republic. This law aims to face the major challenges of education, giving to all students a 'common core of knowledge, competencies and culture',[11] insuring quality of education and equity of access. It wishes to answer to a demand of *transparency* in the process of elaborating new teaching programmes; to a need of *coherence* between the contents taught, the common core, the assessment process and teacher education.

Its missions are defined on the Ministry website:

It provides advices and makes proposals on:

- The general design of education provided to students in schools, colleges and high schools, and the introduction of digital technology in teaching methods and the construction of knowledge
- The contents of the common core of knowledge, skills and culture, and school programmes, ensuring their consistency and their articulation in cycles, and the validation procedures for the acquisition of this base
- The nature and content of the tests of examinations leading to national diplomas of secondary education and the baccalauréat and the adaptability and development of these tests for students with a disability or disorder Health invalidating
- The nature and content of the tests of teacher recruitment exam for first and second degrees, the possibilities of adaptation and development of these tests for candidates with a disability or a disabling health disorder, as well as goals and overall design of initial and continuing training for teachers

Three major features appear: the will to articulate the teaching of the different disciplines (from a disciplinary programme approach to a more integrated curricular approach); the importance given to the digital technologies for rethinking the teaching methods and the construction of knowledge, and the will to establish a real coherency between teaching methods and modes of assessment.

---

[11] This common core, proposed by the CSP to the discussion among teachers, is composed of five domains: languages for thinking and communicating; methods and tools for learning; the education of the person and of the citizen; observing and understanding the world; Representing the world and human activity.

New programmes are currently under discussion, following these principles. This process of renewing begins in pre-primary school and primary school. The programmes are organised by 'cycles'. For example, the fourth cycle concerns grades 6, 7 and 8. It is composed of three parts: the common competencies targeted by this cycle; the contributions of each discipline to the achievement of these competencies; and the structured programme for each discipline. Regarding mathematics and tools: the first part of the programme underlines the importance of creativity, abstraction and modelling, using ICT tools; the second part underlines the responsibility of the different scientific disciplines for learning to move from one language to another one (including programming languages), to move from one representation to another one, to develop inquiry-based approach; the third part of the programme underlines the major components of mathematical activity (searching, modelling, representing, reasoning, computing and communicating), needing the use of DGS, spreadsheet, and calculators. A major change is the introduction of 'programming and algorithmic', giving occasions for individual and collective student work.

The following step will be the renewing of the high school programmes, combined with the renewing of baccalauréat, reconnecting perhaps, 15 years after, with the pioneering experimentation of the mathematics practical test... It is certainly difficult to decide to analyse this in terms of 'loss of time', or, at the contrary, in terms of 'insuring the conditions of the success of an innovation': beginning by the basis (the pre-primary school) and conceiving the teaching system as a whole (changing in an articulated way the curricula and modes of assessment for all the disciplines), and the preliminary experimentation, in 2007, of this practical test paved probably the way for a—later—'scaling-up' (cf. 12.1.3).

The second institutional decision impacting mathematics curricula is the implementation of a 'mathematics strategy'. This strategy[12] is motivated by the 'French mathematics divide': good results of the French researchers at an international level (see for example the Fields medals) and real difficulties for mathematics teaching (middling results in the international comparison test, decreasing number of students engaging in mathematics studies).

This strategy leans on three axis: an updating of teaching programmes (including new links to be established with research in mathematics and mathematics education), improving teachers' education (including an encouragement for student teachers to engage in mathematics courses and teaching careers) and promoting a new image for mathematics (including the development of a national portal for mathematics teaching and learning resources).

Regarding the teaching programme, the mathematics strategy proposes, particularly, the introduction of 'algorithmic', a renewing of mathematics teaching thanks to the introduction of computer science, the use of DGS and programming, the

[12] See the presentation of the mathematics strategy on the French Ministry website: http://cache.media.education.gouv.fr/file/12_Decembre/30/2/DP-l-ecole-change-avec-vous-strategie-mathematiques_373302.pdf.

development of open problem solving and learning games, the development of links with the other sciences, all elements fully coherent with the current work of the CSP.

This strategy comes with an effort to mobilise the mathematics and mathematics education communities[13] via a national initiative 'Living Mathematics Forum', which aims to make more visible a number of initiatives promoting an alternative image of mathematics (including a large place to experimental aspects in technologically rich environments). This common mobilisation recalls the productive atmosphere around mathematics teaching developed in the 1970s (see Sect. 10.2). The challenge is to maintain this mobilisation which depends on both political choices (keep the promise of supporting students' engagement towards mathematics teaching careers) and of collective choices: the development of a portal for mathematics resources appears crucial for making available new resources, meeting the new programme requirement and supporting teachers' engagement in uncertain zones of their practices. It is too early to know if a synergy can be realised between the renewing of the teaching programme and the mathematics strategy but these two processes seem to be a development from the promising parenthesis of 2007, being grounded in cumulative reflections on ICT in mathematics which have developed in the field.

Finally, I have qualified this French period as a *tumultuous* one due to the successive regulation changes, and the brutal ending of a promising experience. But, with some hindsight, it appears mainly as a productive period, evidencing the profound dialectic between experiencing innovative assessment modes and generalising innovation related to the integration of ICT in mathematics.

During this period, ICT usages have developed. The issues of their position in the final assessment at the end of high school appears as sensitive ones. The calculator's regulation during the examination reveals the educational policy regarding mathematics teaching and ICT, echoing the discussions in the community of mathematics education and its evolution with the curriculum. Using the full potential of ICT in this final assessment leads to deep changes in the mode of examination. These changes lead to modification of the attitudes of both teachers and students towards the evaluation, the students' activity during the examination and the nature of what is evaluated. Consequently, it has the potential to influences future modes of teaching and learning mathematics to become more problem oriented. The institutional hesitations to scale-up the experimentations, even in the case of positive results, reveal institutional awareness of the complexity of the new equilibrium to be reached: what is possible with volunteer teachers is not necessary possible with the whole population of teachers. This experimentation raises a range of questions of research about competencies in using ICT vs. competencies in doing mathematics with ICT; ways of assessing them; teacher education; teacher's role in an evaluation process.

[13] Particularly in the frame of the French Commission on Mathematics Teaching (CFEM).

12.5 Conclusion/Discussion

Aiming to draw on this historical and geographical travel through curricula and tools in school mathematics, I would like to underline some main tensions:

- A first tension between the evolution of mathematics curriculum and the integration of tools. The digital metamorphosis conditions the knowledge itself, the way it may be taught; the curriculum may also constrain the tools themselves, see the development of calculators-to-the-test following the French institutional prescription (Sect. 12.4.2)
- A second tension between the intended and the implemented curriculum: there is no easy prescription for using digital tools and their real integration in classrooms practices
- A third tension between changing the ways to teach a given mathematics programme, and the ways to change the programme to be taught. Actually, a real integration of ICT leads both to change in the curriculum (rebalancing experimental phases and more formal phases, towards mathematics laboratory, Sect. 12.2.2) and to introduce elements of computer sciences articulated to mathematics
- A fourth tension between training teacher *in* using ICT, and training teacher *for* using ICT, both objectives being more and more present in the new way of developing teacher education (Gueudet & Trouche, 2011b)
- A fifth tension between the evolution of curriculum and the evolution of assessment. The real curriculum is strongly conditioned by the way students are evaluated, particularly for the final examination… and this final evaluation has to take into account the real curriculum conditioning what the students have learnt; the dual evolution of curriculum and assessment needs to be carefully managed
- A sixth tension between experimentation of new way of assessment and scaling-up towards a generalisation of this innovation. The evolutions, on such a sensitive subject, need time and the creation of a set of conditions which allows positive innovation to 'take root'

Concerning the role of technology in national mathematics examinations, Drijvers (2009) distinguishes between four assessment policies: (1) technology is (partially) not allowed; (2) technology is allowed, but offers no advantage; (3) technology is recommended and useful, but its use is not rewarded; and (4) technology is required and its use is rewarded. With the fourth of these policies, Trouche, Drijvers, Gueudet, and Sacristan (2013, p. 764) underline that 'conceptual skills, such as interpretation, reasoning, mathematisation, justification and modelling are examined. However, designing appropriate examination tasks for such goals is not trivial'.

Designing *appropriate examination tasks* constitute one aspect of a larger issue that is designing *appropriate tasks*, both coherent with the curriculum and taking profit of the full potential of tools (see Chap. 17). One can observe, at least in

France, an increasing institutional awareness of the urgent need of teachers for new resources (see the French 'mathematics strategy'). Here stands a seventh tension between a top-down process, providing teachers with official resources, and a bottom-up process, sharing resources between teachers (Pepin, Gueudet, & Trouche, 2015).

These tensions may be productive ones, leading to a reconsideration of the essence of mathematics teaching. A challenge for the stakeholders, the communities of mathematicians and mathematics educators. The French experience evidences the interest of a joint reflection, combining the analysis of innovations (limited in space and time) and the design of new curricula.

References

Artigue, M. (2011). *Challenges in basic mathematics education.* UNESCO. Retrieved from http://unesdoc.unesco.org/images/0019/001917/191776e.pdf

Buisson, F. (Ed.). (1911). *Nouveau dictionnaire de pédagogie et d'instruction primaire.* Paris: Hachette. Retrieved March 2015, from http://www.inrp.fr/edition-electronique/lodel/dictionnaire-ferdinand-buisson/

Chevallard, Y. (1987). A theoretical approach to curricula. Communication to the seminar on *Comparative Studies of Mathematical Curricula in Different Countries,* held in Frascati (Italy).

Chevallard, Y. (1992). Intégration et viabilité des outils informatiques : le problème de l'ingénierie didactique. In B. Cornu (Ed.), *L'ordinateur pour enseigner les mathématiques* (pp. 183–203). Paris: PUF.

Cornu, B., & Ralston, A. (Eds.). (1992). *The influence of computers and informatics on mathematics and its teaching.* Paris: UNESCO. *Science and Technology Education* 44 (second edition of the 1985 ICMI study).

Drijvers, P. (2009). Tools and tests: Technology in national final mathematics examinations. In C. Winslow (Ed.), *Nordic Research on Mathematics Education, Proceedings from NORMA08* (pp. 225–236). Rotterdam, The Netherlands: Sense.

Drijvers, P., Monaghan, J., Thomas, M., & Trouche, L. (2014). *Use of technology in secondary mathematics* (Report for the International Baccalaureat), London, IB.

Fort, M. (2007). *Expérimentation d'une épreuve pratique de mathématiques au baccalauréat scientifique.* Inspection générale de l'éducation nationale. Ministère de l'éducation nationale, de l'enseignement supérieur et de la recherche. Retrieved from http://media.education.gouv.fr/file/98/3/4983.pdf

George, A. R. (2005). In search of the é.dub.ba.a: The ancient Mesopotamian school in literature and reality. In Y. Sefati, P. Artzi, C. Cohen, B. L. Eichler, V. A. Hurowitz, & A. Experienced (Eds.), *Scribe who neglects nothing. Ancient near eastern studies in honor of Jacob Klein* (pp. 127–137). Bethesda, MD: CDL Press.

Gueudet, G., & Trouche, L. (2011a). ICT in France: Development of usages, institutional hesitations and research questions. In A. Oldknow & C. Knights (Eds.), *Mathematics education with digital technologies* (pp. 23–29). London: Continuum Press.

Gueudet, G., & Trouche, L. (2011b). Mathematics teacher education advanced methods: An example in dynamic geometry. *ZDM, The International Journal on Mathematics Education, 43(3),* 399–411. doi:10.1007/s11858-011-0313-x. Retrieved from http://www.springerlink.com/content/13733h7321658734/

Gutfreund, H., & Rosenber, Y. (Eds.). (2012). *Knowledge and training of secondary school mathematics teachers.* Jerusalem: The Israel Academy of Sciences and Humanities.

Hoyles, C., & Lagrange, J. B. (2010). *Mathematical education and digital technologies: Rethinking the terrain. The 17th ICMI study*. New York: Springer.

Hoyles, C., Noss, R., Vahey, P., & Roschelle, J. (2013). Cornerstone mathematics: Designing digital technology for teacher adaptation and scaling. *ZDM, The International Journal on Mathematics Education, 45*(7), 1057–1070.

Kahane, J.-P. (2002). *L'enseignement des sciences mathématiques. Commission de reflexion sur l'enseignement des mathématiques*. Paris: Odile Jacob.

Kahane, J.-P. (2008). The ICMI studies: their launch. In F. Furinghetti, L. Giacardi, & F. Arzarello (Eds.), *The first century of the International Commission on Mathematical Instruction (1908-2008): Reflecting and shaping the world of mathematics education* (pp. 18–19). Rome: Instituto della enciclopedia italiana.

Kelly, A. V. (1977). *The curriculum, theory and practice*. London: Sage publication.

Lagrange, J. B. (2005). Curriculum, classroom practices, and tool design in the learning of functions through technology-aided experimental approaches. *International Journal of Computers for Mathematical Learning, 10*(2), 143–189.

Lagrange, J.-B., Artigue, M., Laborde, C., & Trouche, L. (2003). Technology and mathematics education: A multidimensional study of the evolution of research and innovation. In A. J. Bishop, M. A. Clements, C. Keitel, J. Kilpatrick, & F. K. S. Leung (Eds.), *Second international handbook of mathematics education* (pp. 239–271). Dordrecht, The Netherlands: Kluwer Academic Publishers.

Lavoie, P. (1994). *Contribution à une histoire des mathématiques scolaires au Québec: l'arithmé tique dans les écoles primaires (1800–1920)*. Thèse de doctorat, Faculté des sciences de l'éducation, Université de Laval, Québec.

Maschietto, M., & Trouche, L. (2010). Mathematics learning and tools from theoretical, historical and practical points of view: The productive notion of mathematics laboratories. *ZDM, The International Journal on Mathematics Education, 42(1)*, 33–47. doi:10.1007/s11858-009-0215-3. Retrieved from http://www.springerlink.com/content/48045470220u4073/

Pepin, B., Gueudet, G., & Trouche, L. (2015). Comfortable or Lost in Paradise? Affordances and constraints of mathematics E-textbooks in/for curriculum enactment, communication to the symposium. *Mathematics Curriculum Contingencies: From Authoring to Enactment via Curriculum Resources*, chaired by D. Clarke, K. Ruthven & M. K. Stein, in the frame of the AERA 2015 meeting, Chicago, 16–20 April.

Trouche, L. (2005). Calculators in mathematics education: A rapid evolution of tools, with differential effects. In D. Guin, K. Ruthven, & L. Trouche (Eds.), *The didactical challenge of symbolic calculators: turning a computational device into a mathematical instrument* (pp. 11–40). New York: Springer.

Trouche, L., Drijvers, P., Gueudet, G., & Sacristan, A. I. (2013). Technology-driven developments and policy implications for mathematics education. In A. J. Bishop, M. A. Clements, C. Keitel, J. Kilpatrick, & F. K. S. Leung (Eds.), *Third international handbook of mathematics education* (pp. 753–790). New York: Springer.

Chapter 13
The Calculator Debate

John Monaghan

13.1 Introduction

Questions about the use of calculators in the learning and/or assessment of mathematics have been raised for many decades and this, in the opinion of us three authors, warrants a chapter on the calculator in this part of the book devoted to issues with regard to tool use in mathematics. I chose the slightly emotive title 'The calculator debate' for this chapter, over a more neutral title such as 'hand-held computational technology', because the digital artefact known as 'the calculator' does, it appears, stir the emotions of many people (on both sides of the debate on the affordances and constraints of this artefact for school students learning mathematics). This affective dimension of tool use in mathematics is not an issue that should be ignored in this book. This artefact, the calculator, is, by my Sect. 1.3.1 definition of a tool, only a tool when it is used to do something and it can be used for non-arithmetic purposes including such as drawing a straight line or as a paperweight. The often extreme valuations (positive and negative) of this artefact as an arithmetic tool are interesting and one of the aims of this chapter is to explore these valuations. Extreme valuations of artefacts for the study of arithmetic are not, historically, restricted to the calculator debate. Buisson (1911), a dictionary of pedagogy and instruction, in a section on the use of the abacus in primary mathematics,[1] reports on a late nineteenth century abacus debate and cites a professor at the Polytechnic of Zurich:

> Le boulier corrompt l'enseignement de l'arithmétique. La principale utilité de cet enseignement est d'exercer de bonne heure, chez l'enfant, les facultés d'abstraction, de lui apprendre à voir de tête, par les yeux de l'esprit. Lui mettre les choses sous les yeux de la chair, c'est aller directement contre l'esprit de cet enseignement.

[1] See http://www.inrp.fr/edition-electronique/lodel/dictionnaire-ferdinand-buisson/document.php?id=2204.

J. Monaghan et al., *Tools and Mathematics*, Mathematics Education Library,
DOI 10.1007/978-3-319-02396-0_13

> The abacus corrupts the teaching of arithmetic. The main purpose of this teaching is to exercise in early childhood the faculties of abstraction, to teach him to see in the head, through the eyes of the mind. It put things in front of the body, goes directly against the spirit of this teaching.[2]

I structure this chapter in three sections. In Sect. 13.2 I position 'the calculator' within 'portable hand-held computational technology' and I briefly review: calculator use; the research literature on the use of the calculator; and the 'calculator debate'. In Sect. 13.3 I consider the calculator with regard to Wertsch's (1998) ten *properties of mediated action*. The last section speculates on a possible future of the calculator debate.

13.2 Hand-Held Computational Technology

When I say 'calculator' in this chapter I use an everyday term to refer to an everyday object and I start by attempting to put some precision on a tacit understanding of this term. Two ways to position the calculator within portable hand-held computational technology are: diachronically, over time; synchronically, at a moment in time. I shall attempt to do both and I start with the synchronic description so as to have a working definition of what I mean by 'the calculator'.

By 'hand-held computational technology' I refer to digital electronic artefacts with batteries that can be held in a human hand. There are an awful lot of these around: electrical probes; weighing scales; etc. These are calculators but they are not what one regards as 'calculators' so I limit the domain to such artefacts which can execute arithmetic operations. This still leaves a lot of artefacts: mobile phones; small tablet computers; etc. These again are calculators but they are not really what one regards as 'calculators' so I further limit the domain to artefacts in which the main purpose is for the user to execute/evaluate mathematical operations/functions. This last definition captures, I feel, what is meant by the term 'calculator' but the definition has lost its extensionality, by which I mean that the first two definitions referred only to the external properties of the artefact but the last one, which captures the meaning of the everyday word, includes human motives for using the artefact.

I thus take a calculator to be a hand-held digital electronic artefact with batteries whose main purpose is for the user to execute/evaluate mathematical operations/functions. This gives us a set of plastic and metal artefacts: arithmetic calculator; scientific calculator; graphic calculator; and symbolic calculator. The calculator debate concerns all types of calculator but it is usually centred on the first two. There are two main differences between arithmetic and scientific calculators. (1) Arithmetic calculators execute arithmetic operations as they are written whereas scientific calculators execute the expression entered prior to pressing the 'enter' or

[2] Translated by Luc Trouche.

'=' key. This makes no difference for expressions such as $2 \times 3 + 4$ but it does for the expression $2 + 3 \times 4$ where the arithmetic calculator will give the wrong answer (20 instead of 14) according to the conventions of mathematics. (2) Scientific calculators have a number of mathematical and statistical functions embedded within keys that arithmetic calculators do not have. Calculators can be viewed as composite tools instead of unitary tools, by which I mean: an arithmetic calculator can be viewed as a tool for four-function arithmetic, for storing intermediate results (via the memory key), for calculating percentages, etc.; the scientific calculator can be viewed as including all the tools of an arithmetic calculator as well as being a tool for doing trigonometry and a tool for statistical calculations, etc. The graphic calculator can be viewed as including all the tools of a scientific calculator as well as being a tool for displaying graphs of functions, etc. The symbolic calculator can be viewed as including all the tools of a graphic calculator as well as being a tool for algebraic manipulation and for calculating derivatives and integrals, etc. There are many interesting issues specific to the use of graphic and symbolic calculators such as their potential for multiple forms of representation but this chapter mainly focuses on arithmetic and scientific calculator.

Diachronically the calculator is simply a recent, in historical time, aid to calculation. Our species has, for millennia, used tools for calculations and in Chap. 4 I considered abaci, the method of prosthaphaeresis, logarithms and slides rules. The quote above from Buisson (1911) concerning abaci shows that calculators are not alone in generating debate on the use of tools for doing arithmetic. I now move on to consider calculator use.

A consideration of calculator use could be lengthy but my purpose here is a 'broad brush stroke' account of the dimensions of calculator use. I start by distinguishing in-school and out-of-school use. In an educational institution, a school, a learner may be directed by another (a teacher) to use or not use a calculator. This is control of the use of an artefact in a public place and there are many examples of this: no smoking; no parking; no ball games; etc. A student has greater personal agency on the use of artefacts outside of school (where they can ignore a demand 'do not use a calculator' in a homework assignment). In adult work activity, restrictions on calculator use are rare but the object of the activity in a workplace setting is rarely on learning mathematics. In the late twentieth century calculator use in the workplace was common in shops, offices and engineering sites. This continued into the twenty-first century but calculator use is increasingly replaced by the use of more advanced digital technology.

I now consider in-school use of calculators. I begin with my own impressions of use (seven dimensions) and then report on a recent survey.

1. Age of the learners: There are often restrictions on the use of calculators with young children. When calculator use is allowed young children are often presented with an arithmetic calculator. Scientific calculators are the norm in secondary schools; implicit and/or explicit restrictions on use may still be applied but these are generally less stringent than they are in primary schools.

2. Tasks: Many tasks from the pre-calculator period have not changed as a direct result of the introduction of calculators. Where calculators have replaced tables of logarithms, the numbers used in tasks are not so likely to be designed for ease of calculation. Some tasks are specifically designed for calculator use (see 6 and 7 below).
3. Attributes of the teacher: Apart from school and/or curricula exhortations to use or not use a calculator, the use of a calculator in a classroom is largely determined by the teacher. There is great variation over mathematics teachers in their attitude to and their understanding of calculator use.
4. Regions: Calculator use varies over nations and, in federations, over states (see Tables 13.1 and 13.2 below).
5. Date: There is some variation in calculator use over time in different regions (restrictions on use may be relaxed at a later date).
6. Curriculum: Although there have been a large number of school-based projects in which the use of calculators is a prominent feature, national curricula have incorporated few new areas of mathematical content to account for calculator use. Numerical analysis (e.g. trial and improvement methods) is an exception.
7. Assessment: Calculator use in high-stakes assessment varies over nations and, in federations, over states. There is variation in grade restrictions (e.g. allowed after a certain age/grade) and examination papers (allowed, not allowed and two levels of papers, one allowed and one not). In calculator allowed papers there is variation as to whether calculators are permitted (no designed change of tasks) or expected (some tasks are changed to facilitate use of calculator).

Table 13.1 Example: fourth-grade calculator policies

Australia	Statements/policies vary by state. In most cases calculator use is encouraged but not mandated during mathematic instruction ...
Austria	Calculators are not used until grade 5
Belgium (Flemish)	Students learn to use the calculator effectively in mathematics instruction
Botswana	Calculator use is permitted in examinations but not encouraged by teachers
Denmark	Calculator use is permitted in examinations

Source: Mullis et al. (2012, pp. 78–79)

Table 13.2 Example: eighth-grade calculator policies

Syrian Arab Republic	It is forbidden to use calculators on exams because mental arithmetic keeps student intellect alive. ... Most students and schools do not have calculators
Tunisia	The use of calculators is permitted to solve simple problems or mathematical operations in instruction
Turkey	Calculator use is allowed in instruction for some objectives in the curriculum
United Arab Emirates	Calculators are allowed during mathematics classes but not during examinations
United States	Statements/policies vary by state. Some states have standards for calculator use in instruction and most states have standards for use in assessments ...

Source: Mullis et al. (2012, pp. 80–81)

TIMMS 2011 Encyclopedia (Mullis et al., 2012) provides the most recent survey available at the time of writing. It provides tables (called Exhibits) on 'National Policies Regarding Use of Calculators in Mathematics Instruction and assessment' at grades 4 and 8. Exhibit 21 (fourth grade, pp. 78–79) reports on 52 countries of which 17 have no policy. Exhibit 22 (eighth grade, pp. 80–81) reports on 45 countries of which 8 have no policy. Both show variation with regard to my dimensions 1, 3, 4 and 7. For example, the stated policies (other than 'no policy') of the first five countries in Exhibit 21 and the last five countries in Exhibit 22 are replicated in Tables 13.1 and 13.2.

Calculator use in instruction and assessment globally clearly shows variation. This variation could cause us to question the meaningfulness of the term—if there are many different calculator uses, then is the term/construct 'calculator use' a valid construct? From the perspective on artefact and tools developed in Chap. 1 of this book, the calculator is an artefact that becomes a tool when it is used to do something (in a particular way). From this perspective the calculator (be it arithmetic, scientific, graphic or symbolic) becomes a myriad of tools through different usages and the term 'calculator use' becomes a collective term for these different tools. I now move on to the research literature on calculator use.

A thorough consideration of the research literature on the use of the calculator would be lengthy and my aim, as with calculator usage above, is to provide a 'broad brush stroke' based on my knowledge of the literature (which is open to claims of bias). I first note two trends in research on calculator use and then consider different types of research. The first trend is that most research on calculator use has been on school-based use. The second trend concerns the date of school-based research. Most research: on arithmetic and scientific calculators was conducted in the 1980s and 1990s: research on graphic and symbolic calculators was largely conducted in the 1990s and the early years of the twenty-first century. I conjecture that reasons for this are (1) graphic and symbolic calculators appeared after arithmetic and scientific calculators and (2) the twenty-first century has witnessed a marked increase in the availability and use of computers in school and researchers have largely turned their attention away from calculator use.

'Research on calculator use', like 'calculator use', is a term/construct that has several meanings ranging from randomised control experiments to scholarly (or -experience-informed) inquiry. Taking research as including all these forms leads to a range of types of publications on calculator use. Within this range I report on: experiments; descriptions of classroom activities and their consequences; meta-studies and explorations of student–calculator actions/activity.

Hedren (1985) reports on a longitudinal study of eight form 4 to form 6 (students developed from age 10 to age 12 over the course of the project) classes in Sweden. The quasi-experiment aspect of the study was only a part of this research which also included questionnaires, interviews with teachers and classroom observations. These eight classes, who could use calculators at any point in their lessons, formed the experimental group; three similar classes formed the control group. The form 4 pre-test items covered mental arithmetic, algorithms and word problems; the difference between the results of experimental and control classes was very small.

The form 6 post-test items included similar content but the items were designed by an independent group. The experimental classes scored significantly better in 47 items but significantly worse in 9 items. Hedren (1985, p. 175) states that experimental 'pupils did not get worse results overall in mental arithmetic [and] ... achieved a better ability to solve word problems'. With regard to the latter, 'We maintain that these positive results are due to our pupils' greater opportunities to concentrate on the process of problem solving when they used their hand-held calculators for calculations'. In the conclusion Hedren (1985, p. 178) states 'we have drawn conclusions on the basis of the observed differences in test results, we cannot eliminate the possibility that the results might have been caused by factors other than the use of calculators'.

Quesada and Maxwell (1994) report on a USA pre-calculus college study conducted over three semesters. Students in the experimental group used a graphic calculator and a textbook written for the use of this calculator. Students in the control group used a scientific calculator and a traditional textbook. Data consisted of student responses to four tests, a final examination and weekly quizzes. 90+% of final examination questions were identical over the groups. One hundred and ninety-nine students in the experimental group and 335 students in the control group were included in the qualitative analysis. 'Statistical results obtained in this study indicated that the test scores of the experimental groups were significantly higher than those of the control groups' (Quesada & Maxwell, 1994, p. 212). However, 'It is not clear what really causes the improvement in scores when the graphing calculator is used' (Quesada & Maxwell, 1994, p. 214). Both Hedren (1985) and Quesada and Maxwell (1994) end with caveats concerning the explanation for the results. This is a common feature of research which relies on statistical methods.

Shuard, Walsh, Goodwin, and Worcester (1991) report on the calculator-aware number (CAN) curriculum and teacher development project, a British project that worked in collaboration with primary teachers in the 1980s. It is an example of a report on classroom activities and their consequences; the research element of this work is a by-product of the development work. It was a radical project in that, at a time when it was estimated that 80 % of primary mathematics time in Britain was devoted to pencil-and-paper practice with standard written algorithms. CAN advocated that: standard written algorithms should not be taught; children should have a calculator at all times; and they should be the ones who decide when calculator use is appropriate. Project teachers developed new tasks for children. The following is a teacher-designed task focused on place value for 6-year-old children.

Put a number inside a square. Then put a number at each corner so that the four 'corner' numbers add up to the number in the square.

Examples of children's work include:

173 in the square and 100, 70, 3 and 0 at the corners; 44444444 in the square and 11111111 at each corner (CAN project children started working with large numbers early in their education).

The CAN project did not simply downgrade one set of tools (traditional algorithms) and upgrade another set of tools (calculator methods) but also placed a great deal of emphasis on investigational work and mental calculations. CAN, its leaders proclaimed, witnessed 'a great flowering of mental calculation' (Shuard et al., 1991, p. 12) and non-experiment test results purported to show CAN project children outperforming non-CAN similar age children on a host of items including many for which a calculator did not appear useful (see Shuard et al., 1991, pp. 59–63 for details). Ruthven (1998), considered below, suggests reasons for the 'great flowering of mental calculation'.

Hembree and Dessart (1986) is a meta-study of 79 research reports which focuses on the effects of calculators on student achievement and attitude. The paper attends to criteria for the selecting and coding of the studies it considers, and also to the methods of analysis. The conclusions are calculator enthusiasts' dream and include:

> In Grades K-12 (except Grade 4), students who use calculators in concert with traditional instruction maintain their pencil-and-paper skills without apparent hard. Indeed, a use of calculators can improve the average student's basic skills with paper and pencil, both in basic operations and in problem solving ... Students using calculators possess a better attitude toward mathematics and an especially better self-concept in mathematics than noncalculator students. This statement applies across all grade and ability levels. (Hembree & Dessart, 1986, p. 96)

These conclusions are, I believe, one reason why Hembree and Dessart (1986) remains a highly cited paper on calculator use. Whilst I do not doubt the scientific integrity of the researchers, I feel it should be pointed out that: my comments on difficulties in explaining results of research which relies on statistical methods applies to this meta-study; the conclusion report on statistically significant results and there are a number of 'results' which are not reported as they are not statistically significant (including results on conceptual understanding); this meta-study reports on research reports and such data can be subject to the *Hawthorne effect* (that the novelty of being involved in a research project may have encouraged teachers of classes using calculators to put extra effort into their lessons).

Ruthven (1998) is a micro-study of the mental, written and calculator strategies of a sample of students' upper primary (10–11 years of age) schoolchildren and is an example of research which explores student–calculator actions/activity. The study also throws some light on CAN's 'great flowering of mental calculation'. Students were drawn from six schools, three of which had participated in the CAN project but CAN terminated before the student sample started at the schools. The contextual information on the schools makes it clear that the legacy of CAN remained in the post-CAN schools and not in the other schools. Ruthven (1998) reports on student responses to four arithmetic word problems such as S*tamps*:

A second-class stamp costs 19p.
How much would 5 second-class stamps cost?
How much change would you get from £5?

Table 13.3 Students use of written and calculator media, from Ruthven (1998)

Medium and level of use	Non-project students	Post-CAN students
No written use	15	24
Multiple written use	5	3
No calculator use	13	14
Multiple calculator use	9	4
No written or calculator use	5 (19 %)	11 (38 %)
Multiple written or calculator use	14 (52 %)	7 (24 %)
All pupils	27	29

The paper catalogues students' strategies. For example, *Stamps* resulted in six mental strategies, two written strategies and one calculator strategy (noting, however, that the trichotomy mental-written-calculator is simplistic and written strategies serve two cognitive functions, recording and spatially schematising). Ruthven (1998) then proceeds to quantitative analysis and my account skips to the part of this which aims:

> To explore more systematically how pupil characteristics might be associated with use of written column or calculator methods, aggregate levels of use were collapsed into three categories—no use, single use (on one problem only) and multiple use (on two or more problems)—and modelled using logistic regression . . . (Ruthven, 1998, p. 35)

Table 13.3 replicates part of the raw data from Ruthven (1998, p. 36) on students strategies in all problems. I have not included indices from logistic regression but I have marked 'interesting' percentages.

Twice the proportion (38 %) of students from post-CAN schools, compared to students from school which were not in the CAN project, employed only mental strategies in all four problems. Over twice the proportion (52 %) of students from schools which were not in the CAN project, compared to students from post-CAN schools, made multiple use of written or calculator strategies in all four problems. Ruthven comments on these figures:

> The greater use of mental strategies by pupils in the post-project schools is of particular interest, as it is consistent with the more positive attitude to mental calculation found amongst such pupils in the macro-study. (Ruthven, 1998, p. 37)

From my position, as someone with a particular interest in tool use in mathematics, this result is interesting because it shows that familiarity with a mathematical tool (a calculator) does not necessarily lead to a reliance on this tool in problem solving. Indeed, it can lead to a proclivity for mental arithmetic methods and perhaps even a 'great flowering of mental calculation'. It also questions policy statement such as 'It is forbidden to use calculators on exams because mental arithmetic keeps student intellect alive' (see Table 13.2 above).

Whilst research on calculator use is 'mixed' along several dimensions and does not 'prove results', the vast body of research points to in-school calculator use does more 'good' in terms of learning mathematics than 'bad'. But if one has a strong opinion on a subject (global warming, the economy or a tool such as the calculator) research may have little or no influence on that opinion. On this note I turn to the calculator debate.

What I call 'the calculator debate' is a series of questions or, more often, statements about calculator use in education and, in particular, in mathematics lessons and examination. The statements are sometimes categorical, 'calculators should not be used in examinations', and sometimes qualified, 'calculators should not be used until students have mastered written methods'. The national policy statements (other, perhaps, than 'no policy') in Tables 13.1 and 13.2 above are premised on national debates on calculator use and Exhibits 21 and 22 in Mullis et al. (2012) show wide variation across countries in the 2011 resolution of the calculator debate. The statements made in this debate: sometimes appear to be made in ignorance but are often carefully constructed arguments; sometimes focus on the calculator alone but often focus on the calculator in concert with innovative (or traditional) means of teaching mathematics. A local version of the latter hit the world news in the late twentieth century as *California's math wars*. Jackson (1997) reports on these 'wars' that centred on *Reform Mathematics*, of which calculators were only a part. Jackson (1997) reports that the anti-reformers believe 'the reformers have swung too far in the direction of "discovery learning" in which students discover mathematical ideas on their own rather than the teacher telling them' (Jackson, 1997, pp. 695–696). The debates/wars continue to rage (though with less publicity than the California's one). Indeed, as I write (2014) a ban (initially announced in 2012) in my country (England) on calculator use in mathematics examinations for 11-year-olds, has just been enacted. The reason that the Education and Childcare Minister Elizabeth Truss gave was that 'children were not getting the rigorous grounding in mental and written arithmetic they needed to progress' (Department for Education, 2012) but this rationale is tied up with concern for England's performance and ranking in international mathematics tests

- Tests for 10- and 12-year-olds in Massachusetts do not allow calculators. In Hong Kong, calculators are not allowed in tests for 9- and 11-year-olds. Elementary students learn how to perform basic arithmetic operations without using a calculator.
- Pupils in Massachusetts, Singapore and Hong Kong outperform pupils in England in international league tables at age 10 and age 14 ... (Department for Education, 2012)

But constructed arguments can also lead to polar position as we shall now see. Gardiner (1995) and Ralston (1999) have a number of similarities: they were written at about the same time; they are written by mathematicians with a keen interest in school mathematics education; they are accounts based on premises which value the culture of mathematics. But these two papers represent polar positions in the calculator debate. I now outline the argument in each paper.

Gardiner (1995) starts by listing four unresolved issues: what should be taught; why is it important; how should it be taught; what level of fluency is expected? Mathematics rests on the *language of expressions* and the fact that the objects and methods of mathematics are *absolutely exact*. A curriculum which embraces calculator use abandons exact forms/objects such as π and $\sqrt{2}$ and turns them into '*algorithms* to be *evaluated*' (Gardiner, 1995, p. 528). This does not mean that mathematics must not use $\sqrt{2} \approx 1.414$ but that 'what should be taught' should differentiate between $\sqrt{2}$ and 1.414, 'the "=" symbol conveys a *moral* message ... not only... *exactly equal*' (Gardiner, 1995) but the person who writes it should be able to explain the transformation. The '=' button on the calculator is a 'completely different animal ... [it is] like the magician's utterance [*abracadabra*], to focus attention on the effect, and hence distract the observer from looking for the true cause' (Gardiner, 1995). The calculator ceases to be a calculating aid and becomes a tool which 'controls, obscures, and distorts the meaning of the symbols and the operations' (Gardiner, 1995, p. 529). The calculator, to Gardiner, is not the sole culprit for the perceived sins of the curriculum he attacks, hand-in-hand with the calculator are artefacts which reduce the need for mathematical thought (tasks which consistently require only one step to obtain a solution) and artefacts which undermine the need for fluency in the language of expressions (formula books).

Ralston (1999) is in part I reaction to the *math wars* and Ralston claims that paper-and-pencil arithmetic (PPA) should 'no longer be a goal of elementary school mathematics' (Ralston, 1999, p. 173). Ralston argues that:

- PPA is not a useful life skill 'in a world where arithmetic is almost universally done using calculators' (Ralston, 1999, p. 177)
- PPA is not useful for professional mathematical pursuits because multi-digit arithmetic in these pursuits is done using calculators or computers
- Expertise in PPA methods does not promote expertise in calculator methods (and vice versa), so 'halfway houses are almost certain to be ineffective' (Ralston, 1999, p. 176)

So 'the argument in favour of learning PPA stands or falls insofar as this skill is necessary to learn subsequent mathematics' (Ralston, 1999, p. 177).

Ralston notes the importance of mental arithmetic and of algorithms for elementary mathematics but he argues that skill in PPA does not help mental arithmetic and the latter 'require that (personal) algorithms be developed and learned' (Ralston, 1999, p. 183). Ralston goes on to outline an elementary curriculum, which he considers mathematically challenging where 'mental arithmetic and calculators should not be the only tools ... Manipulatives and other arithmetic models ... should continue to play an important role' (Ralston, 1999, p. 185).

I expect that there are many unstated opinions (based on the authors' past experience) under the surface of the arguments that Gardiner and Ralston construct that lead to their polar positions but reading these two papers at face-value I am struck by their valuations of algorithms (which are artefacts). Gardiner clearly values specific algorithms above others and his dislike of calculators is partly due to the fact that calculators appear to undermine the algorithms he values. Ralston, on the other hand, values algorithms per se and 'it may be doubted that any real flavour of algorithms is imparted by the teaching of most PPA' (Ralston, 1999, p. 183).

I end this opening section of this chapter here, with a dispute amongst mathematicians. Perhaps mathematicians are too close to our subject to be neutral on the calculator debate. With this I turn to the writings of James Wertsch, who is not a mathematician.

13.3 Properties of Mediated Action

Chapter 2 of Wertsch (1998) examines ten 'basic claims that characterise mediated action and cultural tools' (Wertsch, 1998, p. 25). As I said in Chap. 1, I see the prefix 'cultural' as unnecessary as I do not see how any tool can be a cultural but the term 'cultural tool' has widespread use and I am content to regard it as a synonym for 'mediational means' or just 'tool'. I also do not see Wertsch's ten claims as exhaustive but Wertsch does not claim they are. But I do consider them well-considered and apply to all forms of mediated action. It is thus interesting to view the calculator, as a mediational means, in terms of general properties of any mediational means; the claims provide a means to view the particular (the calculator) in terms of the general (mediational means). I position Wertsch's book and his Chap. 2 before applying it to the calculator debate.

Wertsch is an educator but not a mathematics educator. His theoretical framework is sociocultural and he was an important figure in the 1980s in introducing Soviet activity theory in the West; see Wertsch (1981) which was considered in Chap. 9 of the book you are reading. Wertsch (1998) has six chapters: Chaps. 1 and 2 are introductory; Chaps. 3–6 consider narrative as a cultural tool for representing the past and this leads to a sociocultural analysis of official (Soviet) and unofficial histories of Estonia. Chapter 1 considers the task of sociocultural analysis, which he states is 'to understand how mental functioning is related to cultural, institutional, and historical context' (Wertsch, 1998, p. 3).Wertsch argues that this task is holistic and should go beyond the confines of individual disciplines. Ironically, given the focus on tools in the book you are reading, he writes:

> Dissatisfaction has grown with analyses that limit their focus ... various traditions in the human sciences have had different and incommensurable ideas about the essence of human nature. Some traditions have viewed humans as political animals, others have argued that our essence lies in tool-using activities, still others define us as symbol-using animals (Wertsch, 1998, p. 3)

He takes academic inspiration for his quest from the writings of Vygotsky, Bakhtin and Kenneth Burke. In Chap. 2 he reformulates the task of a sociocultural approach to be:

> to explicate the relationships between human *action*, on the one hand, and the cultural, institutional, and historical contexts in which this action occurs ... this involves focusing on agents and their cultural tools—the mediators of action (Wertsch, 1998, p. 24)

In the remainder of this section I state each claim in bold italics; summarise the claim; consider the implications for calculator use in mathematics education.

In stating Wertsch's claims I repeat some claims about artefacts and tool use made in earlier chapters of the book you are reading to retain the integrity of Wertsch set of claims.

13.3.1 Mediated Action Is Characterised by an Irreducible Tension Between Agent and Mediational Means

Wertsch's analysis of mediated action focuses on humans (agents) and mediational means (cultural tools). This is not to say that there are not other aspects which can/should be considered but this dyad, agent(s)-and-mediational-means, is, in the language I used in Chap. 9, his *unit of analysis*. A focus solely on either part of this dyad loses does not permit an analysis of human actions.

Consider the learner action of keying in $123 \times 45 =$ on a calculator and getting 5535. Did the child get the answer 5535?—Wertsch would answer 'no'. Did the calculator get the answer 5535?—Wertsch would answer 'no'. Did the child–calculator dyad get the answer 5535?—Wertsch would answer 'yes'. Wertsch would give similar responses if 'calculator' was replaced by 'standard written algorithm' or 'tables of logarithms' or . . . From this child–calculator dyad position it is meaningless to say that a calculator is either beneficial or detrimental for the learning of mathematics because such statements consider just one part of the essential dyad. It is, however, meaningful to speak of 'banning calculator use' because, with regard to this claim, this statement effectively means 'banning learner-with-calculator actions'. It could be (and is!) argued that some learner-with-calculator actions are beneficial or detrimental for learners at specific stages in their mathematical development and a calculator ban amounts to 'throwing out the good with the bad'.

Gibsons' *affordances* (considered in Sect. 7.3.1) is a relevant construct to introduce under this claim because the affordances of the environment are what it *offers* the agent (for good or bad); Wertsch focuses down to the agent-mediational means aspect of this environment. In mathematics similar (but not identical) artefacts can offer the child different affordances (dependent on the task). Consider the following task:

Copy and complete the following equations:
$1 + 2 = \_$
$4 + 5 + 6 = 7 + \_$
Write the next three equations.
Write down any patterns you notice.

The visual affordances of most arithmetic and scientific calculators for this task are different. In keying in, for example, $4 + 5 + 6$ on an arithmetic calculator the terms are lost as one key in the numbers and the final display is just 15. This does not happen on a scientific calculator, 15 is displayed but the expression $4 + 5 + 6$ is also displayed. This may or may not be important for a child attempting the task but it does indicate that the child–calculator dyad can be extended to the child–calculator-task triad and the type of calculator used in the task is worthy of note. This is related to what Bartolini Bussi and Mariotti (2008) call the *semiotic potential of an artefact*, the potential of an artefact to focus learners on relationships between signs.

13.3.2 Mediational Means Are Material

Wertsch (1998) notes that many mediational means have 'a clear-cut materiality in that they are physical objects that can be touched and manipulated ... and they continue to exist as physical objects even when not incorporated into the flow of action' (Wertsch, 1998, p. 30) but 'In some instances, mediational means do not have materiality in the way that prototypical primary artifacts do' (Wertsch, 1998, p. 31). Language is the prime example of the second kind but the materiality of language is evident in its acoustic properties (language can be recorded). Establishing the materiality of mediational means is important to Wertsch because:

> The external, material properties of cultural tools have important implications for understanding how internal processes come into existence and operate. Such internal processes can be thought of as skills in using particular mediational means. The development of such skills requires acting with, and reacting to, the material properties of cultural tools. Without such materiality, there would be nothing to act with or react to, and the emergence of socioculturally situated skills would not occur. (Wertsch, 1998, p. 31).

This claim is important as terms such as 'conceptual tool' and 'psychological tool' are not uncommon in the mathematics education literature (e.g. Douady, 1985); my point here is not to proscribe such terms but to make their link the material world clear. The claim is not important to the calculator debate as the materiality of the calculator is not questioned. I would, however, add a rider to Wertsch's claim: behind the use of any material form of a tool there is also an *ideal* form of the tool.[3] Before an agent uses a tool, the agent must have an idea, which may be quite rudimentary, of what the tool is to be used for and how to act with the tool. The ideal form of a tool is not a Platonic ideal form but simply what an agent conceives prior to action with an artefact at a particular time. The ideal form of a calculator will vary across children and, in an individual child, will likely vary over the course of their mathematical development with a calculator. See Cole (1996, pp. 117–118) for further considerations about ideal and material forms of tools.

---

[3] This rider has similarities to (but is not identical to) Luc Trouche's distinction between an artefact and an instrument—see Luc's definition of a tool in Sect. 1.3.

13.3.3 Mediated Action Typically Has Multiple Simultaneous Goals

The main point behind this claim is that 'mediated action typically serves *multiple* purposes ... [which] are often in conflict ... [and] mediated action cannot be adequately interpreted if we assume it is organised around a single, neatly identifiable goal' (Cole, 1996, p. 32). This point is related to Leont'ev's distinction between 'goals' and 'motives' considered in Chap. 9. Wertsch considers the case of pole-vaulting where the obvious goal is to cross the bar but other goals may include impressing an audience or beating a particular opponent or many other things. Beyond the individual, a pole-vaulting event is set in a social setting which brings in collective goals.

In this section Wertsch considers calculations. The goal of performing a multiplication may be to get the right answer but this goal varies and/or splits into distinct artefact specific goals in different contexts. In a workplace context the right answer is usually an end-in-itself but in an educational context the goal is usually related to obtaining the goal with or without a specific artefact, such as using a standard algorithm or not using a calculator. Further to this, 'the goal of obtaining the right answer needs to be coordinated with other aspects of the sociocultural setting' (Cole, 1996, p. 34) such as, in an educational context, a test situation or practicing an algorithm in a classroom or conceptual understanding. In such settings 'the goal of the agent and the affordances of the mediational means [may] come into conflict' (Cole, 1996, p. 33). For example, the calculator does not afford developing the skills required to perform standard written algorithms.

With this claim Wertsch also illustrates that the calculator debate is but an instance of many disputes in society about the use of specific mediational means. For example, in sporting events, 'excellence' is often really 'excellence with regard to a specific and standard artefact'. For example, in the shot put, the weight of the shot is 7.26 kg for men and 4 kg for women and there are many other rules (e.g. the athlete must not wear gloves). Calls for bans on specific artefacts are fairly common over sporting history: fibreglass poles in pole-vaulting; specific types of tennis rackets; and specific types of golf balls. These calls for bans on the use of specific mediational means are associated with the affordances and constraints of the mediational means and the goals of the mediated action. Quite often the calls for bans celebrate the constraints of the old and castigate the affordances of the new.

13.3.4 Mediated Action Is Situated on One or More Developmental Paths

This claim is an elaboration of Vygotsky's (1978, pp. 64–65) assertion that, 'the historical study of behaviour is not an auxiliary aspect of theoretical study, but rather forms its very base'; the irreducible tension between agents and mediational means 'always have a particular past and are always in the process of undergoing further change' (Wertsch, 1998, p. 34). Wertsch contrasts aircraft design in the 1960s (with slide rules and drafting equipment) and in the 1990s (with computers) and asks 'What developed?' (Wertsch, 1998, p. 35). His answer is threefold: any developed intelligence goes hand-in-hand with development of mediational means; we cannot interpret development without some idea of a *telos* (end point); but development is not determined by a preordained end point, development is contingent on all sorts of things.

This claim provides means to understand the views of some players in the calculator debate. As we saw in the first part of this book, written methods, including 'standard algorithms', are simply a part of the history of human methods of calculating. Oral means and semiotic tools (abaci, tables of logarithms, . . .) have always been a part of our means of calculating and these have, over the centuries, been in a constant state of development. Some players (those who suggest banning calculators in some form) in the calculator debate do not appear to appreciate this historical development and/or implicitly consider that the end point 'arrived' with the standard written algorithms for calculation. Some other players in the debate appear to be taken with a sort of positive sense of the *telos*, that calculators are a positive force for development. I have been careful in the above to say 'some', I do not loosely attribute naivety to players in the debate. My own view is that this *telos*/development aspect of the calculator debate is important but complicates matters. It is important to know where we came from to understand where we are now (with regard to means to perform calculations) and to consider where we might be going. But we are simply at one point along a developmental path and the complication is that (1) there is no pre-determined *telos* but (2) we need to consider a *telos* to consider where the path might go.

13.3.5 Mediational Means Constrain as well as Enable Action

This claim is an elaboration of the Gibsons' construct of affordances and constraints to mediated action, 'even if a new cultural tool frees us from some earlier limitation of perspective, it introduces new ones of its own' (Wertsch, 1998, p. 39). Academics, Wertsch argues, who consider mediated action, 'can often be seen as falling into one of two basic camps, depending on whether one takes a "half-full" or "half-empty" perspective' (Wertsch, 1998). Regardless of one's perspective, 'the constraints imposed by cultural tools are typically recognised only in retrospect through a process of comparison' (Wertsch, 1998, p. 40).

The half-full and half-empty perspectives are often clearly marked in the calculator debate: Ralston is a half-full author and Gardiner is a half-empty author though I do not believe either fails to see the other perspective (they simply do not value the other perspective). Regarding constraints being recognised only in retrospect, I do not feel this really applies to the calculator debate; the accuracy constraints of log tables and, especially, slide rules were recognised long before the digital calculator arrived. But drawing attention to the Gibsons' construct of affordances and constraints is, however, almost always relevant in debates on mediated action. The Gibsons', in their many writings on affordances and constraints of agent–environment dyads, often note that these can be regarded as 'bad' or 'good'. 'Bad' and 'good' are, clearly, value-laden terms and not common terms in academic writing but they are common terms in public debates.

13.3.6 New Mediational Means Transform Mediated Action

Wertsch cites Vygotsky (1981, p. 137), 'by being included in the process of behaviour, the psychological tool [sign] alters the entire flow and structure of mental operations' and argues that this can be understood via consideration of the different *genetic domains* of phylogenesis, sociocultural history, ontogenesis and microgenesis. But, regardless of the genetic domain, 'the introduction of a new mediational means creates a kind of imbalance in the systematic organisation of mediated action, an imbalance that sets off changes in other elements . . . Indeed, in some cases an entirely new form of mediated action appears' (Wertsch, 1998, p. 43). I consider mental operations, genetic domains and imbalance in turn with regard to the calculator.

With regard to the flow of mental operations, consider teaching a class of children aged 7–8 how to write a fraction as a decimal. Shuard et al. (1991, p. 21) report on one CAN project teacher doing this with ¼. The teacher said 'The way we write it, it contains the numbers 1 and 4. What can you do with 1 and 4 on a calculator?' One child wrote:

$4+1=5\ 1+4=5$
$4\times 1=4\ 1\times 4=4$
$4-1=3\ 1-4=-3$
$4\div 1=4\ 1\div 4=0.25$

He then said 'I think a quarter is 0.25' and checked it in two ways:

$0.25+0.25+0.25+0.25=1$
$0.25\times 4=1$

This example is unspectacular inasmuch as the mathematics education literature teems with examples of students' idiosyncratic ways of tackling tasks with a wide variety of tools. But the example nicely illustrates the Vygotsky-Wertsch point that a new tool alters the flow of mental operations. Shuard et al. (1991) do not provide details of the child's actions beyond those I replicate above but it is reasonable to think that the child combined a systematic 'operation search' on the calculator, using the signs 1, 4, =, +, −, × and ÷, and then focused on $1\div 4=0.25$. We have no idea what he was thinking at this point but whatever it was the focus of the thought was produced by child–calculator dyad. His subsequent 'double check' on his hypothesis also used the calculator. I say 'used the calculator' but it appears to me that this may be a weak term, the calculator appears to not only influence the flow of mental operations but to be an integral part of the mental operations. This phenomena often goes by the name 'distributed cognition' in the literature (Hutchins & Klausen, 1992), cognition in action is distributed over human and non-human agents. Notice that I have now edged into ascribing agency (in some form) to a tool. I shall put a stop on this speculation at this point as the example does not support further exploration but we shall come back to this issue at various points in the remainder of this book.

I now consider genetic domains and the imbalance which sets off changes in other elements. The first appearance of the digital calculator was in the genetic domain of sociocultural history—a new artefact/tool[4] appeared in society. Mathematics (used by workers in the field of electronics) contributed to this appearance but the appearance was not a part of mathematics education of the time. Before long this influenced the microgenetic actions of students engaged in arithmetic tasks, such as the actions of the student considered immediately above. Such moves from sociocultural history to microgenetic actions are not unique to the calculator, the introduction of tables of logarithms (considered in Chap. 4) led to a similar movement over genetic domains. This move from sociocultural history to microgenetic actions will occur as long as the new artefact/tool is used (in the case considered here, is used in mathematics) but the influence of this use on ontogenetic development is only possible if the use is sustained over a considerable period of time. The CAN project is an example of sustained calculator use over time and Ruthven's (1998) study appears to provide evidence of students' ontogenetic development, the use of the calculator appears to have influenced the development of students' mental calculation strategies. One could speculate on the influence on phylogenetic development, as Prenksy (2001) does with the ascription of twenty-first century children as *digital natives*, but this is a speculation too far for me at this moment in time.

The imbalance that Wertsch talks about, with regard to calculators, can be viewed as a *fallout* from the move from sociocultural history to microgenetic actions but it can also be viewed in terms of Leont'ev's (1978) triple (operations, actions, activity; considered in Chap. 9), calculator use by students transforms not only students' actions but their operations (keystrokes in place of written signs) and the activity of learning arithmetic itself. This imbalance can lead to fundamental 'what are we doing?' questions amongst those concerned with mathematics instruction—and we see differing responses to this question in Gardiner (1995) and Ralston (1999). The introduction of the calculator into mathematics classes also created curriculum and assessment imbalances. At the time the CAN project started the primary mathematics curriculum at the school level (almost all focused on

[4] I use the term' artefact/tool' in compliance with my distinction between artefacts and tools, an artefact becomes a tool in use.

arithmetic facts and algorithms) in England was organised with regard to age-related constraints on the size and type of numbers involved (e.g. learn whole number addition facts up to 20). The CAN project simply did away with this organisation, 'Most children enjoy using the largest numbers they can handle confidently. Many CAN activities encourage children to use numbers of their own choice' (Shuard et al., 1991, p. 13). In assessment the imbalances border at time on being absurdities such as 'Calculators are allowed during mathematics classes but not during examinations' (see Table 13.2 above). This says, in other words, the tool used for learning is not allowed in the examination of this learning. If a similar rule was applied to learning to drive, using the indicator for signalling turning in learning to drive a car but using hand signals when sitting a driving test, it would strike many people as strange.

13.3.7 The Relationship of Agents Towards Mediational Means Can Be Characterised in Terms of Mastery

This claim (and the next) address Wertsch's interpretation of mediated action with regard to 'internalisation':

> ... the process of internalisation consists of a series of transformations:
>
> (a) *An operation that initially represents an external activity is reconstructed and begins to occur internally ...*
> (b) *An interpersonal process is transformed into an intrapersonal one ...*
> (c) *The transformation of an interpersonal process into an intrapersonal one is the result of a long series of developmental events ...*Vygotsky (1978, pp. 56–57).

With obvious respect towards Vygotsky, as a founder of sociocultural analysis, Wertsch (1998) is nevertheless critical of much discourse which invokes the term 'internalisation':

> It encourages us to engage in the search for internal concepts ... [and] entails a kind of opposition, between external and internal processes, that all too easily leads to the kind of mind-body dualism that has plagued philosophy and psychology for centuries ... it seems that many different interpretations [of internalisation] clutter the conceptual landscape and that these are tied to different exemplars. (Wertsch, 1998, pp. 48–49)

Wertsch uses the word 'exemplars' in the sense of Kuhn (1970) in a passage where Kuhn is reacting to the view that, in students learning, 'Scientific knowledge is embedded in theory and rules; problems are supplied to gain facility in their application' (Kuhn, 1970, p. 187). Kuhn argues that this view is wrong, 'at the start and for some time after, doing problems is learning consequential things about nature. In the absence of such exemplars, the laws and theories [the student] has previously learned would have little empirical content' (Kuhn, 1970, p. 188). The exemplars that Wertsch considers concern mediated action and he finds the word 'mastery' (know how) appropriate to this domain as it comes with less *conceptual*

baggage than 'internalisation'. Further to this, it is not clear to Wertsch how many (most) forms of mediated action could be fully internalised (though some internal transformation may, of course, accompany mediated action).

I include this claim of Wertsch for completeness; it does not appear particularly relevant to the calculator debate as I am not aware of participants in the debate making positive or negative claims with regard to the internalisation of calculator use. For example, it is difficult to imagine how multiplying 343×822 with a calculator (or, indeed, with a standard algorithm) could be internalised. Mastery does appear to be a better word as we can be clear whether or not a student has mastered an arithmetic algorithm with a specific tool. With regard to clarity of exposition the 'anti-calculator' subset in the calculator debate appear to have the upper hand; the minister's statement cited above, that 'children were not getting the rigorous grounding in mental and written arithmetic they needed to progress' (Department for Education, 2012) may be open to criticism but it is a clear statement with regard to mastery.

13.3.8 The Relationship of Agents Towards Mediational Means Can Be Characterised in Terms of Appropriation

> In addition to being characterized by level of mastery, the relationship of agents to mediational means may be characterized in terms of "appropriation." In most cases, the process of mastering and appropriating cultural tools are thoroughly intertwined, but ... this need not be the case. The two are analytically and, in some cases, empirically distinct. (Wertsch, 1998, p. 53).

Wertsch uses the term 'appropriation' in the sense of Bakhtin, 'taking something that belongs to others and making it one's own' (Wertsch, 1998, p. 53) Bakhtin's interest was language:

> The word in language is half someone else's. It becomes "one's own" only when the speaker populates it with his own interpretation, his own accent, when he appropriates the word, adapting it to his own semantic and expressive intention. Prior to this moment of appropriation, the word does not exist in a neutral and impersonal language ... but rather it exists in other people's mouths, in other people's contexts, serving other people's intentions ... (Bakhtin, 1981, p. 293)

But appropriation is also a relevant construct with regard to non-linguistic artefacts. There are many such artefacts that people appropriate: bicycles; clothes; guns; musical instruments; ...; and calculators. Appropriation of an artefact can be viewed as a continuum, from non-appropriation to complete appropriation and all points between these poles. A cyclist may be someone who simply uses a bicycle but a positive assertion 'I am a cyclist' is likely to reflect a person who enjoys cycling, is proud that s/he is not polluting the atmosphere and may make modifications to their bicycle—being a cyclist to such a person is a part of their identity. I have never heard anyone say ‘I am a calculatorist’ but there are people who are proud of their calculator, set up special modes and have preferred sequences of keystrokes for certain types of problem. But it is difficult to imagine using a calculator being a part of someone’s identity, this artefact, unlike bicycles or computers (with ‘geeks’), does not seem to lend itself to intense positive personal ownership. I have, however, in the course of classroom-based research over the decades encountered students (generally high attaining students) who pride themselves on avoiding calculator use. Appropriation of (or aversion to) artefacts is not just an individual ‘choice’ as trends (peer pressure) in clothes, music and gadgets amongst different age group shows. In the case of the calculator a student may be influenced by the calculator debate itself, ‘My teacher/parents say that calculators make you bad at maths’.

I posit that appropriation and mastery of calculators (and other artefacts) are, in general, interrelated in as much as low/high appropriation is often paired with a low/high level of mastery. An artefact that has been widely appropriated and mastered in the twenty-first century is the mobile phone. This artefact is not just a phone but a source of games, *apps* and enables the owner to access the internet—it is widely used and used for different purposes. The calculator is used for mathematics and this use is often restricted to a small set of its possible functionalities. I have seen students use an arithmetic calculator for arithmetic operations but they do not know what the memory button does, and students use a scientific calculator for trigonometric questions but they do not know what the statistical functions on the calculator are for. Mastery of their calculators in these students is localised to specific uses/functionalities and this is not likely to generate appropriation of the calculator as a useful tool for doing mathematics. Sheryn (2005), which reports on research which monitored the use of graphic calculators (GC) by six senior school students studying academic stream mathematics over 9 months, found similar behaviours, ‘None of the six students were extremely proficient with their GC at the end of the year although some were confident using a very limited selection of features of the GC’. (Sheryn, 2005, p. 106) and goes on to say ‘I have seen that only a few students appropriate their GC’ (Sheryn, 2005, p. 107).

Low mastery and appropriation of calculators by students appears to set a serious problem for the ‘pro-calculator’ subgroup in the calculator debate. It may be that without large-scale integration of calculators coupled with the virtual abolishment of standard written algorithms (such as those enacted in CAN or proposed by Ralston), the majority of students will fiddle with keys they know to be useful for specific types of problems.

13.3.9 Mediational Means Are Often Produced for Reasons Other Than to Facilitate Mediated Action

Wertsch’s eight claims above concern how mediational means are taken up and used. In this claim he considers how and why they are produced. Sometimes they are produced for the purpose for which they are used but this is not always the case, sometimes they are a *spin-off*. Wertsch cites fibreglass pole-vaulting poles. Fibreglass was developed by the military for reasons that had nothing to do with pole-vaulting. But once the material was produced it was available to be made into poles for pole-vaulting. Werstch also cites the QWERTY keyboard which was designed to slow down typists on manual typewriters so that the keys did not stick (a common phenomenon in nineteenth century pre-QWERTY keyboard typing). In mathematics education two of the most commonly used digital technologies have similar design histories: spreadsheets, which were designed for finance; the calculator which, as we have seen in Chap. 7, was produced because the technology to produce it was available.

It is tempting to throw out Wertsch's claim with 'So what, the technology is available, let's use it' but a consideration of the production of mediational means can provide insights into old and new mediational means. I now consider the standard written method of adding positive integers and contrast it with a calculator method and mental methods. A statement which many people on both poles of the calculator debate would agree with, though the valuations behind the agreement would differ, is 'calculator use does not reinforce skill with the traditional written method'. Let us take 363 and 448. The standard written method starts with the least significant digits, adds these (11), records the least significant digit of this addition (1) in the unit column and 'carries' the most significant digit of this addition (1) into the tens column, etc. The calculator method is key in 363, +, 448, =. This clearly does not reinforce the written method because it was not designed to do this (the technology to reproduce the visual design was not available when the first calculators were produced). Now let us consider performing 363 + 448 without writing anything down. Where do you start? Most likely with the most significant digits, which is the opposite to the standard written method. The standard written method appears to work against the mental method whereas the calculator method could be said to be neutral. The methods children use will, in general, be tied to the context of use. Selter (2001) investigates primary children's use of mental, informal written and standard algorithms in addition and subtraction tasks up to 1000 and found 'The written algorithms became the main method after they had been introduced not least because a high amount of time was devoted to them during the lessons' (Selter, 2001, p. 166). But Threlfall (2002), which focuses on mental calculations of similar aged children, reports on an orally presented task to 53 children, 45 + 48. Threlfall records eight solution types, none of which starts with the least significant digit. The point of relaying the information above is to point out differing perceptions of mediational means with regard to the time of their production. The written method itself was produced for a specific technology, pen and paper. It was produced so long ago that it is tempting to think that it is, somehow 'natural' but it is not, it is 'artefactual' just like the calculator method. The gulf between initial production of this written method and current consumption may also lead to a belief that it encourages skill in 'mental methods' but it appears that this is not so. The production of the calculator is more recent than the standard written method and there

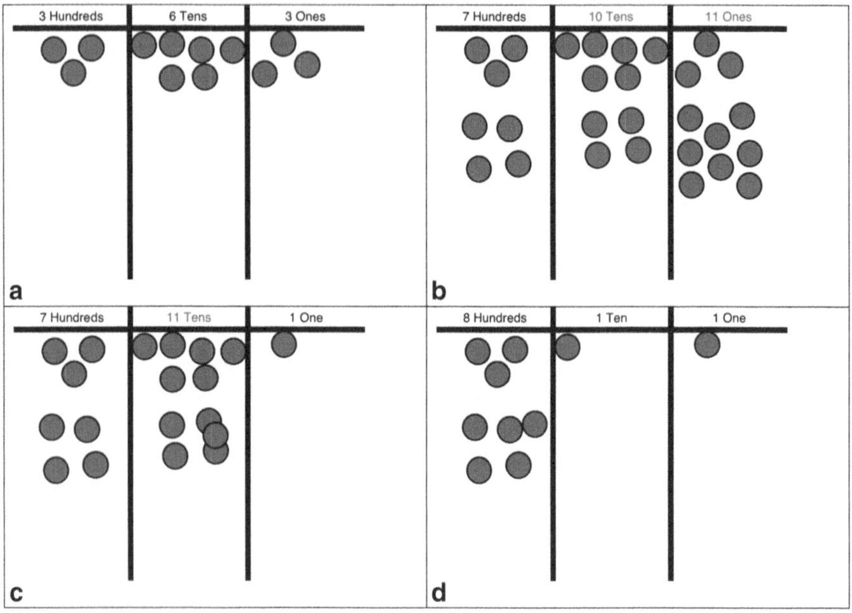

Fig. 13.1 Four images from the *app* Place Value Chart. *Instructions* Tap the screen to create tokens. Move tokens from field to field by dragging. Moving from a lower place to a higher place is only possible if you have enough tokens. Moving tokens from a higher place to a lower place is always possible. Moving one token, one field to the right will result in ten tokens there. Remove tokens by shaking the device or moving single tokens to the top

appears to be a belief that it is detrimental for skill in 'mental methods' but it appears that this is not so.

The production of mediational means develops over time. The production of twentieth century calculators resulted in linear output that did not highlight place value as both the written and mental methods above do. I write this in 2014 and I do not have a calculator that will highlight place value in the calculation of $363+448$ but I do have an *ipod* (which is smaller than most calculators) with an *app* that will do this (see Fig. 13.1). The explanation of Fig. 13.1 is as follows: in frame (a) I have tapped the screen to obtain a visual representation of 363; in frame (b) I have tapped the screen to obtain a visual representation of 448 below the representation of 363; frame (c) shows the visual representation after I have 'swiped' the units column to the tens column; frame (d) shows the visual representation after I have 'swiped' the tens column to the hundreds column; By the time you are reading this such a calculator may exist. The production of future calculators may change the tenor of the calculator debate.

13.3.10 Mediational Means Are Associated with Power and Authority

A gun is associated with power and authority though, returning to Wertsch's first claim, it is the dyad man-with-a-gun that may wield power and authority. But Wertsch is interested in more subtle examples and cites scholars who 'have argued that the rise of print media and literacy have had a transformatory effect on how power is organised and exercised in society' (Threlfall, 2002, p. 65). In the field of education Freire (1993), amongst others, has attacked 'transmission teaching' as a means to control a population's thoughts and actions. Wertsch again looks to Bakhtin and language and how words can encourage or silence communication. The phrase 'knowledge is power' is not just appropriate to government level spies, instrumental sheet music often comes with suggested fingering but instrumental virtuousi often keep their fingerings to key pieces a secret. Mathematics is not immune from power relations: the linguistic artefact initiation-response-evaluation (what is $x+x$?-2-good) can be used to wield power in the classroom; and mathematicians, not school children, determine what counts as an elegant solution to a problem.

A classical text on authority is Weber (1947) who posited three types of authority (and responses to that authority): legal (rational obedience to the law); traditional (loyalty); and charismatic (devotion). Perhaps if Weber was alive today he would add a fourth type, 'artefactual' (obedience to digital technology). Amit and Fried (2005) examined two 8th grade mathematics teachers and their classes through observations and interviews. This study found evidence of the authority of friends and shared authority but 'The teacher's tremendous authority, in every sense of the word, was evident in all of the student interviews in both Danit's and Sasha's class' (Amit & Fried, 2005, p. 155).

Calculators in the mathematics classrooms have the potential to level out some power relations with regard to knowledge. The teacher is traditionally the possessor of knowledge in the mathematics classroom but, in a subset of mathematical problems, a calculator can give a student equal authority to the teacher as both can press a few keys and obtain the solution to $363+448$. In the case of symbolic calculators, the student-with-a-calculator may sometimes have more authority than the teacher with traditional tools, e.g. solving $\int \frac{1}{1+x^2}dx$ requires about the same number of keystrokes as solving $363+448$. This issue is not, to my knowledge, raised in the literature around calculator use in the classroom; perhaps this aspect of power associated with this mediational means is invisible to many.

One aspect of the calculator debate is about the authority associated with different mediational means for doing mathematics though 'authority' is usually implicit in the normative language of the debate, 'Once the moral imperative of the "=" symbol (*exactness in principle*) is lost, mathematics becomes no more than an experimentally based bag of tricks' (Gardiner, 1995, p. 529). A more overt aspect of the calculator debate associated with authority and mediational means are the

various proposals to ban calculator use (legal authority). In the language of mediational means this is using one mediational means, a law, to proscribe the use of another mediational means, a calculator. This is a very common relationship between mediational means and accounts for a great many laws in every nation (laws on vehicles, on firearms, . . .). It is interesting that a digital artefact used in the mathematics classroom is singled out for this treatment. A compass is potentially more dangerous to the physical well-being of students but the perceived 'mental dangers' for students appear to be seen by some as more dangerous than these physical dangers. In stressing this point I am not suggesting that it is wrong for people to suggest a ban on calculators but not compasses, I am merely noting that it appears to class calculator use a mental hazard like pornography.

13.4 A Future for the Calculator Debate?

Wertsch's claims do not, of course, resolve the calculator debate but they help us to see the calculator debate in the wider perspective of the influence of tools on and in practices. 'Tool-X' in the following could be 'calculators' or 'fibreglass poles' or many other tools.

> Tool-X came along at a certain period in time and people started doing things with it. Tool-X was, over time, incorporated into a specific practice but it came into practice that pre-dated the arrival of tool-X. When used, tool-X transformed that practice. Some people did not value the new practice as they valued the old practice though some preferred it.

When an artefact appears and someone does something with it, they do it for a reason but that is not to say that this reason for using the artefact was built into the artefact; it does not even mean there is a reason for the existence of the artefact other than it could be produced. When the artefact is used this use may be ascribed as 'good' (or 'bad') for, say, developing mathematical understanding but in reality: all artefacts have affordances and constraints; expecting an artefact used in doing mathematics to have any 'natural' link to the way the mind works is probably expecting too much. The use of a new artefact transforms mediated action. From the point of view of mathematics education, this transformation is at the heart of the calculator debate, whether the transformation is appreciated or not (and the forms of this appreciation will differ, even on one side of the debate, say the 'anti-calculator' side, over the participants, e.g. mathematicians and politicians).

Wertsch's ten claims were not intended to be exhaustive. I would add three claims about mediational means relevant to the calculator debate: (i) a given mediational means enables action only in concert with other mediational means; (ii) mediational means exert agency; (iii) many mediational means have a finite useful lifespan. With regard to (i), calculator use in a classroom is coordinated with the use of other artefacts: tasks; black/whiteboards; computers; textbooks; pencil and paper; the structure of the lesson (the time sequence and the spatial arrangement of desks). All of these things have the potential to interrelate. With regard to (ii) I

first note that all artefacts exert agency of some form. In some cases this is minimal and takes the form of 'resistance' (Latour, 2005), an artefact offers resistance when the artefact won't let me do what I want to do. At the other extreme, computer-management learning (CML), which, incidentally 'have a minimal effect on student mathematics achievement' (Cheung & Slavin, 2011, p. 17) are often programmed to give students a sequence of tasks which is determined by the students' test result; a CML (in concert with the design team) can exert a strong agency over the students. The calculator falls between these two extremes. My third claim (considered in the paragraph below) is, perhaps, the most relevant for the future of the calculator debate as it implies that the debate may simply go away.

There are mathematical tools that have withstood the test of time since they were first produced. The compass and the straight edge are examples. But tools that aid calculation have been replaced, at least at the global level (the abacus is still used in some, generally Far Eastern, educational practices and tables of logarithms and slide rules are occasionally used by history of mathematics enthusiasts). Given this history it may be reasonable to assume that the calculator will be replaced—but by what? In my discussion of Wertsch's fourth claim, mediated action is situated on one or more developmental paths, I stated that there is no pre-determined *telos* but we (mathematicians and mathematics educators) need to consider a *telos* in order to take the issue of calculator use in mathematics education forward. This is complicated for reasons related to what Engels (1894/1968, p. 694) states, 'Men make their history themselves, only they do so in a given environment, which conditions it, and on the basis of actual relations already existing'. But existing artefacts can give us a vision for future artefacts. I refer to Wartofsky's (1979; considered in Sect. 7.3.2) 'tertiary artefacts', 'artefacts of the imaginative construction of "off-line" worlds' (Wartofsky, 1979, p. 208). Even though our thoughts are bounded by our experience, these experiences allow us to imagine a world beyond our experiences. Figure 13.1 shows the output of an artefact that allowed me to imagine a touch-screen future artefact with visual output that could replace current calculator technology. The artefact which produced Fig. 13.1 was produced by Ulrich Kortenkamp, and is a feature of Ladel and Kortenkamp (2013, discussed in Chapter 9) who was inspired to produce the *app* on the basis of his work (his 'online praxis' in the language of Wartofsky) with young children engaging with mathematics on touch-screen technology. Such an artefact could (only could) bring the pole divisions in the calculator debate closer together or even make the calculator debate in its current form a thing of the past.

References

Amit, M., & Fried, M. N. (2005). Authority and authority relations in mathematics education: A view from an 8th grade classroom. *Educational Studies in Mathematics, 58*(2), 145–168.

Bakhtin, M. M. (1981). *The dialogic imagination: Four essays by M.M. Bakhtin* (Edited by M. Holquist and C. Emerson & M. Holquist, trans.). Austin: University of Texas Press

Bartolini Bussi, M. G., & Mariotti, M. A. (2008). Semiotic mediation in the mathematics classroom: Artefacts and signs after a Vygotskian perspective. In L. English et al. (Eds.), *Handbook of international research in mathematics education* (pp. 750–787). Mahwah, NJ: LEA.

Buisson, F. (1911). Nouveau dictionnaire de pédagogie et d'instruction primaire. Paris: Hachette. Retrieved from http://www.inrp.fr/edition-electronique/lodel/dictionnaire-ferdinand-buisson/

Cheung, A. C. K., & Slavin, R. E. (2011). *The effectiveness of educational technology applications for enhancing mathematics achievement in K-12 classrooms: A meta-analysis.* Baltimore: Johns Hopkins University, Center for Research and Reform in Education. *The Best Evidence Encyclopedia.* Retrieved from www.bestevidence.org.

Cole, M. (1996). *Cultural psychology: A once and future discipline.* Harvard, MA: Harvard University Press.

Department for Education. (2012, November 9). Press release, *Government bans calculators in tests for 11-year-olds*, Elizabeth Truss MP.

Douady, R. (1985). The interplay between different settings. Tool-object dialectic in the extension of mathematical ability. In *Proceedings of the Ninth International Conference for the Psychology of Mathematics Education* (Vol. 2, pp. 33–52).

Engels, F. (1894/1968). Letter to W. Borgius. In K. Marx & F. Engels (Eds.), Selected works in one volume. Moscow: Progress Publishers.

Freire, P. (1993). *Pedagogy of the oppressed.* New York: Continuum.

Gardiner, T. (1995). Matters for Debate: Back to the Future. *The Mathematical Gazette*, 526–532.

Hedren, R. (1985). The hand-held calculator at the intermediate level. *Educational Studies in Mathematics, 16*(2), 163–179.

Hembree, R., & Dessart, D. J. (1986). Effects of hand-held calculators in precollege mathematics education: A meta-analysis. *Journal for Research in Mathematics Education, 17*(2), 83–99.

Hutchins, E., & Klausen, T. (1992). Distributed cognition in an airline cockpit. In Y. Engestrom & D. Middleton (Eds.), *Communication and cognition at work.* Cambridge, England: Cambridge University Press.

Jackson, A. (1997, June/July). The math wars: California battles it out over mathematics education reform (Part I). *Notices of the AMS*, pp. 695–702.

Kuhn, T. S. (1970). *The structure of scientific revolutions.* University of Chicago press.

Ladel, S., & Kortenkamp, U. (2013). An activity-theoretic approach to multi-touch tools in early mathematics learning. *International Journal for Technology in Mathematics Education, 20*(1), 3.

Latour, B. (2005). *Reassembling the social: An introduction to actor-network-theory.* Oxford, England: Oxford University Press.

Leont'ev, A. N. (1978). *Activity, consciousness, and personality.* Englewood Cliffs, NJ: Prentice-Hall.

Mullis, I., Martin, M. O., Minnich, C. A., Stanco, G. M., Arora, A., Cenurino, V. A. S., et al. (Eds.). (2012). *TIMMS 2011 encyclopedia: Education policy and curriculum in mathematics and science* (Vol. 1 and 2). Boston: TIMMS & PIRLS International Study Centre.

Prenksy, M. (2001). Digital natives, digital immigrants. *On the Horizon, 9*(5), 1–6.

Quesada, A. R., & Maxwell, M. E. (1994). The effects of using graphing calculators to enhance college students' performance in precalculus. *Educational Studies in Mathematics, 27*(2), 205–215.

Ralston, A. (1999). Let's abolish pencil-and-paper arithmetic, *Journal of Computers in Mathematics and Science Teaching, 18*(2), 173–194.

Ruthven, K. (1998). The use of mental, written and calculator strategies of numerical computation by upper primary pupils within a 'calculator-aware' number curriculum. *British Educational Research Journal, 24*(1), 21–42.

Selter, C. (2001). Addition and subtraction of three-digit numbers: German elementary children's success, methods and strategies. *Educational Studies in Mathematics, 47*(2), 145–173.

Sheryn, S. L. (2005). Getting an insight into how students use their graphical calculators. In *Proceedings of the British Society for Research into Learning Mathematics* (Vol. 25, no. 2,1 pp. 3–108).

Shuard, H., Walsh, A., Goodwin, J., & Worcester, V. (1991). *Calculators, children and mathematics: The calculator-aware number curriculum*. Hemel Hempstead: Simon & Schuster, for the NCC.

Threlfall, J. (2002). Flexible mental calculation. *Educational Studies in Mathematics, 50*(1), 29–47.

Vygotsky, L. S. (1978). *Mind in society*. Cambridge, MA: Harvard University Press.

Vygotsky, L. S. (1981). The instrumental method in psychology. In J. V. Wertsch (Ed.), *The concept of activity in Soviet psychology* (pp. 134–143). Armonk, NJ: M. E. Sharpe.

Wartofsky, M. (1979). *Models: Representation and the scientific understanding*. Dordrecht, The Netherlands: Reidel.

Weber, M. (1947). *The theory of social and economic organization* (A. R. Henderson & T. Parsons, Trans.). London: William Hodge.

Wertsch, J. V. (1998). *Mind as action*. Oxford, England: Oxford University Press.

Wertsch, J. V. (1981). *The concept of activity in Soviet psychology*. Armonk, NY: Sharpe.

Chapter 14
Tools and Mathematics in the Real World

John Monaghan

14.1 Introduction

This chapter has two main foci: (1) the use of mathematics in out-of-school[1] mathematical practices; (2) making school mathematics relevant to activities beyond mathematics classrooms (which I shall call 'out-of-school' mathematics/practices). Both foci are important issues in mathematics education and both are problematic issues. Both foci, of course, will be investigated with special regard to tool use. This chapter has four sections. The two central sections address the two main foci. The opening section sets the scene with an historical account of ways that mathematics has been subdivided with regard to its application(s). The last section considers problem issues.

14.2 Can Mathematics Be Subdivided with Regard to Its Application(s)?

Mathematics has been subdivided in various ways over two millennia. A subdivision, with regard to the application of mathematics, that has been used in Western mathematics for over 100 years, is that between pure and applied mathematics. This sounds like a promising way into the two main foci of this chapter so I start by considering this division: to what extent is it a real division; does tool use vary over this division?

There is a sense in which the division pure and applied mathematics is a real division between mathematical activity for intrinsic or extrinsic purposes. To take an elementary example, if I am teaching (or writing about the teaching of) 456 + 78 using the standard written algorithm, then I will pay careful attention to the fact that

[1] I use the term 'school' loosely to denote any educational institution.

J. Monaghan et al., *Tools and Mathematics*, Mathematics Education Library,
DOI 10.1007/978-3-319-02396-0_14

the ‘1’ I carry into the tens column (when I add 6 and 8) is not the number 1 but 1 unit of 10. But if I am doing this addition for a purpose other than doing mathematics, say, in checking my accounts, then this explicit attention to place value within a calculation is not important (in activity-theoretic terms, see Sect. 9.2, the object of the activity is different). This has immediate implications for tool use in mathematical activity. In the first activity the focus is on the correct use of a specific mathematical tool (a specific algorithm). In the second activity the focus is on obtaining the correct mathematical result and the tool I use to get the answer, as long as it gives the correct answer, does not matter a great deal (an abacus or a calculator or the standard written algorithm will do).[2] But there is also a sense in which the division between pure and applied mathematics is not a ‘natural’ division but a cultural–historical division; and this leads to a brief tour into the history of mathematics and mathematics education: the Ancient Greeks to the nineteenth century (Sect. 14.2.1); the twentieth century (Sect. 14.2.2).

14.2.1 Subdivision of Mathematics over Time

Fauvel and Gray (1987, p. 56) write of the ancient Greek *quadrivium*, ‘the four-part classification of mathematical sciences … into arithmetic, music, geometry and astronomy’ and claim that this ‘came to constitute a major part of the liberal arts curriculum of medieval universities’. This four-part classification, however, did concern ‘pure’ mathematics for Plato writes (see Fauvel & Gray, 1987, p. 69) of arithmetic, ‘what a subtle and useful instrument it is for our purpose, if one studies it for the sake of knowledge and not for commercial ends’. In seventeenth century Europe the division of mathematics was between pure and mixed mathematics. Francis Bacon wrote in 1603 (see Fauvel & Gray, 1987, pp. 290–291):

> To the pure mathematics are those sciences belonging which handle quantity determinate, merely severed from any axioms of natural philosophy; and these are two, geometry and arithmetic … Mixed hath for subject some axioms or parts of natural philosophy … For many parts of nature can neither be invented with sufficient subtility … nor accommodated unto use with sufficient dexterity, without the aid and intervening of mathematics; of which sort are perspective, music, astronomy, cosmography, architecture, enginery, and divers others … there cannot fail to be more kinds of them, as nature grows more disclosed.

This division was a part of Bacon’s tree of knowledge where the natural sciences were divided into physics and metaphysics and metaphysics divided into pure and mixed mathematics. Academic debate in the succeeding two centuries, according to Brown (1991), was subject to local variation as new areas of mathematics developed but largely retained Bacon’s distinction. For example, in mid-eighteenth century France, D’Alembert, writing in Diderot’s *Encyclopédie*, placed the new

[2] I think this ‘use of a specific tool in pure mathematics’ and ‘use of a range of tools in applied mathematics’ is a fairly common phenomena but I do not claim that it is a universal phenomenon.

field of probability (analysis of games of chance) into mixed mathematics but he placed the new field of calculus within pure mathematics.

The decline of the term 'mixed mathematics' occurred in the nineteenth century; Brown (1991) notes that the eighth edition of the *Encyclopedia Britannica* (1853–1860) used the pure-mixed classification but in the ninth edition (1875–1889) this was changed to 'pure' and 'applied' mathematics. Behind these classifications is ideology. There is a strong elitist ideology of 'learned men' behind writings from Plato to D'Alembert. Brown (1991, p. 84) writes:

> The mathematician was concerned with doing mathematics; the philosophe with analysing its importance to society. Who best to write about "mixed mathematics" than the scholar who was both a "geometer" and a "philosopher"? Neither Daniel Bernoulli, Euler, Lagrange, nor Laplace could be considered men of letters. That left only Condorcet and D'Alembert.

But ideology and mathematics shifted their foundations in Europe during the nineteenth century. Non-Euclidean geometries emerged which eroded geometry's claim as an a priori constructive field and science was viewed through positivist empirical eyes. 'By 1875 theories were no longer "mixed" with experience, they were "applied" to experience' (Brown, 1991, p. 102).

14.2.2 Subdivisions of Mathematics in the Twentieth Century

So we enter the twentieth century with a division, in the West, between pure and applied mathematics. In the Soviet Union, however, Vygotsky (cf Sect. 7.3) introduced a division between *everyday* and *scientific* concepts. Vygotsky did not introduce this distinction with mathematics in mind but it is, from a Vygotskian perspective, applicable to mathematics; in practical everyday mathematical activity an addition such as $456 + 78$ will likely involve 'things' (such as units of currency) but in *mathematics* addition comes with a history (the culture of mathematics) and mathematicians add numbers, not things. As noted in Sect. 7.3, Scott, Mortimer, and Ametller (2011, p. 6), in writing of Vygotsky's distinction, note 'scientific concepts are taken to be the products of specific scientific communities and constitute part of the disciplinary knowledge of that community'; 'the world is flat' was once a scientific concept. In a cultural vein similar (but not identical) to Vygotsky's, Bishop's (1988) study of mathematical enculturation differentiated between 'mathematics' and 'Mathematics':

> the mathematics which is exemplified by Kline's *Mathematics in Western Culture* is a particular variant of mathematics, developed through the ages by various societies. I shall characterise it as 'Mathematics' with a capital 'M'. (Bishop, 1988, p. 19)

These cultural approaches do not directly address the distinction between pure and applied mathematics but are concerned with the division between types of mathematical activity. By the end of the twentieth century, with mathematics education established as an academic discipline (see Sect. 7.2), scholars in this field made further divisions within mathematical activity. Blum and Niss (1991) is an interesting example of this because it represents the collective thoughts of a conference working group and presents 'a pragmatic attempt to give some working definitions' (Blum & Niss, 1991, p. 37). It considers two types of mathematical problems:

> It is characteristic of an applied mathematical problem that the situation and the questions defining it belong to some segment of the real world and allow some mathematical concepts, methods and results to become involved. By real world we mean the "rest of the world" outside mathematics, i.e. school or university subjects or disciplines different from mathematics, or everyday life and the world around us. In contrast, with a purely mathematical problem the defining situation is entirely embedded in some mathematical universe. This does not prevent pure problems from arising from applied ones, but as soon as they are lifted out of the extra-mathematical context which generated them they are no longer applied. (Blum & Niss, 1991, pp. 37–38)

The starting point for Blum and Niss (1991) is a 'real problem situation':

> This situation has to be simplified, idealized, structured, . . . This leads to a *real model* of the original situation . . . [which has to be] has to be mathematized, i.e. its data, concepts, relations, conditions and assumptions are to be translated into mathematics . . . [and] results have to be re-translated into the real world . . . real problem situations can also be called applications. . . mathematical models . . . can be seen as belonging to applied mathematics. Of course, this definition does not imply a strict separation between "pure" and "applied" mathematics. (Blum & Niss, 1991, pp. 38–40)

So, we are back to the division between pure and applied mathematics but the division is not a strict one and there is also a slight difference between 'applications' and 'modelling'. The translation and re-translation that Blum and Niss speak of is often presented in a diagram, like the two leftmost columns in Fig. 14.1, in mathematics education literature on applications and modelling of mathematics. With reference to these two columns Fig. 14.1, the left column represents 'reality' (the real world) and the right column represents 'mathematics' (the mathematical world). The diagram represents a cycle: situation → mathematical model → mathematical results → real result → compare with the situation and

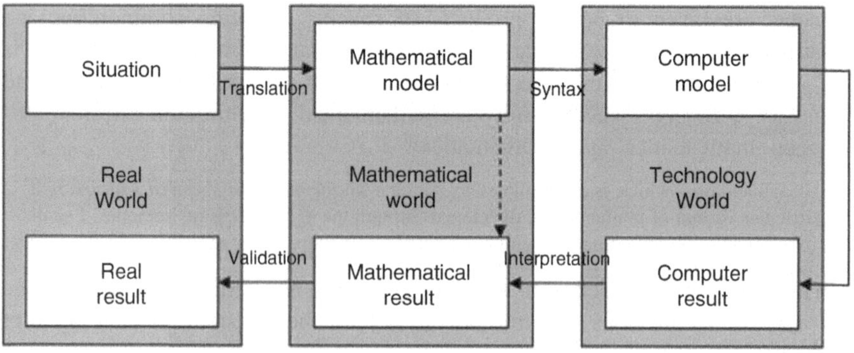

Fig. 14.1 Siller and Greefrath's (2010) extended modelling cycle

note discrepancies → adjust mathematical model → mathematical results → etc. Of the many comments that can be made on this modelling cycle, I make three. First, it does appear (albeit in an oversimplified way) to approximate what goes on in applied mathematical problem solving. But, second, what does it represent? Is it supposed to be descriptive (of the work of mathematicians) or prescriptive (for educational purposes)? Blum and Niss (1991) appear to regard it as descriptive, 'This leads to a *real model* of the original situation' (Blum & Niss, 1991, p. 38), but my own experience of the first step is more akin to 'situation ↔ mathematical model' than it is to 'situation → mathematical model' (i.e. there is a lot of 'fiddling' with the mathematical model). My third point regards tools and is simply that the two column modelling cycle, which is what is usually offered, does not include tools and I consider this further below.

It is curious, from the point of view of tool use in mathematics, that the word 'tool' does not feature in Blum and Niss' account except within the context of talking about computers as a tool. I use the word 'curious' in the sense that Arthur Conan Doyle ascribes to Sherlock Holmes in the *Memoirs of Sherlock Homes* where the detective is talking to a police inspector:

> 'Is there any point to which you would wish to draw my attention?'
> 'To the curious incident of the dog in the night-time.'
> 'The dog did nothing in the night-time.'
> 'That was the curious incident,' remarked Sherlock Holmes.

Computers are very interesting tools but it is curious that tools, other than computers, are not mentioned in Blum and Niss (1991) when tools (measuring artefacts, machinery for experiments, formulas as tools, etc.) are clearly important in modelling and applications of mathematics. I shall mention similar omissions at other points in this chapter, so I give it a name, 'tool blindness'—not seeing something until it *hits you in the face*. Computers are mentioned in the second part of Blum and Niss (1991) in relation to 'trends' and 'obstacles':

- With regard to professional modelling 'For several years it has been evident that computers form a highly powerful tool for the numerical and graphical treatment of mathematical applications and models' (Blum & Niss, 1991, p. 52).
- With regard to mathematical education, computers allow 'More complex applied problems ... relief from tedious routine ... Problems can be analysed and understood better by varying parameters ... [and] Problems which are inaccessible from a given theoretical basis ...may be simulated numerically or graphically' (Blum & Niss, 1991, p. 58).

Siller and Greefrath's (2010) also focus on the place of computers in modelling (for educational purposes). They offer the first representation of the modelling cycle (to my knowledge) that includes tools (computers), see Fig. 14.1.

Whilst it is nice, from my tool perspective, to see a recognition of the place of technology in modelling, I am sceptical that there are three distinct worlds (as the presentation may suggest). The practice of modelling (be it in-school of out-of-school) is a reality. These three worlds seem to impose an 'unreal' partition of this practice.[3] The resolution of this problem issue may be simply to jettison the modelling cycle and look to real practice (in which tool use will be an integral part of the practice). I return to this point in the final section of this chapter and now consider out-of-school mathematical practices.

14.3 Out-of-School Mathematical Practices

Out-of-school mathematical practices cover an extensive field of activities and I must pare this field to keep this section manageable. Section 14.3.1 sets the scene by mapping the field. This mapping includes constructors (people who design technology) and operators (people who use technology) and Sects. 14.3.2 and 14.3.3 consider constructors and operators in turn. The final subsection looks at the place of computers in out-of-school mathematical practices because computers hold a prominent position in many of these practices in the twenty-first century.

14.3.1 Varieties of Out-of-School Mathematical Practices

There are many out-of-school mathematical practices—certainly too many to list. I will first attempt a map of the field and then consider a subset of Western workplace practices which have been a focus of research and address tool use and mathematics. My map of the field includes three divisions: leisure and work practices; levels of involvement with tools; and Western vs. 'other' practices.

The distinction between leisure and work practices is not a precise one since there are instances where such practices overlap (e.g. voluntary work). Leisure, considered as non-paid activity, includes domestic and recreational activity. Domestic activity includes practices which can have mathematical aspects such as: cooking, following a new recipe (which is an artefact which is used and is thus a tool by my Sect. 1.3 definition)—cooking also involves using a great many utensils (tools), some of which (e.g. weighing scales) are 'pre-mathematicised'; monitoring household accounts, which is often facilitated by the tools available in e-banking; and domestic repairs such as drilling a hole (finding the right drill size and 'feeling the right angle' with your body). Recreational activity includes practices which can have mathematical aspects includes: travelling, buying e-tickets online and co-ordinating rail and flight schedules (artefacts); programming the recorder on your TV-media unit; performing music; and playing games (see considered in Chap. 19). Gameplay always has a mathematical aspect as games have rules (which are artefacts, 'mediational means' in the language of Wertsch—see

[3] Siller and Greefrath's (2010, p. 2138) note, 'The three different worlds shown in Fig. 2 are idealized; they influence each other.'

Sect. 13.3) and these rules include sequencing actions. Although some games (e.g. soccer) can be played with only an epsilon of mathematical activity in a similar way to which they were played 100 years ago, gameplay has increasingly been influenced by digital technology. This is certainly so in the case of digital games but it is also sometimes the case in games such as soccer where even some amateur teams use performance analysis software, which provide statistics on video-recorded motion analyses, to improve their gameplay.

My second division concerns levels of involvement with tools and I employ the language of Skovsmose (2005). Skovsmose is interested in 'critical mathematics' and technology and distinguishes between 'constructors', 'operators' and 'consumers'. With regard to technology, constructors are professionals who design/develop technology, operators are those who use/manipulate the technology and consumers are people not involved in the construction or operation of the technology but are affected by it. For example, a manager and a computer scientist (constructors) may design/implement a new system of calculating wages, computer operators run the wage system and the consumer is affected with wages and a pay statement. My consideration of leisure practices above concern the consumer level of involvement with tools but this level is also common in workplace practices, especially amongst low paid workers. People's encounters with tools/technology at the consumer level is typically as 'black-boxes', a term originally from cybernetics that refers to artefacts where the input–output relationship is hidden from the user. I write at a time when an international banking crisis is having a profound negative effect on the quality of life of many consumers of banking technology. Enabling people to critically engage with black-boxes is important to critical mathematics. This is one reason why mathematical modelling is considered important.

My third division concerns Western vs. 'other' practices. Western research in mathematics education is dominated by Western researchers researching Western contexts. A partial exception is what is commonly referred to as 'ethnomathematics'. This was a new but rising area of research at the turn of the Millennium but it met problems. Ethnomathematics investigates 'indigenous, socio-, informal, spontaneous, oral, hidden, implicit, and people's mathematics' (Gerdes, 1996, p. 909). Activities investigated, such as basket weaving in Mozambique, are characterised as being both highly context bound and highly creative (Gerdes, 1997). Ethnomathematics is not a non-Western phenomenon but concerns traditions in any locality, though reports on ethnomathematical activities are often written by people with a Western education reporting on the practices of those who have not received a Western education. Dowling (1998, p. 14) considers that these studies succeed in 'celebrating non-European cultural practices only by describing them in European mathematical terms, that is, by depriving them of their social and cultural specificity'. Pais (2011) considers this and other criticisms of ethnomatics as it has been researched. This and my Western background lead me to leave an account of tool use in non-Western out-of-school practices to a more capable author.

14.3.2 Tool Use in Constructors' Mathematical Practices

Frejd and Bergsten (2016) focus on constructors with specific regard to modelling. They interviewed nine professionals (in both the commercial and academic sectors) with Ph.D.s (all in the sciences, five in mathematics) and focused on three phases: pre-construction, the reason for the modelling activity; construction, how the model is developed; post-construction, the consequences of using the model. The analysis reveals three types of modelling which they call 'empirical', theoretical' and 'applicational'; the use of computers was an essential feature in each type of modelling. A defining characteristic of empirical modelling is data from empirical observations though the data, of course, varied over contexts (e.g. financial risk and workforce scheduling). The mathematical model in every case was implemented in a computer system. Issues with the data (to feed into the computerised mathematical model) included: getting sufficient data; cleaning data; dealing with gaps (e.g. for time series analysis); locating errors in the data. In my experience of such things the time actually using the tool (computer) is a tiny proportion of the time preparing the data for tool use but the tool is central to the activity.

Theoretical modelling involves:

> ... setting up new equations based on already theorised and established physical equations. This is followed by the activation of computer resources for computational purposes to solve the new equations with aim to get information about the 'theorised' equations. (Frejd & Bergsten, 2016, p. 24)

Example problems in theoretical modelling include predicting climate change and the design of a new material. At the heart of theoretical modelling is the mathematical model and its implementation on a computer. In the problems cited in Frejd and Bergsten (2016) this computer had to be 'powerful', as the designer of models for new materials said, 'The computer is our big tool, not least when it comes to solving these quantum mechanics equations' (Frejd & Bergsten, 2016, p. 26).

Applicational modelling refers to 'identifying situations where some mathematics or some established mathematical models can be directly applied' (Frejd & Bergsten, 2016, p. 26); this was an aspect of the work of all nine modellers. For example, one of the nine modellers was a biologist who was looking into the spread of diseases between oak trees. His starting point was differential equations:

> Fourier transformations are really good and you can then rewrite anything as a sum of sine functions. [...] This has been used by people at the department of systems control [...] Basically it is knowledge about mathematical methods that do the work, and sometimes you start with the problem and then you add a method [...] It is basically the same thing if bugs fly between oak trees or if animals are transported in trucks. (Frejd & Bergsten, 2016, p. 27).

In the above summary of Frejd and Bergsten (2016) I focused on the construction phase where a mathematical model (an artefact) and various other mathematical tools, especially computers, were central features of the activity in all cases. The discussion of the pre- and post-construction phases in Frejd and Bergsten (2016), however, highlight that mathematical tool use is but a part of the activity of modelling. These modellers serve clients who are not necessarily able to understand the model or computer use. In the pre-construction phase the client needs to be convinced that the model will be useful for his/her purpose and in the post-construction phase the client needs to be made aware, by the modeller, of the potential and the limitations of the model. Communication/dialogue is also an essential feature of professional modelling activity. I now turn to Skovsmose's (2005) 'operators'.

14.3.3 Tool Use in Operators' Mathematical Practices

Operators, people who use/manipulate rather than design/develop technology, are a very large class in themselves which includes technicians (manual, e.g. plumbers, and blue collar, e.g. insurance clerks), social service workers (e.g. nurses and police), sales people and teachers (e.g. a mathematics teacher using mathematical software). Skovsmose's (2005) three categories (constructors, operators and consumers) are wide categories and there are 'grey areas'. For example technicians may adapt given tools to their needs in a specific activity and clerks who operate payroll systems are themselves customers of a payroll system. The categories are nevertheless useful for focusing on tool use in out-of-school mathematical practices.

Noss and Hoyles (1996) distinguish between 'visible' and 'invisible' mathematics in out-of-school activity. Visible mathematics is that which is immediately recognised as being mathematics. This distinction is clearly context/person specific. The mathematics in the work of the constructors considered in Frejd and Bergsten (2016) was visible but it is common, when you ask an operator 'What mathematics is involved in your job?', that they reply 'None' or 'Very little'. Very often there is mathematics in this job but they do not see it as mathematics, it is invisible to them, often 'hidden' in tools they use. I shortly explore these general statements in some detail in the context of research I was involved in but I first outline research by a group that provides themes for a discussion of issues related to operators, mathematics and tools.

Hoyles, Noss and Pozzi focused on operators in a series of publications (see Noss, Hoyles, & Pozzi, 1998, for a summary) which examined mathematics in nursing, banking and flying workplaces. In a report of nursing practice (Pozzi, Noss, & Hoyles, 1998, considered in Sect. 9.2) they focus on drug administration and fluid balance monitoring aspects of patient care. The research team made multiple hospital visits to 12 experienced nurses over 4 months which resulted in 80 h of observation.

They set out to observe activities which involved visible mathematics and the mathematisation of the nurses' professional practice. In all cases they

> attempted to delve beyond simple arithmetic procedures to try to understand more complex, but perhaps less visible parts of decision–making on the ward ... [by separating] out

> episodes involving routine behaviour from those involving a *breakdown* in the normal habits of nursing practice (Pozzi et al., 1998, pp. 107–108, italics added).

For example, drug administration appears to involve ratio and proportion but proportional reasoning was replaced in routine practice by arithmetic rubrics. They provide an example of a nurse preparing 85 mg of an antibiotic from a vial containing 100 mg of the drug in 2 mL and using the formula (an artefact) $\frac{\textit{Amount you want}}{\textit{Amount you have got}} \times \textit{Volume it is in}$. The formula, however, is not conceived 'mathematically' but as a strategy for calculations for specific drugs, 'It was often heard that "with amikacin you can double it and divide by a hundred" or "with ondansetron, you only need to half it"' (Pozzi et al., 1998, p. 110). I now turn to an example where Pozzi et al. (1998) interpret a 'breakdown' situation. For reasons of space I omit many of the details which can be found in Pozzi et al. (1998).

Two nurses are discussing a fluid balance chart (an artefact) of a patient who has recently had a kidney transplant. The chart is not questioned by the nurses and it comes with a mathematical structure: the rows record times; the columns record fluids in and fluids out. Sam, an experienced nurse who is new to the ward asks 'why are you recording the difference between these two?' and Al, the nurse who is not new to the ward replies 'Because then when I come to add it up, I add my hourly totals. To get this one, that's why I need to know that figure' (Pozzi et al., 1998, p. 113). Further dialogue around the numbers in the chart ensues, basically along the lines of Al providing rationales for the calculations in the given chart and Sam questioning how the numbers relate to the patient's situation. Eventually Al sees Sam's point and concedes 'I suppose you should write down the rate'.

In their summary, Pozzi et al. (1998) note:

> Professional cultures contain a huge number of artefacts which are, like the nurses' chart or the nursing rule, already mathematised . . . workers rarely think mathematically without an artefact to help them to organise or compute the data. In routine use, this mathematics is invisible, and remains so—indeed, the functionality of artefacts often crucially depends on this invisibility. But at times, people will need to understand the models which underlie their artefacts, to sort out what is happening or what has gone amiss . . . As we saw in the fluid balance episode, this typically occurs when there is a breakdown, and in such a situation, people need to represent to themselves how the underlying structures work (Pozzi et al., 1998, p. 118)

I think Pozzi et al. (1998) provides both a well-grounded evidence-base for its claims and insightful comments on tool use in the practice of a group of operators (nurses). But, taken alone, there is a danger that these claims for one practice may be viewed as generic for operators in general. I now consider Magajna and Monaghan (2003) which has similarities and differences to Pozzi et al. (1998).

Magajna and Monaghan (2003) is a case study of the mathematics and tool use of six computer aided design and manufacture (CAD-CAM) technicians. It reports on these technicians' calculations of the internal volume of moulds they produce for glass factories. The six technicians work as a team but three of them (constructors) liaise with clients and three (technologists) liaise with machinists in their factory who produce the metal moulds for glass bottles. The technicians were observed for

60 h over 3 weeks. Constructors evaluate whether a mould for a bottle can be manufactured, define the dimensions of the mould, design the bottle and the mould and make technical drawings. Technologists define the surfaces to be cut, write the programs for computer-numerically-controlled machines and, independently from the constructors, determine the inner volume of the mould.

A mould consists of three parts which close around a piece of molten glass of a given weight/volume. Compressed air is pumped into the molten glass which adheres to the inner shape of the mould. When the glass is cooled, the three pieces of the mould are separated and the bottle is released. The important job is to define the inner shape of the mould and to cut them out using appropriate machines. There are several volumes: of the bottle filled; of the glass; and the inner volume of the mould. The relations between these volumes is obvious to the technicians and the only volume they are concerned with is the inner volume of the mould. When they spoke about a volume related to a bottle they meant the inner volume of the mould. Getting the inner volume correct to a high degree of precision is essential for client satisfaction. The technicians do not distinguish between exact and approximate volumes as all calculations are approximate to them. Six methods of calculating the volume of a shape were observed:

1. The constructors drew the 2D-profile on a computer system and then used a program which automatically calculated the volume of the rotated shape.
2. The constructors represented the shape of the bottle in terms of horizontal cross-sections at various heights and drew a sequence of cross-sections. The volume of the part of the bottle between two horizontal sections with respective areas A and B and the height h between the sections was calculated using the formula $V = h\big(A + \sqrt{AB} + B\big)/3$. The constructors did not know where this formula—it was 'a shop-floor tradition'.
3. The constructors calculated the volume of a bottle using a 3D-solid CAD.
4. The volumes of standard geometric shapes were calculated using school-learnt formulae, e.g. to calculate the volume of a prism of height h, the constructor drew its base on a computer to obtain its area, A, and then used the formula $V = hA$.
5. The technologists obtained the volume of a shape using a 3D-surface modeller integrated into the CAM software they used. The program they used calculated the volume of a polyhedron with the vertices on the mesh points, but the technologists ignored this.
6. Once the mould was made, its volume was measured by weighing the water it held.

I now consider similarities and differences between this research and that of Pozzi et al. (1998). Both studies provide evidence that the operators under scrutiny rarely engage in mathematics without the use of an artefact/tool. Pozzi et al. (1998, p. 115) add, 'the use of artefacts never fully structures activity. People are not necessarily slaves to the tools they use' but Magajna and Monaghan (2003, p. 119) state, of the technicians in their study:

> The mathematics they were really doing, their work mathematics, was inextricably joined with the technology they used. The geometry elements in their designs always represented technological entities and the calculations they performed were grounded in technology. Our practitioners used mathematical tools, including software, as 'black-boxes'. They were not observed to reason about the mathematics hidden in these tools and if a tool-based method did not work, they simply chose another method or overcame the problem by technological means.

A second difference is that Pozzi et al. (1998) concerns 'breakdowns'; as cited above, 'at times, people will need to understand the models which underlie their artefacts ... this typically occurs when there is a breakdown' (Pozzi et al., 1998, p. 118). Magajna and Monaghan do not dispute that this did occur in observations of the nurses but did not find this to be the case in their study. In breakdown situations their practitioners either chose another method to overcame the problem by technological means:

> participants' reactions to 16 cases of non-trivial mathematics-related errors were observed. In 14 out of the 16 cases the error was due to a mistake in a computer generated geometric construction. Analysis revealed the following causes of errors: poor understanding of some detail in a construction command (11 cases), undocumented details about construction in the software (3 cases), a misunderstanding between participants (1 case) and difficulty in visualising the shape (1 case). In such breakdown situations the participants never reasoned about possible mathematics-related causes of the error, e.g. whether they understood the mathematical aspect of the applied construction. In most cases the geometric error was left unresolved and a solution was found by technological means (8 cases). (Pozzi et al., 1998, pp. 113–114)

This comparison of these two studies suggests that the differences observed/interpreted in these two studies are likely to result from the particularities of the different workplaces observed and that further studies on tool use in other workplaces are needed.

14.3.4 Computers in Out-of-School Mathematical Practices

I end this section with a consideration of the place of computers in workplace mathematics. Although a computer is just another tool (or 'set of tools'—see Sect. 1.2), the prominence of computers in the workplace in the twenty-first century merits special consideration. This prominence is evident in the above discussion of constructors. It is also evident in Magajna and Monaghan's (2003) study of CAD-CAM technicians/operators. Studies in trends in workplace skills provide evidence that ICT is an increasingly important part of employment:

> There has been a striking and continued increase since 1986 in the number of jobs in which advanced technology is used. There has also been a marked increase over the last four years in the proportion of jobs in which computing is considered to be an essential or very important component of the work. Over 70 percent of people in employment now make use of some type of automated or computerised equipment, and computerised equipment is seen by 40 percent as essential to their work. (Felstead, Gallie, & Green, 2002, p. 12)

Researchers who have addressed this issue are, again, Celia Hoyles and Richard Noss (with colleagues). Noss and Hoyles (2009) reconsider modelling to address the advance of ICT in twenty-first century work practices, 'With the ubiquity of IT, employees now require new kinds of mathematical knowledge that are shaped by the systems that govern their work' (Noss & Hoyles, 2009, p. 76). Behind this paper are two reports which I now consider.

Hoyles, Wolf, Molyneux-Hodgson, and Kent (2002) reports on research into mathematical skills used/needed in seven areas of employment spanning engineering, financial services and health care. It coined the term 'mathematical literacy' which arose from the required skills and made four recommendations. ICT ('IT' in the language of the Report) is not the sole focus but it is a major focus: mathematical literacy is defined by a list of 12 skills of which the first 2 are 'Integrated mathematics and IT skills; an ability to create a formula (using a spreadsheet if necessary)' (Hoyles et al., 2002, p. 5); all the recommendations bar the last one on communication refer to IT:

> That IT and mathematical skills are interdependent . . .Developing models of new forms of training for all employees which reflect mathematical literacy that is integrated with IT competence . . . To investigate the development of training programmes which will be effective in the workplace by achieving a balance between physical experiences and software packages (Hoyles et al., 2002, pp. 3–4)

The IT dimension of Hoyles et al. (2002) was further developed in Hoyles (2007) under the term, 'Technomathematical literacies (TmL), that is, being able to reason with quantitative or symbolic data processed by information technology as part of decision-making or the communication process' (Hoyles, 2007, p. 16). A construct introduced in Hoyles (2007) is 'technology-enhanced boundary object' (TEBO). The construct 'boundary object' was introduced in Star and Griesemer (1989) and has been widely used in social science research since its introduction. A boundary object is an artefact created in one community of practice and travels to a distinct community of practice. Boundary objects abound in all practices including mathematics (e.g. Sloane's online *Encyclopedia of Integer Sequences*, see Chap. 3) and mathematics education (a new version of a mathematics curriculum devised by Ministry workers and sent to teachers). An interesting feature of boundary objects is that the meanings ostensibly embedded in them by their creators are re-interpreted by members of the receiving community. Hoyles (2007) TEBOs were linked to TmLs and workplace learning opportunities, 'Learning opportunities incorporated interactive software tools that modelled elements of the work process, or were reconstructions of the symbolic artefacts from workplace practice . . . TEBOs . . . involving many cycles of collaborative design' (Hoyles, 2007, p. 18).

Hoyles (2007) provides an example of a packaging factory making plastic film by an extrusion process. The computer control and monitoring system, it is claimed, served as a boundary object between managers, engineers and shop-floor machine operators. The computer system captures data on the stages in the process and presents this data in graphical form but shop-floor operators rarely looked at them. The research team identified a TmL:

> Understanding systematic measurement, data collection and display; appreciation of the complex effects of changing variables on the production system as a whole; being able to identify key variables and relationships in the work flow; and reading and interpreting time series data, graphs and charts (Hoyles, 2007, p. 21)

The research team in collaboration with employees developed a TEBO, a computer simulation of the production process with a goal to achieve stable running of the extrusion process. The hands-on TEBO training was viewed by a process engineer as a superior learning opportunity for operators than prior observational style training. From the perspective of tool use in mathematics it is an interesting case of using a tool (a computer model) in workplace training to simulate another workplace tool (the computer system that monitors the actual process).

14.4 Links Between In-School and Out-of-School Mathematical Practices

This section has two subsections. The first presents a case that linking in-school and out-of-school mathematical practices is an incredibly difficult undertaking. The second looks at research that has sought ways into making links between in-school and out-of-school mathematical practices.

14.4.1 Difficulties in Linking In-School and Out-of-School Mathematical Practices

The application of school mathematics to everyday and work settings is one of the main rationales for the place of mathematics in national curricula: 'This fact in itself could be thought to provide a sufficient reason for teaching mathematics' (Cockcroft, 1982, paragraph 1). Last century there was a perception, that I believe was widespread, that people, as students, learnt mathematics in school and applied this same mathematics, when appropriate, in out-of-school settings. In the UK, for example, the Mathematical Association wrote, concerning the teaching of arithmetic in schools, 'The arithmetic rules and processes needed in the practice of double entry book-keeping are, in the main, those with which the pupils of secondary schools are familiar ... The corresponding arithmetic work may be [there follows a list of topics]' (Mathematical Association, 1952, p. 73).

This perception commonly goes by the name 'transfer' (of knowledge or of learning). In the late twentieth century a number of studies presented data and theories of learning/doing mathematics that ranged from regarding transfer as highly problematic to rejecting it outright as a myth. In this section I first consider two studies/theories that question transfer. I then consider the school mathematics and attempts to make links between in-school and out-of-school mathematical practices.

A book that generated a great deal of interest (ranging from revelation to outrage) in the mathematics education community is Lave (1988). Lave presented data that people could do mathematics 'better' in supermarkets than in a test; her examples were arithmetic, the cost of items in a supermarket and equivalent 'sums' in a paper and paper test. Lave had a theory, which came to be called 'situated cognition', that supported her data, that claimed that how one thinks is tied to the practice one is engaged in. 'Situated cognition' is probably an unfortunate name from Lave's viewpoint as she is scathing of traditional cognitive research on 'knowledge':

> the effect on cognitive research of "locating" problems in "knowledge domains" has been to .separate the study of problem solving from analysis of the situations in which it occurs ... "knowledge domain" is a socially constructed *exoticum*, that is, it lies at the intersection of the myth of decontextualized understanding and professional/academic specializations (Lave, 1988, p. 42)

To Lave (1988) learning in and out of school are different social practices and there is no reason to expect learning in one social practice to influence another social practice. But Lave's, 1988 exposition is not illuminating from the point of tool use in mathematics because tool use in learning does not feature in this account. Indeed, of the wider literature on *communities of practice* (which includes Lave's, 1988 account), Kanes and Lerman (2008) write, 'a theory of mediation is needed ....The nature and ....role of artefacts and tools is hazy' (Kanes & Lerman, 2008, p. 320).

Lave (1988) regards the perception of transfer of learning across social practice as a myth. Around the same time as Lave developed her theory, Saxe (1991) developed an approach that viewed transfer as problematic but not necessarily impossible.

Saxe (1991) uses a model, developed in ethnographic research into the transformation of mathematical practices of Papua New Guinean tribespeople, to examine the candy-selling practices of Brazilian street children, and then to explore links between in-school and out-of-school mathematical practices. Saxe's model has three components: analysis of practice-linked goals; form-function shifts in cognitive development; the interplay of learning across contexts (i.e. 'transfer'). It suffices for this section of this chapter to focus on the first component where of 'practice-linked goals' means 'emergent goals'—'must do' things that arise in practice and can interrupt that practice is they are not resolved. For example, buying something in a shop in a foreign country may induce the emergent goal 'determine the values of these coins in my wallet'. Emergent goals may or may not be related to mathematics; the agent is not necessarily aware of emergent goals. Saxe claims that four 'parameters' impinge on the resolution of emergent goals:

- *Activity structures*, 'general tasks that must be accomplished in the practice- and task-linked motives' (Saxe, 1991, p. 17)
- *Social interactions*, relationships between participants

- *Conventions and artefacts*, 'the cultural forms that have emerged over the course of social history' (Saxe, 1991, p. 18)
- *Prior understandings, which* 'constrain and enable the goals they construct in practices' (Saxe, 1991, p. 18)

I have found Saxe's model useful in examining in-school and out-of-school mathematical practices (see Magajna & Monaghan, 2003; Monaghan, 2004; Monaghan, 2007b) since it affords an analysis of practice to consider the dialectic between mathematics, tool use, networks of artefacts and social activity (which was discussed in Part I of this book). When Saxe (1991) gets to examining learning across contexts (transfer), this model allows him to explore aspects of 'transfer' rather than make general claims as to its existence or not.

I provided these brief accounts of two late twentieth century frameworks related to the perception of transfer to establish a background assumption in twenty-first century academic mathematics education, that transfer, and making links between in-school and out-of-school mathematical practices, is problematic. The issue is ongoing. Although Lave's (1988) 'situated view' of transfer was that it is a myth, Engle (2006) presents a situated view of transfer as 'framing'—'making references to both past contexts and imagined future ones . . .[to] make it clear to students that they are not just getting current tasks done, but are preparing for future learning' (Engle, 2006, p. 456), and forms of learner participation. A research question awaiting a researcher is whether framing tool use in mathematics learning can be used to promote intercontextuality.

Neither Lave (1988) nor Saxe (1991) explore school mathematics classroom practices to any depth and it is appropriate to consider this practice at this juncture. There are differences between countries (Mullis, Martin, Foy, & Arora, 2012), within countries (Noyes, 2012) and within schools (Noyes, 2012) in school mathematics classroom practices but a common feature of mathematics classrooms is that they consist of a set of learners and a teacher (or teachers) who have come together, ostensibly for the teacher to help the learners engage in mathematics. It is important to that the age/experience of the children is taken into account though it should not be assumed that young children cannot engage in 'applied mathematics'. Mathematical practices in classrooms are distinct from those of mathematicians (pure or applied). Sect. 10.4 details Chevallard's notion of 'didactical transposition'; this is neatly encapsulated by Lagrange (2005, p. 69) 'mathematics in research and in school can be seen as a set of knowledge and practices in transposition between two institutions, the first one aiming at the production of knowledge and the other at its *study*.' Strange things such as *the suspension of sense-making* (see, for example, Verschaffel, Greer, & De Corte, 2000) can happen in mathematics classrooms. For example studies in various countries have presented primary school children with 'There are 26 sheep and 10 goats on a ship. How old is the Captain?' and a common answer is '36'. There is, of course, a sense to this answer, 'this is a mathematics class and there are two numbers in this question, I'll add them', but this sense works against the sense needed to link in-school and out-of-school mathematics. School mathematics here is seen as a sort of 'game'.

Verschaffel, Greer, and de Corte (2002, p. 262) cite a 13-year-old student who, when asked by an interviewer why she did not make use of realistic consideration in her solution to a problem, responded:

> I know all these things but I would never think to include them in a math problem. Math isn't about things like that. Its about getting sums right and you don't need to know outside things to get sums right.

And school mathematics is often such a game. In disturbing research by Cooper and Dunne (2000) the researchers presented upper primary and lower secondary school children with 'esoteric' (e.g. $2x+1=17$, find x), realistic mathematics questions and analysed responses with regard to the children's social class. The working class children held their own very well, against children with parents in the professions, in the esoteric questions but performed comparatively poorly in the realistic questions. An interpretation of this data is that working class children took the realistic questions seriously but the other children knew it was *just a game* and this disadvantaged the working class children. For example, a question about the price of a soft drink and a bag of popcorn in a cinema was a disguised simultaneous equations question and children drawing on knowledge of actual cinema prices would get the answer wrong. But even when it is not seen as a game, school mathematics is almost always done in a mathematics lesson and this 'situation' appears to matter. Monaghan (2007b) reports on a study where a company director came into a mathematics class (students aged 14–15) and gave them a problem he was working on (about how to use a GPS position to register when one of his haulage vehicles had arrived at its destination). The research picked up the following exchange between two students:

Student 1 Shall we draw this as a graph?
Student 2 Why?
Student 1 'Cos that's normally what you do with co-ordinates.

The company director wanted a solution to a real problem. He expected that mathematics could be used in the solution but Student 1 expected to use a particular approach due to the mathematical content.

14.4.2 Attempts at Linking In-School and Out-of-School Mathematical Practices

I now focus discussion towards artefact/tool use in attempts to link in-school and out-of-school mathematics. The widespread use of artefact/tools in out-of-school mathematics documented in the first half of this chapter suggests that this focus may have potential to make links between in-school and out-of-school mathematics. I first note, in my experience, a restrictive vision with regard to tools in the applications of mathematics in schools. In 2005, at the outset of a project in my locality concerned with linking in- and out-of-school mathematics, I sent out a questionnaire to local schools with a series of questions on this topic (this is reported in Monaghan & Sheryn, 2006). One question was:

> Does your department use any special resources for linking school mathematics to out-of-school mathematical activity? Anything from surveying equipment, to catalogues to computer software, please specify.

Fifty-two percent of the schools stated that they did not use any special resources. The remainder mentioned occasional use of resources. Twelve percent stated that they had holidays and shopping catalogues. Other resources mentioned were trundle wheels and clinometers. There appears to be a bit of a *tool blindness* (not consciously recognising the use of tools in activity) here as none of the schools mentioned software and I knew that many of them did use spreadsheets in mathematics work. Nevertheless, it does not appear that artefact/tools (resources) are viewed as important in the applications of mathematics. Tool blindness (or, at least, partial vision) appears in research too. Masingila, Davidenko, and Prus-Wisniowska (1996) employs Saxe's framework; it reports on three workplace mathematics studies (dietetics, carpet laying and restaurant management). Selected problems from these contexts were given to pairs of secondary students who were observed and questioned as they solved the problems. They found differences in 'the goals of the activity, the conceptual understanding of persons in each context, and flexibility in dealing with constraints'. Although the paper discusses the role of artefacts/tools in workplace mathematics in its presentation of the theoretical framework, it says surprisingly little about artefacts/tools in its comparison of workplace and in-school problem solving and when it does, it does so in quite general terms, for example:

> For both the restaurant manager and the interior designer, solving the problems were necessary parts of their jobs. They used mathematics as a tool to help them solve problems and not as the goal of the problem. The students, however, seemed to view the problems as mathematical exercises and immediately started using algorithms that they thought would be appropriate. (Masingila et al., 1996, p. 182)

Even when they explicitly consider Saxe's parameter concerned with artefacts they merely mention, with regard to carpet laying, 'students may invent notation to indicate when objects are the same size and shape, in the course of working in a measurement context, before they have formalised the concept of congruence' (Masingila et al., 1996, p. 196).

Two school-based studies that do focus on genuine artefacts are Lowrie (2011) and Bonotto (2013), though the artefacts in question in both papers are not mathematical tools.[4] Lowrie's focus is twofold, the use of genuine artefacts and collaborative learning in solving realistic mathematics problems. The children were a Grade 6 class (11–12 years old) from an Australian primary school. The artefacts

[4] This is not meant to belittle the mathematical potential of artefacts that are not mathematical tools. Many artefacts of this kind enable what the Freudenthal school (see Freudenthal, 1991) call 'horizontal mathematization'; mathematics can be extracted from the artefact and the artefact can be mathematically structured by the agent.

used were brochures, menus, bus timetables, photographs and a real map from a local theme park. The children worked in small groups to plan a group trip to the theme park, 'plan the day's events with appropriate details and budgetary considerations . . . use the map as your main reference point' (Bonotto, 2013, pp. 4–5). The artefacts were judged to have learning potential, they:

> Encouraged the children to make connections to real-life experiences . . . [children] sourced a great deal of visual, spatial and graphical information from the artefacts . . . established a strong motivational intention for the open-ended task. (Lowrie, 2011, pp. 7–8)

With regard to collaborative learning, however, there was considerable variation in the: quality of the solution; the authenticity of the solution; the manner in which the group work (collaboratively or with one student dominating). Lowrie's interpretation of this is interesting:

> These artefacts establish a sense of problem solving 'integrity' . . . helps to establish meaningful engagement between peers . . .However, as the students accessed and used personal knowledge to solve problems, they were less likely to monitor and manage collaborative group goals. (Lowrie, 2011, p. 14)

So the use of genuine artefacts has great potential for applied problem solving in schools but the solutions by individuals in groups are often rich, complex and varied and many of these students found it difficult to simultaneously focus on the complexity of their own solution and that of their peers. It is useful when research alerts us to matters such as these which may not be obvious.

Bonotto (2013) has similarities to Lowrie (2011), the age of the children and the types of artefacts, but focuses on artefacts as a source of real-life problem-posing (I will only refer to problem-posing when necessary as it is not my focus in this chapter). The study was in two parts. The first part was exploratory, 'to evaluate . . . the products of the problem-posing process when it is implemented in situation involving the use of suitable artifacts, with its related mathematics, and particular teaching methods' (Bonotto, 2013, p. 42). The evaluation was largely positive:

> children had no difficulty translating typical everyday data, present in the artefacts, into problems suitable for mathematical treatment . . . [but] it was decided to modify some of the data of the problem in order to render the resolution of the problem more straightforward. (Bonotto, 2013, p. 43)

The second study had three phases: presentation of the artefact (a brochure for an amusement park); a problem-posing activity; a problem-solving activity. Two classes from different primary schools participated. There were similarities and differences between these classes. Whilst all but one of the 189 mathematical problems posed were mathematical one school generated 58 problems whilst the other generated 131 problems. About three quarters of problems from each school were solvable. The school which generated more problems also had a greater variety of types of problems and more 'original problems', 'Original problems include inverse problems, and problems containing almost all the information on the artefact' (Bonotto, 2013, p. 50). This may suggest that problem-posing from an artefact is related to the nature of the teaching children have experience (and this seems a reasonable hypothesis).

Lowrie (2011) and Bonotto's (2013) study provide evidence that artefacts can be useful in generating links between in-school and out-of-school mathematics, though both raise issues for further research. A further aspect of artefacts in generating such links is the production of artefacts. This, as we saw in Chap. 8, ties in with constructionist thinking, 'we extend the idea of manipulative materials to the idea that learning is most effective when part of an activity the learner experiences as constructing is a meaningful product' (Papert, 1987, abstract). Monaghan (2007a) reports on the production of artefacts within secondary school mathematical activities designed to link in-school and out-of-school mathematics activities. In this 2-year study eight teachers worked with the researchers (and often an out-of-school expert) to co-design school-based projects on out-of-school themes. Of 20 project designs 13 were implemented and in 4 of these the production of artefacts was the student outcome: 'designing a mathematical garden' involved transforming a garden including making a sundial; 'designing shelf-ready packaging' involved making a cardboard template of the packaging which was suitable for assembly on a production line; 'writing a rap song' with specialist music software resulted in electronic music; 'setting up your own business' involved producing a business plan on a spreadsheet which was suitable to send to a bank. In all of these projects, artefacts were used to produce new artefacts, which is the case in many out-of-school practices.

14.5 A Consideration of the Issues

The distinction between pure and applied mathematics, the use of mathematics in out-of-school practices and linking in-school and out-of-school mathematical practices are substantial and ongoing issues and it would be foolish of me to expect my tool-focused consideration of these issues in this chapter to bring a resolution to any of them. But a lot of detail has been presented in the three sections above and it is appropriate to consider the *punch line* of this scholarship and research. I structure this section by considering the 'problem issues' (problems of interpretation and problems of apparent gaps in understanding/research).

I started this chapter suggesting that there is a sense in which the division between pure and applied mathematics is a real division between mathematical activity for intrinsic or extrinsic purposes but also a sense in which the division between pure and applied mathematics is a cultural–historical division (and then I took a quick historic tour of ways of conceptualising divisions in mathematical activity). This is a problem issue—what, if anything, is the distinction between pure and applied mathematics?

I think the key to understanding this problem is recognition that there exists the practice of doing mathematics and interpretations of this practice. If we return to Jon's Chap. 3, an account of the practice of doing mathematics, we can see what might be called 'pure' (Case Study Ia: Iterative Reflections) and what might be called 'applied' (Case Study Ib: Protein Confirmation) mathematics but Jon does not employ these terms in his discussion of these case studies, he just reports on mathematics research (and the significant use of tools in this practice). This appears to be true of non-experimental mathematicians as it is for experimental mathematicians. This 'problem issue' is a problem issue for (some) interpreters (philosophers, historians and mathematics educators), not for (most) practitioners. But the interpreters, it seems, do play a role in determining the education (and thus the practice) of future practitioners in as much as they play a significant role in determining the structure of mathematics curricula (from Greek to medieval to modern times). With regard to tool use in mathematics, there appears, to use a term I coined above, *tool blindness* in many practitioners and interpreters. This does not appear important in the case of practitioners as they will use tools whether they realise it or not but it is important in the case of interpreters who have a say in structuring curricula (if their interpretations of the appropriate tools for doing mathematics are out of synch with the tools future practitioners need).

The second problem issue I raise is 'the modelling cycle' discussed in Sect. 14.2. I outlined my problems with this cycle above: it is oversimplified; it is not clear what it represents; and the usual presentation of this cycle does not attend to tool use. The Siller and Greefrath (2010) version of this cycle partially attends to tool use but raises complications by positing three worlds. I suggest that it may be useful to ignore the modelling cycle and simply look to practice and this appears to be an approach of current research, for example Noss and Hoyles (2009). I hope that this book will contribute to a focus on tool use (including computers but not just computers) in these practices. But if this leads to a new interpretation of out-of-school mathematical practices, then we should not view this as the final interpretation. Any interpretation will, from a cultural–historical perspective, be an interim interpretation in the developmental path of our understanding of the divide between 'pure' and 'applied' mathematics, a step on the way from the ancient Greek *quadrivium*, to pure and mixed mathematics, to pure and applied mathematics to . . . another understanding.

The third problem issue I raise is the difficulty of characterising tool use in out-of-school mathematical practices. Contributing to this problem issue are: the sheer number of out-of-school mathematical practices and the variation in both the tools used and the way tools are used in these practices; research into these practices requires contextual data, often obtained by time consuming ethnographic methods, so surveys of tool use may have limited value; mathematical tools are often invisible to practitioners and researchers. Section 14.3 only considered three studies in any depth. These studies all pointed to the importance of tool use in out-of-school mathematical practices but they do offer differing interpretations of tool use in practice. I am not worried about these differences but they suggest that we have barely scratched the surface of understanding tool use in practice.

Finally I raise a set of problem issues related to school mathematics. School mathematics is/can be viewed as a game and when this game is applied to linking mathematics to the real world it often results in *the suspension of sense making* and can disadvantage certain classes of children (Cooper & Dunne, 2000). Schools and classrooms are institutions and we should not expect 'real-life' reasoning to arise 'naturally' (Monaghan, 2007b) in them or that learning will 'naturally' transfer out of them. Tools and genuine artefacts appear to hold some hope that school mathematics can be related to real-life activities but many teachers and some researchers appear to have a form of *tool blindness*. But an awareness of problem issues can be a prelude to attempts to address problem issues.

References

Bishop, A. (1988). *Mathematical enculturation: A cultural perspective on mathematics education*. Dordrecht, The Netherlands: Kluwer.

Blum, W., & Niss, M. (1991). Applied mathematical problem solving, modelling, applications, and links to other subjects—State, trends and issues in mathematics instruction. *Educational Studies in Mathematics, 22*(1), 37–68.

Bonotto, C. (2013). Artifacts as sources for problem-posing activities. *Educational Studies in Mathematics, 83*(1), 37–55.

Brown, G. I. (1991). The evolution of the term "mixed mathematics". *Journal of the History of Ideas, 52*, 81–102.

Cockcroft, W. H. (1982). *Mathematics counts*. London: HM Stationery Office.

Cooper, B., & Dunne, M. (2000). *Assessing children's mathematical knowledge: Social class, sex and problem-solving*. Buckingham, England: Open University Press.

Dowling, P. (1998). *The sociology of mathematics education: Mathematical myths, pedagogic texts*. Washington, DC: Falmer Press.

Engle, R. (2006). Framing interactions to foster generative learning: A situative explanation of transfer in a community of learners classroom. *The Journal of the Learning Sciences, 15*(4), 451–498.

Fauvel, J., & Gray, J. (1987). *The history of mathematics: A reader*. Milton Keynes, England: The Open University.

Felstead, A., Gallie, D., & Green, F. (2002). *Work skills in Britain 1986–2001*. London: DfES.

Frejd, P., & Bergsten, C. (2016). Mathematical modelling as a professional task. *Educational Studies in Mathematics, 91*(1), 11–35.

Freudenthal, H. (1991). *Revisiting mathematics education: China lectures*. Dordrecht, The Netherlands: Kluwer.

Gerdes, P. (1996). Ethnomathematics and mathematics education. In A. J. Bishop, K. Clements, C. Keitel, J. Kilpatrick, & C. Laborde (Eds.), *International handbook of mathematics education* (pp. 909–943). Dordrecht, The Netherlands: Kluwer.

Gerdes, P. (1997). Survey of current work on ethnomathematics. In A. Powell & M. Frankenstein (Eds.), *Ethnomathematics: Challenging eurocentrism in mathematics education* (pp. 331–372). Albany, NY: SUNY Press.

Hoyles, C. (2007). *Understanding the system: Techno-mathematical literacies in the workplace* (ESRC End of Award Report, RES-139-25-0119). Swindon, Scotland: ESRC.

Hoyles, C., Wolf, A., Molyneux-Hodgson, S., & Kent, P. (2002). *Mathematical skills in the workplace*. London: Science, Technology and Mathematics Council.

Kanes, C., & Lerman, S. (2008). Analysing concepts of community of practice. In A. Watson & P. Winbourne (Eds.), *New directions for situated cognition in mathematics education* (pp. 303–328). New York: Springer.

Lagrange, J. B. (2005). Curriculum, classroom practices, and tool design in the learning of functions through technology-aided experimental approaches. *International Journal of Computers for Mathematical Learning, 10*(2), 143–189.
Lave, J. (1988). *Cognition in practice*. Cambridge, England: Cambridge University Press.
Lowrie, T. (2011). "If this was real": Tensions between using genuine artefacts and collaborative learning in mathematics tasks. *Research in Mathematics Education, 13*(1), 1–16.
Magajna, Z., & Monaghan, J. (2003). Advanced mathematical thinking in a technological workplace. *Educational Studies in Mathematics, 52*(2), 101–122.
Masingila, J., Davidenko, S., & Prus-Wisniowska, E. (1996). Mathematics learning and practice in and out of school: A framework for connecting these experiences. *Educational Studies in Mathematics, 31*(1–2), 175–200.
Mathematical Association. (1952). *The teaching of arithmetic in schools: A report prepared for the Mathematical Association*. London: G. Bell & Sons.
Monaghan, J. (2004). Teachers' activities in technology-based mathematics lessons. *International Journal of Computers for Mathematical Learning, 9*(3), 327–357.
Monaghan, J. (2007a). *Linking school mathematics to out-of-school mathematical activities* (ESRC End of Award Report, RES-000-22-0739). Swindon, Scotland: ESRC.
Monaghan, J. (2007b). Linking school mathematics to out-of-school mathematical activities: Student interpretation of task, understandings and goals. *International Electronic Journal of Mathematics Education, 2*(2), 50–71.
Monaghan, J., & Sheryn, L. (2006). How do secondary teachers make mathematics applicable? *Mathematics in School, 35*(4), 13.
Mullis, I. V., Martin, M. O., Foy, P., & Arora, A. (2012). *TIMSS 2011 international results in mathematics*. Chestnut Hill, MA: TIMSS & PIRLS International Study Center, Boston College.
Noss, R., & Hoyles, C. (1996). The visibility of meanings: Designing for understanding the mathematics of banking. *International Journal of Computers for Mathematical Learning, 1*, 3–31.
Noss, R., & Hoyles, C. (2009). Modeling to address techno-mathematical literacies in work. In R. Lesh, C. Haines, P. Galbraith, & A. Hurford (Eds.), *Modeling students' mathematical modeling competencies* (pp. 75–86). New York: Springer.
Noss, R., Hoyles, C., & Pozzi, S. (1998). ESRC end of award report: Towards a mathematical orientation through computational modelling project. Mathematical Sciences Group, Institute of Education, London University, London.
Noyes, A. (2012). It matters which class you are in: Student-centred teaching and the enjoyment of learning mathematics. *Research in Mathematics Education, 14*(3), 273–290.
Pais, A. (2011). Criticisms and contradictions of ethnomathematics. *Educational Studies in Mathematics, 76*, 209–230.
Papert, S. (1987). *Constructionism: A new opportunity for elementary science education*. DRL Division of Research on Learning in Formal and Informal Settings. Retrieved from http://nsf.gov/awardsearch/showAward.do?AwardNumber=8751190
Pozzi, S., Noss, R., & Hoyles, C. (1998). Tools in practice, mathematics in use. *Educational Studies in Mathematics, 36*(2), 105–122.
Saxe, G. B. (1991). *Culture and cognitive development: Studies in mathematical understanding*. Hillsdale, NJ: Laurence Erlbaum Associates.
Scott, P., Mortimer, E., & Ametller, J. (2011). Pedagogic link-making: A fundamental aspect of teaching and learning scientific conceptual knowledge. *Studies in Science Education, 47*(1), 3–36.
Siller, H. S., & Greefrath, G. (2010). Mathematical modelling in class regarding to technology. In *Proceedings of the Sixth Congress of the European Society for Research in Mathematics Education* (pp. 2136–2145).
Skovsmose, O. (2005). *Travelling through education. Uncertainty, mathematics, responsibility*. Rotterdam, The Netherlands: Sense.

Star, S. L., & Griesemer, J. R. (1989). Institutional ecology, 'translations' and boundary objects: Amateurs and professionals in Berkeley's Museum of Vertebrate Zoology, 1907–39. *Social Studies of Science, 19*, 387–420.
Verschaffel, L., Greer, B., & De Corte, E. (2000). *Making sense of word problems.* Lisse, The Netherlands: Swets & Zeitlinger.
Verschaffel, L., Greer, B., & de Corte, E. (2002). Everyday knowledge and mathematical modelling of school word problems. In K. Gravemeijer, R. Lehrer, B. van Oers, & L. Verschaffel (Eds.), *Symbolizing, modelling and tool use in mathematics education* (pp. 257–276). Dordrecht, The Netherlands: Kluwer.

Chapter 15
Mathematics Teachers and Digital Tools

John Monaghan and Luc Trouche

Abstract This chapter considers mathematics teachers' appropriation and classroom use of digital tools. The first section considers teachers—who are they, how are they conceived in the literature and what aspects of teachers have been studied? The second section examines twenty-first century research on mathematics teachers using digital tools. This sheds light on the complexity of mathematics teachers' appropriation and classroom use of digital tools but what we find is that our focus is too narrow and we need to consider digital tools within the range of resources use in planning and realising their lessons, which leads us to the third section, mathematics teachers using resources. We end with a review of the current state of understanding and an agenda for future research.

15.1 Introduction

This chapter tackles a complex issue, mathematics teachers' appropriation and classroom use of digital tools. We say 'complex' because we have both seriously applied ourselves to understanding this issue over two decades and we have both found it to be complex (in terms that will emerge in this chapter). It is fairly easy for someone to say what they think teachers should do but analysing the reasons[1] for what teachers actually do is a different and difficult task. The first section below does not begin with tools but with teachers—who are they, how are they conceived in the literature and what aspects of teachers have been studied? We then look at twenty-first century research on mathematics teachers using digital tools. This sheds light on the complexity of mathematics teachers' appropriation and classroom use of digital tools but what we find is that our focus is too narrow and we need to consider digital tools within the range of resources use in planning and realising their lessons, which leads us to the third section in this chapter. We end

[1] 'Reasons' may not be the best term if it suggests some explicit logic behind actions; 'agencies at play' in an institutional setting may be a better term, suggesting a range of factors influencing teachers' actions.

J. Monaghan et al., *Tools and Mathematics*, Mathematics Education Library,
DOI 10.1007/978-3-319-02396-0_15

with a review of the current state of understanding and an agenda for future research.

15.2 Constructing the Mathematics Teacher

Almost everyone reading this chapter will have spent hundreds of hours in school mathematics lessons led by specialist teachers. Given this, it is not unreasonable that we may think there is some essential quality to being a mathematics teacher. Our view, however, is that this is not the case, mathematics teachers are mathematics teachers by virtue of their employment—they are (usually) paid to instruct a set of people in the subject we call mathematics. This employment usually takes place in an institution called a classroom within an institution often called a school within a larger context (for example secondary teaching), etc. The locations of their employment, each with an ensemble of social relations and, we hasten to add, an ensemble of artefacts, contribute to the social construct known as 'mathematics teachers'. Chevallard's anthropological theory of didactics (considered in Sect. 10.4) sheds further light on our view of mathematics teaching: teaching is a praxeology[2] (an idiosyncratic practice) with a logos and a praxis specific to an institution; the mathematics taught is not the mathematics Jon writes about in Chap. 3 but is a *transposed* form of this mathematics (Chevallard & Bosch, 2014), mathematics which has been adapted for study in an institution.

The adjective 'professional' is often linked with teachers (e.g. teacher professional development) but the status of 'a teacher in a given institution' is not equivalent to a lawyer's status in the judiciary, or a doctor's status in a medical institution. Etzioni (1969) refers to teaching (as well as nursing and social work), as *semi*-professions: 'Their training is shorter, their status is less legitimated, their right to privileged communication is less established, there is less of a specialised body of knowledge, and they have less autonomy from supervision or societal control than "the" professions' (1969, p. v). This may ring less true in 2015 than it did in 1969 but, in the context of mathematics teachers' classroom use of digital tools, there is a sense in which, say, when a teacher faces a given problem (for example, integrating calculators in her teaching), s/he looks, in general, neither for solutions relevant to all colleagues, nor for an institutional solution; the solution to the problem is likely to be an individual one. In this sense teaching is more a craft than a profession.

Academics have attempted to describe teachers (in general, not just mathematics teachers) within the paradigmatic constructs of their age (e.g. behaviourism, constructivism, etc.) as we are doing in this chapter. Olson (1992, p. 1) outlines Western attempts in the second half of the twentieth century:

[2] Actually a set of praxeologies realised in different institutions.

> The *systems* model used organisational theory to understand and manage change. It is concerned with the techniques of change. The *ecological* model studies the work environment of the teacher and recommends how that environment should be changed so that teachers can work effectively. The *cognitive* model concentrates on how teachers process information from their environment—it searches for schemes teachers can follow to bring about learning. In each model, predictive social science is the basis for the quest for the effective teacher.

In the same period academics in the field of mathematics education investigated teachers' attitudes but, as Hannula (2012), in writing about attitude in mathematics generally (not just teachers' attitudes), says with regard to two classic texts from 1989 to 1992 discussing attitudes, beliefs and emotions, if we try to combine these two views, 'we see that attitude is at the same time a parent and a sibling to emotions and beliefs... This apparent mismatch highlights the different usages of terminology in the field' (Hannula, 2012, p. 140). A study of teachers' attitudes to using *Logo* that rejects classic views is Moreira and Noss (1995) who looks at the evolution of two teachers' attitudes towards *Logo* and the use of *Logo* in an INSET course and in their classrooms. They argue that there is a dialectic between attitudes towards the use of computers in the classroom and the classroom situations themselves: that the situations of use structure the attitudes and the attitudes structure the situations of use. Moreira and Noss' position resonates with our view of teachers' attitudes to the use of digital tools in mathematics teaching; these attitudes are not well formed pre-existing entities that somehow reside in the teachers' minds but are a part of the unit of analysis (cf. Sect. 9.2) of teachers' activities.

An important touchstone in academic study of teachers is two papers by Shulman (1986, 1987) which introduced the construct 'pedagogical content knowledge' (PCK), which has had a considerable influence on the academic study of teaching and teachers (both in general and in mathematics education). PCK is a powerful construct but one that is open to many interpretations: at a trivial level it is the intersection of 'content knowledge' and 'pedagogical knowledge'; at a deeper level it describes Chevallard's didactical transposition in a positive light, as the transformation of personal content knowledge into knowledge that can be appropriated by learners. Mishra and Koehler (2006) added technology[3] knowledge to the mix in their construct technological pedagogical content knowledge (TPCK); like PCK, TPCK is open to many interpretations (see Ruthven, 2014, for a critique). PCK, is an artefact and, as an artefact becomes a tool when it is used to do something (see Sect. 1.3), so PCK becomes a range of tools for the academic study of teachers and teaching. For example, Williams (2011, p. 161), in an activity theory framework, regards PCK as a 'boundary object between reflection on

---

[3] Mishra and Koehler (2006), like many authors, use the term 'technology' (short for 'digital technology'). We prefer, in this book, the term 'digital tool' but realise that we are 'nit picking'. We adopt the following usage in this chapter: when we are developing our own line of thought we shall use the term 'digital tool' but when we are discussing literature that uses the term 'technology', then we shall use the term 'technology'.

Fig. 15.1 A classification of knowledge for mathematics teaching (from Hill, Schilling & Ball, 2008)

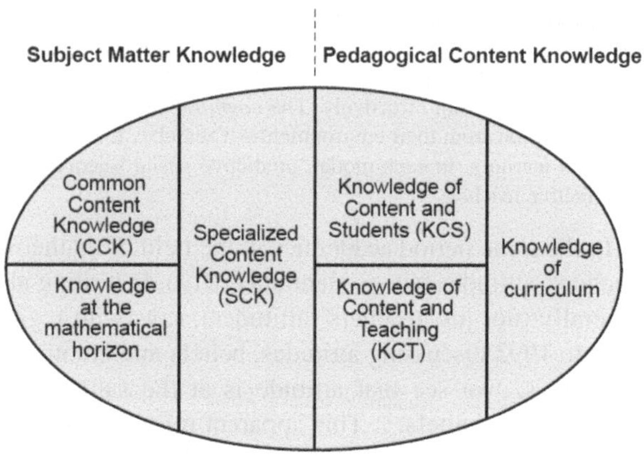

teaching and the practice of teaching' and Williams uses this in his analysis of audits of teachers' understanding of their students' knowledge.

Deborah Ball (with various colleagues) is, at the time of writing, the mathematics educator most closely associated with 'teacher knowledge' and her work, with colleagues, builds on that of Shulman. Figure 15.1 shows a multi-part classification of knowledge for mathematics teaching (from Hill, Schilling & Ball, 2008, p. 377). This division of knowledge, however, is not without its critics, for example Watson (2008, p. 1) states 'I try to think about mathematical knowledge in teaching as a way of being and acting, avoiding categorisation and acquisition metaphors of knowledge'. Further to this, though not as a direct attack on the division in Fig. 15.1, Rowland and Ruthven (2011, p. 2) state:

> … much work in this field has treated mathematical knowledge for teaching as residing solely with the classroom teacher. We aim to follow an approach which recognises the part played by textbooks and other tools and resources in classroom teaching and learning

An explicit focus on knowledge and tools which is strangely absent in the most current discourse on teacher knowledge is 'instrumental genesis' introduced by Guin and Trouche (1999). This has been discussed in Chap. 10 but it is worth revisiting here. 'Instrument' here, as Luc said in Sect. 1.3, is a composite entity composed of the artefact and the associated knowledge (both the knowledge of the artefact, and the knowledge of the task constructed when using this artefact). The artefact and the agent (or agents) are interrelated: the artefact shapes the actions of the agent, *instrumentation*; the user shapes the use of the artefact, *instrumentalisation*. The process of turning an artefact into an instrument is called 'instrumental genesis'. The agent brings her/his knowledge and the artefact brings its potentialities and constraints to the artefact agent interaction. Important constraints are: internal (linked to hardware); command (linked to the syntax required for use); organisation (linked to the artefact-user-interface). Instrumental genesis is a process that students and teachers will go through whenever they meet a new tool (digital or not); it is, moreover, a process without a unique end point, the process occurs in a zone of proximal development. Instrumental genesis can be used to provide a critique of Mishra and Koehler's (2006) TPCK when we consider software that is explicitly mathematical since content knowledge resides in both the agent and the artefact and technological knowledge is knowledge that develops in instrumental genesis. Instrumental genesis also leads to a model of analysing teachers' use of digital tools in classrooms, which leads us to the next section.

15.3 Twenty-First Century Research on Mathematics Teacher Using Digital Tools

Monaghan (2004, p. 329) advances the view that 'the 1990s witnessed a progressive if somewhat uneven realisation that teachers teaching with technology is a complex issue'. He cites professional and academic literature, *circa* 1990, that appears to view technology as somehow miraculously transforming teachers' classroom roles when they use technology:

> Teacher exposition leads naturally to discussion between teacher and pupils with the computer display as a focus. (Mathematical Association, 1992, p. 18)
>
> . . .classroom roles must and do shift. It is no longer possible for teachers to serve as ex cathedra authorities . . . Teachers and students must and do learn to listen carefully to and assess the qualities of one another's arguments. (Schwartz, 1989, p. 57)

Monaghan's view is complemented by a meta-study of ICT technologies in mathematics education in the period 1992–1998 (Lagrange, Artigue, Laborde, & Trouche, 2003, see also Sect. 10.5.2) which notes that the majority of papers reviewed in the meta-study were optimistic and few focused on the everyday practice of integrating digital technology into teaching and learning. The paper, however, noted:

> a long term motion towards awareness of a more complex integration and the subsequent necessity of new dimensions of analysis. It is confirmed by what we know of the institutional and instrumental dimensions in today's research studies and of the emerging reflections on the teacher (Lagrange et al., 2003, p. 260)

In 2001 three papers appeared which addressed ordinary teachers' practices with digital technology in their classrooms. Kendal and Stacey (2001) reports on work with Australian teachers to implement the same high school calculus module but found they did this in different ways. The paper claims teachers privilege (a construct from Wertsch, 1991) techniques and uses of digital technology which resonate with their pre-use-of-technology ways of working in the classroom. Laborde (2001) documents changes (made over several years) that teachers make in the tasks they use in classrooms with a dynamic geometry system. Monaghan (2001) details teachers' ways of working in classrooms with technology and states that changes in routines from classroom without digital technology are often the result of the material conditions of their work, e.g. teachers spent more time (respectively less time) speaking to two or more students (rather than individuals) in technology lessons (respectively lessons without digital technology) but this could be explained by the fact that the availability of computers (for the teachers he observed) forced students to work with two or more to a computer.

These papers, along with papers emanating in French research (see Chap. 10 and the papers by Artigue, 2002; Guin & Trouche, 1999 and in particular) and socio-cultural studies (see Chap. 9), helped establish an international dialogue which influenced a subset of papers on teachers' practices with technology in their classrooms in the first decade of the twenty-first century. Sections 15.3.1–15.3.5 trace five of these strands: a pedagogical model; time; emergent goals; zones of free movement and promoted actions; and instrumental orchestration. We choose these strands because they represent complementary aspects of mathematics teachers working with digital tools. We present the ideas in these strands, including the theoretical frameworks, by considering representative papers in each strand.

15.3.1 A Pedagogical Model

Ruthven and Hennessy (2002) construct what they call a practitioner model of the use of computer-based tools and resources to support mathematics teaching and learning from teachers' accounts of their practices and experiences in using technology in their classrooms. The theoretical framework is sociocultural and the writings of Wertsch and Bakhtin (see Chap. 13) are cited as influential. The data thematically analysed are transcripts of group interviews with mathematics teachers in each of seven state schools. The interviews sought to illicit 'how teachers conceive their incorporation of use of computer tools and resources into mainstream mathematics teaching' (Ruthven & Hennessy, 2002, p. 50). The main mathematics software teachers used were graphing packages, *Logo*, spreadsheet and software designed to teach or test a topic. Ruthven and Hennessy (2002) isolated ten interrelated themes associated with teachers' use of software, which they grouped into three groups:

> Four themes depend most directly on exploiting affordances of ICT: *Ambience enhanced* in changing the general form and feel of classroom activity; *Tinkering assisted* in helping to correct errors and experiment with possibilities in carrying out tasks; *Routine facilitated* in enabling subordinate tasks to be carried out easily, rapidly and reliably; and *Features accentuated* in providing vivid images and striking effects which highlight properties and relations. Three further themes depend in turn on these processes: *Restraints alleviated* in mitigating factors inhibiting student participation such as the laboriousness of tasks . . .; *Motivation improved* in generating student enjoyment and interest, and building student confidence; and *Attention raised* in creating the conditions for students to focus on overarching issues. Three final themes depend again on preceding processes: *Engagement intensified* in securing the commitment, persistence and initiative of students in classroom activity; *Activity effected* in maintaining the pace and productivity of students within classroom activity; *Ideas established* in supporting the development of student understanding and capability through classroom activity. (Ruthven & Hennessy, 2002, p. 81)

The model is not based on observations of teachers' practice but does seek a tentative model of practice. The first group of four themes, which focuses on affordances of the digital tools for teaching and learning mathematics, is clearly an academic interpretation of perceived practice.

15.3.2 Time

Assude (2005) has links with Laborde (2001) in as much as it is focused on teachers' use of the dynamic geometry system *Cabri* and 'time' but, whereas Laborde focused on changes in tasks teachers set over chronological time, Assude focuses on time management. It is difficult to describe the theoretical framework in a few words it approximates to a fusion of the anthropological theory of didactics and sociocultural theory. The paper is based on the work of two primary teachers teaching geometry (plane figures) in two consecutive years (the first year without *Cabri*, the second year with *Cabri*). Data collected included teachers' lesson preparation notes, students' exercise books and recorded classroom observations. Two constructs, 'didactic time' and 'time capital' are central to the paper. Didactic time is related to 'the process of transformation of a body of knowledge into a knowledge which can be taught' (Laborde, 2001, p. 185), i.e. the 'didactical transposition' (see Chap. 8). Didactic time is linear and sequential (i.e. knowledge is broken down and taught in some order). Time capital is time as it is measured by a clock and the suffix 'capital' suggests that teachers have a fixed curriculum time allocation to teach a given topic. The pace of a lesson/course can be viewed as 'the rate at which didactic time advances relative to the time capital allotted to it' (Laborde, 2001, p. 186) and the statement 'that was a fast-paced' lesson can be understood in terms of this ratio. Assude (2005) presents tables which show the number of hours devoted to different parts of the geometry work in the 2 years (with and without *Cabri*). The aim of the work in these years was the same but the content was not identical: the second year included, for example, 3 h for students to become familiar with *Cabri*; the number of hours spent on specific topics varied over the 2 years. Overall, however, the tasks given to students in both years were similar and teachers exercised control over didactic time though there were changes in the pace of the lessons. The use of *Cabri* did influence the pace of parts of the geometry work. For example, in a unit on quadrilaterals:

> A construction activity was much slower than an activity in which previously constructed figures were analysed . . .the relation between the pupils' working time and didactic time is not economical relative to time capital, because didactic time hardly moves forward; pupils must have sufficient command of the software in order to do the constructions, which is not necessarily the case for analysis (Assude, 2005, p. 194)

Assude (2005) goes on to describe teachers' strategies which 'allowed teachers to save their time capital in working with Cabri and, as a result, this software could be integrated in the day-to-day work of the class' (Assude, 2005, p. 201).

15.3.3 Emergent Goals

Monaghan (2004) is an interpretation of the Monaghan (2001) in terms of Saxe's (1991) sociocultural four parameters model of practice linked goals.[4] The context was a project where 13 state secondary school mathematics teachers with limited prior experience of using technology in mathematics lessons attempted extensive use of technology in their lessons for a school year. Over the 13 teachers the following digital tools were used: spreadsheet; graphic packages and calculators; computer algebra systems (CAS); and dynamic geometry systems. The stated aim of Monaghan (2004) is 'to find an holistic way to examine teachers' practice ... because of a conviction that the whole undertaking involves a fusion of many factors and analyses suffer if these factors are taken in isolation' (Monaghan, 2004, p. 327). Data collected included teacher journals with lesson plans, various interviews and four video-taped lessons of each teacher spread out over the year (the first of a lesson without digital tools). Monaghan (2004) goes through each of Saxe's parameters in turn. With regard to activity structures, all the initial lessons had a similar format: teacher exposition and examples followed by students doing questions from a textbook but here was considerable variation in the format in the lessons incorporating the use of digital tools and the tasks given to students in these lessons differed from the textbook tasks in the initial lessons. Student work on some of these tasked appeared, to the teacher, to be focused on the technology rather than 'on maths' and this caused some teachers to question 'is this maths?'. With regard to prior understandings teachers had a clear understanding of how a 'normal lesson' and lesson plans were, more often than not, notes on mathematical content to be covered. But in lessons with digital tools:

> Back to being like a student teacher to be honest because I felt happier writing things down. I think it's a kind of a confidence thing. When you're just teaching a normal lesson without ICT you don't, you're prepared for any eventuality without realising it but when you're doing ICT and it's the first time you've done a topic like that you're not. (Monaghan, 2004, p. 337)

With regard to conventions and artefacts, we restrict the summary to software and hardware. Mathematical software is not a uniform category to teachers. For example, a teacher who is a graphing software and a CAS viewed graphing software as a tool to do straightforward tasks that fitted in with his 'normal' lesson ideas but the CAS was a 'monster' that could do virtually anything and forced him to rethink his lesson planning. Hardware had a differential effect on the organisation of teaching. For example, teachers who did not have computers in their class (and so had to book a computer room) tended to have 'all or nothing' computer-based lessons. Monaghan (2004) notes many changes with regard to social interactions as the result of incorporating digital tools but, as noted above, these are often the result

[4]This was outlined in Chap. 14. A brief resume is: the parameters activity structures, social interactions, conventions and artefacts and prior understandings interact and influence, and can interrupt, emergent goals arising in practice.

of the material conditions of their work. Monaghan (2004) then looks at interrelationships between these parameters, which are extensive. The following quote from a project teacher (one of a pair in one school) shows a dialectic between activity structure (task), prior understandings (start with structured tasks) and artefacts (printers):

> We started by giving the students structured tasks to do such as solving simultaneous equations graphically. We then moved on to less structured work—usually an investigation. . . . These investigations were much more successful with set 1 than with set 2 . . . As a result of these difficulties we adapted the assignments to reduce the number of printouts and to make the tasks more prescriptive. (Monaghan, 2004, p. 349)

Monaghan (2004) concludes that the integration of technology into classroom practice is a complex transformation of practice and that teachers varied considerably in the extent to which they could successfully adapt (in their own evaluations) to this transformation.

Lagrange and Erdogan (2009) elaborate on Monaghan's use of Saxe's frameworks by linking (networking) it to the anthropological theory of didactics (cf. Sect. 10.4) and 'values'.[5] Lagrange and Erdogan (2009) focus on two experienced secondary teachers using spreadsheet in their secondary school classes. One of these teachers, Mrs. P_{EX} (experienced in the use of technology), is positively disposed towards using spreadsheet, the other, Mrs. P_{SCEP} (sceptical), is not. The purpose of the paper, 'is to highlight episodes marked by improvisation and uncertainty as a central feature of teachers' classroom activity involving technology' (Lagrange & Erdogan, 2009, p. 65). Like Monaghan (2004) they note the influence of Saxe's four parameters on each teacher's work. We focus on one situation, the *Birthday Situation* (see below), from Lagrange and Erdogan (2009) which both teachers chose to make into a task from a set of situations offered to them (though in Mrs. P_{SCEP} this 'choice' had 'strings attached'—the curriculum was such that she was expected to use technology in her lessons and it seems that she chose this situation as *the best of a bad lot*).

> Birthday Situation: Sabine has just been born; Her grandmother opens a credit account for her, makes a first 100€ deposit and decides to make each year a new deposit of the same amount plus double Sabin's age.

Both teachers turned this situation into a task with sub-tasks leading to 'How much money will Sabine have on her 18th birthday?' Mrs. P_{EX}'s version of the task does not refer to spreadsheets. Mrs. P_{SCEP}'s version of the task (from a textbook) makes specific reference to 'spreadsheet algebra', e. g, 'Which of the following formulas should we write in cell B3 . . . 1) = B2 + 2*A3 . . .' (Lagrange & Erdogan, 2009, p. 73). Lagrange and Erdogan's (2009) account of Mrs. P_{SCEP}'s lesson shows that Mrs. P_{SCEP} was often surprised with the student work resulting from the task and 'became aware that her activity format of individual interaction with the

[5] Pragmatic values which concern the range of application of a technique and epistemic values which concern the role of techniques in promoting mathematical understanding.

students was not adapted for this question' (Lagrange & Erdogan's, 2009, p. 75). In Mrs. P_{EX}'s class the work the students worked, worked in two teams and could decide what tool to use to do the task. The idea of using a spreadsheet emerged in whole class discussion on the first sub-task and the students suggested using a spreadsheet.

Lagrange and Erdogan's (2009) provide a wealth of detail regarding the development of each lesson which we leave the interested reader to pursue though we do note that the development of work in Mrs. P_{EX}'s class was not without tensions. We now jump to their summary comments in terms of Saxe's model, the anthropological theory of didactics and 'values'. The activity structure, following the different task, in each class were different and this difference appeared to be accentuated by the two teachers' views of 'mathematics with technology': Mrs. P_{SCEP}—learn the notions and notations and then apply them; Mrs. P_{EX}—explore, discuss and synthesise. For Mrs. P_{SCEP} the spreadsheet was a tool for applications but for Mrs. P_{EX} it was a 'tool to introduce exploration and modelling' (Lagrange & Erdogan's, 2009, p. 79). Regarding social interactions, Mrs. P_{EX} privileged teamwork whereas Mrs. P_{SCEP} privileged individual work which led to unwanted emergent goals of keeping individual students on the execution of the task in the way she determined it should be done. With regard to techniques Lagrange and Erdogan's (2009) note that:

> The teachers attached great importance to techniques for their epistemic values, but speaking to the students, they rather insisted on a supposed pragmatic value. Students were not convinced. It seems that teachers prefer to give superficial reasons rather than discuss in depth the interest of a technique. (Lagrange & Erdogan's, 2009, p. 81)

15.3.4 Zones of Free Movement and Promoted Actions

We now consider work on teachers and technology that employ Valsiner's sociocultural constructs zone of free movement (ZFM) and zone of promoted actions (ZPA). Valsiner (1987) introduces these zones in relation to Vygotsky's zone of proximal development (ZPD) to explain how adults shape an environment for a child's expected proximal development. The ZFM characterises the child–environment relationship, what the child can and cannot do in an environment. The ZPA concerns the 'set of activities, objects, or areas in the environment, in respect of which the child's actions are promoted' (Valsiner, 1987, pp. 99–100). These two zones interact and 'work jointly as the mechanisms by which canalisation of children's development is organised' (Valsiner, 1987, p. 101). Valsiner is a child psychologist but these constructs do appear useful for conceptualising both students and teachers in mathematics classrooms with or without technology: what actions are available (or constrained) in a classroom? What actions are promoted or not?

Goos (2005) applies Valsiner's zone theory to novice teachers, and their identities as teachers, in the context of using technology in their early teaching experiences. Goos views novice teachers': ZPD as including their experience in working with technology and their knowledge on how to integrate technology into mathematics teaching; ZFM as including access to hardware, software and teaching materials, support from colleagues and curriculum and assessment requirements; ZPA as including university-based education, practicum and professional development. Much of Goos (2005) considers the case of Geoff who was encouraged to use technology in the university element of his course. Geoff was an experienced computer user and was enthusiastic towards integrating technology into his teaching, 'make things easier to understand because ... it's dynamic and not static' (Goos, 2005, p. 47). In his practicum placement it appeared that Geoff had a number of apparent constraints on his movements and actions with regard to the integration of technology into his lessons:

> the ZPA offered by this supervision excluded technology, and thus was not well matched with the ZPD that defined the direction in which Geoff hoped his teaching would develop ... In Figure 3 we see that the university ZPA is distinct from the school ZPA and thus largely outside the school's ZFM (Goos, 2005, p. 48)

Nevertheless, Geoff was able to reconfigure the ZFM with a low attaining class, 'a class that nobody cares about' (Goos, 2005, p. 49), to include the university ZPA (which he wanted to enable). Goos (2005) adds that Geoff, 'was able to construct his practice as a pre-service teacher of low status mathematics students to develop further his emerging identity as a teacher for whom technology was an important pedagogical resource'.

Goos (2005) follows Geoff into his first year of teaching in a school which, unlike his practicum school, promoted the use of technology. Goos (2005) reports on a successful lesson where Geoff presented a graphing task that involved capturing student data in 'walking contests' using a motion detector linked to a graphics calculator and displayed using a view-screen. Goos (2005, p. 51) comments:

> Geoff's ZFM appeared to afford teaching actions consistent with his beliefs about mathematics ... Furthermore, the ZPA offered by his teaching colleagues seemed to be consistent with both his development as a teacher (i.e., his ZPD) and the ZPA offered by his pre-service course, in that new graduates teaching at the school were actively supported in integrating technology into their practice ... Geoff's pedagogical identity was afforded by the apparent relationships between his ZPD, ZFM, and ZPAs.

However, when Geoff attempted to extend his use of technology to computers, the school constrained Geoff's practice because the computer rooms were not available for him to use with his class:

> These components of Geoff's ZFM tended to undermine his goal of infusing technology as a partner in assessment tasks as well as learning activities ... these factors led to a contraction of Geoff's ZFM in ways that tended to exclude some of the pedagogical practices promoted by his pre-service course (Goos, 2005, p. 52)

There is a sense in which studies, such as Goos (2005), of the ZFM and the ZPA in relation to a teachers' ZPD provide a meta-level analysis of the study of the Gibson's affordances and constraints with regard to the use of digital tools by teachers. We now move on to instrumental orchestration, which ends our consideration of papers and ideas which influenced scholarship on teachers' practices with technology in their classrooms in the first decade of the twenty-first century.

15.3.5 Instrumental Orchestration

The term 'orchestration' is a widely used metaphor in academic literature concerned with digital technology and education. Kennewell (2001), for example focuses on the Gibsons' affordances and constraints, 'The role of the teacher is to orchestrate the affordances and constraints in a setting in order to maintain a gap between existing abilities and those needed to achieve the task outcome' (Kennewell, 2001, p. 107). But in the field of mathematics education, at the time of writing, 'orchestration' is most widely associated with 'instrumental orchestration', which is an extension of Guin and Trouche's (1999) instrumental genesis applied to teaching and learning. Trouche (2004) introduces the term:

> to point out the necessity . . . of *external steering* of students' instrumental genesis . . . An instrumental orchestration is defined by *didactical configurations* (i.e. the layout of the artifacts available in the environment . . .) and by exploitation modes of these configurations (Trouche, 2004, p. 296).

Instrumental orchestrations can be enacted at different levels: of the artefact; of the instrument; of the relationship between the user and the instrument(s). Trouche (2004) focuses on modes of use of an algebraic calculator with regard to a specific configuration which he calls *Sherpa-student-orchestration* (see also Sect. 19.2.1). This configuration includes individual students with calculators linked to a view-screen and involves one student (the Sherpa), displaying his/her calculator work to the class and 'Sherpa-student-orchestration' was, for a number of years, the prime example of instrumental orchestration.

Drijvers, Doorman, Boon, Reed, and Gravemeijer (2010) extend the repertoire of instrumental orchestrations. The context of this extension is a project which explored teaching mathematics in technology-rich environments with the teachers who had limited experience in using digital tools in their secondary school classrooms. The mathematics to be taught centred on constructing function within a Realistic Mathematics Education framework. Tasks were presented in Java applets embedded in an e-learning environment. The research focuses on describing the types of instrumental orchestrations the teachers enacted in their classrooms. Drijvers et al. (2010) add 'didactical performance' to Trouche's (2004) didactic configurations and exploitation modes:

> A didactical performance involves the ad hoc decisions taken while teaching on how to actually perform in the chosen didactic configuration and exploitation mode: what question to pose now, how to do justice to (or to set aside) any particular student input, how to deal with an unexpected aspect of the mathematical task or the technological tool, or other emerging goals. (Drijvers et al., 2010, p. 215)

Didactical performance constitutes a critical enrichment of the instrumental orchestration model, allowing to see an orchestration 'as an artefact for a teacher, evolving through successive phases of design and implementation in classroom situations' (Trouche & Drijvers, 2010, p. 676).

Data collection included analysis of videotapes of 38 lessons by 3 teachers in 5 classes. Qualitative data analysis focused on the whole class and the unit of analysis was the whole class use of technology in the execution of a task. Six orchestration types were identified:

> *Technical-demo* orchestration concerns the demonstration of tool techniques by the Teacher ... *Explain-the-screen* orchestration concerns whole-class explanation by the teacher, guided by what happens on the computer screen ...In the *Link-screen-board* orchestration, the teacher stresses the relationship between what happens in the technological environment and how this is represented in conventional mathematics of paper, book and blackboard ...The *Discuss-the-screen* orchestration concerns a whole-class discussion about what happens on the computer screen ...In the *Spot-and-show* orchestration, student reasoning is brought to the fore through the identification of interesting DME student work during preparation of the lesson, and its deliberate use in a classroom discussion ...In the *Sherpa-at-work* orchestration, a so-called Sherpa-student uses the technology to present his or her work, or to carry out actions the teacher requests. (Trouche & Drijvers, 2010, pp. 219–220)

The six orchestration types are not seen as exhaustive (simply those observed) and 'are not isolated, but part of orchestrational sequences' (Trouche & Drijvers, 2010, p. 220). Drijvers et al. (2010) present a table showing orchestration types used by three teachers which shows that different teachers have orchestration types which they appropriate and others which they never or rarely used (and these vary over the teachers). Drijvers et al. (2010) comments on the didactical configuration and exploitation mode associated with each orchestration type and didactical performance is considered later in the paper in a detailed analysis of a classroom episode involving *Spot-and-show* orchestration; we do not summarise these descriptions for reasons of space. Drijvers et al. (2010) notes that the teacher dominates the discourse in first three orchestration types listed above and calls these 'teacher-centred' whilst students have opportunity to contribute in the last three orchestration types listed above and calls these 'student-centred'.

This analysis of how teachers use technology is clearly important but, as the researchers would surely agree, they capture the teachers' ways of classroom work with technology in specific contexts. Specifics that apply to Drijvers et al. (2010) include: secondary school mathematics; the software is 'expressive' in the sense defined in Chap. 8. It is likely that primary school teachers using mathematical software which is not expressive will orchestrate learning in ways not listed by Drijvers et al. (2010).

15.3.6 On the Complexity of Integrating Digital Tools into Classroom Practice

We opened this chapter stating that we consider mathematics teachers' appropriation and classroom use of digital tools to be a complex issue and think that Sects. 15.3.1–15.3.5 reveal different dimensions of this complexity. We also think all of the above approaches to viewing mathematics teachers using digital technology are important because they focus on real issues in teachers' actual practice in using digital tools in ordinary classrooms. An argument could be made that the constructs introduced in Sects. 15.3.1–15.3.5 are complementary but different. Should we be looking to 'network[6]' them? It may be possible and Lagrange & Monaghan (2009) consider this with Saxe's approach and the ATD. But, as is often the case in research, a line of inquiry leads to new lines of inquiry rather than a resolution of the original line of inquiry. Work on mathematics teachers using digital technology continues at the time of writing but Gueudet and Trouche (2009) opened a new line of enquiry, which we address in the next section.

15.4 Mathematics Teachers Using Resources

Opening a new line of enquiry, if not artificial, supposes a certain number of *raisons d'être*. We present in this section these 'raisons d'être', then this new line of inquiry, namely the documentational approach, and conclude by a short visit of the work in progress: as mentioned above, the emerging of a new line of enquiry opens new questions to be explored.

15.4.1 Some 'raisons d'être' of a New Line of Inquiry

In the case of the documentational approach, we mention, a posteriori, three *raison d'être*: a dramatic evolution of learning environments; a maturation of reflection carried by a research community; and the crossing of several scientific fields favoured by the existence of boundary objects.

The first reason is the dynamics of the technological changes. We have presented (Chap. 5) the dramatic evolution of the practice, teaching and learning of mathematics after the invention of writing. The development of scribal schools and the transmission of knowledge through clay tablets create the conditions for new

[6] Networking theoretical approaches is a term used by the Congress of European Research on Mathematics Education (CERME), see, for example, Kidron, Bikner-Ahsbahs, Monaghan, Radford, and Sensevy (2012).

equilibrium between language and written signs, between memorisation and use of written supports. The digital metamorphosis of the supports of knowledge, today, creates also the conditions for a dramatic evolution of the practice (cf. Sect. 4.5), of the teaching and the learning of mathematics. Students and teachers are permanently exposed to an abundance of 'machines' (cf. Chap. 5) having the potential to answer—and sometimes to pose—complex mathematics questions and to 'show' representations of mathematical objects and processes. These machines do not necessarily have the appearance of a mathematical artefact: you can use Google to compute a multiplication or to search for a proof. Hyperlinks open a new multidimensional space for thinking, and they modify deeply the usual relationship with the 2D organisation of a paper page. Reading and writing become two interrelated actions (meaning that usage and design of resources are completely interrelated). Last, and not least, each Internet resource gives access to a number of artefacts, of different levels, for a teacher. Look at, for example, the front page of the *GeoGebra* tube (Fig. 15.2). This resource gives access to software (Geogebra), to a great number of mathematical activities (98,484 at the beginning of the afternoon whilst writing this section and 98,510 at the end of the afternoon) and to a community (forum, technical help...).

Traditional artefacts supporting teachers' work, such as textbooks, are evolving towards e-textbooks (Pepin, Gueudet, Yerushalmy, Trouche, & Chazan, to appear), giving access to a lot of dynamic resources. Such textbooks are not only conceived by specialists (designers, teachers educators, ...) but also by teachers themselves, gathered in associations of free designers (see in France, for example, the experience of Sésamath in Sabra & Trouche, 2011, Sect. 19.3.2). In this context, the mediation of teachers' activity is not only supported by *artefacts* (such as compasses, calculators, ...), but also by a lot of 'things' (a colleague's e-mail, a comment on a forum...) that are currently named 'resources' (we will elaborate further on this word). And the right level of analysis is not a closed set of resources (for example a textbook and a calculator) that could be identified for supporting teacher activity, but rather a *resource system* which is constantly evolving through interactions of the teacher with her/his environment.

The second *raison d'être* is the internal evolution of a community of research (which, of course, is not independent of the technological evolution). We have presented in Chap. 10 the emergence of the instrumental approach, in the context of the French school of thought. At the beginning, the instrumental approach was mainly considered with regard to the genesis of instruments for students in the course of their situated activity. Rapidly (cf. Sect. 15.3.5), however, the crucial role of teachers for monitoring students' instrumental geneses appeared, and this led to the conceptualisation of instrumental orchestration. At the beginning, teachers' work was considered as a sequence of choices and design: choice of a didactical goal, choice of a *didactical situation*, in the sense of Brousseau (Chap. 10), choice of a technological environment, design of an orchestration. In this new situation, however, the analysis of teachers' work leads to a more flexible view:

Fig. 15.2 The front page of Geogebra Tube, a resource... full of resources for a teacher as well as for a student

- An Internet resource often proposes a didactical situation, a technological environment, and elements of orchestration, in various orders (in the case of Geogebra, Fig. 15.2, the choice of a technological environment opens the way for finding ideas of mathematical activities and instrumental orchestrations)
- Instrumental orchestrations have to be conceived as living entities, permanently renewed by didactical performances (Sect. 15.2)

This view leads to a conception of a teacher working with/for resources, the word 'resource' having to be understood in a broad sense.

Finally understanding teachers' work requires analysing their *resource system* (Ruthven, 2012), a complex and living entity, made of 'things' that do not constitute a uniform category: artefacts; didactical situations; environments; seeds of instrumental orchestrations and resources coming from interactions with colleagues and students. These evolutions are evident if we compare two chapters dedicated to technology in mathematics education in two successive Handbooks, both written by a team having the instrumental approach as a main reference: the words 'resource', 'collective' or 'community' never appear in Lagrange et al. (2003); in Trouche, Drijvers, Gueudet, and Sacristan (2013), the same words appear 40 times (for 'resource'), 15 times (for 'collective') and 20 times (for 'community').

The schema (Fig. 15.3), published in the Third Handbook on Mathematics Education, summarises major evolutions of the research, giving more attention to the *creative* role of the teacher (resources from top-down to bottom-up), to the collective context of her/his engagement, the essential question becoming the *integration* of new resources in the whole resource system of a teacher.

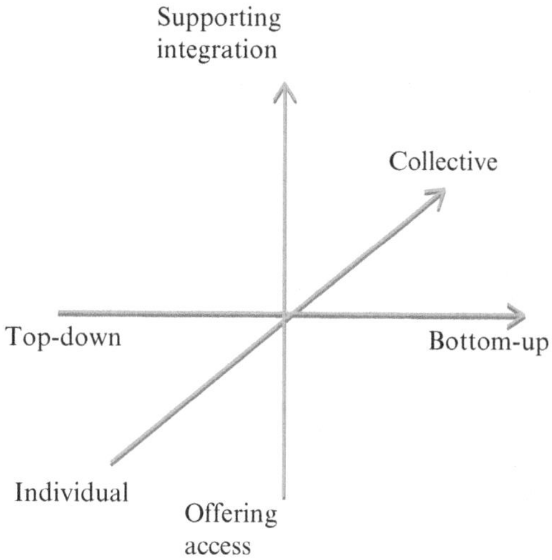

Fig. 15.3 The evolution of the major trends of the research in ICT in mathematics education (Trouche et al., 2013)

The third *raison d'être* is the interaction with three other conceptualisations, or lines of enquiry, not exactly in a networking perspective, but in a perspective of mutual enrichment. The first one is the conceptualisation of a resource (Adler, 2000), understood as all that have the potential *to re-source* teacher activity, enlarging the consideration of teacher environment from artefacts to cultural, social and human resources. The second one is the conceptualisation of curricular resources (Remillard, 2005), analysing teachers' work with mathematics curricula as 'following, interpreting, subverting the text'. This view of a dialectical relationship between teachers and mathematical content through a curriculum appears clearly in the two schemas proposed by Ball (Fig. 15.4).

The third one is the conceptualisation of a *document*, in the field of information architecture (Pedauque, 2006), as a social construct, dedicated to a given usage, opening a possible distinction between what is available for supporting activity (a resource) and what is appropriated for achieving this activity (a resource). We expand on this 'document' construct in the next section.

15.4.2 The Documentational Approach of Didactics

The mere presence of *raisons d'être* is not enough for something to appear. An effort is needed for drawing on the consequences of theoretical evolutions and establishing links with other theoretical frames. An occasion to do this was provided by a French summer school on didactics, in 2007, when Gueudet and Trouche were asked to give a lesson on 'didactical situations and documents for the teacher'. The reflections for preparing this course led to the so-called documentational

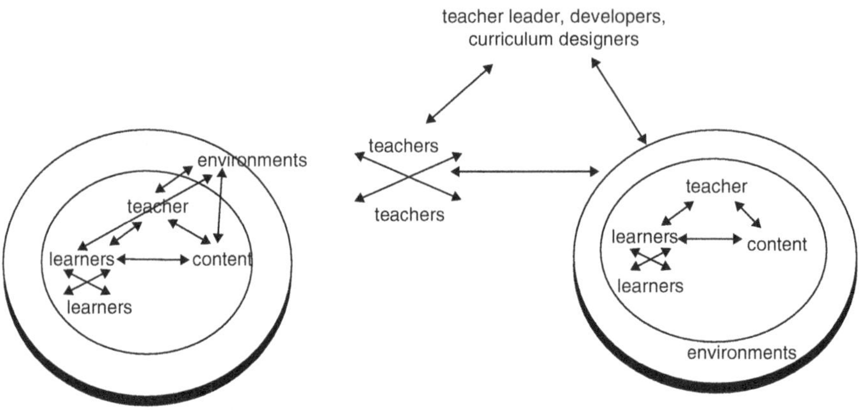

Fig. 15.4 A first view (*left*) on the relationships between a teacher and the curricular resources, and a second view (*right*), more complex, taking into account interactions between teachers and with curriculum designers

approach of didactics, presented to an international audience in Gueudet and Trouche (2009) (Fig. 15.5).

The documentational approach, enlarging the instrumental approach (Sect. 15.2) mainly conceptualises teacher's work as an interplay between a teacher and a set of resources, guided by a teaching goal (for example: introducing the notion of derivative for a tenth grade class). This goal actually constitutes a class of situations: such an introduction could mobilise different kinds of examples, animations, exercises… This interplay combines instrumentation and instrumentalisation (Sect. 15.2). Through successive preparations, implementations in class and revisions, a hybrid entity emerges, composed of the resources (adapted and recombined) and a scheme (a way of using these resources). This hybrid entity is named a *document* (as something documenting teacher's activity), and the process leading, from a set of resources to a document, is named a *documentational genesis*. The notion of scheme has been introduced by Piaget (Sect. 7.3), and defined by Vergnaud (1998) as the invariant organisation of activity (Sect. 10.3). The document is structured by knowledge—called operational invariants—explicit and implicit knowledge about mathematics, pedagogy and technology and knowledge obtained by a teacher's initial trials in using the resources gathered for achieving a given goal. Giving a short example of a documentational genesis is not easy, as such a genesis takes time, and mobilises a number of resources and actors (mainly students, colleagues), and needs to be understood to have a number of elements of contexts. Such developed descriptions can be found in Gueudet and Trouche (2009). The extract below provides a brief summary of a case of documentational genesis.

> Frédéric has taught for 15 years to students from grades 6 to 10. During the interview, he presented to us a mathematical task designed to introduce the square root in grade 9. This task deals with the areas of squares: several side-lengths are given, students must compute the areas, then place the points with coordinates (length, area) on a graph, draw a curve

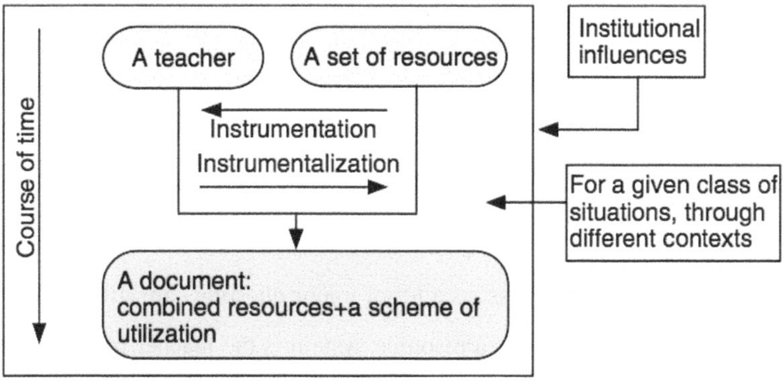

Fig. 15.5 A schematic representation of a documentational genesis (Gueudet & Trouche, 2009)

> through these points, and use the curve to find approximate values of the side-length for given areas. These approximate values are then compared with the values obtained with a calculator, using the square root key. Frédéric declared that he had used this task for more than 10 years (it had changed along the years, and was initially used in grade 8 before a curriculum change; but the features described above were always present). For the class of professional situations: 'Preparing the introduction of the square root in grade 9', Frédéric now draws on a set of resources comprising the original textbook extract; the student sheet proposed the year before, with notes on it about changes he thought of when using it in class; a slide with the points and the curve joining them; but also the students' calculators. We claim that Frédéric developed a scheme of utilization of this set of resources for this class of situations. This scheme entails general operational invariants: 'a new notion must be introduced through a mathematical task that yields evidence of the meaning of this notion'; and invariants linked with the mathematical content: 'searching for the side-length of a square for a given area gives evidence of the meaning of square root'; 'the square root is the reverse process of squaring'; 'the calculator square root key supports the introduction of the symbol' (Gueudet & Trouche, 2009, p. 205)

This extract reveals only a part of the underlying model and it should be kept in mind that:

- Unlike an ordinary birth, documentational genesis is never ending, because a document is always evolving through implementations in new contexts. This evolution impacts the resources themselves, and the scheme, i.e. the way the usages are used and the teacher knowledge itself. Thus following a documentational genesis along the time is exactly following a teacher's development from a novice level to an expert level
- Mentioning only 'a teacher' is restrictive as each documentational genesis involves several actors, and interactions *between* teachers are certainly essential (Gueudet, Pepin & Trouche 2013). These interactions intervene at different levels: for enlarging the possible choice of resources as well as for enlarging the possible choice of usages

In this model, the teacher resource system is constituted by all the resources that a teacher has integrated and worked with, to constitute the matter of her/his teaching.

15.4.3 Work in Progress

This approach is still in progress, as a lot of major questions are still open:

- Question of methodology: as a resource system is the teacher property, assembling very different things, how is it possible to have access, at least to a part of it, to describe and analyse it? A methodology of *reflective investigation* (Gueudet & Trouche, 2009) has been designed, mobilising teachers' views of their own resources, through different tools (mainly logbooks, guided tours and schematic representations of the resource system, see an example Fig. 15.6)
- What about the structure of a resource system? Are there particular resources playing a particular role (as pivotal resources)?
- What relationships hold between a teacher's resource system and the resource system of a collective s/he is involved in? (see Sabra & Trouche, 2011)
- How does the resource system vary over different phases of education, for example primary, secondary and tertiary teaching? (see Gueudet, Buteau, Mesa, & Misfeldt, 2014 in the case of higher education)

15.5 Retrospect and Prospects

Mathematics teachers' use of digital tools is an area in need of much further research. Section 15.2 suggests this research must go beyond studies focused on teachers' knowledge and teachers' attitudes. Section 15.3 provides many frames and constructs which shed light on teachers' use of digital tools. Section 15.4 makes a case that research should view teachers' use of digital tools in the wider perspective of documentational genesis but there is still a case for a consolidation (a 'networking') of the frames and constructs outlined in Sect. 15.3. In this closing section we consider two matters central to the purpose of this book. The first addresses the question 'is there anything special about teachers' use of digital tools?' The second looks to the future and speculates on new forms of teacher education and teacher professional development in the 'digital age'.

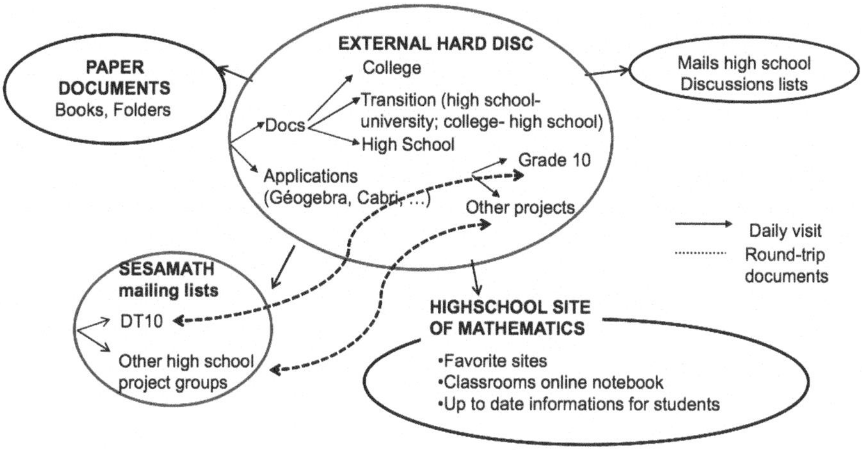

Fig. 15.6 A schematic representation of her resource system drawn by a teacher (Sabra & Trouche, 2011)

15.5.1 Is There Anything Special About teachers' Use of Digital Tools?

Our response to this question is 'no and yes'. We begin by expanding on the 'no' aspect of our response by summarising Larry Cuban's historical work on teachers and technology in the USA in the twentieth century. Cuban (1986) traces the introduction and establishment of film and radio, television and computers in American classrooms from 1920 to the 1980s. This book opens with a 1927 photograph[7] of an aerial geography lesson. This is a fascinating picture with regard to artefacts/tools: with a new artefact (an airplane) in which to study physical geography (by looking out of the windows), the individual wooden student desks of the period are arranged so that the students face the teacher, who is pointing to a globe. Our interpretation of this opening is to make the point that the artefacts of the past shape the instructional use of the artefacts of the future. Cuban (1986) traces the earlier use of film (1910), optimism for the potential of the artefact for education and the establishment of the use of films in classrooms (1930s). By the 1950s, Cuban's evidence suggests, 'most teachers used films infrequently in classrooms' (Cuban, 1986, p. 17). Cuban's interpretation of reports on classroom use of films points to four 'obstacles': teachers' lack of skills; costs; accessibility; fitting a film to a lesson. Similar patterns later emerged for radio and then television; his comments on computers are less well formed as the educational use of computers was in the 'establishment' stage at the time of writing. Cuban (1986, p. 109) concludes, 'The search for improving classroom productivity through technological innovations has yielded very modest changes in teacher practice'. We see this in

[7] The photograph is available on https://iamliterate.wikispaces.com/Social+Studies+IRP.

some (not all!) of the mathematics lessons we have observed, for example a teacher presenting a PowerPoint 'demonstration' on a topic to a class. Such a demonstration uses a computer in place of a board on which one may write but the only significant difference between the two may be the presentation medium. In noting this we are not blaming these teachers, like the teacher in the 1927 photograph of an aerial geography lesson, they are using a new artefact in a manner that makes sense to them in terms of their past use of artefacts. So the 'no' part of our response is simply that there need not be anything special about teachers' use of digital tools.

We now address the 'yes' aspect of our response. Our argument that there is something special about teachers' using digital tools basically boils down to 'the opportunity for there to be something special; digital tools are there if teachers have a vision for them to be special'. Constructionists (amongst others) have a vision that there is something special about digital tools and we preface our 'yes' response by returning to three statements made by constructionist in Chap. 8. Two statements from Sect. 8.4: a microworld is a place where 'the teacher and the learner can be engaged in real intellectual collaboration' (Papert, 1980, p. 115). Noss (2001, p. 22) point about Papert's vision, that the computer 'can be changed (even change itself) into any number of forms' The third statement (from Sect. 8.5) concerns Noss and Hoyles (1996) distinction between media to present mathematics to a learner and media which encourages a learner to express mathematical relationships.

Taken together these statements provide a vision for new roles for teachers when expressive tools are used by their students. We illustrate this with a fairly pedestrian example, constructing a kite in a dynamic geometry system. Our experience with this task is that students often start by simply drawing a shape (using the software) which looks like a kite (a bit of dragging their points to destroy the kite-like shape convinces students that this is not sufficient). A little prompting on the properties of a kite allows students to construct a kite that remains a kite under any amount of dragging. In a large class several distinct constructions are produced and this can lead to the teacher asking 'Can you construct a kite like Jean's one?'[8] and later a challenge 'Construct a kite in as many different ways as you can'. This challenge has, in our experience, produced legitimate constructions (using different tools on the DGS tool-bar) that we had not expected and we have sat down with students saying 'How did you do that'. Figure 15.6 shows some different ways to construct a kite. From left to right the construction are made by: reflecting the point *A* in the line *BC*; constructing the perpendicular bisector *KL* to the line segment *HI*; marking two points (*E* and *F*) on the circumference of a circle (centre *D*) and constructing the angle bisector of $\angle EDF$; reflecting a triangle in one of its sides (Fig. 15.7).

Classroom teachers do not need to be aware of the constructionist statements or an advanced knowledge of mathematics to work with a class in this way but they need a willingness to work in this way. We stop short, however, of saying that teachers should have a willingness to work in this way; Cuban (1989) says,

---

[8] Comments such as this are provided as possible comments which can be provided, not as exemplar comments.

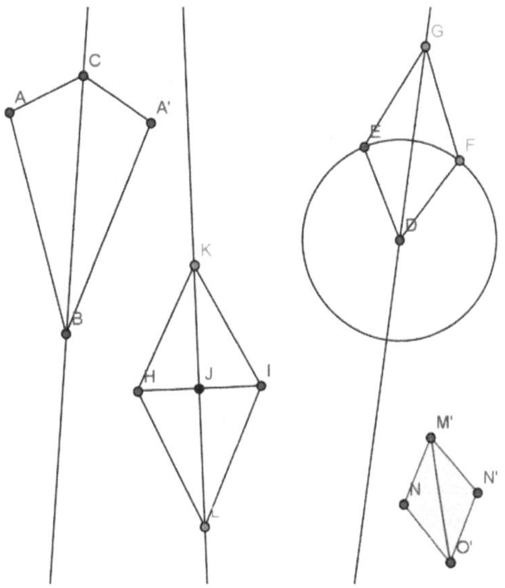

Fig. 15.7 Different ways of constructing a kite using a dynamic geometry system

in response to exhortations for teachers to work this was in their classrooms, such visions 'are out of synch with organisational realities and go well beyond what ordinary, well-intentioned, good-hearted people can do' (Cuban, 1989, p. 221).

15.5.2 New Forms of Teacher Education and Teacher Professional Development

Going beyond the ordinary exhortations for teachers to work differently supposes one to be aware of the complexity of ICT integration (Guin & Trouche, 1999) and of the institutional conditions for helping teachers to deal with. *Teacher education*, appears then as a *key issue* (Artigue, 1998) for the integration of computer technologies. Gueudet and Trouche (2011), a consideration of in-service teacher education and ICT, underline that usual training strategies were essentially based on the transmission of 'expert resources', mostly organised in a short period (about 3 days), isolated from school practice and they do not allow for *continuous* support to be provided during the necessary adaptation of resources to each teacher's usage context. To overcome these problems, innovative programmes had been developed, from the 1990s, in the US (Allen, Wallace, Cederberg, & Pearson, 1996), relying on teachers' *networks*, designing, with the help of experts, situations of use for dynamic geometry software (DGS); this perspective, of *teachers empowering teachers* appeared powerful, both for integrating DGS and for promoting new inquiry-based teaching practices. Trouche & Guin (Allen et al., 1996) have taken up this idea, conceptualising *collaborative work on resources* as a way of teacher education on technology in developing an innovative teacher training programme named SFoDEM (Sect. 19.3.1).

The development of ICT itself provides new means for providing this continuous support for ICT integration, based on teacher collaboration, design, implementation and revision of resources. From this perspective Gueudet and Trouche (2011) analyse a French innovative program, Pairform@nce.[9] Grounded on the work on/for resources, this experiment is presented as an illustration of the documentational approach (Sect. 15.4).

Pairform@nce, aims to develop in-service teachers' skills in using ICT in class and proposes *training paths* for all topics, from primary to secondary school levels. These training paths are available on an *online platform*. Each path is structured in seven stages: (1) Introduction to the training session, (2) Selection of teaching contents and organisation of teacher teams, (3) Collaborative and self-training, (4) Collaborative design of a lesson, (5) Test of this lesson in each trainee's class, (6) Shared reflection about feedback of class tests, (7) Evaluation of the training session. Gueudet and Trouche (2011) focus on trainees, analysing their collective involvement in a training path dedicated to DGS and its effects on practice and knowledge. The trainees are lower secondary school teachers, teaching from grade 6 to 9.

The training takes place during 13 weeks and consists of three face-to-face one-day workshops. Between these workshops, continuous work is done using e-mail and the distant training platform (see the agenda Table 15.1).

Gueudet and Trouche (2011, pp. 409–410) draw some lessons from the implementation of the DGS training session

> In the [DGS] path, the collective documentation work is the key feature retained in an objective of professional development. The resources proposed, in particular: the scenario, observation and report grids, and the examples of lessons aim at supporting the documentation work of the teams of trainees. The implementation of the scenario elaborated by each team as lesson in class, and the revision of this scenario, following a design-in-use principle, are also essential features of the path. All this could be organised in a distance training, using a platform. The work in presence was nevertheless also essential in the [DGS]-training; it permitted indeed to discuss about important features of an inquiry-based lesson using dynamic geometry, in particular: presentation of the task, appropriation of the problem by the students; design of an experiment by the students using the DGS, formulation and test of conjectures; argumentation, organisation of a debate; articulation between experimentation and proof. These elements were discussed during each one of the three workshops: about the examples of lessons presented by the trainers (workshop 1), about the scenarios built by the teams (workshop 2), about the lessons designed by the teams (workshop 3). We consider that the intertwining of such discussions and of scenarios design by the teams of trainees contributed to the development of the trainees practices.

Regarding the development of the trainees practices, Gueudet and Trouche noticed the first use of the computer lab (thanks to the collective work); the coordination, through the use of software, of multiple representations for learning the concept of function; the careful presentation of the task to students; and teachers taking into account the appropriation of the DGS by the students. What is

[9] Pairform@nce stands for 'training (*formation* in French) based on collaborations with colleagues (pairs in French)'. The symbol @ evokes the importance of Internet in this program.

Table 15.1 Agenda of training as planned by the DGS path

Week 1	Week 2	Weeks 3 and 4	Week 5	Weeks 6–12	Week 13
Questionnaire sent via e-mail by the trainers, filled by the trainees: equipment, experience with DGE, expectations	*Workshop 1*	Design, by the teams of trainees, of a scenario for one of the problems texts proposed	*Workshop 2*	Design and experimentation of the lessons by the teams of trainees	*Workshop 3*
	Presentation of the training (principles, agenda, platform)		Presentation by the teams of trainees and discussion of the scenarios		Presentation by the teams of trainees and discussion of the lessons
	Constitution of the teams (4 teachers in 2 different schools)		Choice of a lesson theme	At least one implementation in class by a member of the team, observed by another member	
	Presentation by the trainers and discussion of two examples of lessons		Beginning of the lesson design	The three grids (scenario, observation, report) are filled and uploaded on the platform	
	Presentation of three grids: description of scenario, observation of a lesson, final report				
	Proposition of two problem texts by the trainers				

The weeks mentioned in grey correspond to distant work of the trainers and teams of trainees (Gueudet & Trouche, 2011, p. xx)

interesting also to note is that the support of ICT (here a distant platform) is not enough for integrating ICT; the combination of face-to-face and distant collaboration appears as crucial. The nature of ICT supporting teachers' collaboration is also important: 5 years after the Pairform@nce experience, the French Ministry is developing (in 2014) a new program (named M@gister), based on a new platform, more flexible, and giving more responsibilities to teams for organising their own paths (varying the successive stages, beyond the initial seven stages).

New forms of activity with digital tools enabled by connecting these tools also provide teachers with opportunities to engage in new forms of action in their classrooms, and with their colleagues, giving new opportunities for teacher professional development. We return to these issues in Chap. 19.

References

Adler, J. (2000). Conceptualising resources as a theme for teacher education. *Journal of Mathematics Teacher Education, 3*, 205–224.

Allen, R., Wallace, M., Cederberg, J., & Pearson, D. (1996). Teachers empowering teachers: Vertically-integrated, inquiry-based geometry in school classrooms. *Mathematics Teacher, 90* (3), 254–255.

Artigue, M. (1998). Teacher training as a key issue for the integration of computer technologies. In J. D. Tinsley, & D. C. Johnson (Eds.), *Information and communication in school mathematics* (pp. 121–129). London: Chapman and Hall.

Artigue, M. (2002). Learning mathematics in a CAS environment: The genesis of a reflection about instrumentation and the dialectics between technical and conceptual work. *International Journal of Computers for Mathematical Learning, 7*(3), 245–274.

Assude, T. (2005). Time management in the work economy of a class. A case study: Integration of Cabri in primary school mathematics teaching. *Educational Studies in Mathematics, 59*(1–3), 183–203.

Chevallard, Y., & Bosch, M. (2014). Didactic transposition in mathematics. In S. Lerman (Ed.), *Encyclopedia of mathematics education*. Dordrecht, The Netherlands: Springer.

Cuban, L. (1986). *Teachers and machines: The classroom use of technology since 1920*. New York: Teachers College Press.

Cuban, L. (1989). Neoprogressive visions and organizational realities. *Harvard Educational Review, 59*, 217–222.

Drijvers, P., Doorman, M., Boon, P., Reed, H., & Gravemeijer, K. (2010). The teacher and the tool: Instrumental orchestrations in the technology-rich mathematics classroom. *Educational Studies in Mathematics, 75*, 213–234.

Etzioni, A. (1969). *The semi-professions and their organization: Teachers, nurses, social workers*. New York: The Free Press.

Goos, M. (2005). A sociocultural analysis of the development of pre-service and beginning teachers' pedagogical identities as users of technology. *Journal of Mathematics Teacher Education, 8*, 35–59.

Gueudet, G., Buteau, C., Mesa, V., & Misfeldt, M. (2014). Instrumental and documentational approaches: From technology use to documentation systems in university mathematics education. *Research in Mathematics Education, 16*(2), 139–155.

Gueudet, G., Pepin, B., & Trouche, L. (2013), Collective work with resources: An essential dimension for teacher documentation, *ZDM, The International Journal on Mathematics Education, 45*(7), 1003–1016. doi:10.1007/s11858-013-0527-1

Gueudet, G., & Trouche, L. (2009). Towards new documentation systems for mathematics teachers? *Educational Studies in Mathematics, 71*(3), 199–218.

Gueudet, G., & Trouche, L. (2011). Mathematics teacher education advanced methods: An example in dynamic geometry. *ZDM: The International Journal on Mathematics Education, 43*(3), 399–411.

Guin, D., & Trouche, L. (1999). The complex process of converting tools into mathematical instruments: The case of calculators. *International Journal of Computers for Mathematical Learning, 3*(3), 195–227.

Hannula, M. S. (2012). Exploring new dimensions of mathematics-related affect: Embodied and social theories. *Research in Mathematics Education, 14*(2), 137–161.

Hill, H. C., Ball, D. L., & Schilling, S. G. (2008). Unpacking pedagogical content knowledge: Conceptualizing and measuring teachers' topic-specific knowledge of students. *Journal for Research in Mathematics Education*, 372–400.

Kendal, M., & Stacey, K. (2001). The impact of teacher privileging on learning differentiation with technology. *International Journal of Computers for Mathematical Learning, 6*(2), 143–165.

Kennewell, S. (2001). Using affordances and constraints to evaluate the use of information and communications technology in teaching and learning. *Journal of Information Technology for Teacher Education, 10*, 101–116.

Kidron, I., Bikner-Ahsbahs, A., Monaghan, J., Radford, L., & Sensevy, G. (2012). CERME7 Working Group 16: Different theoretical perspectives and approaches in research in mathematics education. *Research in Mathematics Education, 14*(2), 213–214.

Laborde, C. (2001). Integration of technology in the design of geometry tasks with Cabri-Geometry. *International Journal of Computers for Mathematical Learning, 6*(3), 283–317.

Lagrange, J.-B., Artigue, M., Laborde, C., & Trouche, L. (2003). Technology and mathematics education: A multidimensional study of the evolution of research and innovation. In A. J. Bishop, M. A. Clements, C. Keitel, J. Kilpatrick, & F. K. S. Leung (Eds.), *Second international handbook of mathematics education* (pp. 39–271). Dordrecht, The Netherlands: Kluwer.

Lagrange, J.-B., & Erdogan, E. O. (2009). Teachers' emergent goals in spreadsheet-based lessons: Analyzing the complexity of technology integration. *Educational Studies in Mathematics, 71* (1), 65–84.

Lagrange, J. B., & Monaghan, J. (2009). On the adoption of a model to interpret teachers' use of technology in mathematics lessons. In *Proceedings of the sixth Conference of the European Society for Research in Mathematics Education* (pp. 1605–1614).

Mathematical Association. (1992). In W. Mann, & D. Tall (Eds.), *Computers in the mathematics curriculum: A report of the mathematical association*. Leicester, England: The Mathematical Association.

Mishra, P., & Koehler, M. (2006). Technological pedagogical content knowledge: A framework for teacher knowledge. *The Teachers College Record, 108*(6), 1017–1054.

Monaghan, J. (2001). Teachers' classroom interactions in ICT-based mathematics lessons. In M. van den Heuvel-Panhuizen (Ed.), *Proceedings of the 25th Conference of the International Groups for the Psychology of Mathematics Education* (Vol. 3 pp. 383–390).

Monaghan, J. (2004). Teachers' activities in technology-based mathematics lessons. *International Journal of Computers for Mathematical Learning, 9*(3), 327–357.

Moreira, C., & Noss, R. (1995). Understanding teachers' attitudes to change in a logomathematics environment. *Educational Studies in Mathematics, 28*, 155–176.

Noss, R. (2001). For a learnable mathematics in the digital culture. *Educational Studies in Mathematics, 48*, 21–46.

Noss, R., & Hoyles, C. (1996). *Windows on mathematical meanings: Learning cultures and computers*. Dordrecht, The Netherlands: Kluwer.

Olson, J. (1992). *Understanding teaching: Beyond expertise*. Milton Keynes, England: OU Press.

Papert, S. (1980). *Mindstorms: Children, computers, and powerful ideas*. Brighton, England: The Harvester Press.

Pedauque, R. T. (Ed.). (2006). *Le document à la lumière du numérique (Document under digital light)*. Caen, France: C & F éditions.

Pepin, B., Gueudet, G., Yerushalmy, M., Trouche, L., & Chazan, D. (to appear). E-textbooks in/for teaching and learning mathematics: A potentially transformative educational technology. In English, L., & Kirschner, D. (Eds.), *Handbook of international research on mathematics education*. Taylor & Francis.

Remillard, J. T. (2005). Examining key concepts in research on teachers' use of mathematics curricula. *Review of Educational Research, 75*(2), 211–246.

Rowland, T., & Ruthven, K. (2011). *Mathematical knowledge in teaching* (Vol. 50). Dordrecht, The Netherlands: Springer.

Ruthven, K. (2012). Constituting digital tools and materials as classroom resources: The example of dynamic geometry. In G. Gueudet, B. Pepin, & L. Trouche (Eds.), *From text to 'Lived' resources: Mathematics curriculum materials and teacher development*. New York: Springer.

Ruthven, K. (2014). Frameworks for analysing the expertise that underpins successful integration of digital technologies into everyday teaching practice. In A. Clark-Wilson, O. Robutti, &

N. Sinclair (Eds.), *The mathematics teacher in the digital era* (pp. 373–393). Dordrecht, The Netherlands: Springer.

Ruthven, K., & Hennessy, S. (2002). A practitioner model of the use of computer-based tools and resources to support mathematics teaching and learning. *Educational Studies in Mathematics, 49*(1), 47–88.

Sabra, H., & Trouche, L. (2011). Collective design of an online math textbook: When individual and collective documentation works meet. In M. Pytlak, T. Rowland, & E. Swoboda (Eds.), *Proceedings of CERME 7, 9–13 February 2011* (pp. 2356–2366). Rzesów, Poland.

Saxe, G. (1991). *Culture and cognitive development: Studies in mathematical understanding* (pp. 15–18). Hillsdale, NJ: Lawrence Erlbaum. Chapter 2.

Schwartz, J. L. (1989). Intellectual mirrors: A step in the direction of making schools knowledge-making places. *Harvard Educational Review, 59*(1), 51–62.

Shulman, L. S. (1986). Those who understand: Knowledge growth in teaching. *Educational Researcher, 2*, 4–14.

Shulman, L. S. (1987). Knowledge and teaching: Foundations of the new reform. *Harvard Educational Review, 57*(1), 1–23.

Trouche, L. (2004). Managing the complexity of human/machine interactions in computerized learning environments: Guiding students' command process through instrumental orchestrations. *International Journal of Computers for Mathematical Learning, 9*, 281–307.

Trouche, L., & Drijvers, P. (2010). Handheld technology for mathematics education, flashback to the future. *ZDM, The International Journal on Mathematics Education, 42*(7), 667–681.

Trouche, L., Drijvers, P., Gueudet, G., & Sacristan, A. I. (2013). Technology-driven developments and policy implications for mathematics education. In A. J. Bishop, M. A. Clements, C. Keitel, J. Kilpatrick, & F. K. S. Leung (Eds.), *Third international handbook of mathematics education* (pp. 753–790). New York: Springer.

Valsiner, J. (1987). *Culture and the development of children's action: A cultural-historical theory of developmental psychology*. John Wiley & Sons.

Vergnaud, G. (1998). Toward a cognitive theory of practice. In A. Sierpinska & J. Kilpatrick (Eds.), *Mathematics education as a research domain: A search for identity* (pp. 227–241). Dordrecht, The Netherlands: Kluwer.

Watson, A. (2008, March 18). Developing and deepening mathematical knowledge in teaching: being and knowing. In *MKiT 6, Nuffield Seminar Series*. University of Loughborough.

Wertsch, J. V. (1991). *Voices of the mind: A sociological approach to mediated action*. Cambridge MA: Harvard University Press.

Williams, J. (2011). Audit and evaluation of pedagogy: Towards a cultural-historical perspective. In T. Rowland & K. Ruthven (Eds.), *Mathematical knowledge in teaching* (pp. 161–178). Dordrecht, The Netherlands: Springer.

Chapter 16
Interlude

John Monaghan, Luc Trouche, and Jonathan M. Borwein

These pages are, after Chaps. 6 and 11, the third 'space for reflection' in this book and our focus is on issues in mathematics education related to tool use; issues that we address and issues which we have not addressed (at least as the focus for a chapter in the book). Part III focused on the curriculum, the calculator debate, mathematics in the real world and the mathematics teacher. Part IV, with an eye to the future, focuses on task design, games and connectivity. We structure our reflections in two unequal sections: a short section on futurology and the selection of issues in Part IV; a consideration of issues on which we have not explicitly focused.

We approach 'futurology' with caution—we hold that we can only imagine from that of which we have current experience[1] (anticipating even medium terms future is a fools' game). In a similar vein Borwein (2015) cites David Bailey:

> How could anyone, even ten years ago, have predicted that by now almost everyone above the age of eight would now be on Facebook for minutes (or hours!) of every day? How could anyone have predicted twenty years ago that almost every person above the age of eight would have their own personal supercomputer, systems far more powerful (and useful!) than the Cray supercomputers of the time?

Of the future scenarios which we could imagine we settled on three issues which feature in (some) current practice: tasks; games; and connectivity. We selected these because:

- Task design in mathematics education is very important but as a subject of study it is in its infancy. Using different tools for what appears to be the same task involves the person doing the task in different mathematical processes. We explore this in Chap. 17.
- Although games are not new (in themselves or as means to enable learning), digital tools offer new ways to exploit games for learning. Chap. 18 explores a

[1] Wartofsky (1979) said basically the same thing but expressed this in terms of 'tertiary artefacts' (imaginative constructions, see Sect. 7.2.2) which have their basis in real (non-imagined) praxis.

J. Monaghan et al., *Tools and Mathematics*, Mathematics Education Library,
DOI 10.1007/978-3-319-02396-0_16

future potential of games and to provide opportunity for engaging in mathematical activity, with a specific focus on the role of artefacts in gameplay.
- 'Connectivity', with regard to internet connection, is a widely used term but quite often with extreme (good/bad) valuations. In Chap. 19 we explore various meanings of the term and the potential of connectivity to support students' learning of mathematics and teachers' professional development.

Of issues on which we have not explicitly focused we comment, in turn, on four: the future of mathematics; equity; identity; and who uses tools. In the first two we summarise recent texts which have addressed this issue.

Borwein (2015) addresses the future of mathematics (with extreme caution!). After noting that mathematics will continue to be important and that some things will not change much he notes that the need to make our subject accessible is more pressing than 50 years ago. With regard to tools he notes:

- INRIA's prototype *Dynamic Dictionary of Mathematical Functions*, http://ddmf.msr- inria.inria.fr/, points to a future in which mathematical knowledge is generated algorithmically and extensibly, rather than from lookup tables.
- The future of mathematics is intimately coupled to computing and more emphasis on algorithms and constructive methods.
- Automatic theorem proving is already important and one can anticipate a time in the distant future when all truly consequential results are so validated; interactive theorem proving will become increasingly important.

Activity around mathematics is also likely to change. Collaborative research (which can be exaggerated) is likely to increase. Experimental mathematics raises numerous issues of computational reproducibility and there will be increased need for researchers to keep a record of workflow, computer hardware and software configuration, or parameter settings. In a hopeful (rather than an expectant) mood he envisions the development of comprehensive computation and publishing system with features that allow one to manipulate mathematics while reading it and which ensures published mathematics is rich and multi-textured, allowing for reading at a variety of levels.

Jon's view on the future of mathematics is, of course, informed by his experimental mathematics. Reading the reflections of Cédric Villani, the 2010 Fields Medalist we realise how the work of each mathematician is sensitive to the tools available for writing and communicating. Speaking of TEX, the software developed by Donald Knuth for writing mathematics, Villani (2012, p. 62) notes: 'thanks to this software, Knuth is probably the living person who changed the most the daily life of mathematicians'. Villani also writes on the extensive use of email in advancing, in a community of mathematicians, ideas towards a proof of a conjecture and how this seems to foster the emergence of a new kind of collective intelligence. Digitalisation impacts indeed all the spheres of the mathematics community.

Section 4 of Hoyles and Lagrange (2010) is devoted to issues of access and equality. With regard to regional differences with regard to access to digital technology they note progress with regard to a 1986 report but note differences both across and within countries. They note similarities at the policy level of recognising the importance for curricula to address the use of digital technologies in school mathematics but conclude that:

> . . .conditions of schooling, social stratification, even within early developing countries, and political conditions mitigate strongly against all students benefiting from the possibility for meaningful learning of mathematics which digital technologies have to offer. (Julie et al., 2010, p. 381)

With regard to gender and the learning of mathematics with technology, research shows disparities, 'in some countries the gender gap favouring males may be closing, while in other countries, where there have been little or no gender differences in the past, the gap may be widening' (Forgasz et al., 2010, p. 385).

Who we are (our identity) is interrelated with the tools we use. This can be seen in the need of many young people to get the latest mobile phone. It can also be seen in us three authors: John and Luc wrote their chapters for this book in Word but Jon's chapter was written in TeX (mathematicians write in TeX). With regard to tools for mathematics John and Luc are both proponents of graphic calculators (partly due to the appearance of these artefacts at a formative stage of their professional lives) but students today may view graphic calculators as 'old technology' (such a perception even appeared at the turn of the millennium, see Rodd & Monaghan, 2002), not 'cool' things to be seen with. Issues of identity emerge in mathematics lessons too and 'who one is' in a mathematics class has a tool dimension. John recalls a high achieving calculus class he visited many years ago. The class was doing work using a computer algebra system. There was one girl who was clearly not enjoying the work. John asked the teacher about her, who told him that she was the best mathematician in the class. John talked to the girl. She did not like using *Derive* and did not use it except when the classwork was directed towards *Derive* use. This disposition towards *Derive*, we hold, is interrelated with her identity as a mathematician. She was a 'high flying' mathematics student. Her judged ability was not mathematical ability in a vacuum, it was ability in using traditional paper and pencil mathematical tools and signs and *Derive* was able to do what she was good at—who likes a tool that threatens who you are?

The final issue we broach in this interlude is 'who uses tools?' This is actually a non-question as everyone uses tools. Our focus in this section concerns the use of specific tools in mathematical activity in specific contexts. This was raised, in non-academic contexts, in Sect. 14.2.1 with regard to Skovsmose's (2005) distinction between 'constructors', 'operators' and 'consumers'. All of these groups of people use tools (related to specific activities) but the control they have over the tools used can, and often does, vary: operators in production engineering, for example, may be forbidden to change specific settings of tools they operate.

In school contexts a limited set of tools for mathematics (ruler, compass, protractor, calculator, specific algorithms) is currently the norm in most countries and student use (or not) of these tools is often dictated by the teacher (who themselves are often constrained by curricular regulations). These comments are 'common knowledge' to those in education but we are not aware of research on this issue.

Anecdotal evidence from England suggests that the increased use of digital tools in classrooms is largely teacher-only use of digital tools. John reports on two experiences to support this hypothesis. The first arises from classroom observations since Interactive White Boards (IWBs) became common in classrooms. In John's opinion, a very large proportion of the use of IWBs is teacher use of IWBs with PowerPoint (rather than interactive mathematics software) and the result is 'teacher demonstration'. The second experience is an ongoing project in which teachers of academic stream, 16–19-year-old students, have been observed. A summary sheet to each lesson observation records information on whether mathematics software was used and, if so, then who used it. In the first round of observations 122 lessons were observed. Remarkably, mathematics software was used in exactly half (61) of these lessons but in 45 of the 61 lessons, the software was used by the teacher alone. Teacher demonstration, of course, is important for student learning but, if these figures reflect a wider phenomenon, then students are not being granted wide access to tools to explore mathematical relationships.

Observations like those above challenge us to explain this practice. It could be that there are, in specific situations, tools that are in the hands of the teacher (like the IWB)) and tools that are in the hands of the student. The first ones may stay in the teacher's hand even though their purpose is to direct student activity. If, however, teachers use these tools to collaborate with their peers (see Chap. 19), they may be more keen to share their tools with the students.

References

Borwein, J. M. (2015). The future of mathematics: 1965 to 2065. In *Celebrating a century of advancing mathematics* (MAA Centenary Volume, pp. 313–330).

Forgasz, H. J., Vale, C., & Ursini, S. (2010). Technology for mathematics education: Equity, access and agency. In *Mathematics education and technology-rethinking the terrain* (pp. 385–403). Springer US.

Hoyles, C., & Lagrange, J. B. (2010). *Mathematical education and digital technologies: Rethinking the terrain*. The 17th ICMI study. New York: Springer.

Julie, C., Leung, A., Thanh, N. C., Posadas, L. S., Sacristán, A. I., & Semenov, A. (2010). Some regional developments in access and implementation of digital technologies and ICT. In *Mathematics education and technology-rethinking the terrain* (pp. 361–383). Springer US.

Rodd, M., & Monaghan, J. (2002). Graphic calculator use in Leeds schools: Fragments of practice. *Journal of Information Technology for Teacher Education, 11*(1), 93–108.

Skovsmose, O. (2005). *Travelling through education. Uncertainty, mathematics, responsibility*. Rotterdam, The Netherlands: Sense.

Villani, C. (2012). *Théorème vivant*. Paris, France: Grasset.

Wartofsky, M. (1979). *Models: Representation and the scientific understanding*. Dordrecht, The Netherlands: Reidel.

Part IV
The Future?

Chapter 17
Tasks and Digital Tools

John Monaghan and Luc Trouche

17.1 Introduction

'Tasks' in this chapter refers to what teachers plan and design for triggering and supporting learners' activity. Given the focus of this book we are, of course, interested in mathematical tasks and the roles of tools in the design and execution of tasks. To keep this chapter manageable we focus on 'scholastic tasks' (as opposed to vocational training tasks which have a mathematical element) and do not attend to tasks designed to instruct students in the use of a tool (though such tasks are often necessary before a mathematical tasks can be presented). The design of tasks with digital tools often involves the design or the adaption of a digital tool (for example, a teacher configuring a spreadsheet for the purpose of mathematical instruction). Tool design and/or adaption is not an explicit focus of this chapter but it will be considered when it is an essential aspect of task design.

In the language of mathematics teaching the 'tasks' go by a number of names: exercises, investigations, problem solving, modelling. Tasks are also subdivided into, for example: open and closed tasks; tasks for the consolidation of skills and for concept development. Further to this there is a dialectic between task, tool and person in learning mathematics. We attended to the task-tool aspect of this dialectic in Chap. 2 of this book which considers how one task, bisecting an angle, can be done using four tools: a straight edge and compass, a protractor, a dynamic geometry system and a book. We noted that each tool has affordances and constraints with regard to what might be learnt in using each tool to do the task. The person-task aspect of this dialectic has several dimensions including the person's engagement in the task and the ability of the person to realise the task (i.e. not too easy or too hard). The person-task aspect of this dialectic also has several dimensions including what Wertsch (1998) (see Sect. 13.2) refers to as mastery and appropriation of the mediational means.

Tasks are extremely important in mathematics learning and teaching. Watson et al. (2013, p. 10) state that tasks 'are the mediating tools for teaching and learning

J. Monaghan et al., *Tools and Mathematics*, Mathematics Education Library,
DOI 10.1007/978-3-319-02396-0_17

mathematics and the central issues are how tasks relate to learning, and how tasks are used pedagogically.' The use of the word 'tool' in this quotation fits with Monaghan's Sect. 1.3.1 definition of a tool, an artefact that is used to do something; a task is an artefact that is used by a teacher to promote learning. It is, then, surprising that is was only recently (at the time of writing) that the first ICMI study on task design (Margolinas, 2013) in mathematics education took place. There is a danger that tasks and tool-use are seen in isolation; tool-use in a task is a part of an often extended period of 'instrumental genesis' (see Sect. 10.4) and time spent by a student on any task is but a moment in a longer term instructional sequences. That said, our focus in this chapter is on tasks and tool-use, with the caveat that tasks are parts of teaching sequences and tool-use is a part of instrumental genesis being taken as read.

This chapter is in the last part of our book where we look to the future of learning and teaching mathematics. Digital tools will, without doubt, be important tools in this future and are the major focus in this chapter. In the first section we provide a set of examples that illustrate a range of issues related to tasks and digital tools. In the second section we review academic literature which sheds light on the dialectic between task, tool and person in learning mathematics. The first two sections are somewhat biased towards tasks in 'ordinary' classrooms (tasks for learning) and issues relating to tasks using mathematical software. In the final section we look at task-tool issues in larger-than-the-individual classroom research and in assessment; we also comment of avenues for further development.

17.2 Tasks and Digital Tool: Examples

The purpose of this section is to raise issues to consider with regard to tasks and digital tools. We begin with two tasks and two (sets of) tools which show that a *mismatch* between tasks and tools can exist. We follow this with further examples which show: a move from procedural to conceptual tasks with digital tools; differences with regard to the mathematics in a task over different digital tools; and an increase in the solution methods with digital tools.

17.2.1 A Mismatch Between Tasks and Tools Can Exist

The two tasks below (Fig. 17.1) could be suitable for pre-calculus work on functions. John selected these tasks to illustrate a *mismatch* and they are, perhaps, 'English' classroom tasks. The first task is made up for the purposes of this example but is generic of many 'Sketch the graph of ...' tasks typically given to students. The second task, designed by a teacher, was observed in a classroom using computer graph plotters to explore the graphical effect of the parameters in $y = ax^2 + bx + c$.

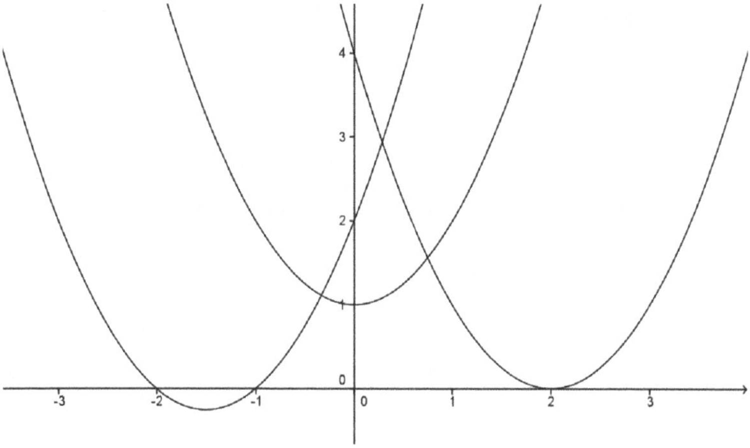

Fig. 17.1 Two tasks on the graphs of quadratic functions

Task 1: Sketch the graph of $y = x^3 - x^2 + 2x - 1$.
Task 2: Reflect the quadratic functions (in Fig. 17.1) in the grid below in the x-axis.

We outline possible student actions (ignoring errors that could be made) in doing each task using digital technology and pencil and graph paper and make reference to 'techniques' (see Sect. 10.3.1).

Task 1 can be approached in a number of ways using pencil and paper. A sophisticated approach is to rewrite and factorise the expression $x^3 - x^2 + 2x - 1$, $x^3 - x^2 + 2x - 1 = x^3 - (x^2 - 2x + 1) = x^3 - (x - 1)^2$ and then use the known graphical representation of $y = x^3$ and $y = (x - 1)^2$ as references from which to sketch $y = x^3 - x^2 + 2x - 1$. The actions here are algebraic (rearranging and factorising) and then graphic (sketching and then coordinating known functions). Students may adopt this approach if techniques which privilege these algebraic and graphic actions have been taught recently (or have been internalised, see Sects. 7.2 and 13.2). A less sophisticated approach students may use is to construct a table of x and y values (a technique), select suitable scales, draw axes on graph paper, plot the points and sketch the graph by interpolating values between points. The techniques here involved in this second approach are generic to much that goes by the name of sketching graphs.

In Task 1, using graphing software, students may start by inputting the functional equation into the input bar (typing digital-mathematics) and then pressing an appropriate key (keying). Further actions (dragging and zooming in or out) may be needed to ensure the graph can be seen. The actions (in brackets) require techniques (quite different techniques to the techniques used in the pencil and paper approach).

In Task 2, using pencil and graph paper, students may simply draw the reflection (drawing).

In Task 2 using graphing software students may start by determining the functional equation of each graph (attending to salient features of each graph). They may then find a means to express the reflection as a function (expressing a geometric feature algebraically). They are then likely to input the functional equation into the input bar (typing) and then press an appropriate key (keying). If the student cannot immediately determine the functional equation of a graph, then the student may recognise salient features (for example, that a in $y = ax^2 + bx + c$ is 1), 'try out' a specific function of this form and use the computer-generated graph as feedback from which to try out another specific function of this form.[1]

It seems to us that there are a lot more mathematical actions (and arguably mathematical thinking) in Task 1 when it is done using pencil and graph paper but that the opposite is the case for Task 2. This is what we mean when we say that a *mismatch* between tasks and tools can exist. Lagrange has argued in various papers (see, for example, Lagrange, 1999), techniques do not disappear when tasks are done with digital technology, they are, rather, transformed and replaced (in Task 1 using graphing software the technique of inputting a functional equation appears). We shall consider the relationship between techniques and tasks in further detail in Sect. 17.3.2 below but for now simply state that this transformation of techniques presents problems for classroom teachers who are charged with ensuring that a specific curriculum is taught (as most are) because of a *mismatch* in techniques required for tasks with digital tools and techniques privileged in mathematics curricula.

17.2.2 From Procedural to Conceptual Tasks with Digital Tools

We now move on to our other three examples. The first concerns a need, at times, to move from procedural to conceptual tasks with digital tools and this need spans the years of schooling. Johnson (1981) illustrates this with regard to arithmetic and calculators. He considers tasks when a concept has already been taught and 'the calculator activity is planned to provide an opportunity to practice or apply the concepts and/or relationships which have been studied' (Johnson, 1981, p. 28). For example, the first of a set of ten exercises on estimation is:

(37 ○ 21) ○ 223 = 1000, where the circle represents a missing operation: +, −, ×, ÷.

A similar task which could be given to learners is 'estimate $(37 \times 21) + 223$'. A calculator is pretty pointless for this task because using a calculator negates the need for estimation. Johnson's task is designed to highlight operations: (37×21)

[1] This is an example of instrumentalisation (see Sect. 10.4.1).

$+223$ is very different to $(37+21) \times 223$, and the calculator provides feedback to the students on the effect of operations. The design principle implicit in Johnson's task is 'move from procedures to obtain a result (which becomes trivialised with the digital technology) to a focus on the mathematical relationships which produce a given result'. Once this principle is appropriated by a teacher this principle can be applied to tasks related to other areas of mathematics and tools:

- Scientific calculators $\square 30° + \cos 44° = 1.2967$ (to 4dp)
- Symbolic calculators $\frac{dy}{dx}(\sin \square \cdot \cos x) = 2\cos x \cos 2x - \sin x \sin 2x$
  $\square$ denotes a missing symbol

17.2.3 *Differences with Regard to the Mathematics in a Task Over Different Digital Tools*

The second example concerns differences with regard to the mathematics in a task over different digital tools.[2] Perks, Prestage and Hewitt (2002), for example, consider the task, 'find a line that passes through the point (4, 3) and has a gradient of 2' with a graph plotter, a dynamic geometry system and a spreadsheet. Suppose the students know that the graph has the form $y=2x+c$. With a graph plotter students obtain practice typing $y=2x$ ... into the computer/graphic calculator. They are likely to use a trial and improvement approach ($y=2x+1$ is too high, try $=2x-1$). The task may focus students' attention on the parallel lines and/or that different values of c translate the lines. A 'lucky guess' of $y=2x-5$ would effectively diminish the educational value of the task and it might be better to turn the task into a two-person game of 'find my line'.

Perks et al. (2002) then illustrate a dynamic geometry system solution to the task with the software *Geometer's Sketchpad*. They click on two points and they choose the menu option that draws the straight line between these two points, which displays the equation of the line in the form $y=2x+c$. They then move the points around the screen and note:

> The dynamic quality of the package gives a very different feeling to the task ... the mathematics here is about the relationship between the way the line changes and the numbers in the equation. It allows:
>
> - A feel for rotation' and gradient in relation to a particular point;
> - A feel for the constant in relation to a particular point;
> - Working on small changes in m and c. (Perks et al., 2002, p. 31)

Perks et al. (2002) then illustrate a spreadsheet solution to the task where the task consists in setting up a table of values (as shown in columns A and B in Fig. 17.2) where the B column cells are calculated from A, C and D column class as shown in the formula line in Fig. 17.2. The student highlights the table and inserts a graph.

---

[2] This subsection has a similar focus to Chap. 2 but here the focus is on different digital tools.

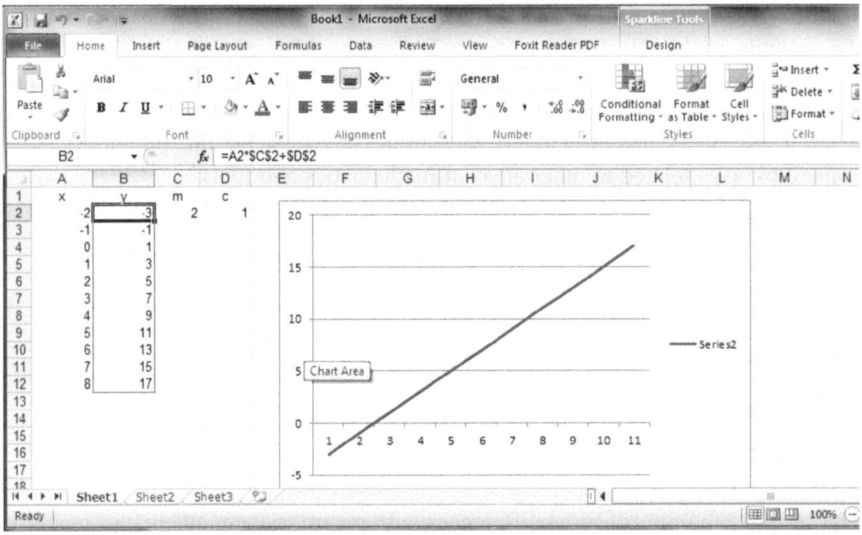

Fig. 17.2 A spreadsheet method of attacking the task

Perks et al. (2002) stress that different areas of mathematics are highlighted in the different types of software with the same task. This point is reinforced when we consider that different software exhibits different mathematical forms. For example, if you connect the points (4, 3) and (2, −1) with a line in *Geometer's Sketchpad* it displays the equation of the line in the form $y = 2x - 5$ but if you do this in *GeoGebra* (which includes a dynamic geometry system) it displays the line in the form $2x - y = 5$.

17.2.4 An Increase in the Solution Methods with Digital Tools

The third example concerns an increase in possible solution methods with digital tools. Consider solving a quadratic equation, say $x^2 + 4x - 1 = 0$. Traditional methods of solution include factoring (and solving the factors), completing the square, using the formula and graphical means. Arithmetic/scientific calculators add trial and improvement methods. Graphic calculators (or packages) add new graphical means. Symbolic calculators (or packages) add new algebraic means. These tools allow tasks-which-suit-the-tool to be given to students (as in Sect. 17.2.1). For example, the task below could be set for students to do with *GeoGebra* (see Fig. 17.3).

1. Write $f(x) = x^2 + 4x - 1$ in the form $f(x) = (x - 1)^2 + s$. Hence or otherwise draw the graph of $f(x)$.

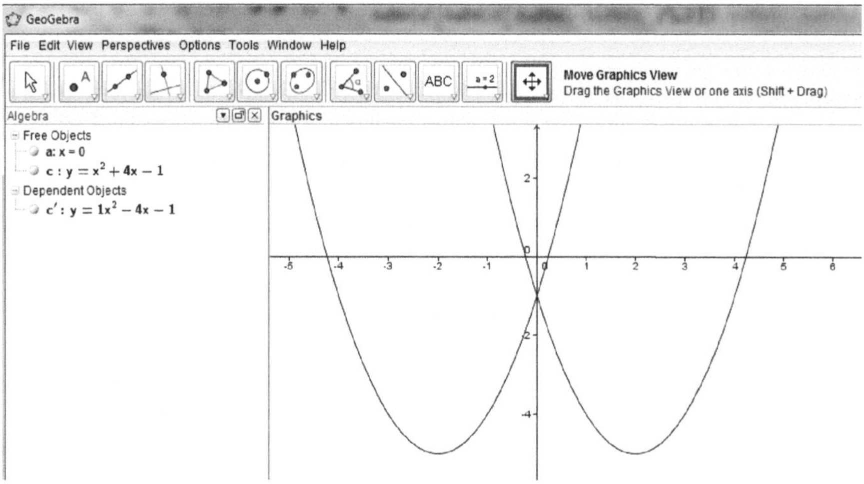

Fig. 17.3 A screenshot from *GeoGebra* related to the task above

2. Reflect the graph of $f(x)$ in the line $x = 0$. Call this $g(x)$. Sketch the graph of $g(x)$.
3. Write $g(x)$ in the form $g(x) = (x - p)^2 + q$ and compare the completing the square forms of $f(x)$ and $g(x)$.
4. Make an hypothesis about the reflection of a function $h(x) = (x - m)^2 + n$ in the line $x = 0$.

17.3 Tasks and Tools: The Development of Ideas and New Opportunities

This section explores scholarly work on tasks and tools in mathematics education. Although we expressed surprise in the opening of this chapter that we had to wait until 2013 for an ICMI study on task design, the academic literature on tasks is far from non-existent. This section has three subsections: the first two explore the development of ideas in the literature; the third applies ideas from the literature to explore new opportunities for task design with digital tools. Sect. 17.3.1 sets the scene with an overview of theoretical and practical approaches to task design. Tools are not always an explicit feature in these approaches but if you delve into these approaches you will find tools at the root of tasks. Sect. 17.3.2 explores a subset of the approaches outlined in Sect. 17.3.1 further with an explicit focus on digital tools. Sect. 17.3.3 explores new opportunities for task design with digital tools with regard to: students' contributions; teacher collaboration; exploitation of resource systems (cf. Sects. 10.5 and 15.3); and teacher assessment.

17.3.1 Academic Literature on Tasks: An Overview

In this subsection we briefly revisit Chaps 8–10 (constructionism, activity theory and French didactics), and visit work using variation theory, Realistic Mathematics Education (RME) and tasks suitable for small group learning.

Chapter 8 considered constructionism in which task design is a central feature. Constructionist tasks are invariably set in a *microworld* (cf. Sect. 8.3) and software that enables the user to *express* (cf. Sect. 8.4) mathematical relationships. The task itself is often only partially specified so that the user can pose, as well as solve, problems. Tasks are designed to stimulate 'situated abstractions':

> Situated abstraction describes how learners construct mathematical ideas by breathing life into the web using the tools at hand, a process which, in turn, shapes the ideas. Tools are not passive: in a microworld, for example, the designer's intentions are constituted in the software tools. These tools wrap up some of the mathematical ontology of the environment and form part of the web of ideas and actions embedded in it. Yet it is students who shape these ideas ... A microworld comprises tools to construct objects. But these tools are themselves objects which encapsulate relationships. (Noss & Hoyles, 1996, p. 227)

Chapter 9 considered activity theory (AT) where task was a constant theme though not a focal point. Tasks are important in AT because AT focuses on the object of activity, and tasks are a part of this object (and sometimes the object in school mathematics lessons). AT is also a theory that views context as paramount, so tasks in AT are appropriately viewed in context. AT theory also pays homage to Vygotsky's zone of proximal development (ZPD) which, in terms of tasks, can be viewed as the zone in which a task is challenging for the learner but not impossible; a task in which the learner, in mathematics, must be assisted (mediation) to appropriate cultural forms of reasoning.[3] This said, however, AT does not give mathematics education an explicit theory of tasks and task design though the human–computer interaction strand of AT does explicitly address artefact design related to tasks:

> Artefacts should be designed to enable efficient transformation into instruments... the importance of designing flexible, open artefacts that can be modified by users and adjusted for various tasks, including unanticipated tasks and the need for designers to take into account the actual transformation of practices and the real needs of users over the course of appropriating an artefact.[4] (Kaptelinin & Nardi, 2006, p. 110)

Chapter 10 considered French schools of thought in mathematics education and the schools of thought associated with Brousseau, with Chevallard and with Rabardel all have important things to say about tasks. Tasks are a prominent feature in Brousseau's Theory of Didactical Situations (TDS) and the *didactical contract* (where the teacher hands over responsibility to the student) depends on an appropriate task, 'The teacher must see to it that the student solves the problems set for

---

[3] The ZPD is a much deeper construct than this sentence suggests. Abdul Hussain, Monaghan, and Threlfall (2012) discuss complexities in the ZPD.

[4] This approach has many similarities to' instrumentalisation' (see Sect. 10.4).

him, so that both of them can assess whether he has accomplished his task' (personal communication from Brousseau cited in Warfield, 2006, p. 34). Brousseau designed his tasks (by iterative refinement of classroom experiments) so that students could fully engage in the situations of action, formulation and validation. Brousseau evidences how the choices of some elements of the task (parameters, type of material)—the *didactic variables*—can deeply modify students' activity and learning. Tasks are a named part of Chevallard's anthropological theory of didactics (ATD). The central features of ATD are praxeologies (idiosyncratic practices which reside in institutions) which consist of four elements (T, τ, θ, Θ) in two pairs: T/τ (tasks/techniques) concerns the practice part of the praxeology and θ/Θ (technology/theory) is related to the theory and the discourse which describes, justifies and interprets the practice. By this view tasks are set by mathematics teachers because they allow students to engage in techniques privileged in the institution (school). Task design by this view is not a simple matter of making a 'nice task' but of analysing the praxeology and the types of tasks that can be set in an institution. Bosch and Chevallard (1999) draw attention to the role of tools (within the wider category of 'ostensifs') in this approach (see Sect. 10.3.2). Rabardel's contribution to work on tasks and tools was the basis for the instrumental approach in mathematics education (see Sect. 10.4) provides artefactual insights on Brousseau's construct of *didactic variables* (considered further in the next subsection).

A comparably new (last 10 years at the time of writing) approach to task analysis and design is the use of variation theory (which focuses on variation in people's perceptions of phenomena). John Mason and Anne Watson (various papers) are the mathematics educators most closely associated with this approach. An early paper on this theme encapsulates the appeal of this theory for task design:

> Marton's identification of 'dimensions of variation' offers a way to look at exercises in terms of what is available for the learner to notice . . . This approach offers a structured and structural approach to exposing underlying mathematical form. We find it useful to consider dimensions of possible variation as experienced by a learner or by a teacher in any given situation (what could change and still the situation remains much the same), since this varies both between learners and even within one person at different times. . . . By asking the highly mathematical question 'what changes and what stays the same?', and by examining the nature of the changes offered, we can be precise about what an exercise affords the learner. (Watson & Mason, 2004, p. 108)

The 'exercise' becomes the designed artefact in this approach. An exercise can be thought of (and often is) as a series of mathematical tasks related to a technique. The 'dimensions of variation' approach requires that the individual tasks in the exercise are not merely related but, sequentially, provide variation.

RME is both a theoretical framework and a practical approach to designing mathematical learning activities that entered the public domain with Freudenthal (1973). The 'realistic' in the title refers to mathematics education being realistic to the learners' experience, not to real-life applications and modelling (though in practice there is likely to be a large overlap in these two realities). Task design from these principles should endeavour to, 'invite students to develop "their own"

Table 17.1 A classification of mathematical tasks with regard to their suitability for collaborative student group work

Category	Type of exchange model	Nature of task
Drilling basic skills I	Very limited	Closed in terms of method and outcomes
Applying a formula or algorithm II	Limited in how to proceed and checking results	Typically closed in terms of outcomes and also the methodology
Measuring and collecting data I—IIA	Provides some opportunities	Some openness in terms of methodology but rather closed in terms of outcomes
	For equal exchange of ideas and opinions	
Real problem solving IIB–III	Equal exchange	Real problems are those encountered in everyday life. They may or may not involve mathematical models. Openness of these tasks may vary from rather closed to open
Mathematical modelling III	Several opportunities for rich equal exchange	Modelling tasks are typically real problems that require mathematical principles and formulas in order to be solved. Tasks are open in terms of procedures and outcomes
Mathematical investigations IVA	Several opportunities for rich equal exchange	Basic investigations are often closed in terms of outcomes but open to various methods. Extended investigations are typically open tasks
Designing projects and studies in mathematics V	Rich equal exchange	Projects and studies are the most open mathematical tasks. The openness includes the setting of questions and selection of methods

From Sahlberg and Berry (2003), p. 74

mathematics … [with] guidance from the teacher … find such meaningful phenomena that beg to be organised and structured by the targeted mathematical knowledge' (Drijvers, Boon, Doorman, Bokhove, & Tacoma, 2013, p. 56). Further to this the tasks should gradually mathematicise an experientially real (for the student) situation to become a mathematical object, 'the challenge is to find suitable situations that ask for the development of such models, and allow for a process of progressive abstraction' (Drijvers et al., 2013).

Sahlberg and Berry (2003) address the question 'what kinds of mathematical tasks are suitable for small group learning?' Table 17.1 is an extract from this book. Prior to the appearance of this table in Sahlberg and Berry (2003) a number of concrete mathematical tasks are analysed resulting in the categories in the left column and the descriptions in the right column. The middle column, 'equal exchange', describes the extent to which the task affords equal contributions from all members of a student group tackling the task. For example 'Solve $2x - 3 = 17$' would be in their 'drilling basic skills' category and the statement 'very limited' refers to the situation where one or more students know(s) the answer and tell(s) the other group members.

Sahlberg and Berry (2003) does not consider tools but tools and artefacts can be 'read into' the categorisation (e.g. algorithms in the first two categories, measuring/data collection in the third category, designing artefacts in the last category).

The above focuses on the academic literature but, as discussed in Sect. 15.3, tasks are not only designed by academics but also by teachers. Such tasks rarely have a theoretical foundation but this, of course, does not relegate them as somehow inferior to tasks which have a theoretical underpinning. As Sect. 15.3 further shows, advances in digital tools (such as *GeoGebra Tube*) provide a medium in which teachers can publish and exchange tasks. There are also publications, such as Mason and Johnston-Wilder (2006), which aim to help teachers develop their skills in writing tasks for their students.

17.3.2 Academic Literature on Tasks with Digital Tools

In this subsection we revisit themes from the previous subsection with particular regard to digital tools. We begin with activity theory and French didactics and end by relating these approaches to task design to other approaches.

Chapter 9 concluded by saying that although AT has contributed much to our understanding of tool-use in mathematics, it offers us nuanced understandings and a similar statement, in our opinion, holds with regard to tasks-with-tools. We illustrate this by considering two 'camps' within AT, one represented by Chiappini (2012) and the other by two papers, Robert (2012) and Abboud-Blanchard and Vandebrouck (2012).[5]

Affordances are central in Chiappini's account. The affordances of a tool is regularly referred to by mathematics educators in relation to software but Chiappini is interested in cultural affordances because he is a cultural historical activity theorist and he is interested in how tool-use can bring learners to focus on the cultural ideas of mathematics. His attentions, then turn to the design of mathematical software which, through its affordances and constraints, promotes the emergence of student solutions to tasks. The task to Chiappini is not an end in itself but, when tackled on the software provide output to student input and some of this output should surprise the student and produce cognitive conflict. Tasks should be designed to allow the student to exploit the affordances of the software 'to allow students to explore the conditions, causes and explicative mechanisms of conflicts' (Chiappini, 2012, p. 139) and a role of the teacher is to assist in the process of raising students' consciousness of these conflicts. There is much more to Chiappini (2012) than I have stated here[6] but this summary suffices to contrast this approach to that below.

---

[5] All three papers were summarised in Sect. 9.3.

[6] See Sect. 9.3 or, better still, Chiappini (2012), for details.

Central to Robert (2012) and Abboud-Blanchard and Vandebrouck's (2012) activity theoretic approach is a production loop in which students' actions/activity is focused on the solution of a task and a construction loop concerned with the construction of knowledge in the student. This leads to a focus on the description of the ways that students work on tasks-with-tools in classrooms and to design, via analysis of these descriptions (over time through iterative refinement) classroom tasks with regard to both production and construction. Many different types of tasks are possible but exercises, the resolution of which involves specific knowledge to be learned, are the main interest. But the focus is not just student-task as the teacher is central to students completing tasks (the productive component) and student understanding resulting from this (the constructive component) and an important aspect of the design of tasks-with-tools is the development of the teacher to enable productive and constructive student activity.

Both approaches can be said to adhere to AT principles and there are some similarities (e.g. both include a concern with tasks whose resolution involves specific knowledge to be learned) but the range of possible tasks described by each camp differs significantly. Our point in arguing this is merely to note that there is not a single AT approach to task design in mathematics education. We now move on to consider three approaches within French didactics.

We start with Brousseau's TDS. There is a sense in which TDS (the fundamental situations, the didactical contract, the student-mathematics-milieu dialectic) continues to permeate almost all French mathematics education research (even when the researchers explicitly align themselves to another theoretical framework). For that reason we consider Joubert's (2013) appropriation of TDS as a framework for tasks-with-digital-tools design, as Joubert is not French.

Joubert (2013) uses TDS as a basis to formulate principles of task design for mathematical tasks which expect the use of mathematical computer software. She notes that TDS requires that the intended learning (by doing the task) is clear and 'the design of the task should attend to the mathematics embodied in the tools, and to working out how this might mediate the learning' (Joubert, 2013, p. 73). She notes that TDS views epistemological obstacles are key to learning and should be built into tasks-with-digital-tools. Further to this, TDS regards feedback from the milieu to the student as crucial for learning and when the computer software is part of the milieu, then the task designer must focus on student feedback from the software including how the student might engage with this feedback. It is also important for Joubert that task designers explicitly attend to the situations of action, formulation and validation: in the situation of formulation in, say, using the computer to draw graphs, 'students may be able to notice characteristics of families of graphs, develop conjectures based on their observations and test their conjectures on the computer' (Joubert, 2013, pp. 75–76); in a situation of validation students may provide 'an explanation to justify a prediction that a computer-generated graph will have particular intercepts by noticing a relationship between the intercepts and the equations of graphs generated previously' (Joubert, 2013, p. 76) and the task designer should work to ensure that students' attempts at validations are based on mathematical relationships and not just screen images. We now move on to consider Rabardel's theory.

It is important, here, to situate the role of Rabardel' s theory (see Sect. 10.4) on the French school of thought. Rabardel intervened at the French summer school of didactics of mathematics and his lecture (Rabardel, 1999) deeply influenced subsequent conceptualisations. In line with his conceptualisation of 'what an instrument is' (see Sect. 10.4), as a developing structure appropriate to a given learner, he extended the Vygotsky's notion of proximal zone, evidencing the need for a teacher to know the potential of the artefacts to be used for the targeted activities, and the stages of students' instrumental geneses, before designing the task (Fig. 17.4).

Trouche (2004), following this perspective, proposes a task aiming to introduce what is called in France 'the theorem of the compared increasing of the power and the exponential functions'. His work was tried out on a group of students each using a symbolic calculator (a TI-92). An *a priori* analysis of the constraints of the tool revealed that, both in the approximate computation mode and in the graphical mode (see Fig. 17.5), the calculator 'shows' that the equation $e^x = x^{10}$ has three solutions, but $e^x = x^{20}$ has only two solutions (while there is a third solution, as e^x will

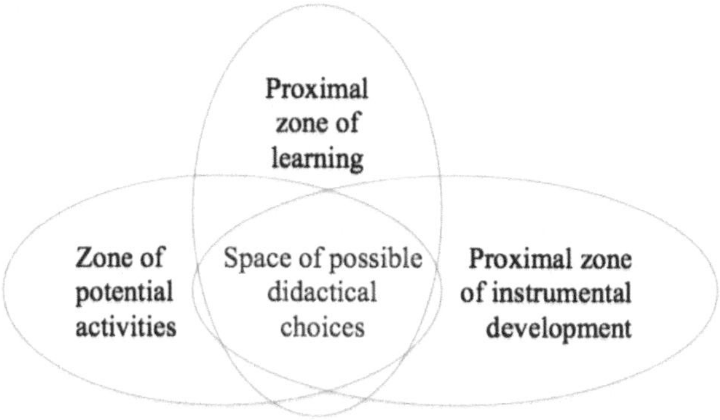

Fig. 17.4 What are the conditions bearing on the teacher's didactical choices, i.e. the task design, according to Rabardel (1999), p. 212, our translation

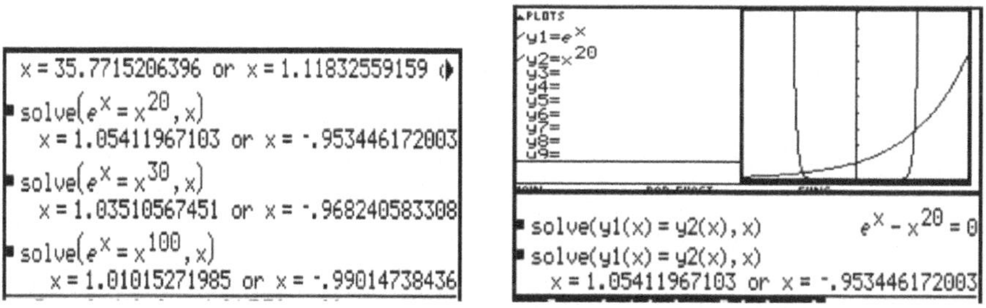

Fig. 17.5 Some screen copies of the calculator used by the students, opening some windows on the mathematical task (Trouche, 2004, p. 302)

necessarily become greater than x^{20}). In the phase of instrumental genesis that the students were in at the time, the students privileged the graphic mode to search for a function limit to solve an equation. In this phase of their learning of functions, students imagined that what will occur in the behaviour of two functions outside of the graphic window of the calculator would simply be an extension of the behaviour visible in the window; thus the students imagined that the graph of x^{10n} will 'always' be above the graph of the exponential function for integers $n \geq 2$. The task given to the students was: How many solutions has the equation $e^x = x^{10n}$? The task clearly exploits the limitations of the calculator, giving the right solution for $n = 1$, then wrong solutions for $n > 1$.

This subtle reflection on the didactic variables (in this case, the value of the integer n), depending on the didactical situation and of the available tools, is at the heart of didactical engineering that has been developed in the TDS. This didactical engineering could be defined as an experimental basis for task design. Artigue (2005) developed two cases of didactical engineering, for introducing the equivalence of algebraic expressions and for introducing the derivative and details of three levels that are used in this process:

- The *surprise level*: this plays on the effect of surprise produced by unexpected results so as to destabilise erroneous conceptions, to promote questioning, to motivate mathematical work;
- The *multiplicity level*: this plays on the potential offered by technology for producing a great number of results very quickly, so as to promote the search for regularities and invariants and to motivate mathematical work aiming to understand these;
- The *dynamic level*: this plays on the dynamic potential of graphic representations to overcome the evident limitations of paper-and-pencil work and to promote a dynamical way of approaching mathematical concepts and problems, the potential of which is now widely acknowledged (Artigue, 2005, p. 286).

We now move on to the ATD. Michèle Artigue and Jean-Baptiste Lagrange, amongst others, have worked on the implications of ATD for learning and teaching with digital tools and we focus on their work on the task-technique praxeology pair.

> Techniques help to distinguish and reorganize tasks. For instance different techniques exist for the task "find the intervals of growth of a given function" depending on what is known about the function. If the function is differentiable the task can then be related to the task "find the zeroes" of another function. In other cases, a search based on a more direct algebraic treatment can be more effective. (Lagrange, 2005, p. 116)

They view the relationship between techniques and conceptual understanding as complex and argue that digital tools do not provide a means to bypass techniques in our endeavours to construct situations which lead students to conceptual understanding (see Lagrange, 2000, for the details of this argument). Lagrange (1999) notes that: technical work does not disappear when doing mathematics with digital tools, it is transformed; novices progressively become skilled in techniques by doing, talking about, and seeing the limits of techniques; and diminishing the role of techniques encourages teachers to avoid talking about them. Artigue (2002) distinguishes between the pragmatic and epistemic values of techniques. Pragmatic values concern the range of application of a technique and epistemic values concern the role of techniques in promoting mathematical understanding. Lagrange (2005) provides an example with expressions of the form $\frac{a+b\sqrt{2}}{c+d\sqrt{2}}$, where a, b, c and d are integers. The standard technique of multiplying top and bottom by $c - d\sqrt{2}$ has pragmatic value in writing any such expression in a form that has a rational denominator and potential epistemic value in developing students' knowledge of properties of quotients and radicals. The lessons for tasks design is that designers need to take techniques seriously and consider the potential values of techniques inherent in a task.

Thomas and Lin (2013) addresses designing tasks-with-digital-tools where the techniques students employ have potential epistemic value for the students. They note that tasks such as Task 1 in Sect. 17.2 above, 'have little or no epistemic value,[7] since solving these with "black box" technology use does not assist students to focus on, or understand, the constructs of mathematics' (Thomas & Lin, 2013, p. 112) and they ask 'What should be considered when constructing tasks employing technology that have epistemic value?' (Thomas & Lin, 2013). They suggest that tasks should require 'dynamic multiple representations... and versatile interactions between representations ... integrating technological and by-hand techniques' (Thomas & Lin, 2013, p. 113). An application of this principle to Task 1 (Sect. 17.2.1) could result in the following task:

1. Show that $x^3 - x^2 + 2ax - a^2$ can be written in the form $x^3 - (x - a)^2$.
2. In *GeoGebra*, set up a slider for a and produce the graph of $y = x^3 - x^2 + 2ax - a^2$. By changing the values on your slider, investigate conditions on a, such that the resulting graphs have two local extrema.
3. Using calculus, or otherwise, explain why the conditions on a hold.

We end this subsection by noting that academic work on task design under the theoretical frameworks considered in this and the preceding subsections 'feed off' each other. Although one must 'connect theories' with care (see Prediger, Arzarello, Bosch, & Lenfant, 2008), theories do not exist without people who interpret these theories. Academics involved in task design work together on 'issues'—a case in point being this chapter, written by two academics who have 'a lot in common' but do not have identical theoretical frameworks.

[7] Earlier in the paper Thomas & Lin refer to the 'epistemic value of techniques'. Their reference here to the 'epistemic value of a task' can be taken as shorthand for the 'epistemic value of the techniques required to solve the task'.

17.3.3 New Opportunities for Task Design with Digital Tools

We present here what appears as three main opportunities for task design with digital tools: *involving students* in this design, *involving teachers* in a new way for designing textbooks, *involving users* in the improving of tasks.

Involving students in the design of task was the main objective of 1-year work of Trouche (1998), resulting in a book written with his students: '*Practicing mathematics with symbolic calculators, conjecturing and proving, 37 variations on a given problem*'. 37 variations stands for the ways the 37 students engage in a given task, contributing actually to its design. This creativity is clearly supported by the potential of the technological environment for varying the representation of mathematical objects, doing complex computations, as evidenced in Fig. 17.6.

Finally, the various students' contributions lead to the emergence of an extended possible task. What is always the case in a solving problem process is all the more true in technological enriched environment,[8] needing of course a set of didactical

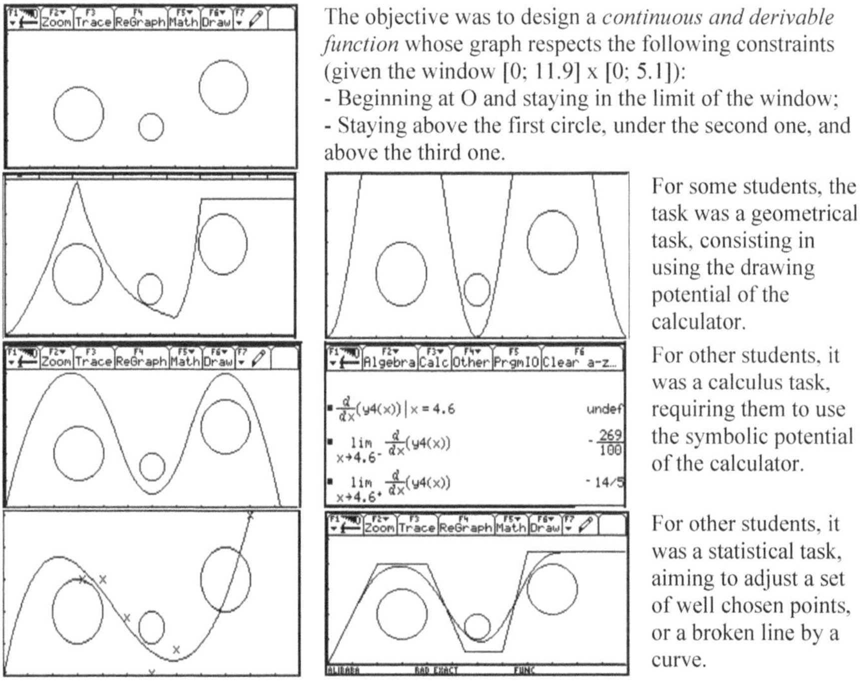

Fig. 17.6 Various calculators screen copies illustrating different students' works for solving the same task (Trouche, 1998, pp. 33–43)

[8] See the European project MC2 (http://www.mc2-project.eu), focusing on social creativity in the design of digital media intended to enhance creativity in mathematical thinking.

conditions, in terms of new orchestrations, for allowing students to develop their own creativity (Trouche, 1998).

The second opportunity involved teachers in new ways of designing textbooks. The digital environments afforded new means for connecting teachers: we will come back to this essential feature in Chap. 19. It allowed for the development of teacher associations, sharing in a first step their resources, and co-designing in a second step their resources and associated tasks, until a common construct of a whole set of resources: a textbook.

Gueudet, Pepin, and Trouche (2013) compare two modes of the design of a textbook: the first one is a 'classical one', by a small team of experts (teacher trainers and researchers); the second one is a new one by a large team of teachers gathered in an online association, drawing their expertise from their joint experiences. They underline differences at six levels:

The modes of design: more *collaborative* in the first case, more *co-operative* in the second;

The nature of the structure: the first book is a *single whole*, with an organised structure (organised by the team of experts); the second is an *atomistic system* that can be arranged differently by different users;

The organisation of the content: more *didactically original*, linked to the didactical choices of a 'homogeneous' team in the first case; more *aligned with* the institutional instructions in the second;

The content: more *open* to a variety of ways for solving a given problem in the first case; more *driven* by an expert solution in the second;

The integration into the whole grades 1–9 mathematics curriculum; links with primary school more taken into account in the first case than in the second; and

The links to the users; the textbook provided as a *final product* given to the teachers in the first case; and as a *proposal* to be enriched by teachers' contributions in the second.

These modes of design evidences, with regard to digital environments, the emergence of new modes for designing tasks, with their strength and weakness. Linking designers and users appears as new means for improving tasks, as we wish to evidence now, as an essential potential of digital environments.

Improving the quality of mathematical tasks given to students in dynamic geometry environments was one of the main goals of the *Intergeo* European project,[9] renewing the usual contract between a designer of a mathematical task and a user:

> An implicit contract binds the needs of the author and the needs of the teacher. The later needs resources of good quality and the former needs recognition for the work that was put in creating the resource.

[9] Interoperable and Interactive Geometry for Europe http://i2geo.net/xwiki/bin/view/Main/About

> Users are bound by interest, not interest for money, but interest for knowledge. The use of the platform, although provided for free, in the sense that you don't have to pay for it, is not provided without moral obligations. *Respect* for the work of others is the main pillar of our philosophy. Tokens of respect is the currency that we exchange in the intergeo project. That means that:

- If you upload an educational resource, it should have some *interesting* facets; half-baked resources are ok if it is explicitly labelled as an idea begging for improvement by others. It means as well to respect fellow teachers' opinions on your work; if they invested time reporting on your resource, please consider it in good faith as venues for improvement for your resource. If you don't act on it, or if you disagree with the opinions of users, don't be upset if others modify them whenever the license allows them to do so. The project is here to foster your resources like growing evolving organisms.
- If you use an educational resource in your classroom, you are *expected to report* on that use. We are expecting this quality report from users, in order for the project to be useful for everybody; and authors are expecting this feedback from their fellow teachers, as a sign of respect. It is especially the case for groups releasing their work. Please consider that we *value* your feedback and you should value it as well, not giving it lightly but giving it eagerly.

> Respect for good work means as well that quality reviews are not always praises. When reporting, you should assume *good faith* from the author and should be constructive in your critics but weak points should be pointed out for the author to be able to iron them out (http://i2geo.net/xwiki/bin/view/About/Quality)

The digital tools are used here at different levels (mainly dynamic geometry software and the Intergeo platform, allowing designers and users to communicate). Trgalová and Jahn (2011) analyse the process of evaluating resources via a questionnaire that the users had to fulfil, taking into account different levels of quality (ergonomics, technical, mathematical as well as didactical). They evidence both the potential of this process, and the conditions for making this potential effective:

> Importantly, the review from peers comes usually only after some delay, so the feedback provided by the review might not match the author's need for improving his/her resource. We can thus conclude that the quality assessment process has the potential to improve the resource quality, but it seems not to be fully effective. [...] These results show that the repository and the tools developed within the Intergeo project can be used efficiently in the framework of teacher training initiatives in which their appropriation by the participating teachers can be accompanied and facilitated by tutors' interventions (p. 985).

The opportunities provided by digital tools open new questions and new means for connecting communities in mathematics education (we return to this in Chap. 19).

17.4 Tasks and Digital Tools: Issues

In this final section we look at task-tool issues in larger-than-the-individual classroom research and in assessment. We also comment of avenues for further development.

We have, over the decades, been involved in various mathematics curriculum and assessment development projects that have involved the creation of new kinds of tasks (see, for example, Monaghan, Pool, Roper, & Threlfall, 2009). The consideration of tasks in such work requires careful consideration of where teachers and students are, with regard to tasks and tools, at the outset of the project and their target states at the end of the project. It would be improper to expect teachers and students to work effectively with new types of tasks with new tools in a short period of time; teachers and students need time to accommodate new tasks and tools into their daily mathematical routines and the time needed can run into years. Laborde (2002) documents the evolution of teachers' tasks with a dynamic geometry system (*Cabri*) over several years. The teachers' initial classroom tasks-with-*Cabri* were not connected with the 'traditional' tasks (prior to and running into the project work) but over the project the 'role' of *Cabri*, in the tasks teachers designed, evolved: tasks in which *Cabri* facilitates the material aspects of the task; *Cabri* as facilitating the mathematical task; tasks modified when given in *Cabri*: tasks only existing in *Cabri*.

In addition to the 'evolutionary time' required for new tasks-with-tools to 'settle in', if the project is extended beyond the initial project schools, then one should not expect the new tasks-with-tools work in the next wave of schools to mirror the development of schools/teachers/students in the original project schools. Apart from expected individual variation in patterns of work, it is often the case that less 'development work' goes into next wave schools. Extending project work beyond classrooms and/or schools involved in one stage of a project is referred to as 'scaling up'.

Roschelle, Tatar, Shechtman, and Knudsen (2008) examine scaling up in relation to Jim Kaput's *SimCalc* project work which focuses on the mathematics of change and variation. *SimCalc* is designed to engage students in tasks ('realistic' in the sense of RME) that enable students to make connections between representations and to use software to facilitate this. Teacher professional development is an integral part of the *SimCalc* project. Roschelle et al. (2008) view scaling up as more than involving a larger number of schools in the project and a central construct is 'robustness', 'the consistency of the innovation's benefits for student learning when deployed consistently to a wide variety of students, teachers, and settings' (Roschelle et al., 2008, p. 151). Robustness includes ensuring that the instrumental genesis students and teachers experience provides them with tools which enable them to tackle challenging tasks. In discussing scaling up in the state of Texas they comment:

> we realized that the existing Texas curriculum only presented students with proportionality tasks in which they were given three numbers and asked to find a fourth. If we simply taught students to use a formula with three slots instead of a formula with four slots, we would have accomplished nothing. (Roschelle et al., 2008, p. 159)

Hoyles, Noss, Vahey, and Roschelle (2013) discusses scaling up the work described in Roschelle et al. (2008) in England and focus on a unit of work, which was expected to last 8–10 h of classroom time, focused on coordinating algebraic, graphic, tabular and natural language representations of linear functions with regard to motion (position, time and speed/velocity). An independent evaluation of the project was positive. With regard to tasks Hoyles et al. (2013), p. 1066 comment that 'teacher ownership fall short of the re-design of tasks themselves to exploit the use of technology to give sense to mathematical concepts' though 15 of the 17 teachers in the project 'used their professional judgement to repackage the material by choosing to teach disparate pieces together, or to decompose one idea into many' (Hoyles et al., 2013). Teacher appropriation of tasks in scaling up projects, including teacher re-design of initial project tasks, is clearly an area for future work in the professional development of mathematics teachers. We now turn to assessment.

There are many types of assessments and divisions within assessment. We will consider: formative (feedback to students on their learning intended to enhance learning) and summative (which grades students' work at the end of a learning sequence) assessment; portfolio work and timed 'unseen' examinations; assessment by technology and assessment by people (teachers or examiners) of student work in which they use technology.

Formative assessment in the mathematics classroom will always be of students using tools and quite often it will make little difference whether these tools are digital or not. In cases where there is a difference this can be positive or negative: students working with, say, a calculator, may only record a result, arguably making formative assessment more difficult; however, students may be working with digital technology that records their digital actions and this may benefit formative assessment. The size of the digital technology can be important as it is easier for a teacher to view, and thus comment on student work on a large screen than on a small screen. Formative assessment by technology will vary with the sophistication and the pedagogic design of the technology. There is a range of software for automated feedback on student work on a computer that ranges from simply stating whether an answer is correct or not to that with a sophisticated design. Bokhove (2013) reports on an example of the latter type in tasks focused on the algebraic solution of quadratic equations. Task design is informed by Watson and Mason's (2004) dimensions of variation (considered in Sect. 17.3.1 above) and the constructivist idea that learning happens when the learner finds a task very difficult (which Bokhove refers to as a 'crisis'). Task items are of three types:

> Pre-crisis items: In the initial items students are confronted with equations they have experience with ... Crisis item: Students are then confronted with an intentional crisis ... Post-crisis items: After the crisis item students are offered help ... feedforward information (Bokhove, 2013, p. 22)

The degree of post-crisis help gradually decreases.

Students' mathematical project work (portfolio work) is only, as far as I know, marked by people. A wide range of mathematical software can be used in students' portfolio work and in statics the internet is a source of large data sets and the use of these is highly likely to enhance the authenticity of the work (and in turn brings a need to use statistical software to analyse the data sets). Lumb, Monaghan, and Mulligan (2000) report on students' portfolio work in a numeric analysis module where the use of a computer algebra system was encouraged but most of the students chose to use spreadsheets, graph plotters or graphics calculators. Lumb, Monaghan and Mulligan's example raises two immediate issues; this highlights the importance of student agency in determining the software to use in portfolio work. When portfolio work is used for summative assessment of the computer, in the form of the internet, raises problems for potential plagiarism.

Tasks which make use of digital technology by students in summative assessment raises a host of issues, many of which are problematic (see Chap. 11 for a consideration of problems perceived with the use of calculators in examinations). There is also a danger in higher level mathematics examinations that the use of symbolic calculators may make examination more difficult as procedural tasks (such as 'differentiate $\sin 2x \cos x$') which most students can do, may be removed from examinations (see Monaghan, 2000). The tasks set in a summative assessment which allows digital technology may be as before or may change. In the 1990 the *International Baccalaureate Organisation* (IBO) incorporated graphic calculators into the curriculum and assessment of the Diploma Programme. They instigated a two-stage strategy: graphic calculator allowed; graphic calculators expected. Assessment tasks in the second stage were designed to make use of the affordances that graphic calculators offered students in these items but Brown (2010), a study of graphic calculator, questions in three high-stakes examinations including IBO, did not find major changes in the examination questions due to technology. In defence of the people and organisations behind innovations that do not appear to be greatly innovative in digital technology in mathematics assessment we note that it is difficult for people who have matured mathematically with one set of tools to turn to a new set of tools. The problematic issue of examination tasks is supported by Drijvers (2009) review of the use of digital technology in national mathematics examinations in Western Europe.

Some students are sitting entire examination on computers and this is likely to increase. An interesting study in this field with regard to tasks is Threlfall, Pool, Homer, and Swinnerton (2007) which reports on a comparison of pencil and paper (P&P) with 'equivalent item' computer-based (ICT) mathematics assessment on standard test items for children aged 11 and 14 years in work sponsored by the English government assessment agency. 400 students at each age level were given 24 items (12 P&P and 12 ICT). Facilities (% correct) for each question/format were calculated. 17 of the 24 items had very similar facilities in P&P and ICT formats but seven items differed. I report on two of these, *Circles* and *Shapes*. The table below shows the age of the students and the facilities.

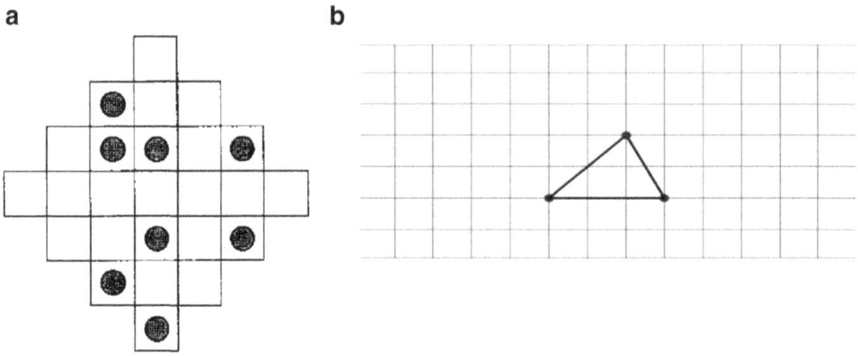

Fig. 17.7 (**a**) The pattern displayed in *circles*; (**b**) the pattern displayed in the ICT version of *Shapes*

Age of student	Question	P&P facility (%)	ICT facility (%)
11	Circles	64.4	88.1
14	Shapes	49.0	14.6

Circles states 'Here is a grid with eight circles on it'. The P&P item states 'Draw two more circles to make a symmetrical pattern'. The ICT item states 'Move the two extra circles on to the grid to make a symmetrical pattern'. The same pattern was displayed in both P&P and ICT items (see Fig. 17.7a).

In *Shapes* the P&P item presents a blank 1 cm^2 square grid paper and states 'Draw a triangle that has an area of 9 cm^2'. The ICT item presents Fig. 17.7b and states, 'Move the red dots to make a triangle that has an area of 9 cm^2'.

Threfall et al. explain response differences between the P&P and ICT formats in terms of affordances and attunements:

> On paper, the circles cannot actually be drawn until after a decision has been made about where they should go . . . The pupil needs to decide that it will look right without being able to try it . . . On computer, the pupil can put the two circles on and make a judgement by recognition—does this arrangement look symmetrical? . . . Here the affordance of the computer medium enables easier success—by recognition of symmetry . . .
>
> [in *Shapes*] It seems that the computer affordance to enable exploratory action was not as useful as might be supposed. . . . On the paper and pencil question . . . most pupils began by drawing a horizontal line, and then building a triangle up from it . . . and many pupils evaluated size by counting squares. . . . The computer version of Shapes seems to require a more analytic and strategic approach to the problem than the paper version does. On paper . . . the affordances of the medium, starting with a plausible line, then seeing what it leads to. In the absence of a similar attunement to the computer affordances . . . pupils probably had to consider the problem in terms of the formula for the area of a triangle. (Threlfall et al., 2007, p. 345)

More studies of this type would be useful for mathematics educators to assess the impact of different forms of tasks on students' responses in P&P and ICT media. We now move towards closing considerations in this chapter.

Designing tasks-with-digital-tools which enable learning which we, as mathematicians, value is a relatively new field with a great many issues and, indeed, problems; it is a field in which we have a lot to learn. Research in this area has been broad and sometimes it is good to narrow down but perhaps now is not the time to do that. Consider, for instance, the research reported in Sect. 17.3.2. There are many different frameworks represented in this research. Sometimes different frameworks taken together can be a problem but the different frameworks reported on in Sect. 17.3.2 appear to be feeding off each other and providing new insights. We are far from having 'the answer' and we want as many insights as possible. Two areas that we would like to see more research on is 'tasks which enable students to go under the bonnet' and cultural affordances.

In Sect. 3.3.5 Jon writes:

> Below the hood, Maple is optimizing polynomial computations using tools like Horner's rule, running multiple algorithms when there is no clear best choice, and switching to reduced complexity (Karatsuba or FFT-based) multiplication when accuracy so demands. Though, it would be nice if all vendors allowed as much peering under the bonnet as Maple does.

And it would be nice if students could go at least a little 'under the bonnet' with their software. This is effectively a design principle in tasks by constructionist educators. Kynigos (2007) even goes as far as building bugs into his 'half-baked microworlds' so that the user needs to go under the bonnet to rectify the bugs. Going under the bonnet is a way of viewing the mathematical functions, the mathematical relationships, embedded in the software. The opposite of going under the bonnet is just accepting what is and this seems contrary to the nature of being a mathematician who wants to know why it is. Going under the bonnet as Jon does require a great deal of experience and insight but 'how does the computer do ___' tasks could generate a great deal of mathematics whilst simultaneously leading students to a greater understanding of the tools they use to do mathematics.

The cultural affordances of software for mathematics learning is a new area opened up by Chiappini (2012) which appears important but not greatly appreciated. Mathematics has a culture. $a^2 - b^2$has a cultural significance in mathematics and there is a sense in which Maple recognises this cultural significance but a spreadsheet (with $C1 = A1 \times A1 - \mathrm{B}1 \times \mathrm{B}1$) does not. This has implication for the kinds of tasks we can ask students to do in different systems. In Maple I can design a task, as Kieran and Drijvers (2006) do, to investigate factorisations of $x^2 - 1,\ x^3 - 1,\ x^4 - 1,\ x^5 - 1, \ldots$ and get a lot of mathematics out of this task; in a spreadsheet the task is pretty pointless, you just get a lot of numbers.

References

Abboud-Blanchard, M., & Vandebrouck, F. (2012). Analysing Teachers' Practices in Technology Environments from an Activity Theoretical Approach. *International Journal for Technology in Mathematics Education, 19*(4), 159–163.

Abdul Hussain, M., Monaghan, J., & Threlfall, J. (2012). Teacher-student development in mathematics classrooms: Interrelated zones of free movement and promoted actions. *Educational Studies in Mathematics, 82*(2), 285–302.

Artigue, M. (2002). Learning mathematics in a CAS environment: The genesis of a reflection about instrumentation and the dialectics between technical and conceptual work. *International Journal of Computers for Mathematical Learning, 7*(3), 245–274.
Artigue, M. (2005). The integration of symbolic calculators into secondary education: Some lessons from didactical engineering. In D. Guin, K. Ruthven, & L. Trouche (Eds.), *The didactical challenge of symbolic calculators: Turning a computational device into a mathematical instrument.* New York: Springer.
Bokhove, C. (2013). Using crises, feedback and fading for online task design. In C. Margolinas (Ed.), *Task Design in Mathematics Education. Proceedings of ICMI Study 22.* Oxford. Retrieved from http://hal.archives-ouvertes.fr/docs/00/83/74/88/PDF/ICMI_STudy_22_proceedings_2013-FINAL_V2.pdf
Bosch, M., & Chevallard, Y. (1999). La sensibilité de l'activité mathématique aux ostensifs. Objet d'étude et problématique. *Recherches en didactique des mathématiques, 19*(1), 79–124.
Brown, R. G. (2010). Does the introduction of the graphics calculator into system-wide examinations lead to change in the types of mathematical skills tested? *Educational Studies in Mathematics, 73*(2), 181–203.
Chiappini, G. (2012). The transformation of ergonomic affordances into cultural affordances: The case of the Alnuset system. *International Journal for Technology in Mathematics Education, 19*(4), 135–140.
Drijvers, P. (2009). Tools and tests: technology in national final mathematics examinations. In C. Winslow (Ed.), *Nordic Research on Mathematics Education, Proceedings from NORMA08* (pp. 225–236). Rotterdam: Sense.
Drijvers, P., Boon, P., Doorman, M., Bokhove, C., & Tacoma, S. (2013). RME principles for designing online tasks. In C. Margolinas (Ed.) *Task Design in Mathematics Education. Proceedings of ICMI Study 22.* Oxford. Retrieved from http://hal.archives-ouvertes.fr/docs/00/83/74/88/PDF/ICMI_STudy_22_proceedings_2013-FINAL_V2.pdf
Freudenthal, H. (1973). *Mathematics as an educational task.* Dordrecht, The Netherlands: Reidel.
Gueudet, G., Pepin, B., & Trouche, L. (2013). Textbooks' Design and Digital Resources. In C. Margolinas (ed.), *Task Design in Mathematics Education* (pp. 327–337). ICMI Study 22, Oxford. Retrieved from http://hal.archives-ouvertes.fr/docs/00/83/74/88/PDF/ICMI_STudy_22_proceedings_2013-FINAL_V2.pdf
Hoyles, C., Noss, R., Vahey, P., & Roschelle, J. (2013). Cornerstone mathematics: Designing digital technology for teacher adaptation and scaling. *ZDM, The International Journal on Mathematics Education, 45*(7), 1057–1070.
Johnson, D. C. (1981). Calculator exploration for concept reinforcement. *Mathematics Teaching, 95*, 28–29.
Joubert, M. (2013). Using computers in classroom mathematical tasks: revisiting theory to develop recommendations for the design of tasks. In C. Margolinas (Ed.), *Proceedings of ICMI Study 22* (pp. 71–79). Oxford, UK: ICMI
Kaptelinin, V., & Nardi, B. (2006). *Acting with technology: Activity theory and interaction design.* Cambridge, MA: MIT Press.
Kieran, C., & Drijvers, P. (2006). The co-emergence of machine techniques, paper-and-pencil techniques, and theoretical reflection: A study of CAS use in secondary school algebra. *International Journal of Computers for Mathematical Learning, 11*(2), 205–263.
Kynigos, C. (2007). Using half-baked microworlds to challenge teacher educators' knowing. *International journal of computers for mathematical learning, 12*(2), 87–111.
Laborde, C. (2002). Integration of technology in the design of geometry tasks with Cabri-geometry. *International Journal of Computers for Mathematical Learning, 6*(3), 283–317.
Lagrange, J.-b. (1999). Complex calculators in the classroom: Theoretical and practical reflections on teaching pre-calculus. *International Journal of Computers for Mathematical Learning, 4*(1), 51–81.
Lagrange, J.-B. (2000). L'intégration d'instruments informatiques dans l'enseignement: une approche par les techniques. *Educational Studies in Mathematics, 43*(1), 1–30.

Lagrange, J.-B. (2005). Using symbolic calculators to study mathematics. In D. Guin, K. Ruthven, & L. Trouche (Eds.), *The didactical challenge of symbolic calculators: Turning a computational device into a mathematical instrument* (pp. 113–136). New York: Springer.

Lumb, S., Monaghan, J., & Mulligan, S. (2000). Issues arising when teachers make extensive Use of computer algebra. *International Journal of Computer Algebra in Mathematics Education, 7* (4), 223–240.

Margolinas, C. (Ed.) (2013). *Task Design in Mathematics Education. Proceedings of ICMI Study 22*. Oxford. Retrieved from http://hal.archives-ouvertes.fr/docs/00/83/74/88/PDF/ICMI_STudy_22_proceedings_2013-FINAL_V2.pdf

Mason, J., & Johnston-Wilder, S. (2006). *Designing and using mathematical tasks*. York, England. QED Press.

Monaghan, J. (2000). Some issues surrounding the use of algebraic calculators in traditional examinations. *International Journal of Mathematical Education in Science and Technology, 31*(3), 381–392.

Monaghan, J., Pool, P., Roper, T., & Threlfall, J. (2009). Open-start mathematics problems: An approach to assessing problem solving. *Teaching Mathematics and its Applications, 28*(1), 21–31.

Noss, R., & Hoyles, C. (1996). *Windows on mathematical meanings: Learning cultures and computers*. Springer.

Perks, P., Prestage, S., & Hewitt, D. (2002). Does the software change the maths? *Micromath, 18* (1), 28–31.

Prediger, S., Arzarello, F., Bosch, M., & Lenfant, A. (Eds.). (2008). Comparing, combining, coordinating—networking strategies for connecting theoretical approaches. *ZDM, The International Journal on Mathematics Education, 40*(2), 163–327.

Rabardel, P. (1999). Éléments pour une approche instrumentale en didactique des mathématiques. In M. Bailleul, *Actes de la dixième université d'été de didactique des mathématiques* (pp. 203–213). ARDM, Caen.

Robert, A. (2012). A Didactical Framework for Studying Students' and Teachers' Activities when Learning and Teaching Mathematics. *International Journal for Technology in Mathematics Education, 19*(4), 153–157.

Roschelle, J., Tatar, D., Shechtman, N., & Knudsen, J. (2008). The role of scaling up research in designing for and evaluating robustness. *Educational Studies in Mathematics, 68*(2), 149–170.

Sahlberg, P., & Berry, J. (2003). *Small group learning in mathematics: teachers' and pupils' ideas about groupwork in school*. Painosalama Oy.

Thomas, M. O., & Lin, C. (2013). Designing Tasks for Use With Digital Technology. *Task Design in Mathematics Education Proceedings of ICMI Study 22*, 109.

Threlfall, J., Pool, P., Homer, M., & Swinnerton, B. (2007). Implicit aspects of paper and pencil mathematics assessment that come to light through the use of the computer. *Educational Studies in Mathematics, 66*(3), 335–348.

Trgalová, J., & Jahn, A.P. (2011). Quality issue in the design and use of resources by mathematics teachers. *ZDM – The International Journal on Mathematics Education 45*, 973–986.

Trouche, L. (1998). *Faire des mathématiques avec des calculatrices symboliques, conjecturer et prouver. 37 variations sur un thème imposé*, IREM, Université Montpellier 2.

Trouche, L. (2004). Managing the complexity of human/machine interactions in computerized learning environments: guiding students' command process through instrumental orchestrations. *International Journal of Computers for Mathematical Learning, 9*, 281–307.

Warfield, V. (2006) *Invitation to Didactique*. http://www.math.washington.edu/~warfield/Inv%20to%20Did66%207-22-06.pdf (accessed 13 December 2015).

Watson, A., & Mason, J. (2004). The exercise as mathematical object: Dimensions of possible variation in practice. In *Proc. 24th Conf. of The British Society of Research in Learning Mathematics* (Vol. 2, pp. 107–112).

Watson, A., Ohtani, M., Ainley, J., Bolite-Frant, J., Doorman, M., Kieran, C., Leung, A., Margolinas, C., Sullivan, P., Thompson, D., & Yudong Yang, Y. (2013). Task Design in Mathematics Education – Introduction, *Proceedings of ICMI Study 22*, Oxford, England.

Wertsch, J.V. (1998). *Mind as action*. Oxford: Oxford University Press.

Chapter 18
Games: Artefacts in Gameplay

John Monaghan

18.1 Introduction

Games, and in particular digital games, are perceived by some to be an important part of the future of mathematics education. This chapter explores the potential of games as a resource for learning mathematics. This book focuses on tool use in mathematics, so the place of tools (or, rather, 'artefacts', as shall be discussed later) in games permeates this chapter. The chapter is in four sections: the first section considers the range of games. The second section considers artefacts in games and gameplay. The third section addresses games in mathematics education. The final section looks to possible future development.

18.2 The Range of Games

Games are not new. Dice and board games are known to have been played 5000 years ago (Schwartz, 2006). But new forms of games are developing with technology and this co-development (of games and technology) is an interest of this chapter.

Before launching into the range of games I would like to give 'my take' on two terms, 'games' and 'gameplay'. In Sect. 14.2 I said 'games have rules . . . and these rules include sequencing actions'. The first part of this statement has links to Wittgenstein's (1953) idea of characterising games via rules. I think this characterisation is insightful but it neglects ontogenesis and the origin of games in playful activity. I now turn to the second part of this statement, that rules have sequencing actions. The sequencing (which do not need to be linear) of actions according to the rules, to me, makes playing a game, at least, a proto-mathematical activity. Apart from the sequencing of rules, games have a range of mathematical features built into them. The card game 'poker', for example, requires that players recognise: five

J. Monaghan et al., *Tools and Mathematics*, Mathematics Education Library,
DOI 10.1007/978-3-319-02396-0_18

cards per person; similarity of numbers/pictures or suits; and the linear sequence of winning hands. The card game 'snap', however, merely requires that players recognise similarity of numbers/pictures. But beyond the game itself there is the gameplay, how one plays a game.[1] Poker, for example, can be played by picking cards up and placing them down at the appropriate point in the game but it can be played by estimating the probability of drawing a specific card after discarding a card in the player's hand.[2] Much of the mathematics in a game, I contend, will be in the gameplay. With this distinction between a game and gameplay made I move on to the range of games with the rider that I will, for brevity, often refer to 'games' when, according to this distinction, I should refer to 'games and gameplay'.

There are too many games to list (even by 'type of game') in an intentionally short chapter but I list some types of games. Sport has many games, most of them are competitive games (a player/team's objective is to win the game and the rules may then determine that the game is over). The artefacts used in these games may, like the soccer ball, have interesting mathematical features but there appears to be little scope for developing mathematics in the gameplay of a game like soccer though mathematics may be developed about the gameplay—the physicist Stephen Hawking used logistic regression to calculate the probability of England winning the 2014 World Cup. At the other extreme, some card, dice and board games (e.g. bridge, backgammon and chess) require significant mathematising during gameplay (at least for intermediate-and-above level players).

Another type of game in the annals of ludology (the study of games) is educational games. The ascription 'educational' to a game has nothing to do with gameplay but concerns the avowed purpose of the game. But in practice, of course, an educational game may not have the desired educational consequences and student activity with non-educational gameplay can, as we shall see, be viewed positively from the point of view of mathematics education. Educational games is a wide category that includes games to reinforce basic skills (at one extreme) and serious games (in professional training) and epistemic games (designed to get users to 'think like' a doctor or architect or . . .) at the other extreme.

Other categories of games are defined by their intended audience (e.g. children's games), their expected venue (e.g. party games), the surface they are played on (e.g. table top games) and the artefact(s) required to play the game (e.g. dice games and video games). And these categories of games branch out into a wide variety of games. Video games, for example, include digitised versions of a large number of pre-video games as well as introducing games. Video games are so called because of their display screen and are played on many electronic devices—an early (1958) game simulated tennis using two analogue control devices linked to an oscilloscope. They vary over: 'platform' (device required to play them); genre; purpose

[1] This distinction between 'game' and 'gameplay' is, perhaps, an English language distinction as the word 'jeu' in French is used for both terms.

[2] Decision making in a wide range of games is studied in the mathematical theory of games but this is not a focus in this chapter.

(ranging from amusement to training); and number of players (ranging from single player to massively multiplayer). Simulation video games is a sub-category of video games but a super-category in itself including life simulation, vehicle simulation and sport simulation—and each of these can be sub-divided according to whether the game is designed for amusement or professional purposes.

I'd like to note, however, that board games are not necessarily dying out as a result of video games—see The Guardian (2007) for a report on an apparent revival of traditional games as well as the establishment of new non-digital games such as *Settlers of Catan.*

This section was intended to 'set the scene' but what a big scene it is. It is so big that a consideration of the conditions under which gameplay may promote the acquisition or construction of mathematical concepts or skills must be addressed in specific contexts with specific games.

18.3 Artefacts and Mathematics in a Sample of Games

I now consider artefacts in games and gameplay. The primary focus on artefacts is because artefacts appear to have a wider application in gameplay than tools, though the Sect. 1.2.1 defines that an artefact becomes a tool when it is used by an agent to do something, means that the artefacts I mention often function as tools in gameplay. I start with a consideration of the types of artefacts in gameplay and then move on to issues of learning mathematics through gameplay. The types of artefacts I consider are: rules; artefacts essential to gameplay; artefacts linked to gameplay; and player constructed artefacts. There is a myriad of interrelations between these categories of artefacts.

Rules are artefacts and are the principal mediational means of gameplay. Rules, as mentioned above, are sequential though not necessarily linearly ordered. The rules may involve numbers (poker: deal five cards), spatial arrangement and movement of pieces (chess) or players (soccer) and/or logic (pontoon: if your total hand totals more than 21, then you lose). In some games the rules unfold during gameplay. An example is checkers (or draughts), when a piece is 'crowned' it has a greater number of possible moves. An extreme example is the card game Fluxx™ where the basic rules are 'pick up a card and place a card down' but cards that are picked up introduce new rules. In zero-player games, those which can be played without human agency, rules take on additional importance: the evolution of Conway's Game of Life is completely determined by the initial state of the cells and the rules; all that snakes and ladders requires of the players is that they follow the rules. When someone (a teacher or student) designs a game, it is essential that the rules are internally consistent (a move is not both allowed and forbidden) and, when the game is competitive, that there are clear criteria which determine a 'win'.

Games often require an artefact for gameplay, a special board or a computer, though, these may have minimum requirements, e.g. some card and dice games merely require a surface that is approximately flat. The boards in board games often have geometric features: squares on chess boards (and the board itself is a square); hexagons for *Settlers of Catan* (and the board itself is a hexagon). In games such as *Monopoly*, gameplay requires paying special attention to mathematical rules which come into play when a player lands on a particular cell on the board, such as calculate the rent to be paid when landing on an opponent's property with two houses on it (and this calculation requires accessing data on an artefact, the card for the property). When the artefact for gameplay is digital (a computer, a game console or mobile phone) it may simply replicate the non-digital artefact for gameplay (i.e. gameplay may take place on a virtual chess board) but the artefact in games for more than one player can often be (or provide) a player itself. This is a significant development in games and gameplay. Games have been played for over 5000 years and the players, until the arrival of digital artefacts, were human. The virtual players, through programming, have access to search strategies and memory far exceeding human players and for this reason game designers often build in difficulty levels (from easy to expert) for virtual players and this, of course, can impinge on the human player gameplay.

Artefacts can also be linked to a game or gameplay. These artefacts can be a part of the game/gameplay or about the gameplay such as books and online advice on how to play games. Table 18.1 is from a site[3] which helps on strategies for the turn-based strategy video game *Civilisation*. Table 18.1 is an extract from a part of the site dealing with terrains to cultivate in the opening moves of the game.

The purpose of the table is to help players optimise the squares they 'settle' at the start of the game. Examples of linked-artefacts that are part of the game/gameplay are betting chips/money in card games such as poker and the doubling cube in backgammon. The use of these two linked-artefacts is related to the difference between 'friendly' and 'aggressive' games of poker and backgammon. In both cases the artefacts are incorporated into the aggressive gameplay and there are special rules for their use. These two linked-artefacts are interesting with regard to the mathematics and gameplay as they can provoke a player to explicitly mathematicise their chances of winning the game and the longer term consequences of accepting a raised bid in poker or a 'doubled game' in backgammon. This is an example of what was called a 'breakdown situation' in Sect. 14.2. Linked-artefacts (e.g. joysticks) are not unusual in digital gameplay but I am not aware of instances in digital gameplay (other than digital games which imitate non-digital games) where linked-artefacts provoke breakdown situations.

There are a variety of gaming situations in which players construct artefacts during gameplay. It is not unusual for people of all ages to construct (make up) their own game and this involves constructing the rules and other artefacts with which to play the game. Some commercial games expect players to construct tools. *Minecraft* is a very popular (at the time of writing) digital sandbox survival game (a game where players construct the virtual environment in which their avatars reside). As the gameplay unfolds over virtual time it is important that the player

[3] http://www.civfanatics.com/content/civ3/strategy/cracker/civ3_starts/index.htm

Table 18.1 Three rows (from 19 rows) on 'powerful terrain squares' in *Civilisation*

Terrain type	Despot raw food	Despot raw shields	Despot raw gold	Despot raw power	Improvement	Improved despot -food	Improved despot shields	Improved despot gold	Improved despot power
Grass bonus + cattle	3[a]	2		5	Irrigate	4[a]	2	1	7
Flood plain + wheat	4[a]	0	1[b]	5	Irrigate	5[a]	0	2[b]	7
Grass bonus + wheat	3[a]	1		4	See note 1	3[a] (4[a])	2 (1)	1	6

[a]In the table above indicates that the output from the square has been reduced by 1 due to despotism
[b]In the gold columns in the table above almost any terrain square that is next to a river will gain one extra gold unit of production. Flood plains are already next to rivers so this bonus is always built into flood plains

constructs virtual tools, such as axes to cut down trees, in order to survive. The *Playground* project (considered in the next section) combines both of the above aspects of tool construction, it provides 'a computational 'world' on top of ToonTalk [a programming language] where children can find the resources and tools they need to build games' (Adamson, 2002, p. 9). Tool construction can occur during games in which tool construction is not an expectation. The board game *Settlers of Catan* has resources on hexagons which are assigned a number (from 2 to 12) and I have seen players create a table, with numbers as rows and resources as columns, to keep a record of their resources; in the next section I report on research where a boy created a tool to ensure that the virtual house he was constructing was symmetrical. A further, and common, player constructed tool is the gameplay strategy. In some games the strategy is explicitly mathematical. For example, there is a game in the 1980s *MicroSmile* series of computer games (designed for use in school mathematics lessons) in which players win by being the first to obtain a cumulative score of 100 by rolling a virtual D6 die. Players may roll the die as many times as they like but if they get a 6, then their cumulative score for the series of rolls the 6 occurs in is void. Players must play one strategy (such as 'roll the die 3 times' or 'stop when the total for a series of rolls is 12') throughout a game. The aim of the game, then, is to find a good strategy.

The die game above is an example of focusing students towards developing a strategy that is effective. Focusing students towards specific or restrictive use of tools in gameplay can be done in a number of ways. I illustrate this with a game whose origins I do not know. The context of the gameplay is a mathematics class with the focus on multiplication (or division) of a number by a number between 0 and 1. The game is to be played by pairs of students with one calculator between them. A two digits prime number is entered into the calculator. The players may only multiply by a number and in each turn the players cannot change the number that is already displayed on the calculator. The winner is the first person to get to 100 ± 0.1. The table below displays initial possible opening moves of two players:

Input	Player 1's input	Player 2's input	Output
71	$\times 1.2$		85.2
85.2		$\times 1.2$	102.24
102.24	$\times 0.98$		100.195

Digital tool use (a calculator) is an essential feature of this game but the functionality of the tool is intentionally restricted. Limiting the functionality of a tool can sometimes be effected by hiding features of the tool. *Green Globs* is a computer graphics from the 1980s (see Dugdale, 1982) where points (green globs) are marked on a Cartesian grid and the players must 'hit' the points with linear or quadratic (or whatever) graphs. The game can be played on the mathematical software *GeoGebra* (see Fig. 18.1) but a challenge for students would be to play the game without the *Line through two points* tool (which can be 'hidden').

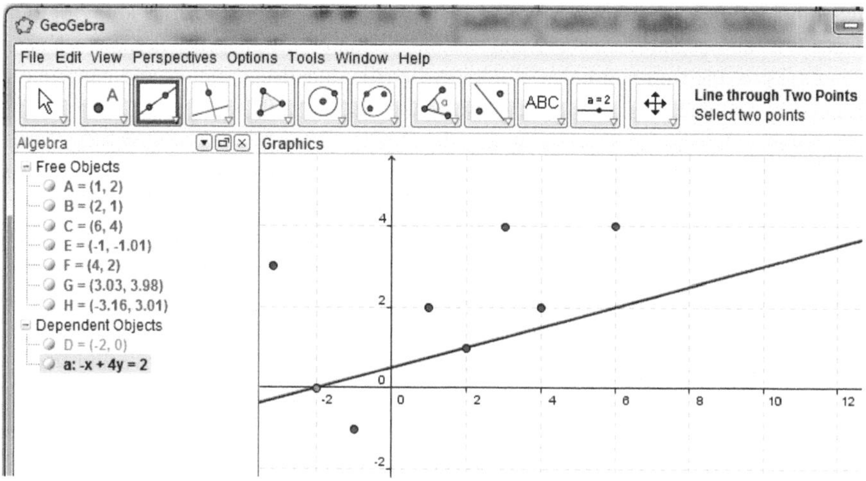

Fig. 18.1 *Green globs* on *GeoGebra*

18.4 Games in Mathematics Education

The previous section provided some examples of the use of games in school mathematics but these were provided to illustrate the artefacts in gameplay. The purpose of this section is to provide an overview of games used (or having the potential to be used) in school mathematics in order to locate the potential (and the potential problems) of games, and digital games in particular, as a resource for learning mathematics. This section is divided into two subsections for presentational reasons: the first subsection presents a selective chronological overview (including, but not exclusively, research); the second subsection presents specific research findings and considers 'understandings' that specific types of research can provide us with.

18.4.1 The Use and Potential of Games in Mathematics Education

'Educational games' in the twenty-first century is a software category but non-digital games have been used in education for some time: in nineteenth century Germany Froebel's kindergarten 'included singing, dancing, and games' (Kidwell, Ackerberg-Hastings, & Roberts, 2008, p. 142)[4]; in my experience (my own

[4] Froebel also made extensive use of mathematical artefacts: wooden cubes and 'tables marked with a grid of lines, much like graph paper; each square in the grid was the size of the face of one of the small cubes. Arranging cubes on the grid produced pleasing patterns' (Kidwell et al., 2008).

teaching and many classrooms where I have been an observer) games are a regular feature of some (and an occasional feature of many) English mathematics classrooms. I provide four examples of approaches to educational games and gameplay, related to mathematics, in the literature in the last 50 years.

Zoltan Dienes was considered in Sect. 7.3 as an influential early non-behaviourist mathematics education researcher. The first chapter of Dienes (1963) is 'On the function of play in mathematical thinking' and provides 'a number of examples to illustrate ways in which play-energy may be transformed into higher cognitive activity' (Dienes, 1963, p. 54). He considers a distinction between primary (related to instincts) and secondary play (related to planned activity) but sees this as a 'gradually changing degree of awareness, rather than of a dichotomy' (Dienes, 1963, p. 22). Secondary play can be rule-bound or manipulative and '*Rule-bound play* is essentially 'playing a game' (Dienes, 1963, p. 24). The use of rule-bound play is illustrated in a series of educational games in which artefacts are a central feature.

Brousseau's theory of didactical situations was outlined in Sect. 10.2. Brousseau designed 'didactical situations', classroom activity based on three 'situations': of action; of formulation; of validation. An early activity which he wrote a great deal about was the *Race to 20*, which is a variant of the game Nim:

> the first player writes either 1 or 2, then on successive turns players alternate writing numbers, each of which must be either 1 or 2 greater than the one previously written. The objective is to be the one who writes "20". (Warfield, 2006, p. 19)

In the situation of action the students play the game. In the situation of formulation the students develop (winning) strategies. In the situation of validation they attempt to justify their strategies mathematically. *Race to 20* is a game appropriated for educational purposes; it can, by the way, be played using artefacts such as matchsticks.

In contrast to designing or appropriating games for mathematical purposes, the Shell Centre (1987–1989) *Design a Board Game* is a resource box. This is not a game but a set of resources that enable students to play games (in mathematics lessons), reflect on their games, design their own games, and play these games. The focus of the mathematics here is not so much in the gameplay but in the design of games.

My last example, Devlin (2011), looks to the future. It is an impassioned plea from a mathematician[5] to teachers to take the mathematical learning potential of video games seriously; not as an alternative to other forms of education but as an enhancement. He draws on situated cognition (see Sect. 7.3), ideas about identity and Gee's (2003) video game learning principles to argue that video gameplay is motivating and provides challenge and reward. He does not focus on a specific area of mathematics in which video games are particularly suited (though he notes an age range, middle school mathematics) and argues that video games which involve

[5] A co-author of a book (Borwein & Devlin, 2008) discussed in Sect. 3.1 with a co-author of this book.

mathematical thinking can be designed and can provide players with mathematical challenges and rewards, 'the video game ... gives the mathematics meaning by embedding it in a real context' (Devlin, 2011, p. 175).

18.4.2 Research on the Use of Games in Mathematics Education

The purpose of this section is not to provide a comprehensive review of research on games in mathematics education but to point to the types of research and the consideration of artefacts in gameplay in this research. I begin with a few words on research on games in education and games (in general) in mathematics education before focusing on research concerned with digital games.

Research on games in education (educational games and games not designed for educational purposes) is an early twenty-first century boom industry and new academic journals[6] devoted to this field have appeared. Scholarly work in mathematics education often draws on this more general field. For example, Lowrie (2015) reviews research mathematics and visuospatial reasoning in digital gameplay and draws on both mathematics education and general research to reach the conclusion that 'Digital games appear to accommodate ... the visuospatial- reasoning skills to interpret and manage information systems than traditional classroom practices and pedagogies' (Lowrie, 2015, p. 90). However, mathematics has its own culture, and it would be naïve, for example, to assume that a cell in a game in the general shape of a square communicates any of the properties of a 'mathematical square' to game players; this is merely to caution that research on games in education should be viewed critically when regarding mathematics education.

In the field of mathematical education, research into games and learning pre-dates digital games (see, for example, Bright, Harvey, & Wheeler, 1985 as well, of course, Dienes and Brousseau who are considered above). The National Council of Teachers of Mathematics (2004) claim that mathematical games 'can foster mathematical communication...can motivate students and engage them in thinking about and applying concepts and skills' (http://www.nctm.org/fractiontrack/) refers to games in general. But research in mathematics education can, in turn, inform research on games in mathematics education. For example, Nilsson (2007) examines ways in which 12–13-year-old students handle chance in dice game situations. The students play four games with standard cubical dice but not the standard numbers: game 1—the numbers 1, 1, 1, 2, 2, 2 on each dice; game 2—the numbers 2, 2, 2, 4, 4, 4 on one die and 3, 3, 3, 5, 5, 5 on the other; game 3—the numbers 1, 1, 1, 2, 2, 2 on each die; game 4—the numbers 2, 2, 2, 2, 4, 4 on one

[6] For example, *Game Studies: the International Journal of Computer Game Research.* See http://gamestudies.org/0902/about

die and 3, 3, 3, 3, 5, 5 on the other. All games have a board for playing—a cuboid with the numbers 1–12 marked—and a set of counters. Two teams (pairs of students) compete. They are to distribute their markers on the board at the start of the game and throw the dice (in turn) and sum the total on the dice. If the total corresponds to a number with one of their counters on, then the team may remover one counter. The first team to remove all of their counters wins. In the first (respectively second) two games the students get 24 (respectively 36) counters. The research focus was to explore students' probabilistic reasoning, including their (informal) construction of the sample spaces, in each game. The result of student gameplay suggests that students: can discern impossible totals from possible totals; have difficulty in taking the order of the dice into account (e.g. (1, 2) and (2, 1) are viewed as identical); and are prompted to make false assumptions by employing school mathematics (e.g. there are three possible totals in game 1, there are 24 counters and $24 \div 3 = 8$, so put 8 counters on each of the cuboid paces marked 2, 3 and 4). Nilsson (2007) concludes that 'a competitive attitude and confirmative feedback does not always provide for probabilistic explorations' (Nilsson, 2007, p. 312) and this is a conclusion worth keeping in mind when we consider the mathematics that can/cannot be picked up in gameplay that is ostensibly mathematical.

A strand of research on games in mathematics education are quantitative studies of the effects of digital educational games on student achievement and/or attitudes and/or other constructs. Studies that I am aware of to date do not appear to be converging to a consensus of these effects. For example, with regard to attitude/motivation, Lopez-Morteo and López (2007) report on an electronic collaborative learning environment which includes games which was used by high school students and conclude that the environment 'positively affected students' attitudes towards mathematics' (Lopez-Morteo & López, 2007, p. 639). Kebritchi, Hirumi, and Bai (2010) report on the effects of software designed for algebra instruction and 'no significant differences were found in students' motivation between students who played and did not play the games' (Kebritchi et al., 2010, p. 436). This is supported by some reviews of research with regard to achievement, for example Hays (2005). This lack of consensus is not really surprising for, as Bright et al. (1985) stated of mathematical games in general, 'games can be used to teach a variety of content in a variety of instructional settings…there is no guarantee that every game will be effective' (Bright et al., 1985, p. 133). Further to this, 'it appears that assumptions that students will see the usefulness of mathematics games in classrooms are problematic' (Bragg, 2006, p. 233). With regard to the use of artefacts in games, quantitative studies (other than those, if they exist, which study the gameplay itself) are not enlightening simply because artefacts are used in gameplay and quantitative studies examine the effects of gameplay and not the gameplay itself. With this I turn to qualitative studies and present two contrasting (one where tool use was planned, the other where tool use arose from the gameplay) examples related to the use of artefacts (used as tools in these two examples) in gameplay.

In the Playground Project (Adamson, 2002; Goldstein, Kalas, Noss, & Pratt, 2001; Hoyles, Noss, Adamson, & Lowe, 2001) 6–8-year-old children (in schools but not in lesson time) played with, changed and created their own games using child-friendly (non-textual) programming languages, *Imagine* and *ToonTalk*. The programs provide screen robots with possible actions which define how objects in the games behave, e.g. a screen tiger is programmed with the command 'When joystick button 1 is pressed, I jump up 30' (Goldstein et al., 2001, p. 275). A design principle is that children can inspect and change the actions available to the robots.[7] The theoretical framework is constructionism (see Chap. 8) where 'designing, creating and debugging meaningful "external" artefacts is a source of learning' (Adamson, 2002, p. 8). I focus on the aspect of the research concerned with rules, as this has been a recurring theme in this chapter, and the research question 'How is children's understanding and expression of rules mediated by a programming language in which the rules are available for inspection and change?' (Hoyles et al., 2001, p. 170). The concept of a rule here involves three concepts of the game: the Game Formal—the rules that the children have created as computer code; the Game Outside—the game that people see; the Game Inside—the game, with rules, that exists inside the children's heads. These three games are not necessarily identical, 'An essential ability for the programming of games is to transform ideas belonging to the Game Inside into the formal rules of the Game Formal that instantiate the Game Outside' (Goldstein et al., 2001, p. 278), and this 'ability', Hoyles et al. (2001) argue, is developed through gameplay. Children may not be able to predict the consequences of the rules they have programmed but when they observe (and attempt to explain) screen actions they begin, it is conjectured, a developmental path towards this ability though this 'does not necessarily lead to a formal understanding of all the rule's implications' (Hoyles et al., 2001, p. 175) but continued 'playing with the rule and observing its implications, begins the process of appreciating the logic of the implied consequences of a new rule' (Hoyles et al., 2001 p. 176).

There are several types of artefacts/tools available to children in their work in this project: the programming language; screen objects; programmable actions that children program into the screen objects; and the formal rules that ensue from these actions. These interact with the children's developing conceptions of these rules as prototype formal systems. The research focus in this project differs from that in most (all?) quantitative studies in that the focus is the development of these basic formal systems, and not the acquisition of mathematical content or motivation. I now consider a qualitative study where specific tool use arose from the gameplay rather than by design.

Avraamidou, Monaghan, and Walker (2012) report on the game play of an 11-year-old boy, Costas, as he built virtual houses in the commercial-off-the-shelf

[7] This has links to 'peering under the bonnet, as *Maple* does' (Sect. 3.5).

life simulation game *The Sims* in his bedroom; the theoretical framework draws on Saxe's (1991) four-parameter model of emergent goals in practice (see Sect. 14.3.1). In *The Sims* the goal is determined by the player and usually involves providing a virtual family with life's necessities—a house, education, jobs, recreation, etc. The player receives a limited sum of virtual money at the start of the game but must ensure that the family has an income in order to survive. The game has three modes: *live* (watch a family grow up over virtual time), *buy* (houses and artefacts) and *build* (houses and artefacts) and I report on a *build* set of Costas' actions where he has built two trial houses and now wants to build his 'dream house' within a specific budget. *The Sims* has a number of in-built house construction tools, e.g. square tiles that determine the floor area, rectangles for walls, etc. Gee and Hayes (2010) claim that in *The Sims build* mode 'tools require one to use a good deal of geometry to get all the angles and shapes to fit perfectly together' (p. 114). I consider specific aspects surrounding tool use in Costas' third attempt at building his dream house. It should be noted that Costas has a strong desire to make the house symmetric and that he encountered problems in ensuring that his first two houses were symmetric (and these problems resulted in a loss of virtual money). Costas' goals mainly emerged through a combination of house building tools, mathematics and everyday knowledge. For example, he created the sides of the house to be an even number of *The Sims* length units long because virtual doors available were two units long and he discovered that he could not build a central door in a wall an odd number of units long. Costas wanted to make a swimming pool on the side of his 18 by 18 unit house. Costas comments:

> Since the other houses were too big when I added extra rows for the pool, I am thinking of cutting the (unwanted) cubes[8] differently this time. I think I will draw a line in the middle like I did with the cubes before, and then start cutting from left and right. (Avraamidou et al., 2012, p. 14).

The Sims does not allow one to draw a line in the middle so Costas created a two-cube (two square tiles) artefact, that he painted black and placed on the ninth and tenth unit lengths of the 18 by 18 unit house, to act as an indicator of a middle line whilst he 'cut cubes'. Avraamidou et al. (2012) argue that the creation and use of this artefact/tool is a mathematical abstraction in the sense of Hershkowitz, Schwarz, and Dreyfus (2001).

The Playground project and the report by Avraamidou et al. (2012) provide evidence that designed and emergent artefacts can foster the development of mathematical actions and thought in gameplay. However, the mathematics behind these actions and thoughts (in these two studies and, I suspect, quite generally) is not mathematics which is 'privileged' (Wertsch, 1991) in curriculum documents. This is an issue explored in the last section.

[8] 'cubes' is Costas' term for the square floor tiles available in *The Sims build* mode.

18.5 Implications for the Future

Although there is potential for mathematical learning in the gameplay of many games, there are also many problems. Although one might expect that games will motivate learners, research on student motivation is mixed—positive effects in some studies, no significant differences in other studies. Similar mixed results exist in research on student achievement through gameplay and this may simply be down to the games investigated—some games, for reasons we do not (yet) know, do promote measurable mathematical thinking in some learners, other games do not. An additional problem with regard to achievement and mathematical thinking is that mathematical thinking promoted in gameplay is often not related to the mathematics privileged by mathematics curricula. Costas' use of the two-cube artefact he constructed is (for an 11-year-old) creative but it cannot, as far as I know, be mapped onto mathematics in a curriculum document; Bourgonjon et al. (2013) argue that even when teachers recognise learning opportunities promoted by games they are reluctant to use them in their teaching if there is no explicit connection to curricula. This last statement could be phrased 'teachers do not see the curricula warrants' of games but another problematic warrant with regard to many games is the mathematical warrant of the game itself. I unpack what I have just said by returning to Brousseau, Nim and the situations of action, formulation and validation (discussed in Sect. 18.4.1). Brousseau's experimental classroom work was designed to ensure that all three situations were realised by students. The situation of validation is important to the culture of mathematics as it is in this situation that students attempt to justify their formulations mathematically. Mathematically acceptable formulations can be realised, by teacher mediation, in classrooms but in gameplay the warrant for a formulation may be (and, I believe, usually is) 'my strategy worked', which is not a mathematical warrant.

A possible avenue for future work on games as a resource for learning mathematics is the design of games that encourage reflective mathematisation of gameplay by the gamers as this may lead to mathematical warrants for a formulation, strategy or action. A focus on the artefacts/tools involved in gameplay may be useful in such designs. Examples above which approximate to this future scenario are: Brousseau's version of Nim; the use of the doubling cube in Backgammon; the use of external partially mathematicised artefacts such as that displayed in Table 18.1; funnelling students towards specific use of tools such as in the calculator multiplication game; Nilsson's dice games; the explicit focus on rules in the Game Formal/Game Outside/Game Inside in the *Playground* project.

Further development towards the design of games which may promote mathematics with explicit connections to curricula may be provided through a consideration of the affordances and constraints of games (and artefacts/tools associated with gameplay) for formulations that 'fit' with the culture of mathematics. This is related to what Turner and Turner (2002) call 'cultural affordances':

> A cultural affordance (CA) is a feature or set of features which arises from the making, using or modifying of the artefact and in doing so endowing it with the values of culture from which it arises. (Turner & Turner, 2002, p. 94).

The Turners developed this construct in their design of professional training simulation software for maritime and offshore work practices but it has been applied in the design of software for high school algebra (see Chiappini, 2012; see Sect. 9.2.2). The construct is not straightforward to apply to student use of mathematical software (see Monaghan & Mason, 2013) but holds out potential for the design of games that encourage mathematical warrants for strategies and actions in gameplay.

References

Adamson, R. (2002). Playground. *Micromath, 18*(1), 8–13.

Avraamidou, A., Monaghan, J., & Walker, A. (2012). Abstraction through game play. *Technology, Knowledge and Learning, 17*(1–2), 1–21.

Borwein, J., & Devlin, K. (2008). *The computer as crucible: An introduction to experimental mathematics*. Natick, MA: A. K. Peters.

Bourgonjon, J., De Grove, F., De Smet, C., Van Looy, J., Soetaert, R., & Valcke, M. (2013). Acceptance of game-based learning by secondary school teachers. *Computers & Education, 67*, 21–35.

Bragg, L. (2006) Students' impressions of the value of games for the learning of mathematics. In J. Novotná, H. Moraová, M. Krátká, & N. Stehlíková, (Eds.), *Proceedings of the 30th Conference of the International Group for the Psychology of Mathematics Education* (Vol. 2, pp. 217–224), Prague, Czech Republic: PME.

Bright, G. W., Harvey, J. G., & Wheeler, M. M. (1985). Learning and mathematics games. *Journal for Research in Mathematics Education*. Monograph, 1, i-189.

Chiappini, G. (2012). The transformation of ergonomic affordances into cultural affordances: The case of the Alnuset system. *International Journal for Technology in Mathematics Education, 19*(4), 135–140.

Devlin, K. (2011). *Mathematics education for a new era: Video games as a medium for learning*. Natick, MA: A. K. Peters.

Dienes, Z. P. (1963). *An experimental study of mathematics-learning*. London: Hutchinson.

Dugdale, S. (1982). Green globs: A microcomputer application for graphing of equations. *The Mathematics Teacher, 75*, 208–214.

Gee, J. P. (2003). *What video games have to teach us about learning and literacy*. New York: Palgrave Macmillan.

Gee, J. P., & Hayes, E. R. (2010). *Women and gaming: The sims and 21st century learning*. New York: Palgrave Macmillan.

Goldstein, R., Kalas, I., Noss, R., & Pratt, D. (2001). In M. Beynon, C. L. Nehaniv, & K. Dautenhahn (Eds.), *Cognitive technology: Instruments of mind* (pp. 267–281). Berlin, Germany: Springer.

Hays, R. T. (2005). *The effectiveness of instructional games: A literature review and discussion* (No. NAWCTSD-TR-2005-004). Naval Air Warfare Center Training Systems Div Orlando FL.

Hershkowitz, R., Schwarz, B., & Dreyfus, T. (2001). Abstraction in context: Epistemic actions. *Journal for Research in Mathematics Education, 32*(2), 195–222.

Hoyles, C., Noss, R., Adamson, R., & Lowe, S. (2001). Programming rules: what do children understand? In M. van den Heuvel-Panhuizen (Ed.), *Proceedings of the 25th conference if IGPME* (Vol. 3, pp. 169–176). Utrecht, The Netherlands: Freudenthal Institute.

Kebritchi, M., Hirumi, A., & Bai, H. (2010). The effects of modern mathematics computer games on mathematics achievement and class motivation. *Computers & Education, 55*(2), 427–443.
Kidwell, P. A., Ackerberg-Hastings, A., & Roberts, D. L. (2008). *Tools of American mathematics teaching, 1800–2000*. Baltimore: The John Hopkins University Press.
Lopez-Morteo, G., & López, G. (2007). Computer support for learning mathematics: A learning environment based on recreational learning objects. *Computers & Education, 48*(4), 618–641.
Lowrie, T. (2015). Digital games, mathematics and visuospatial reasoning. In T. Lowrie & R. Jorgensen (Eds.), *Digital games and mathematics learning: Potential, promises and pitfalls*. Dordrecht, The Netherlands: Springer.
Monaghan, J., & Mason, J. (2013). Exploring the notion 'cultural affordance' with regard to mathematics software. *Proceedings of the British Society for Research into Learning Mathematics*. Retrieved from http://www.bsrlm.org.uk/
National Council of Teachers of Mathematics. (2004). *Communicating about mathematics using games*. Retrieved from http://standards.nctm.org/document/eexamples/chap5/5.1/
Nilsson, P. (2007). Different ways in which students handle chance encounters in the explorative setting of a dice game. *Educational Studies in Mathematics, 66*(3), 293–315.
Saxe, G. B. (1991). *Culture and cognitive development: Studies in mathematical understanding*. Hillsdale, NJ: Laurence Erlbaum.
Schwartz, D. G. (2006). *Roll the bones: The history of gambling*. New York,: Gotham Books.
Shell Centre. (1987–1989). *Numeracy through problem solving: Design a board game, produce a quiz show, plan a trip, be a paper engineer, be a shrewd chooser*. London: Longman.
The Guardian (2007, August 27). *This time it's personal*.
Turner, P., & Turner, S. (2002). An affordance-based framework for CVE evaluation. In *People and Computers XVII—The Proceedings of the Joint HCI-UPA Conference* (pp. 89–104). London: Springer.
Warfield, V. M. (2006). *Invitation to didactique*. Retrieved from http://www.math.washington.edu/~warfield/Inv%20to%20Did66%207-22-06.pdf
Wertsch, J. V. (1998). *Mind as action*. New York: Oxford University Press.
Wittgenstein, L. (1953/2001). *Philosophical investigations*. Malden, MA: Blackwell.

Chapter 19
Connectivity in Mathematics Education: Drawing Some Lessons from the Current Experiences and Questioning the Future of the Concept

Luc Trouche

19.1 Introduction

'Connecting' is certainly the best representative verb of the Internet era. It gave matter to a lot of constructs, some grounded in research and some purely speculative, 'digital natives' being an example of the later (Helsper & Eynon, 2010). The noun coming from this verb, connectivity, is now used in a number of contexts. The wiki based open content dictionary Wiktionary proposed three meanings: the state of being connected, the ability to make a connection between two or more points in a network in a graph, and a measure of concatenated adjacency (the number of ways that points are connected to each other).

In this Chapter, I focus down on mathematics education issues grounded, as far as possible (because we have not always a sufficient hindsight) in research. In the mathematics education community, the 17th ICMI study (already evoked in Chap. 12) evidenced a strong emergence of the notion of connectivity, constituting the focus of a panel of the conference (Hoyles et al., 2010). The subject Index of this ICMI study proceedings (Hoyles & Lagrange, 2010, p. 486) reveals a number of occurrences, with different meanings: the first one is a *technological* one (the potential, for a given artefact, or an environment, for connecting people to people and/or to Internet); the second one is a *social* one (the state for people, of being connected vs. the ability to make connections) to other people and/or to Internet; the third one is a *cognitive* one (the state vs. the ability, for an individual, of connecting different mathematical representations and meanings); the fourth one is a *theoretical* one (the state, for theoretical frameworks, of being connected vs. the ability to make connections to other theoretical frameworks).

Such a dispersion of meanings is a feature of an emerging concept, and of its potential. As stated by Hoyles et al. (2010), p. 440: '[...] if and how connectivity, in whatever form, transforms mathematical practices in school is a matter of future investigation'. I will conceive this chapter with respect to the emerging situation of this concept, looking at connectivity in the thread of my own experience as a

J. Monaghan et al., *Tools and Mathematics*, Mathematics Education Library,
DOI 10.1007/978-3-319-02396-0_19

teacher and researcher in mathematics education, over the last (at least) 30 years. It seems to me that.

The first section looks at connectivity through the evolution of students' connections to other students, and to mathematics, over their classroom activities. The second section considers connectivity through the evolution of teachers' connections to other teachers, and to mathematics resources, over their *documentation* work (Chap. 15). The third section, back to the ICMI study connectivity panel, questions the notion of cognitive connectivity. The conclusion discusses the dynamics of the concept of connectivity itself.

19.2 Connecting Students and Mathematics Through Digital Artefacts

I present in this section three environments I successively work with, bringing out the strong evolution of students activity according to the available *connecting tools*, keeping in mind that an environment for mathematics learning is not only constituted by sole tools, but also by mathematical problems and teacher' *instrumental orchestrations* (Sect. 15.2.5).

19.2.1 The Sherpa Student Configuration

I could say that the environment based on the sherpa-student configuration (Fig. 19.1) has grounded my reflection about the teacher's role in computerised environments. As a teacher (around 1990), confronted to the usages by students of more and more powerful calculators (see Chap. 13), I wanted to make communicate the small screens, i.e. to go against a spontaneous tendency of each student to keep for him what he was doing, and thinking, with his own calculator. The opportunity for doing that was offered by the calculator manufacturers, providing teachers with a 'view-screen', that is: a tablet with a transparent screen, and a *short cable* connecting it to a calculator (see Fig. 19.1). This device, posed on an overhead projector, allows the calculator screen to be displayed on the classroom whiteboard, or a screen, and then to be visible by the whole classroom. I underline the expression 'short cable', because it was quite obvious, for the manufacturer, that the calculator at stake was the teacher's one: this device was intended to allow the teacher to project *his own calculator* on the screen (it was then in this way that the advertising pictures demonstrates its use). My idea was, instead of connecting to the view-screen *my* calculator, to connect one of my students' calculators.[1] I name

[1] To be noticed, shortly after its first appearance, the cable at stake became longer, allowing a wider use of the view-screen: the material evolves for fitting the usages...

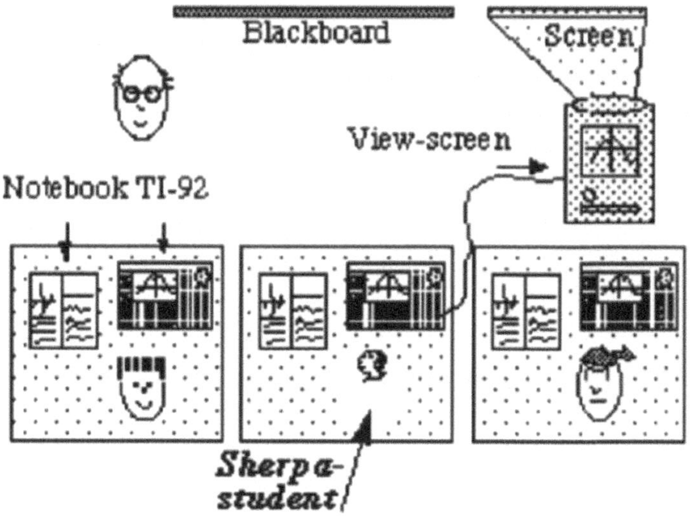

Fig. 19.1 The didactical configuration of the Sherpa-student (Trouche, 2004, p. 298)

him/her the sherpa-student, referring to the person who guides and who carries the load during expeditions in the Himalaya Mountains. My intention was then to underline the *responsibility* of this student helping the class to find its way towards the solution of a given problem, and the *difficulty* of the task.

I presented (Trouche, 2004, p. 299) various exploitation modes of such a configuration, and their possible consequences, in terms of *instrumentation* and *instrumentalisation* processes (Sect. 10.4):

- Sometimes calculators are turned off (and so is the overhead projector): it is then a matter of paper/pencil environment work.
- Sometimes both calculators and overhead projector are turned on and work is strictly guided by the sherpa-student under the supervision of the teacher (students are supposed to have exactly the same thing on their calculator screens as is on the projector screen). Instrumentation and instrumentalisation processes are then strongly constrained.
- Sometimes calculators are on as well as the overhead projector and work is free for a given time. Instrumentation and instrumentalisation processes are then relatively constrained (by the type of activities and by referring to the sherpa student's calculator which remains visible on the big screen).
- Sometimes calculators are on and the projector is off. Instrumentation and instrumentalisation processes are then only weakly constrained.

> These various modes seems to illustrate what Healy (2002) termed filling out and filling in, in the course of classroom social interaction:—when the sherpa-student's initiative is free, it is possible for mathematically significant issues to arise out of the student's own constructive efforts (this is a filling out approach);—when the sherpa-student is guided by the teacher, it is possible for mathematically significant issues to become appropriated during the student's own constructive efforts (filling in approach).

Finally, the usage of such a configuration (Trouche, 2004) evidences that the sole connection of one student's calculator to the common classroom screen, monitored by the teacher, contributes to foster the interactions between the students' instruments and their mathematical thinking. It was, for me, a first occasion of analysing the strong potential of connectivity, understood actually in the three first meanings of this word (see Introduction below): opening opportunities of connections between one student and the whole class through a technological device; opening opportunities of connections between students through the common classroom screen; and opening opportunities of connections between different mathematical representations and meanings, each student having to combine what appears on her calculators screen, and what appears on the classroom screen.

19.2.2 The Calculators Network Configuration

The second occasion for meeting connectivity happens 10 years after (around 2000), when, as a researcher, I analysed the potential of a new device, TI-Navigator, providing wireless communications between students' graphic calculators and the teacher's personal computer (Fig. 19.2). This device consists in an amplification of the view-screen potential, as it allows the teacher to see, on her screen, all the students' screens; she can then decide to connect a given calculator (or some of them), throughout her own computer, to the classroom screen.[2] It leads to a new organisation of the classroom workspace. A manufacturer advertising (Fig. 19.2, left) proposed a configuration attached to a technical constraint: the wireless connection works between hubs, each of them linking four calculators, and the teacher's computer. Then, a natural decision is to split the class into groups of four students. The team of teachers I observed (Hoyles et al., 2010, p. 449) decided

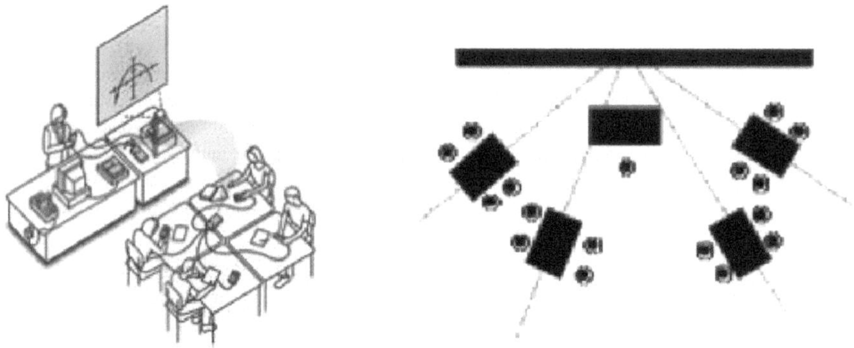

Fig. 19.2 New supports for connecting students' calculators and classroom screen (Hoyles et al., 2010, p. 449)

[2] Actually, in the context of the sherpa-student configuration, I used also to change, during a given mathematical activity, the student playing this role, but, for doing this, I had to plug the cable in another calculator, or to exchange the places occupied by two students. Not so easy to do on the fly.

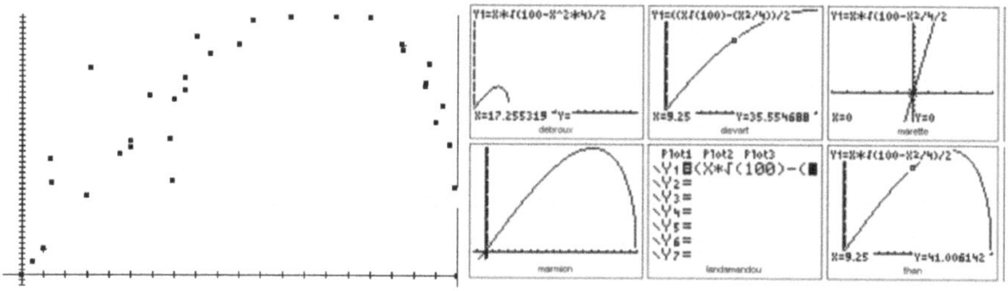

Fig. 19.3 Two main configurations for the TI-Navigator configuration, examples (Hoyles et al., 2010, pp. 447–448)

to slightly adapt this organisation (Fig. 19.2, right), aiming to structure the discussion not only between each students group and the teacher, but between the different groups through the common screen; it appears that the mathematical activity was very sensitive to such an adaptation, that fostered the mathematical discussions in the whole classroom. Actually, the drawing (Fig. 19.1, right) looks like the spatial organisation of an orchestra, evidencing the role of the teacher orchestrating the mathematical situation at stake, taking care of all the students' instruments.

As for the view-screen (Sect. 19.2.1), the technical device does not give matter for a sole teacher's mode of use (the spatial organisation of students, above, constitutes already a strong didactical choice). Actually, Ti-Navigator allows the teacher to use two main configurations: the *common coordinate system* configuration: displaying all of the pupils' data, for example, points or curves, in a single coordinate system (Fig. 19.3, left); the *screen mosaic* configuration: displaying, on the class screen, all (or some) of the pupils' calculator screens in quasi-real time (Fig. 19.3, right).[3]

These two configurations have the common property of connecting all (or some of) the students' calculators to a *common workspace*, situated on the class screen. The orchestration of such a device remains then in the teacher's hands, having to choose the relevant configuration corresponding to her didactical choice, and to select students' calculators to be 'published'.

The mathematical problem giving matter to these screens was (Hoyles et al., 2010, pp. 447–448): an isosceles triangle ABC has two sides AB and AC measuring 4 cm. What is its area? The students tried various values for the third slide BC, drawing the corresponding triangles, measuring their height, and computing their area. Then they send, via their calculators and the hub, the couples (length of BC; area of ABC) to the common screen, obtaining, as a collective result, a cloud of points (Fig. 19.3, left). Then they tried to model this phenomenon with a function,

[3] This application comes actually from the development, by Uri Wilenski, of the HubNet module, see Sect. 19.4.1.

obtaining different curves that the teacher decided to display on the common screen (Fig. 19.3, right).

In Hoyles et al. (2010), p. 447, I draw some lessons of a long term use of TI-Navigator by experienced teachers:

> The work with the TI Navigator was found to foster an emergent real *community of practice* (Wenger, 1998) in the classroom in which we could distinguish three fundamental aspects, *participation*, *reification*, and the existence of shared resources, whose major elements are summarised below:
>
> - Participation with the engagement of students in the mathematical activity and debate.
> - Reification with the collaborative creation of mathematical objects (a good example being the collective creation of the graph of a function that gradually becomes an easily identifiable object) (Fig. 19.3, right).
> - Shared resources most notably the public shared board, which is a place where every student can show her/his mathematical creation. Each student is confronted to her/his production and those of other pupils.
> - In traditional classrooms, speech or writing (when asked or allowed by the teacher) directly on the board are the ways students can express themselves and share with others. With TI Navigator, the situation is very different, for two main reasons:
> - A new interactivity was fostered between the artefact and the student, and between students themselves: students conveys their messages through the artefact; the artefact acts on the students enabling them to extract themselves from their productions thus freeing themselves to become more easily involved in peer exchanges. Thus the common space became a space of debate and exchange that aimed to elaborate a social 'mathematical truth'.
> - Each student becomes detached from his/her production as a distance is created between student and the expression of her/his creation; this distance seemed to improve the reflection on practice. The student became involved in the class activity in a different way as the tool maintained this distance between a student and the results s/he proposed to the class and to the teacher.

As in the case of the sherpa-student configuration, we can notice here the three aspects of connectivity (technological, social and cognitive), with, clearly, didactical difficulties added for orchestrating mathematical activities in such environments: the teacher has to simultaneously manage *all the students calculator screens*, and to take relevant decisions on the fly. She has then to have a deep understanding of the didactical variables of the situation, in order to play on them, according to the dynamics of the classroom activity. Hoyles, Noss, and Kent (2004) give a good description of such a teacher's expertise, based on the collaborative work of a teachers' team. I will go back, regarding connectivity, to teachers work in Sect. 19.3.

19.2.3 Internet as a Connectivity Multiplier

The third occasion for meeting 'connectivity for students' happens 10 years after (around 2010) when I was in charge of a *e-culture teaching unit* for students (third university year) aiming to become mathematic teachers. The name itself of this unit

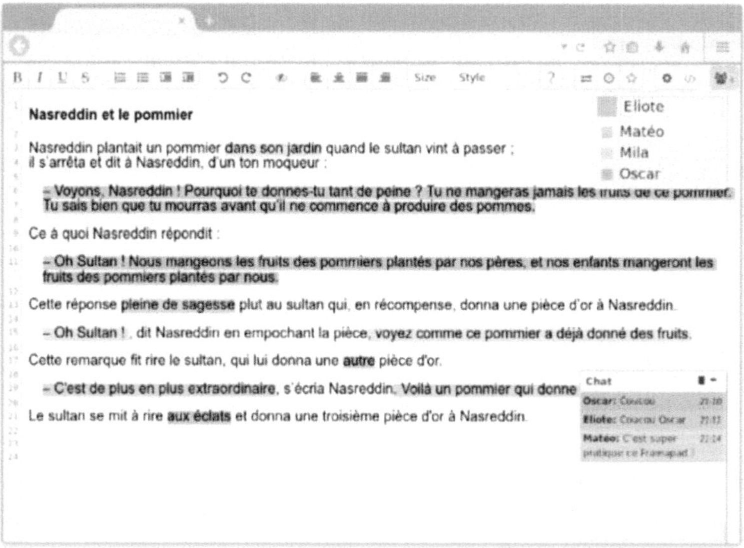

Fig. 19.4 An interface allowing students to share their ideas (screen copy of a designer's advertising)

(e-culture) indicates the major change happening at this time: the integration of Internet for teaching. During the whole sessions, students were connected to Internet and were free to use online software (mainly Geogebra). I integrated also an application[4] providing online collaborative sheets named Pad (Fig. 19.4), allowing students to discuss together (each contributor being identified by a colour, and students using also a chat for commenting their current work). It is up to the teacher: to organise such a sheet for the whole class, or for pairs of students; to be part, or not, of the discussion. Obviously, in such a context, all the tools are not under the teacher's control, as the students can use their own tools for communicating between them… or with somebody else outside!

In this period, I welcomed a Mexican PhD student, aiming to analyse students' work and associated orchestrations in the Internet era (Betancourt, 2014). It was for me a good opportunity to look at connectivity through the eyes of an advanced student.

Betancourt's thesis is related to 'learning of linear algebra supported by digital resources'. In his work (Betancourt, 2014), he related a practical work he proposed, in the context of this e-culture teaching unit, to the students working by pairs: students working together, intentionally, did not seat next to each other, then they had to use the Internet facilities to communicate. The mathematical problem at

[4] Framapad: https://framapad.org

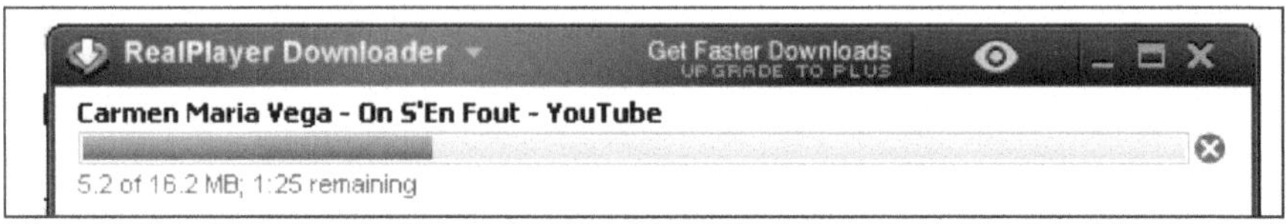

Fig. 19.5 Interface for following the video downloading (Betancourt, 2014, p. 82)

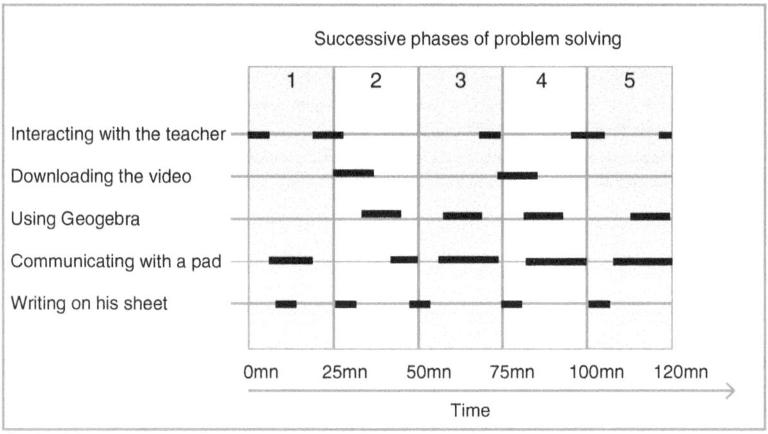

Fig. 19.6 Structure of a student's activity and tools used over 2 h (adapted from Betancourt, 2014, p. 105)

stake consists in modelling the process of a video downloading through Internet. The two questions asked to the students, using an interface for analysing the downloading process (Fig. 19.5), were: is the downloading velocity constant? How, according to you, does the application compute, at any moment, the remaining time for achieving the downloading (for example, Fig. 19.5, the remaining time to download 16.2 Mo is 1.25 s).

The activity proposed to students (Betancourt 2014, p. 143) was divided in five phases (Fig. 19.6): the teacher introducing the problem and the tools to be used; each student downloading the video, getting data and integrating them in Geogebra; discussing within the pair and trying to model the process; downloading again the video for checking the model; final discussion and conclusion.

I cannot, in the frame of this chapter, analyse the content of the students' activity, but I would like to underline some elements of structure over these five phases, following the activity of a given student (Fig. 19.6):

- Phase 1: he lessens to the teacher, then interacted, using the pad, with his pair colleague, expressing some doubts about the constant velocity of the downloading process. At the end of this phase, he interacted with the teacher (Betancourt, 2014, p. 121):

The student S: In my opinion, the velocity of the downloading process is not constant.
The teacher T: Perhaps, but how could you justify this opinion for your colleagues?
S: The downloading velocity, at a given moment, depends on the number of persons downloading the video at the same moment.
T: Perhaps, but how can you evidence this using the data you will pick up from the analysis of the real downloading of the video?

- Phase 2 is dedicated to the analysis of the downloading process and the use of Geogebra for displaying the data.
- During the following phases, the pad and Geogebra seem to be the essential supports of the student's reflection, combining individual mathematics manipulations using the dynamic geometry software, and collective mathematical discussion using the online writing tool.

The use of the Internet, compared to the calculators network configuration, clearly changes the connectivity regulation. The students' interactions are not monitored via the teacher's computer and displayed on the common workspace, but the students' pairs freely organised their work: on their own screen, it is up to them to manage the part of their working *space* dedicated to Geogebra, and the part dedicated to the Pad. With respect to the phases of the mathematics activity orchestrated by the teacher, the two students can negotiate the organisation of their working *time*, and eventually split their work in two parts, one for each of them.

The question at stake—studying the behaviour of Internet through the velocity of a downloading process—illustrates in some way the metamorphosis of the mathematics learning landscape due to the emergence of Internet: Internet appears as a *multiplier of the teacher's orchestration choices* (he can organise students in groups of two, three, or more; with students face-to-face, or at a distance; he can use a common class working space, for showing to the whole class the work of a group, or several groups of students... Internet appears as *rebalancing teacher and students' responsibilities* towards the progress of knowledge in the classroom. Last but not least, Internet appears as a *connectivity multiplier*, opening opportunities for connecting students to students (in this example via a Pad) and for connecting students to Internet resources (in this example Geogebra).

Actually, as underlined Betancourt (2010, p. 127), Geogebra was not the sole resource to be exploited. Students tried also, using their browser, to get direct answers to the problem at stake, with, as I noticed myself, keywords extracted from the teacher's question, as 'modelling downloading process velocity', aiming to find a direct answer... But, doing that, in this case, they did not succeed to find relevant resources.

The Betancourt's experience, with the associated artefacts (essentially a Pad and a dynamic geometry software), could induce the idea of a double connectivity level: for manipulating mathematics objects, students work individually with a given software; for discussing their methods and results, students work collectively.

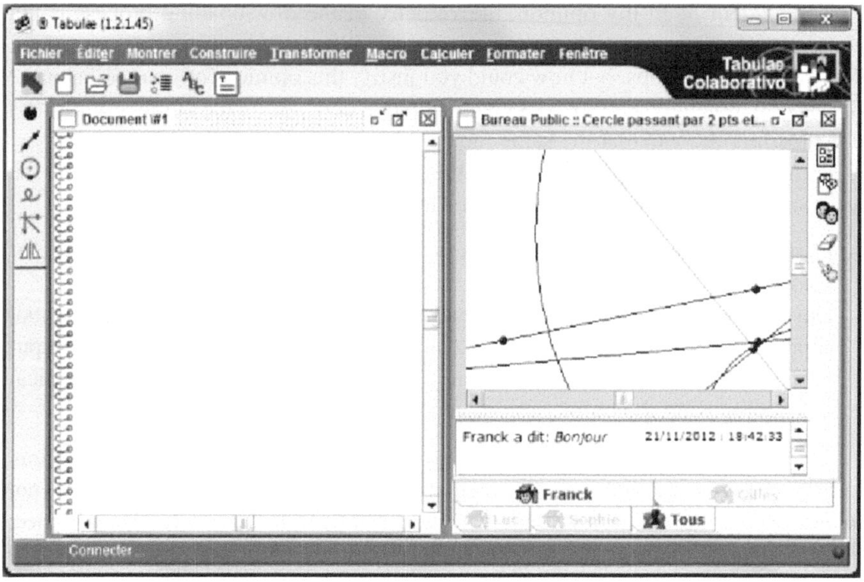

Fig. 19.7 An interface combining individual and collective geometrical work (Bellemain, 2014, p. 31)

What happens effectively in this particular experience is due to the artefacts available for the students. It could work differently with other artefacts. It was the case, in the frame of the same e-culture teaching unit, sometimes after this first experiment, when an invited researcher, Franck Bellemain (2014), proposed a new activity based on a new collaborative dynamic geometry software, Tabulae.[5] This environment (Fig. 19.7) proposes a window combining a space (left) for the individual student's work, and a space (right) for designing collaboratively a mathematical figure and exchanging comments. In such an environment, both mathematical work and associated discussion can be done collaboratively.

In this section, from the sherpa-student configuration to the calculator network configuration, and then to the Internet universe, I evidenced, over 20 years, the emergence of *connectivity* at technological, social and cognitive levels, as a major *potential* factor for renewing students' mathematical activity. This connectivity implies an added complexity for the teacher, that has to conceive and manage orchestrations making profit of these new opportunities. In which way connectivity could also benefit to teachers work? This is the purpose of the following section.

[5] Developed by Luiz Carlos Guimarães at the LIMC laboratory (Laboratório de Pesquisa e Desenvolvimento em Ensino de Matemática e Ciências, http://www.limc.ufrj.br/site/limc_olaboratorio.html) in the Federal University of Rio de Janeiro.

19.3 Connecting Teachers and Mathematics Teaching Knowledge Through Internet Resources

In this section, I follow again the thread of my own experience for analysing the potential and real effect of connectivity for 'teachers working with teachers'. For this purpose, I choose three entries: the experience, from 2000 to 2006, of a teacher training organisation, the SFoDEM; the experience, from 2001, of a teacher online organisation; and the recent experience (2014) of a MOOC aiming to develop the usage of tools in mathematics teaching.

19.3.1 The SFoDEM, Monitoring Teachers for Collaboratively Design Teaching Resources

The SFoDEM[6] was developed in the region of Montpellier, France, from 2000 to 2006, by the local Institute of Research on Mathematics Teaching (IREM, http://www.irem.univ-montp2.fr). The considerations for designing such on organisation were that, in spite of many institutional actions and the enthusiasm of pioneering teachers, in spite of the rapid evolution of technological tools and equipment, integration of ICT into mathematics teaching was slowly increasing in France. Guin and Trouche (2005), pp. 1023–1024 explained the SFoDEM objective, and described its organisation:

> [...] the main objective of SFoDEM was to provide a continuous support for teachers in the *conception, appropriation and experimentation* of pedagogical resources to get over the crucial transition to the pedagogical act. This requires a collaboration to be built between teachers with different teaching experiences aiming to support their day-to-day practice. Various themes were chosen (transition from numerical to algebraic setting and ICT; graphic and symbolic calculators; experiments of teaching sequences towards dynamic geometric diagrams; simulation of random experiences; and cooperative problem solving via Internet) to find invariants in distance training viable beyond the organization and these studied themes [...]
>
> SFoDEM is piloted by a leadership team of three researchers and its platform is managed by an administrator. About 15 trainers are involved in the training network and every year since September 2000, about 100 teachers volunteer to participate in this project. The *training committee* (composed of the leadership team, the administrator and the network of trainers) manages the *coordination* of the five themes: first experiments on distance teaching have pointed out the necessity of compensating distance with an established structured and controlled organization and showed the crucial role of planning and *regulation* [...]. The organization alternates face-to-face meetings and distance periods (the trainers of each theme have a face-to-face meeting each week, the training committee each month, and each theme—trainers and trainees—meets four times a year).

[6] SFoDEM stands for *Suivi de Formation à Distance des Enseignants de Mathématiques*, what could be translated by « Distant follow-up of Mathematics Teachers Training »

The five themes reveal the feature of this period, a transition one between the calculator era and the Internet era. In some sense, I could say that the SFoDEM rested on *social connectivity*, *teaching and training connectivity*, and *mathematics connectivity* to develop the integration of ICT in mathematics teaching. By social connectivity I mean the efforts made for connecting: teachers with trainers on a continuous way (both face-to-face and at distance); the leadership team; and the training committee. By teaching and training connectivity, I mean the efforts made for connecting the day-to-day teaching practice and the training one, the training consisting in designing pedagogical resources to be experimented in each trainees' classroom. By mathematics connectivity, I mean the efforts made for connecting different mathematical fields (calculus, algebra, geometry and statistics) and different artefacts (calculators, dynamic geometry software, Internet) to find invariants of a training organisation aiming to foster teachers' use of ICT.

The SFoDEM objective was quite ambitious, justifying its long time duration. Its pilots draw some main lessons in a CDRom (Guin, Joab, & Trouche, 2006), organised in two parts: a *design path*, and a *library of pedagogical resources*:

- The design path organised in five steps untitled '*Exploring*', '*Defining*', '*Thinking*', '*Exchanging*', '*Revising*', evidenced the central place of Internet for supporting the collaborative design of resources. For example, the first step, 'Exploring', consists, before beginning a new design, in (Fig. 19.2): visiting the main existing repositories, particularly the IREM one and the *Mathenpoche* one (see Sect. 19.3.2); reading already published reports of designing/using resources; searching with a browser and relevant keywords existing resources able to inspire a new design. The four following steps ('*Defining*', '*Thinking*', '*Exchanging*', '*Revising*') needed the use of an online platform dedicated to the interactions between the members of the project. Finally the achievement of the design path leaded to the development of a *technological connectivity* (Fig. 19.8).
- The library of pedagogical resources evidences the importance of a *common model of pedagogical resources* for facilitating both the *design*, the *exchange* and the *appropriation* of a given resource. This common model was composed,

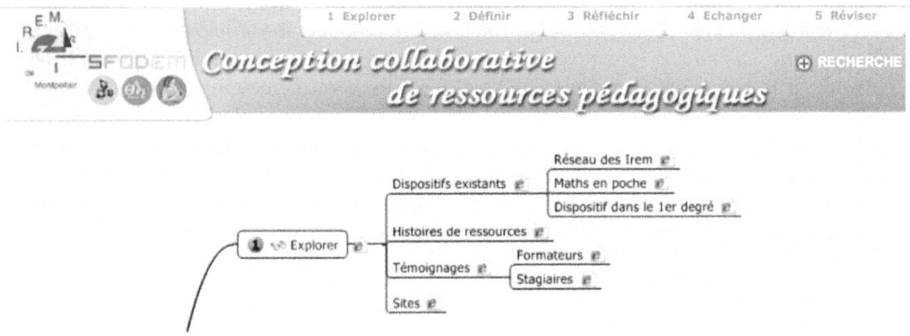

Fig. 19.8 The first step, 'Exploring' of the design path (Guin, Joab, & Trouche, 2006)

at the end of the SFoDEM experiment, of: an identification sheet (including metadata aiming to situate the resource in a larger repository), a student sheet (explaining the mathematical task at stake), a teacher sheet (underlining didactical challenges), scenarios of use and usage reports (enriched over the successive implementation of the resource), traces of students work (evidencing some critical points), a technical sheet (supporting the implementation of the resource in different technological environments) and a CV ('curriculum vitae' of the resource, tracing the main step of its evolution). Finally, the achievement of the library of pedagogical resources leaded to what I could name a *documentation connectivity* (documentation seized in the sense introduced Sect. 15.3.2): the documentation connectivity of a given resource should be defined as its potential for connecting it to different possible usages and associated traces, to different technological possible environments, to different didactical difficulties, and for relying it to its own genesis (where does the resource come from?) and to its different designers.

Such a technological and a documentation connectivity do not develop on a continuous way over the whole life on SFoDEM. Guin and Trouche (2005), p. 2024 underline some major difficulties:

- From a technological point of view: 'this organisation has rapidly revealed that schools equipment [in terms of Internet access] is frequently *inadequate* or *inaccessible*'.
- From the trainers point of view: 'usual trainers' strategies were essentially based on *imitation* strategies where trainees were asked to take the position of a student'.
- From the process of design itself:

> Moreover, initial resources provided by trainers, often expert resources, were too complex for an experimentation by trainees in their own class. Then, there was an evolution towards simpler resources, easier to implement and towards *virtual workshops* of trainees creating resources from initial ideas, named 'germs of resources'. This evolution may be considered as an evolution from a top-down approach towards a bottom-up approach.

Finally, I retain, from the SFoDEM experience, three major lessons: obtaining significant results in terms of integrating ICT in classroom practices needs a strong organisation mobilising over the time researchers and trainers; in this process, social connectivity, technological connectivity, and documentation connectivity seem to jointly develop (other examples can be found in Gueudet & Trouche, 2011); the development of both technological and social connectivity seems to rebalance the responsibilities of trainers and trainees with respect to the design of resources (see the virtual workshops of trainees), recalling the phenomena arising in connected classrooms (Sect. 19.2.3).

Some difficulties encountered seem to be linked to *a period of transition* characterised both by the *emergence of Internet* (just beginning to be a tool available in schools) and the *emergence of online communities*, not so easy among teachers The following section proposes another case study of an online community developing in the same period, but without any institutional support.

19.3.2 Sésamath, Teachers Connecting Teachers

Sésamath is a French association created in 2001. It gathers in-service mathematics teachers, aiming to 'freely distribute resources for mathematics teaching'. Its website front page (http://www.sesamath.net/) claims '*mathematics for everybody*', '*working together, supporting one another, communicating*!'. Its growth has been quite rapid: today, it gathers 100 subscribers, 5000 teachers participating in various projects, and its website proposes 45,000 digital resources for mathematics teaching and welcomes about one million visits, each month. One reason for this growth could be the existence of the French network of IREM, which has, in some sense, paved the way since 1970 (see Chap. 10). But the essential reason seems to be the way this association benefit from the development of Internet and adapt its way of functioning to this development.

The development of Sesamath follows a model (Fig. 19.9) evidenced, in the same period, by other online teachers associations (Gueudet & Trouche, 2012):

- A first group of teachers gathers, for sharing, essentially via Internet, resources.
- Then this founding group, I will call it the kernel, engages in a cooperative work (it is generally the moment of the formal creation of the association), sharing not only resources, but the work for designing them; doing that, it attracts a crown of teachers interested in making profit of these resources, some of them proposing their own resources for the benefit of the whole group.
- At last, the founding group deepens its cooperation for thinking together the whole process of designing the resources and developing the association, moving towards a real community of practice (Wenger, 1998); doing that, it attracts successive crowns of teachers, more or less close to the kernel, according to their engagement in the community project.

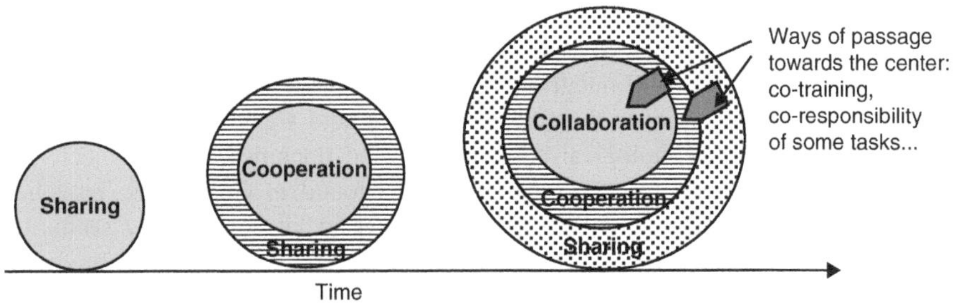

Fig. 19.9 Development of an online teachers community designing and sharing resources (Gueudet & Trouche, 2012, p. 311)

These crowns are not tight: the growing of the 'rolling stone' supposes, for the collaborating kernel, to carefully think ways of passage from the distant crowns towards the centre (Fig. 19.9).

In 2015, the current president of the Sesamath association, Hélène Gringoz, summarised, during a meeting of a research project,[7] the genesis of her association

> The association was established on October 31st 2001, [...]. At the very beginning, it's ten Mathematics teachers who were very, very fan of technology in general, for example calculator, or overhead projector. That was 15 years ago, so the computer was absolutely not as developed as today. And these teachers created websites, and they created resources they put on these websites for their own teaching. [...] They were actually teachers who met because they create resources they missed for their own teaching, they create them together and then put them available to all teachers. [...] For 15 years, it is this spirit that will prevail: the creation of collaborative resources made available to all, it is really the foundation of our association.
>
> The creation of resources took different forms. The best known is Mathenpoche[8]: i.e. the creation, in two years, of a set of interactive exercises that covered the range of teaching level from 6th to 9th grade (the French middle school). And the first printed edition of these exercises occurred in 2002 [...] And since, it works well, it was decided in 2005 to publish the first textbook, for the 7th grade [...] We were seen as precursors, as people a bit wacky, quite innovative but not really serious [...] This situation changed in 2005, since our textbook covered 15 % of the market.[9] And so, it became credible, since we were followed by a number of teachers [...] Today, 15 years after its creation, Sésamath hosts 45,000 resources, addressing all the teaching levels from 1st grade to the University.
>
> At the beginning, we had to face the distribution of resources, it was very complicated, Internet was not working very well. We were just teachers, and therefore, we trained each other so that the distribution of resources goes as well as possible. [...] This led us to create tools as mail servers, list servers, and a collaborative interface. When we began to write in 2003–2004, downloading a file was very heavy and the speed was very low. So, we created in 2003 an interface that can store files and send links automatically via emails, in order to avoid downloading them each time [...].
>
> As the basis of our association is the distribution of resources, gradually came the idea that all our online resources should be free [...].
>
> Our development allows us to propose, to all the mathematics teachers, a new interface, Labomep,[10] a mathematics laboratory where teachers can appropriate Sésamath resources,

---

[7] It was the ReVEA project ('Living resources for learning and teaching', www.anr-revea.fr). The whole interview (audio) is available on the page presenting the ReVEA meeting http://ife.ens-lyon.fr/ife/recherche/groupes-de-travail/revea-collectif. The translation has been made by the author of this chapter.

[8] http://mathenpoche.sesamath.net. The English translation of Mathenpoche should be « Maths in the pocket »

[9] To be noticed: the online version of the Sésamath textbook are, from the beginning, free. Their printed versions are quite cheap (half the price of an 'ordinary' textbook), as the Sésamath authors do not get royalties for their work. The royalties, as low as possible, go to the association, for allowing it to hire the technicians necessary to develop its digital environment.

[10] Labomep (http://www.labomep.net/fiches/fiche26.php), meaning 'Laboratory for math in the pocket', is an interface opened for schools. Once a school is identified, each teacher, individually or collectively with her colleagues, can design her own resources in combining Sésamath resources. Then, she can, through the Sésamath interface, make these resources available for her students.

> combine them with their own resources as they wish. That is the way we develop Sésamath step by step, and that's it: Sésamath offers now a portal, fifteen websites, resources for classroom, for teacher and for student....

The Sésamath president interview is very illuminating, bringing out the way the association develops using the connecting Internet potential, and sometimes anticipating it. A complementary analysis of Sésamath is given by Pepin, Gueudet, Yerushalmy, Trouche, and Chazan (2015), evidencing how Sésamath develops a collaborative design involving a number of teachers—the *social connectivity* point of view—and improve its resources *documentation connectivity* over the design of successive textbooks (Sect. 19.3.1):

> The *mode of design* of these textbooks involves a large number of actors. Many teachers (approximately one hundred, for each textbook) have contributed to its design, in a *collaborative* and *iterative* way, as 'authors of content', or 'designers of didactical scenarios', or 'testers', or 'experimenters' in classes (a single teacher could have several roles, or change roles at different moments). The textbook resulting from this process is expected to fit the wishes and needs of a large number of teachers.
>
> Far from being a simple textbook, the Sesamath textbooks constitute a *hybrid* system of resources for teaching (i.e., including a classical structure in chapters, online supplements, animated corrections). Following their development helps to understand this systemic aspect:
>
> - The first model of Sesamath textbooks was a *single static book*, available both online (under a pdf, but also an odt format, allowing teachers to make modifications) and in hard copy, accompanied by separated animations on line, a set of Mathenpoche exercises, etc. (i.e. a real *resource system*, see Figure 19.1).
> - The second model was a *flexible and dynamic digital textbook*, which a teacher could organize according to his/her needs, with animation and extra exercises integrated in each chapter.
> - The third model was both a flexible and dynamic digital textbook *and* a laboratory for collaboratively adjusting the textbook to the needs and projects of the community (school, team of teachers). This laboratory, named *LaboMEP* allows teachers to develop and share their own lessons, but also to differentiate their teaching according to the results of their students.

As for the SFoDEM case, social connectivity, technological connectivity, and documentation connectivity jointly develop. Besides, some differences between the SFoDEM and the Sésamath cases are clear: nor researcher, or trainers, or institutional support in the second case. SFoDEM designed a limited numbers of resources for a limited number of teachers; on the contrary, Sésamath aimed to cover, with its resources, the whole range of the curriculum needs, and to be in touch with the biggest number of teachers. For guarantying the quality of its resources, SFoDEM relies on a careful didactical analysis by experts of the domain; Sésamath counts upon the contribution of multiple users, allowing the resources to be enriched (and sometimes corrected).

Roughly speaking, I could say that SFoDEM illustrates the *web.1 connectivity* and Sésamath the *web.2 connectivity*, characterised by more interactivity, simplicity and flexibility (O'Reilly, 2005). Is it possible to combine, in developing new forms of connectivity, both the monitoring of experts and the implication of a huge number of resources and users? It is one of the challenges of the MOOCs, I focus on it in the following section.

19.3.3 The MOOC Initiative, as a Connectivity Multiplier

I report in this section on my recent experience in MOOC, opening, for me, new horizon for thinking the connectivity potential in mathematics education.

Opening Wikipedia this morning (10 June 2015), I got this definition for MOOC:

> A massive open online course (MOOC/muk/) is an online course aimed at unlimited participation and open access via the web. In addition to traditional course materials such as filmed lectures, readings, and problem sets, many MOOCs provide interactive user forums to support community interactions between students, professors, and teaching assistants (TAs). MOOCs are a recent and widely researched development in distance education which was first introduced in 2008 and emerged as a popular mode of learning in 2012 (http://en.wikipedia.org/wiki/Massive_open_online_course).

Recently introduced in distance education (Cisel & Bruillard, 2012, Bozkurt et al., 2015), the MOOC have been, at the beginning, mainly developed by the most prestigious universities, benefiting of the well recognised expertise of some of their researchers (see for example the Stanford MOOC on mathematical thinking, taught by Keith Devlin https://www.coursera.org/course/maththink). The rapid development of this very new way of teaching/learning comes with the emergence of a lot of questions (see Fig. 19.10, on the wikipedia), none of them being really solved, at the time where these lines are written.

In 2014, based on the experience in this domain of the IREM network (see Chap. 10) and the IFÉ (French Institute of Education), was launched the MOOC eFAN Math (meaning: Teaching and Training Teachers for mathematics education

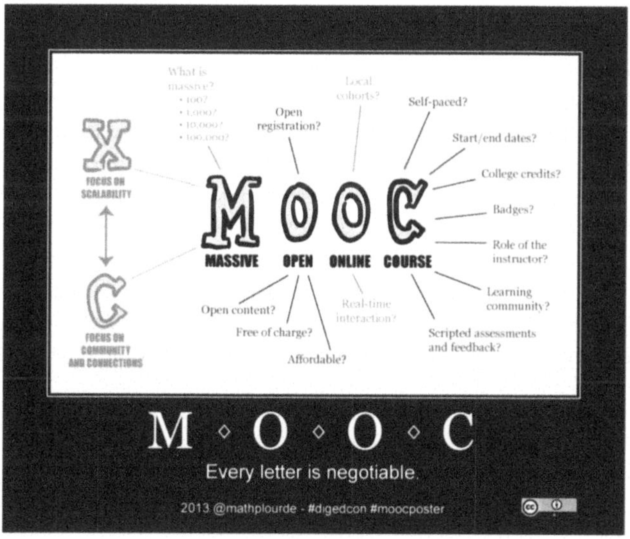

Fig. 19.10 Questioning the true nature of MOOC (2013 @mathplourde)

in digital environments[11]). Its targeted audience was teachers and teacher educators for primary or secondary schools and it aimed to support them for conceiving lessons where instruments and software effectively support students' mathematical activity. For this purpose, it develops a directory of teaching projects, based on the inputs of the participants, and enriched all over the 5 weeks MOOC duration. The orchestration of these 5 weeks clearly expresses the intentions of eFAN maths:

- Week 0: presentation of the MOOC, and constitution of teaching projects teams (the participants were supposed to come into the MOOC with a professional question, as 'how introduce symmetry with a dynamic geometry software?' giving matter to such a team; or to join a team already constituted on a question having sense for them).
- Week 1: presentation of a gallery of possible instruments for doing mathematics (the participants may comment, or/and add new instruments); each teaching project team has to decide which instruments could be used for developing its projects.
- Week 2: presentation of task design processes for reaching a given didactical objective in using a given set of instruments; each teaching project team has to design a relevant task according to its goal and to reflect on the effects of the selected technological environment on students mathematical thinking.
- Week 3: presentation of implementation processes of a given lesson in a given technological environment; each project team has to discuss the teacher's role in term of orchestration.
- Week 4: presentation of processes and tools for sharing a given resource with colleagues, for evaluating and revising it; each project team has to apply/discuss them to the light of its members experiments.

Each week begins with two short videos: a first one summarising the activities and issues of the previous week, the second one presenting the theoretical elements grounding the activities of the week to come, the tools to be used by the teams, the references to go further, and the work to be done. The description of projects in progress were available for all, and opened to comments. All the teachers following the MOOC had to answer, each week, a quiz questioning their understanding of the main notions at stake. Two main tools supported the eFAN Maths activities: the first one, the FUN platform,[12] is dedicated by the French Ministry of Higher Education to the French MOOCs; it hosted the general structure of eFAN Maths,[13] its videos and its quiz. As the FUN platform could not provide tools for collaborative design, a Moodle platform was opened for welcoming the work of the teaching projects teams.

---

[11] The MOOC eFAN Maths was hosted by two French institutions: Ecole Normale Supérieure de Cachan et Ecole Normale Supérieure de Lyon.

[12] The platform FUN (France Université Numérique http://www.france-universite-numerique.fr/moocs.html) is based on the open source technology EdX.

[13] https://www.france-universite-numerique-mooc.fr/courses/ENSCachan/20007/Trimestre_3_2014/about

It is possible to draw some lessons from the point of view of the eFAN Maths team, and from a questionnaire fulfilled by the participants (Gueudet, G. (coord.), 2015, Aldon, 2015). The eFAN pedagogical team was composed of 10 researchers and an engineer coming from the IREM network or from the IFÉ. They all consider this experience as very productive, but very time consuming (the estimate time for the whole process of conception and implementation of this MOOC was, for the whole team, 600 h), and needing to deeply renew the usual teacher training organisation; they estimate also that the available tools (mainly the FUN platform) were not at all adapted to the objective (interactivity and connectivity) of the MOOC. eFAN Math gathered at its beginning 3250 subscribers; the numbers of video downloading decreased from 2800 (first week) to 860 (fourth week); 169 teaching projects were developed and 500 participants were inscribed on the Moodle platform dedicated to the work on these projects. In this sense, eFAN Maths, compared to classical teacher training organisation, appears really as a connectivity multiplier. Finally 161 participants answered the final questionnaire; among them, 75 % estimated that eFAN Maths reached its objectives.

The decreasing number of participants is not surprising: for most of the MOOC, one estimates that the number of participants following the whole training is about 10 %. It was the case for eFAN Maths, if one considers that ‘achieving the training’ corresponds to ‘achieving a teaching project’. The 161 answers to the questionnaire, corresponding more or less to 50 % of the active participants, are then to be considered carefully: 68 % of them wish a more focused training (closer to their teaching, in primary vs. secondary schools, closer to their teaching needs); globally, they wish to have more time for being able to fully conceive, share, experiment, discuss, and revise a teaching project; they wish to have a more effective support from the eFAN team when needed; they wish to dispose of more efficient collaborative tools for designing their projects and a more interactive platform for exchanging with participants and with the pedagogical team.

Some more analyses are certainly needed, for knowing more about the quality of the teaching projects developed during eFAN Maths (their documentation connectivity, particularly from the point of view of ICT integration), the results for teachers knowledge (in terms of cognitive connectivity), and practice. But some results appear critical: the need for time and the complexity of the new equilibrium to be found both in each teacher classroom and in the MOOC itself; the interest to base the training on the design of resources meeting the real teachers needs.

The will for fitting as close as possible the local learners needs and to better monitor their work needs could lead to move towards the notion of SPOC (Small Private Online Courses), as proposes Fox (2013). Effectively, the eFAN experience seems to evidence that, when teachers were working in the same school, they benefit better for the training.

There are also some contradictory tendencies to balance (and decision to be taken, see Fig. 19.10):

- Dillenbourg, Fox, Kirchner, Mitchell, and Wirsing (2014) propose diverse solutions for reinforcing social connectivity within a MOOC, balancing then the interests of MOOC and SPOC:

  > How can we motivate MOOC learners and teachers when the teachers may be both geographically distant and socially disconnected from the learners? MOOCs could, for example, support reciprocal teaching, direct instruction, mastery learning, peer assessment and instruction, small-group/community interactions such as dynamic regrouping of learners to match learning styles and paces, and so on (p. 9).

- Social connectivity and technological connectivity are also to be carefully combined, as notice Dillenbourg et al. 2014 (p. 5):

  > MOOCs take multiple forms. At one end of the spectrum is the xMOOC, which is characterised by a rather tight structure, little social interaction and mainly computer-marked assessments. At the other end is the cMOOC or Connectionist MOOC, which is almost entirely free of pre-provided content and relies instead on very high social interactivity to produce the course content and outcomes. Most current MOOCs lie between these extremes, with some structure (weekly content in the form of video and quizzes) and some important social interactions (discussions, peer-review of work, and so on).

Finally, looking back at the previous sections, the reader may realise how the experiences of SFoDEM and Sésamath were announcing, in some ways, the emergence of MOOCs, under their extreme tendencies, as new forms of fostering teachers professional development on the basis of their collaborative work on teaching resources. For these organisations to work effectively, technological connectivity and social connectivity appear as necessary ingredients.

Internet, both for students (Sect. 19.2) and for teachers (Sect. 19.3) clearly appears as a connectivity multiplier, from a technological as well as a social point of view. To what extent this connectivity improves also teacher documentation, knowledge and practice, as well as students learning and mathematical activity, is not a trivial question. I discuss, in the next section, the way the ICMI study connectivity panel (Hoyles et al., 2010) addressed this issue, mainly from the point of view of students.

19.4 Some Lessons from the ICMI Study Connectivity Panel

The ICMI study connectivity panel, chaired by Celia Hoyles, was based on four presentations (one of them has already been introduced in this chapter, Sect. 19.2.2). I will focus in this section on two of them, the first one concerning the effects of connectivity in a given classroom, the second one across classrooms. I draw then some general lessons from the panel.

19.4.1 Enacting Classroom Participatory Simulations

As I did Sect. 19.2.3, Uri Wilenski (Hoyles et al., 2010, pp. 452–455) exploits 'a neglected affordance of connectivity: the ability to give people a shared interactive experience in classroom contexts'. For this purpose, he presents an outline of his work with NetLogo,[14] using the notion of connectivity in two senses.

The first sense is a *macro-micro level connectivity*:

> In our many years of working with NetLogo in middle and secondary classrooms, we have endeavoured to bring to students descriptions of complex systems at a micro-level and connect those micro-level descriptions to macro-level and observable phenomena. Typically, when we have taught students about systems that can be constructed as complex, we have concentrated on aggregate equations that summarize system behaviour. For example, to describe the behaviour of ideal gases, we rely on equations such as PV = nRT. But agent-based modelling enables students to more directly control and examine the behaviour of elements of the system and connect this behaviour to the system emergent behaviour. Thus in NetLogo's GasLab model suite, students come to understand the ideal gas as composed of Myriad interacting gas molecules and see PV = nRT as an emergent result of these interactions. There are hundreds of NetLogo models we have used in classrooms. Students examine a range of phenomena such as the spread of a disease through a population, or the interaction of predator and prey in an ecosystem [...]

For example, for the interaction predator/prey (Fig. 19.11), students can vary essential parameters as the number of sheep, the number of wolves, the quantity of

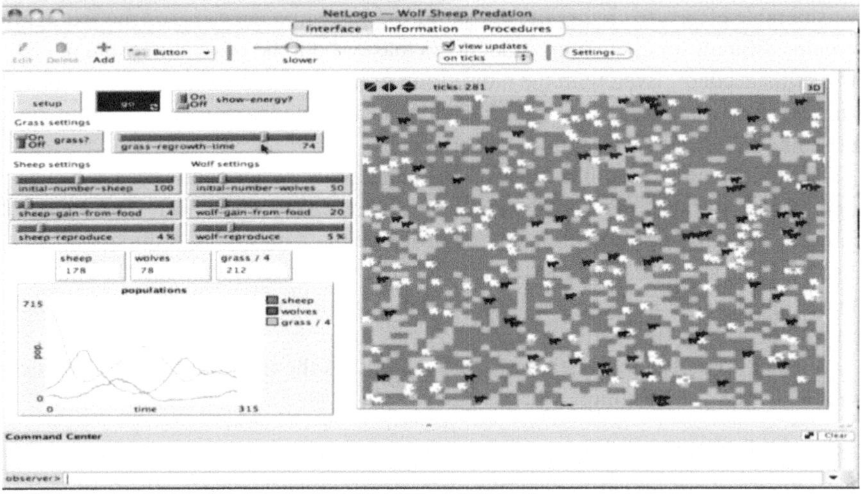

Fig. 19.11 The NetLogo interface for studying a model of predator and prey

[14] NetLogo (https://ccl.northwestern.edu/netlogo/) is a multi-agent programmable modelling environment, developed at the Center for Connected Learning of the Northwestern University. It is an extension of the Logo environment developed by Seymour Papert (http://en.wikipedia.org/wiki/Seymour_Papert)

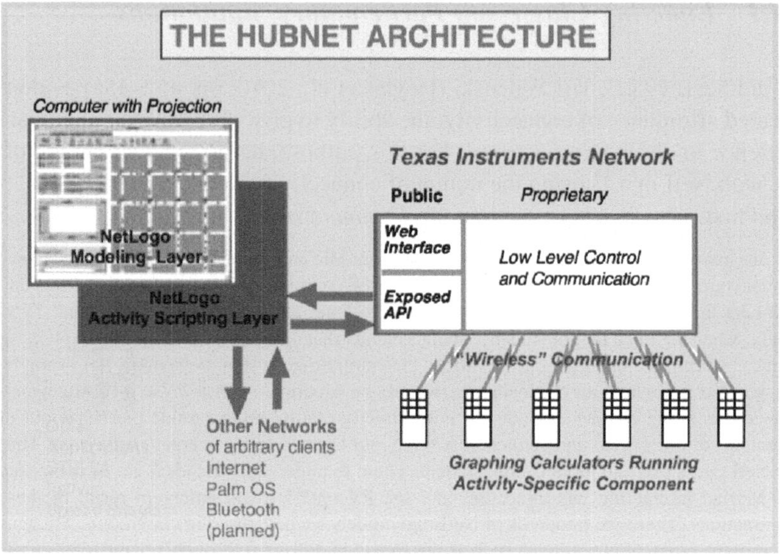

Fig. 19.12 The HubNet architecture (Hoyles et al., 2010, p. 454)

grass, and observe the evolution of the process. The hypothesis is that the micro-level connectivity managed by the application will facilitate students understanding of the system behaviour. Wilenski observes that it was not so simple: 'despite considerable efforts to "lower the threshold" of entry into agent-based modelling, it remains difficult for elementary students to master both the programing and modelling skills needed'. Even with the monitoring of a teacher, this approach 'leaves the student somewhat passive, as only a few can be engaged at any one time and they are limited to discussion of model behaviour'.

These difficulties lead him to develop connectivity in a second sense, a technological one, through the added module HubNet, enabling 'a host of devices to connect to a logo simulation and control agents within that simulation' (see Fig. 19.12 a set of calculators connected to the teacher's computer). The sole modelling activity is then transformed into a participatory simulation, in which each student may take part.[15]

Wilenski (Hoyles et al., 2010, p. 453) underlines the important benefits of such an application for learning:

[...] the modelling activity:

- Becomes more engaging—especially for younger learners. It becomes a social activity and captures much of the same draw as online games.
- Promotes greater student participation. Every student can be actively involved at the same time. Because they often require continuous action on the part of the students, they

[15] This application, through a cooperation with Texas Instruments, gave birth to the TI-Navigator network, that we describe Sect. 19.2.2.

are "in-the moment" motivated to participate. Such universal participation is very hard to achieve in a traditional classroom.

- Enables a shared experience of a complex system. There are very few opportunities, in the classroom or in life, for students to collectively witness the same complex system unfolding. Focal attention to such a system is hard to achieve outside of the virtual and, even when achieved, if the viewing does not connect the micro-level behaviour to the macro-level outcomes, then only the appearance is shared, not the mechanisms of action.
- Facilitates classroom discussion of the system and examination of "what-ifs". Student can suggest experiments with varying critical system parameters and/or agent-rules, hypothesize the observed behavioural change, run the simulation and refine the experiment.
- Scaffolds individual modelling and analysis. Once students have had several opportunities to collectively model and analyse complex systems, they are much better prepared (and motivated) to conduct such inquiry on their own. Often students have suggestions for model experiments that do not get explored in class. These questions are potent seeds of further student inquiry, experimentation and model revision.

What I retained from this rich experiment is the interest of combining different level of connectivity: the *technological connectivity* (HubNet architecture) enables all the students to participate *in the same time* to the construction of a given phenomenon. This collective engagement (*social connectivity*) is stimulated by the student's awareness to be an actor of the *mechanisms of action*, and co-responsible of the final result. The system insures the connection between the micro-level and the macro-level, and the students, being involved in the whole process, incorporates the interrelations between these two levels (*cognitive connectivity*). All over the process, the teacher's orchestration is needed for regulating students' activity. This is made possible by the presence of all the actors in the same time in the same place. I analyse in the following section what could happen when such on activity occurs in different places.

19.4.2 Exploiting Connectivity Across Classrooms

The Noss and Hoyles's presentation (Hoyles et al., 2010, pp. 455–460) in the connectivity panel addresses actually the question of connectivity *within* and *across* classroom, through two projects co-directed by themselves: the Playground Project and the Weblabs Project. I will focus in this section on the first one.

The Playground project,[16] as its name indicates, aims to use the potential of games for stimulating children (from 4 to 8 years old) engagement and learning (about games and mathematics, see Chap. 18). Going beyond the simple 'playing

[16] Its website (http://playground.ioe.ac.uk) points out, on its front page: « The playground project is building computer environments for 4–8 year-olds to play, design and create games. A playground is a place to play with rules not just play by them. We aim to harness children's playfulness, creative potential and exploratory spirit, allowing them to enter into abstract and formal ways of thinking » (see also Chap. 18).

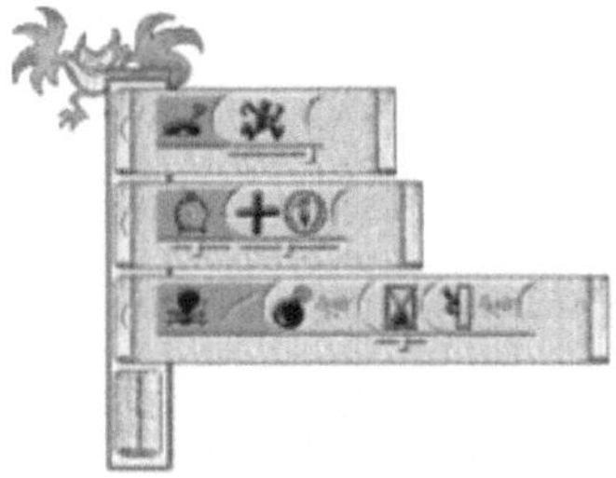

Fig. 19.13 Stones combined for constituting rules defining a monster' behaviour (Hoyles et al., 2010, p. 455)

game', it aims to add a new dimension where children build their own games. Due to the age of the children, the project favours other modalities of interactions than words: mainly speech and direct manipulation. The authors describe the way the project allows children to design their own games:

> Children populated their games with objects which had 'behaviours'—sets of rules that determine their action. Behaviours were defined using collections of iconic rules, which could be viewed by opening a scroll of paper attached to an object (see Fig. 19.13 for rules defining a monster's behaviour). Each rule was expressed as a visible 'sentence' or string of graphic icons, which combined a condition and a series of actions to be executed whenever the condition was true. The icons representing the conditions and actions were represented as 'stones', small concrete manifestations of the concept that could be strung together to constitute a rule. Actions stones had a convex left side so that conditions with their concave right side could naturally fit to their left. Any object could accept any number of these iconic rules, all of which would be executed in parallel whenever the conditions for their execution were satisfied. (Hoyles et al., 2010, p. 456).

When the game starts, I change my speed to 22.5
When a second and half is up, I change my direction plus 25°
When I am shot by a ray, I explode, I wait two second, I appear (every object has an empty rule for making new rules)

The project gives then means for children for *constructing*, *expressing*, and *communicating* their own games. It offered a language allowing them to define rules in a synthetic and no ambiguous way. Once defined a game, the project leads the children to discuss it on two successive phases: sharing the game *through face-to-face interactions* in their own classrooms; sharing the game *using Internet*, either synchronously or asynchronously with a remote classroom. The project findings evidence that, over the two phases, 'children collaboratively came to explain phenomena arising from rules we characterised as either *player* (an agreed regulation), or *system* (a formal condition and action for the behaviour of the game)'

(Hoyles et al., 2010, p. 457). These findings underline also major differences between face-to-face an remote interactions:

> We found that in face-to-face collaboration, the children centre their attention on narrative, and addressed the problem of translating the narrative into system rules, which can be 'programmed' into the computer. This allows the children to debug any conflicts between system rules in order to maintain the flow of the game narrative.
>
> When *we added remote communication* to the system by enabling the sending and receiving of games from within the Playground system, we found that children were encouraged to add complexity and innovative elements to their games, not by the addition of socially-constructed or 'player' rules but rather through additional system rules which elaborate the formalism (games were created using two different kinds of programming systems, neither of which employed textual modalities). This shift of attention to system rules occurs at the same time, and perhaps as a result of, a loosening of the game narrative that is a consequence of the remoteness of the interaction.
>
> This phenomenon was particularly evident *in the case of asynchronous interaction* where, stripped of even the semantics of gestures, our extremely young students found it increasingly natural to try to communicate meaning via the various formalisms we provided. Thus a key historical claim for programming, that is offers a key motivation and model for immersion in a formal system, came to life as children struggled to modify and add rules of their programs that achieved the effects they desired. And it is worth stressing that asynchronous communication, while somewhat less attractive to the students at the time [...] allows students to reflect on, and therefore use more effectively, the formal rules of their games.

The main result I retain here is that 'The shift from narrative to system/formal rules does, in fact, seem to be a direct result of the necessity to formalise, in the absence of all the normal richness of interaction that characterises face-to-face collaboration' (Hoyles et al., 2010, p. 457). In this case, *technical connectivity*, understood as providing children means to communicate via Internet, leads to *cognitive connectivity*, leading the children to relate the implicit rules of the game to formal ones, parts of a system of rules. The discussion among children connects then a level of informal speech and a level of formal system of rules.

I would like to add extra personal comments.

The Playground project concerns a particular part of mathematics, linked to programing. This part will probably strongly develop in the future, supported both by the improvements of the software dedicated to 'children and programing' (see for example Scratch[17]), and by the evolution of curricula, favouring interaction between mathematics and programing (see Sect. 12.3.3).

Noss and Hoyles associate, in this experiment, 'connection to Internet', and 'remote interaction'. Of course, this association is note a necessary one, as connection to Internet and face-to-face interaction may jointly develop (see Sect. 19.2.3 or 19.4.1).

As the authors underline themselves, their project began in the previous century, where peer-to-per connectivity was quite limited. Today, the remote or face-to-face

[17] Scratch: "Create stories, games, and animations, Share with others around the world" (https://scratch.mit.edu), developed by the Massachusetts Institute of Technology, hosting 9,767,423 projects (on 12 June 2005)

interactions could combine texts (under different formats, more or less formal, from SMS to emails), pictures, as well as audio or video interactions.

However, I found useful to present, in the frame of this chapter, some lessons of this experiment, evidencing that even a limited use of connectivity, reduced to peer-to-peer interactions through the exchanges of texts via Internet could have important effects. From this connectivity panel held in the ICMI study, some more general lessons emerge, that I underline in the following section.

19.4.3 The ICMI Study Connectivity Panel, Some Lessons and Perspectives

In this section, I would like to focus on the main lessons drawn by the connectivity panel, then by the ICMI study itself.

Regarding the panel, Hoyles et al. (2010) retain both the potential of *technological connectivity* and the conditions for exploiting it:

- The potential is seen for developing *social and cognitive connectivity* (essentially regarding students):

> Digital technologies are already changing the ways we think about interacting with mathematical objects, especially in terms of dynamic visualisations and the multiple connections that can be made between different kinds of symbolic representations. At the same time, we are seeing rapid developments in the ways that it is possible for students to share resources and ideas and to collaborate through technological devices both in the same physical space and at a distance (p. 439).

The conditions for exploiting technological connectivity are quite largely described: 'Alongside overcoming not inconsiderable technical challenges, establishing an appropriate set of *socio-technical/mathematical norms* that prioritised collaboration [is] crucial regarding connectivity' (p. 460). Some years after, the point of view on technical challenges could seem quite optimistic... But the necessity of changing the socio-technical/mathematical norms clearly appears: the experiences presented during the panel stand at the fringes of the schooling system, and one measures the necessary distance for implementing them in the schooling system.

Among these conditions, even if this question was not addressed by all the panellists, rethinking the teacher's role in terms of new orchestrations appears actually crucial:

> [...] here we are delineating new, even more demanding roles for the teacher, to be aware—across not only her own classroom but those in remote location—of the evolution of discussion, the mathematical substance of what is and what is not discussed, and the need all the while to find ways to keep students on task without removing the exploratory and fun elements of the work. This is, surely, a demanding set of roles for the teacher (p. 460).

Finally, the panellists shared the awareness that connectivity was a promising field of research, specially regarding the cognitive aspect, i.e. implications for learning, quoting Moreno-Armella, Hegedus, and Kaput (2008), suggesting how 'networks can link private cognitive efforts to public social displays thus—potentially at least—enhancing student's metacognitive ability to reflect upon their own work to reference to others'.

The connectivity panel was part of the 17th ICMI study aiming to 'rethink the terrain' of technology and mathematics education. The theme of connectivity appears certainly, within this study, as the one where further research was the most needed, as stated by Artigue in her concluding chapter (Hoyles & Lagrange, 2010, p. 473):

> The way digital technologies can support and foster today collaborative work, at the distance or not, between students or between teachers, and also between teachers and researchers, and the consequences that this can have on student's learning processes, on the evolution of teachers' practices is certainly one essential technological evolution that educational research has to systematically explore in the future [. . .] most of this space is still for us nearly terra incognita. We observe an intense creativity, which very often develops independently of research and this is a very stimulating situation. But we also have to be careful. As stressed by Richard Noss in the panel on connectivity, connectivity does not necessarily imply collaborative work and collaborative work does not necessarily imply better mathematics learning, or, I would add, better mathematics teaching. We are submerged by an avalanche of information, data and possibilities of connection and the way this avalanche can be organized, treated and transformed into knowledge or means for productive action is an open problem.

Since the time of this ICMI study (2006 for the conference, 2010 for its proceedings), connectivity has developed, at least at a technological level, for the students (Sect. 19.2.3) as well as for the teachers (Sect. 19.3.3). Which new lessons and perspectives could be drawn in this new situation? I propose some answers, and some new questions in the next discussion section.

19.5 Discussion

I call this section 'Discussion' rather than 'Conclusion' because the forms of connectivity are evolving so rapidly that I can offer no conclusion. I would like to underline the strong current evolutions, in terms of technologies and usages, then to question the links between connectivity and mathematics, and, last but not least, address the theoretical needs for analysing, in such contexts, mathematics learning and teaching processes.

19.5.1 Internet Uses as a Connectivity Multiplier and a Seamless Learning Tool

I had structured the two first parts of this chapter looking at the evolutions from the students' side, then from the teacher's side. For understanding the processes at stake, I have to embrace in our analyses the two sides in their interrelations. We showed (Gueudet & Trouche, 2012, pp. 313–316), in the case of Pierre, a teacher member of the Sésamath association, the synergy resulting of the interrelations between Pierre's work in/for his association, Pierre's work in/for his class and students' usages:

> He evinces a strong collective involvement both in his school and in Sesamath: he is 'teacher in charge of technology', treasurer of the school cooperative, responsible of the school's chess club. These activities are not all dedicated to mathematics. In Sésamath, as of 2008, he was a member of the board for 5 years. This meant that he spent approximately 1 h a day reading emails and participating in forums 'that engage the association life'. He was also a member of a Sésamath project developing a grade 6 textbook, which is still in progress at this time. He was, finally, the pilot of a new Sésamath project entitled 'mathematics files for primary schools'.
>
> Documentation work takes place within each of these collective involvements and each of them is part of Pierre's work, as he said: 'Consuming time in collective activities is a component of my teaching activity'. He particularly emphasizes the importance of the primary school project ('it gives a better understanding of what my pupils know when arriving at secondary school'), the Sésamath board ('it makes me aware of the questions asked to the profession as a whole') and the 'grade 6 textbook'. It is actually this last project, which appeared as fostering Pierre's documentation. For all the duration of the project (2 years), Pierre decided to have only grade 6 classes (three classes, for 6 h teaching in it), to 'align' his documentation work with the community documentation. Thus, the documentation work that Pierre accomplished in 2008–2009 for the grade-6 level *concentrated* his main efforts, and *connected* individual and community documentation [...]
>
> To this collaborative documentation corresponds a collaborative form of teaching [...]. Using online resources is an important feature of Pierre's documentation work, within or without his students (for preparing his teaching or collaborating in Sésamath projects). Within his classroom, a connected computer, a projector and an interactive whiteboard (IWB) are used to work with online resources. For example at the beginning of each lesson, the teacher opens Pronote, an application allowing displaying the students list, to note the absentees, to memorize what has been done, and what is still to do... Another example of this continuous Internet use: the teacher exploits Google to do any arithmetic operation exceeding students' capacities of mental computation (it was amazing to observe that handheld calculators remain in students' schoolbags!). For continuing to interact with his students outside of the classroom, he developed a collaborative website on which he regularly uploads mathematics problems (that he calls 'enigma'). Students try to solve them and write their solutions on a forum (Gueudet & Trouche, 2012, pp. 313–316).

The use of Google is particularly significant of these interrelations: Pierre uses Google for doing computations, because actually it corresponds to the students' usages. As they are more and more connected to Internet, at home, as well as in school, often through their mobile phones, they tend to use Google as an universal machine: they use the same procedure to answer to a geographical question ('what is the number of inhabitants of such a city?') and to a mathematical one ('what is the result of 45 times 59'?): in the two cases, Google is required to provide the answer. In such a procedure, the constructive aspect of mathematics practices ('yes, *I can* compute 45×59') is lost. It looks like if each result of any question was lying somewhere on the shelves and I had just to go to the relevant shelve and take it: that is the efficiency of Google to do that for us. It is also well known also that, Sesamath providing a wide number of resources covering the whole curriculum, some teachers restrict sometimes their documentation work to 'search on the shelves of Sésamath' what fits their needs of the day. This is of course not the sole aspect of the Internet usages, and there are a lot of productive aspects as I evidenced in Sects. 19.2 and 19.3, but this way of searching direct answers to direct questions is a real economical effect of being continuously connected to a wide repository of resources. The institutions try to control this continuous connectivity (see for example, Sect. 12.3.3, the development of 'machines to the test', excluding during the examination at least any connection between a calculator and 'outside'), but removing connectivity cannot be the sole answer: the development of new environments for communicating based on Internet appeals new kind of orchestrations: the case of MOOC (Sect. 19.3.3) evidences the interest, and the complexity, of such reflections on *Internet as a connectivity multiplier*.

Another aspect appearing in the description of Pierre's work is the dilution of the frontiers between working in school and working out of school, the teacher and his students interacting through a website and Internet resources (as LaboMep, see Sect. 19.3.2). It happens *during* the time of schooling and curriculum knowledge, but I happens also, from a general point of view, *after* the time of schooling, considering lifelong education. It leads to the notion of *seamless learning* (Wong, Milrad, & Specht, 2015),

> [Researchers] propose *seamless learning* as a learning approach characterized by the continuity of the learning experience across a combination of locations, times, technologies, or social settings, (perhaps) with the personal mobile device as a mediator. The basic rationale is that it is not feasible to equip students and knowledge workers with all the skills and knowledge they need for lifelong learning solely through formal learning (or any one specific learning context). Henceforth, student learning should move beyond the acquisition of curriculum knowledge and be complemented with other approaches in order to develop the capacity *to learn seamlessly* (p. xvii).

We are at the beginning of the analysis of this kind of learning: Chaps. 15 on teachers work with digital tools, 17 on the design tasks, 18 on using games open windows on leaning/teaching with Internet.

19.5.2 Connectivity and Mathematics

I have presented in various sections of this chapter (Sects. 19.2.2, 19.4.1 and 19.4.2) the potential of technological connectivity for linking different aspects of the teachers and students' mathematical activities, what I have named documentation connectivity, cognitive connectivity or micro and macro level connectivity. I would like to examine now to which extent mathematics practicing, learning and teaching requires connectivity and in which sense.

We have already met the necessity of 'connecting things' for *learning and teaching mathematics* in two senses: *connecting different representations of mathematical objects* through a specific activity of treatments and conversions, as a central activity for conceptualising (see the work of Duval, Sect. 12.3.3);

connecting ostensives and non ostensives, the last ones guiding the usage of first ones (see the work of Chevallard, Sect. 12.3.2).

More generally, if we consider that *doing* mathematics is solving problem, it appears clearly that doing this needs to connecting different point of views on objects and processes (see examples of proofs in Chap. 6; see Jon's examples on visual theorems, Sects. 3.2.1 and 3.4).

In a discussion between the three authors of the first draft of this chapter John asked Jon:

> I imagine that now, compared to the beginning of your professional life, that you might send out a Maple (or whatever) file to colleagues and say something to the effect "Look at this, there's something strange going on 'under the hood'". Are conjectures more of a shared 'thing' than they were 40 years ago?

Jon replied:

> Life is vastly different than forty years ago. I think to a significant degree this is covered in Chap. 3. It is certainly covered in an article "The Future of Mathematics 1965 to 2065." https://www.carma.newcastle.edu.au/jon/future.pdf. This just appeared as part of the MAA100th anniversary book. When David Bailey and I wrote our book *Mathematics by Experiment* between 2001 and 2004, it was already possible to be scholarly without ever visiting the library and when we revised the book in 2007 this was even more true. The level of connectivity is limited largely by one's imagination and willingness to contact people/ remember what resources may be available. The sociology of this—as with social media—has not yet stabilised. Perhaps it never will. So I am routinely sent stacks of papers by isolated researchers asking me to help them publish them along with even more intrusive requests. Yet on balance this is a wonderful time to be working in a subject which—despite the public image of a solitary researcher staring at a blackboard—has always thrived on and needed human interaction.

Indeed, in responding to one of the questions I posed in Chap. 6, Jon performed an Internet search.

Moving on (but keeping in the domain of mathematics) doing mathematics, since the first written practices (see Chap. 5 and Proust, 2014), has always dealt with highly structured texts. Reading such texts leads one to combine different registers of activity: learning, solving, classifying, archiving, exploring or inventing. Digital tools give us new means for combining these different registers. This links to my construct 'cognitive connectivity' (the internal—in the mind—rather than the external—in action—side of connectivity). Noss and Hoyles (1996), I posit, had similar ideas when, 20 years ago, they compared mathematical connectivity to the functioning of the Web, introducing the notion of webbing:

> Like the web of mathematical ideas, the Web (we will use a capital to denote the electronic network), is too complex to understand globally—but local connections are relatively accessible. At the same time, one way—perhaps the only way—to gain an overview of the Web is to develop for oneself a local collection of familiar connections, and build from there outwards along lines of one's own interests and obsessions. The idea of webbing is meant to convey the presence of a structure that learners can draw upon and reconstruct for support—in ways that they choose as appropriate for their struggle to construct meaning for some mathematics (Noss & Hoyles, 1996, p. 108).

The question is then how the use of the web could support mathematical connectivity. Trouche and Drijvers (2014, p. 6) proposes an approach combining the concepts of webbing and orchestration:

> In the webbing approach, conceptualization appears as a coordination process, 'the process by which the student infers meaning by coordinating the structure of the learning system (including the knowledge to be learned, the learning resources available, prior student knowledge and experience and constructing their own scaffolds by interaction and feedback)' (Hoyles et al., 2004, p. 319). In the instrumental orchestration approach, conceptualization appears as a command process, characterized by the conscious attitude to consider, with sufficient objectivity, all the information immediately available not only from the calculator, but also from other sources and to seek mathematical consistency between them (Guin & Trouche, 1999). 'Very sophisticated artefacts such as the artefacts 25 available in a computerized learning environment give birth to a set of instruments. The articulation of this set demands from the subject a strong command process. One of the key elements for a successful integration of these artefacts into a learning environment is the institutional and social assistance to this individual command process. Instrumental orchestrations constitute an answer to this necessity.' (Trouche, 2004, p. 304). It seems that there is a kind of intended internalization from an instrumental 30 orchestration, seen as an external process of monitoring students' instruments by the teacher, to an internal orchestration, seen as a process of self-monitoring the individual and personal instruments by a student. Coordination and control are certainly two facets of mathematical activity, particularly in a technological rich environment, and the two approaches seem to privilege, each, one of these facets.

With the notion of internal coordination and control comes a new reflection on curriculum resources. Recent analyses of e-textbooks, i.e. textbooks making profit of the digital potentialities, mainly from the point of view of connectivity, underline the necessity, for insuring their quality, to take into account connectivity and coherence (Pepin et al., 2015).

19.5.3 New Theoretical Needs

Describing recent experiences and, regarding for example interactive collaborative mathematical interface (Sect. 19.2.3) or MOOCs (Sect. 19.3.3), I was aware, as I said previously, that we are just at the beginning of the analysis of the connectivity aspects and effects. For developing analyses on new phenomena, sometimes new theoretical frames are needed in order to define new concepts and system of concepts. Taking into account connectivity as a major intellectual challenge has led to the creation of the *connectivism* frame, thus defined by Wikipedia (https://en.wikipedia.org/wiki/Connectivism):

> Connectivism is a hypothesis of learning which emphasizes the role of social and cultural context. Connectivism is often associated with and proposes a perspective similar to Vygotsky's' zone of proximal development' (ZPD), an idea later transposed into Engeström's activity theory (see Chap. 9). The relationship between work experience, learning, and knowledge, as expressed in the concept of 'connectivity, is central to connectivism, motivating the theory's name.

The definition of this theory as 'a learning theory for the digital age' (Siemens, 2005) indicates the emphasis that connectivism gives to technology's effect on how people live, communicate and learn. This is not the choice I had made until now, but I am sure that the development of this new domain will benefit to the other domains of research interested in connectivity.

Actually, for lighting the questions at stake, I am trying a theoretical networking approach as presented by Prediger, Arzarello, Bosch, and Lenfant (2008) (see also Chap. 9), connecting theoretical frameworks for understanding connectivity. In the case of teacher's work, it gave matter to the *documentational approach* (Sect. 15.3.2), crossing the domain of architecture information (Salaün, 2012) and instrumental approach. This approach is used in the frame of a French national project (www.anr-revea.fr) for analysing the evolution of teachers work with resources in a time of digital transition.

In the community of mathematics education, other theoretical approaches should be exploited in order to understand connectivity. My own view is that Sfard's construct 'commognition' is important in this regard. Sfard (2010), p. 432 defines thinking as:

> the *individualized version of interpersonal communication*—as a communicative interaction in which one person plays the roles of all interlocutors. The term *commognition*, a combination of *communication* and *cognition* comes to stress that inter-personal communication and individual thinking are two varieties of the same phenomenon.

According to this perspective, developing social and reflective connectivity is developing opportunities for improving mathematical thinking.

Finally, looking at connectivity in the mathematics education community leads to develop an interdisciplinary program of research, that is before us.

References

Aldon, G. (2015). MOOC, Formations à distance, formations hybrides. *MathemaTICE 46*. Retrieved from http://revue.sesamath.net.

Bellemain, F. (2014). Analyse d'environnements de géométrie dynamique collaborative du point de vue de l'orchestration instrumentale. *Nuances: estudos sobre Educação, 2,* 18–38. Retrieved from http://revista.fct.unesp.br/index.php/Nuances/article/viewFile/2936/2686

Betancourt, Y. (2014). *Uso de recursos didácticos digitales en un primer curso de álgebra lineal,* PhD, CINVESTAV, Mexico.

Bozkurt, A., Akgun-Ozbek, E., Onrat-Yilmazer, S., Erdogdu, E., Ucar, H., Guler, E., et al. (2015). Trends in distance education research: A content analysis of journals 2009–2013. *International Review of Research in Open and Distributed Learning, 16*(1), 330–363.

Cisel, M., & Bruillard, E. (2012). Chronique des MOOC. *Sciences et Technologies de l'Information et de la Communication pour l'Éducation et la Formation,* (19). Retrieved from http://sticef.univ-lemans.fr/num/vol2012/13r-cisel/sticef_2012_cisel_13r.htm

Dillenbourg, P., Fox, A., Kirchner, C., Mitchell, J., & Wirsing, M. (2014). Massive open online courses: current state and perspectives. *Dagstuhl Manifestos* 4(1)1, 1–27. Retrieved from http://ethicalforum2013.fuus.be/sites/default/files/dillenbourg_MOOC.pdf

Fox, A. (2013). From MOOC to SPOC? *Communications of the ACM, 56*(12), 38–40.

Gueudet, G. (coord.) (2015). *MOOC eFAN Maths. Edition 2014. Bilans.* Internal report. ENS de Cachan & ENS de Lyon.

Gueudet, G., & Trouche, L. (2011). Mathematics teacher education advanced methods: an example in dynamic geometry, *ZDM, The International Journal on Mathematics Education, 43*(3), 399–411. doi:10.1007/s11858-011-0313-x. Retrieved from http://www.springerlink.com/content/13733h7321658734/

Gueudet, G., & Trouche, L. (2012). Communities, documents and professional geneses: interrelated stories. In G. Gueudet, B. Pepin, & L. Trouche (Eds.), *From text to 'lived' resources: Mathematics curriculum materials and teacher development* (pp. 305–322). New York: Springer.

Guin, D., Joab, M., & Trouche, L. (2006). *Conception collaborative de ressources, l'expérience du SFoDEM (2000–2006).* IREM, Université Montpellier 2. Retrieved from http://www.math.univ-montp2.fr/sfodem/

Guin, D., & Trouche, L. (1999). The complex process of converting tools into mathematical Instruments: The case of calculators. *International Journal of Computers for Mathematical Learning, 3*, 195–227.

Guin, D., & Trouche, L. (2005). Distance training, a key mode to support teachers in the integration of ICT? In M. Bosch (Ed.), *Proceedings of the Fourth European Conference on Research on Mathematics Education* (pp. 1020–1029), FUNDEMI IQS—Universitat Ramon Llull. Retrieved from http://ermeweb.free.fr/CERME4/CERME4_WG9.pdf

Healy, L. (2002). *Iterative design and comparison of learning systems for reflection in two dimensions*, Unpublished Ph.D. thesis, University of London.

Helsper, E., & Eynon, R. (2010). Digital natives: where is the evidence? *British Educational Research Journal, 36*(3), 503–520.

Hoyles, C., Kalas, I., Trouche, L., Hivon, L., Noss, R., & Wilensky, U. (2010). Connectivity and virtual networks for learning. In C. Hoyles & J.-B. Lagrange (Eds.), *Mathematical education and digital technologies: Rethinking the terrain* (pp. 439–462). New York: Springer.

Hoyles, C., & Lagrange, J. B. (2010). *Mathematical education and digital technologies: Rethinking the terrain.* The 17th ICMI study. New York: Springer.

Hoyles, C., Noss, R., & Kent, P. (2004). On the integration of digital technologies into mathematics classrooms. *International Journal of Computers for Mathematical Learning, 9*, 309–326.

Moreno-Armella, L., Hegedus, S., & Kaput, J. (2008). From static to dynamic mathematics: Historical and representational perspectives. *Educational Studies in Mathematics, 68*(2), 99–111.

Noss, R., & Hoyles, C. (1996). *Windows on mathematical meanings. Learning cultures and computers.* Dordercht, The Netherlands: Kluwer.

O'Reilly, T. (2005, September 30). What Is Web 2.0. design patterns and business models for the next generation of software. *Web 2.0 Conference 2005*. Retrieved June 2015, from http://www.oreilly.com/pub/a/web2/archive/what-is-web-20.html

Pepin, B., Gueudet, G., Yerushalmy, M., Trouche, L., & Chazan, D. (2015), E-textbooks in/for teaching and learning mathematics: A potentially transformative educational technology. In L. English, & D. Kirschner, *Third Handbook of Research in Mathematics Education* (pp. 636–661). New York: Taylor & Francis

Prediger, S., Arzarello, F., Bosch, M., & Lenfant, A. (Eds.) (2008). Comparing, combining, coordinating—networking strategies for connecting theoretical approaches. *ZDM, The International Journal on Mathematics Education, 40*(2), 163–327.

Proust, C. (2014). *Textes mathématiques cunéiformes: des listes pour apprendre, résoudre, classer, archiver, explorer ou inventer.* Retrieved from https://hal.archives-ouvertes.fr/hal-01139604/document

Salaün, J.-M. (2012). *Vu, lu, su. Les architectes de l'information face à l'oligopole du Web.* Paris, France: La Découverte.

Sfard, A. (2010). *Thinking as communicating. human development, the growth of discourses, and mathematizing*. Cambridge University Press.

Siemens, G. (2005). Connectivism: A learning theory for the digital age. *International Journal of Instructional Technology and Distance Learning*. 2(1). Retrieved from http://www.itdl.org/Journal/Jan_05/article01.htm

Trouche, L. (2004). Managing the complexity of human/machine interactions in computerized learning environments: guiding students' command process through instrumental orchestrations. *International Journal of Computers for Mathematical Learning, 9*, 281–307.

Trouche, L., & Drijvers, P. (2014). Webbing and orchestration. Two interrelated views on digital tools in mathematics education, *Teaching Mathematics and Its Applications: International Journal of the Institute of Mathematics and its Applications, 33*(3), 193–209, doi: 10.1093/teamat/hru014, Retrieved from http://teamat.oxfordjournals.org/cgi/reprint/hru014?ijkey=P83FxYUzECbG67e&keytype=ref

Wenger, E. (1998). *Communities of practice: Learning, meaning, and identity*. Cambridge University Press.

Wong, L.-H., Milrad, M., & Specht, M. (Eds.) (2015). *Seamless learning in the age of mobile connectivity*. New York: Springer.

Epilogue

We use these last pages to reflect on matters for further thought and action. Chapters 2–18 describe many aspects of tool use in mathematics and mathematics education and discuss a number of issues surrounding this use but this book, we feel, should be seen as a beginning, not as an end, as there is so much about tools and mathematics which these chapters do not cover, and links between ideas and phenomena which remain to be made and/or clarified. We structure our comments under the themes: history; theory; the nature of mathematics; and action.

We made brief forays into the history of mathematics in Part I: Chap. 4 surveyed prehistory and selected historical topics; Chap. 5 detailed the development of tool use in Mesopotamian scribal schools. But there is so much we have not commented on. An encyclopaedia of the history of tools in mathematics would be a useful publication. We comment on how we would like such an encyclopaedia to be conceptualised by comparing (with a focus on tools) four books on the history of mathematics: Fauvel and Gray's (1987), *The history of mathematics: A reader*; Kidwell et al.'s (2008), *Tools of American mathematics teaching*; Netz's (1999), *The shaping of deduction in Greek mathematics*; and Joseph (2010), *The Crest of the Peacock: Non-European Roots of Mathematics*.

Kidwell et al. (2008) is a wonderful source and the primary focus is on tools for school mathematics but it is restricted (for good reasons) to: the USA; the period 1800–2000; and mathematics teaching. Further to this, it focuses on tools with scant mention of developments in mathematics or conceptions of mathematics. Fauvel and Gray (1987) is also a wonderful source (of original mathematics, with commentary, in English translation) but it focuses on people and concepts with scant (virtually none) mention of tools. Further to this it appears to view 'mathematics' as that which Bishop (1988, p. 19)—see Sect. 14.1.2—calls 'Mathematics' (with a capital M) 'the mathematics which is exemplified by Kline's *Mathematics in Western Culture*'. Netz (1999) is subtitled 'A study in cognitive history'. It approximates more closely than Kidwell et al. (2008) or Fauvel and Gray (1987) do to our vision of an encyclopaedia because, in tracing the shaping of deduction in ancient Greek mathematics it pays close attention to people,

J. Monaghan et al., *Tools and Mathematics*, Mathematics Education Library,
DOI 10.1007/978-3-319-02396-0

concepts and the tools which interacted with people in the shaping of mathematical concepts. Our vision of an encyclopaedia of the history of tools in mathematics would take the positive features of these three books and temper the potential bias towards 'Mathematics' by ensuring that the history that Joseph (2010) writes on is part of mathematics.

In Part II we described a number of theoretical approaches that inform an understanding of tool use in mathematics education. All of these approaches provide insights but none alone provides a definitive means to understand the role of tool use in mathematics education. We hope that this book will be a source for future discussions between theorists of many persuasions as we seek to further understand tool use in mathematics education. There are problems and potential in this hope. Problems are related to the fact that it would be naïve to take a little bit of this theory and a little bit of that theory and hope that something sensible comes out; against there is ongoing work on strategies for connecting theories (see Prediger et al., 2008 for the genesis of this work). Further to this the 'theories' of which we write in Part II are quite different to theories in mathematics or physics, they do not develop 'laws' but are evolving ways of understanding didactical phenomena; and the hope expressed above is for a better understanding of the role of tool use in mathematics education. One possible direction for future connecting of theories towards an understanding of tool use in mathematics education is to consider how different theoretical approaches contribute to what Wertsch (1998) calls 'genetic domain': phylogenesis; sociocultural history; ontogenesis; and microgenesis (these were briefly discussed in Sect. 13.2). Insights on tool use in these domains will differ in emphases (respectively—prehistory and history, cultural dimensions, individual development and individual or group interaction with tools in specific activities). There is a sense in which mathematics education research concerned with tool use has focused on the microgenetic domain and needs to view the 'bigger picture'.

We now turn to what tool use may teach us about the nature of mathematics. When we started writing this book we (or maybe just John) had an expectation that the process of writing would result in us having something substantial to say on the nature of mathematics (with regard to tool use) and the implications of this substantial statement for the learning of mathematics. But we find ourselves at the end of the book being somewhere between having little to say (other than tool use is essential for doing and for learning mathematics) and making a clear statement on the nature of mathematics and learning mathematics with regard to tool use. The following are footnotes (of a sort) to what may become such a statement; the foci are Jon's experimental mathematics (see Chap. 3) and the didactical transposition (see Sect. 10.3).

Experimental mathematics arose through tool use. The word 'through' is important here. Jon used (and uses) tools at hand (digital tools) to attack mathematical problems. Over time his (with other mathematicians) approach got the name 'experimental mathematics'. This approach builds on prior mathematics but digital experimentation adds a 'little something else' that doesn't quite fit with

Mathematics.[1] What we have here is a sort of 'branch line' of mathematics[2] (which could become the 'main line' of mathematics in a possible future) whose establishment was enabled by certain tools but which cannot be reduced to tool use (it includes forms of reasoning including visualising).

We now put experimental mathematics aside for a moment and turn attention to learning mathematics and 'school mathematics'. We accept that there is a didactical transposition, in the words of Lagrange (2005, p. 69) 'mathematics in research and in school can be seen as a set of knowledge and practices in transposition between two institutions, the first one aiming at the production of knowledge and the other at its *study*'. So a problem about addressing 'what tool use may teach us about the nature of mathematics' is that it depends on which (parts or approaches of) mathematics one is talking about. A further problem concerns the culture (or, rather, the cultures) of mathematics. Bishop's 'Mathematics' (with a capital M) is steeped in Western culture but mathematics is, at least, steeped in Eastern and Western culture. We have, in this book, explored tool use in different cultural settings but we have not explored fundamental differences between different traditions in mathematics. *The Nine Chapters* (Chemla & Shuchun, 2004), originating from China first century BCE, evidences strong relationships between mathematical knowledge, problems, solving techniques, and tools. Analysing the differences between different cultural approaches to tool use in mathematics is work that remains to be done.

Our final thought turn to action and a question which Luc posed to John and Jon as we approached the final stages of producing this book: Who is empowered by knowledge on tool use in mathematics? This question challenges us as practical educators. We came to writing this book on tools and mathematics as academics and the book is written for an academic audience. As academics we default to noting the affordances and constraints of specific tools for specific purposes in specific contexts. But we wouldn't have written this book if we didn't have views on 'good' and 'bad' tool use and a vision of mathematics for the masses (we'll come clean, all three of us think, with a caveat here and there, that digital tools that allow people to explore mathematical relationships are a good thing). So why are we writing a book for academics instead of a book for the masses? A good question and a partial response is that books are not always the most effective way to educate/empower people.

So what groups of people could be empowered by knowledge on tool use in mathematics? Valero (2009, p. LV) argues

[1] Mathematics with a capital M here refers to what Bishop (1988, p. 18) calls 'Mathematics', 'the mathematics which is exemplified by Kline's *Mathematics in Western Culture'*.

[2] We are referring to a thing called 'mathematics' here (not with a capital M) but the ensuing discussion (which considers forms of mathematics) makes the idea that there is a thing called 'mathematics' somewhat suspect.

> 'that the time has come to open possibilities of defining both research practices and educational practices in a way that allows tackling in serious, rigorous and systematic ways the social, cultural and political complexity of mathematics education in our contemporary societies'

and mentions: academics (mathematicians and mathematics educators); policy makers; teacher educators; international agencies; technical developers; employers; school staff including mathematics teachers, students and their families. To this we would add online teacher associations and hybrid communities of teachers and researchers (cf. Chap. 18).

We think that all of these groups can be empowered by seeing the importance of tools in mathematical activity (and appreciate the affordances and constraints of specific tools in specific contexts). Most of these groups will not be empowered directly by this book but we hope that this book empowers those who read it to produce materials or activities that can influence people in other groups; and to produce materials or activities under formats that, perhaps, could not be imagined at the time when we wrote these last words.

References

Bishop, A. (1988). *Mathematical enculturation: A cultural perspective on mathematics education.* Dordrecht, The Netherlands: Kluwer Academic.

Borwein, J. M., & Devlin, K. (2008). *The computer as crucible.* Natick, MA: AK Peters.

Chemla, K., & Shuchun, G. (Eds.). (2004). *Les neufs chapitres. Le Classique mathématique de la Chine ancienne et ses commentaires.* Paris: Dunod.

Fauvel, J., & Gray, J. (1987). *The history of mathematics: A reader.* Milton Keynes, England: The Open University.

Joseph, G. G. (2010). *The crest of the peacock: Non-European roots of mathematics.* Princeton, NJ: Princeton University Press.

Kidwell, P. A., Ackerberg-Hastings, A., & Roberts, D. L. (2008). *Tools of American mathematics teaching, 1800-2000.* Baltimore: The John Hopkins University Press.

Lagrange, J. B. (2005). Curriculum, classroom practices, and tool design in the learning of functions through technology-aided experimental approaches. *International Journal of Computers for Mathematical Learning, 10*(2), 143–189.

Netz, R. (1999). *The shaping of deduction in Greek mathematics.* Cambridge, England: Cambridge University Press.

Prediger, S., Bikner-Ahsbahs, A., & Arzarello, F. (2008). Networking strategies and methods for connecting theoretical approaches: First steps towards a conceptual framework. *ZDM, The International Journal on Mathematics Education, 40*(2), 165–178.

Valero, P. (2009). Mathematics education as a network of social practices. In *Proceedings of the 6th Congress of the European Society for Research in Mathematics Education.*

Wertsch, J. V. (1998). *Mind as action.* Oxford, England: Oxford University Press.

Index

J. Monaghan et al., *Tools and Mathematics*, Mathematics Education Library,
DOI 10.1007/978-3-319-02396-0

The manufacturer's authorised representative in the EU is Springer Nature Customer Service Centre GmbH, Europaplatz 3, 69115 Heidelberg, Germany. If you have any concerns regarding our products, please contact ProductSafety@springernature.com

Printed and bound by CPI Group (UK) Ltd, Croydon, CR0 4YY
30/09/2026
02225650-0017